TECHNICAL COMMUNICATION

A Reader-Centered Approach

NINTH EDITION

PAUL V. ANDERSON

Miami University (Ohio)

 CENGAGE

Australia • Brazil • Mexico • Singapore • United Kingdom • United States

Technical Communication: A Reader-Centered Approach, Ninth Edition
Paul V. Anderson

Product Director: Monica Eckman

Product Manager: Vanessa Coloura

Content Developer: Ed Dodd, Alison Duncan

Managing Content Developer: Janine Tangney

Associate Content Developer: Erin Bosco

Product Assistant: Claire Branman

Marketing Manager: Kina Lara

Senior Content Project Manager: Michael Lepera

Senior Art Director: Marissa Falco

Manufacturing Planner: Betsy Donaghey

IP Analyst: Ann Hoffman

Senior IP Project Manager: Kathryn Kucharek

Production Service/Compositor: MPS Limited

Text Designer: Shawn Girsberger

Cover Designer: Beth Paquin

Cover Image: Anatoliy Babiy/Getty Images

For product information and technology assistance, contact us at
**Cengage Customer & Sales Support, 1-800-354-9706
or support.cengage.com.**

For permission to use material from this text or product, submit all requests online at **www.cengage.com/permissions.**

Library of Congress Control Number: 2016953580

Student Edition:
ISBN: 978-1-305-66788-4

Loose-leaf Edition:
ISBN: 978-1-305-67186-7

Cengage
20 Channel Street
Boston, MA 02210
USA

Cengage is a leading provider of customized learning solutions with employees residing in nearly 40 different countries and sales in more than 125 countries around the world. Find your local representative at: **www.cengage.com.**

Cengage products are represented in Canada by Nelson Education, Ltd.

To learn more about Cengage platforms and services, register or access your online learning solution, or purchase materials for your course, visit **www.cengage.com.**

Printed at CLDPC, USA, 02-20

FOR MY FAMILY

Margie

Christopher and Kirsten

Soren and Sigrid

Rachel and Jeff

Anderson

Drew

Mom and Dad

AND FOR MY TEACHERS

James W. Souther and Myron L. White

BRIEF CONTENTS

CONTENTS

PREFACE FOR INSTRUCTORS

Welcome to the Ninth Edition of *Technical Communication: A Reader-Centered Approach*. This edition, like the previous ones, has a single central goal: to help you prepare your students to write effectively in their careers. Once again, I am deeply indebted to the generous suggestions of instructors and students—my own and others.

Key Features

While this edition includes many new features, it retains the features that instructors and students have found most helpful.

Reader-centered approach. No matter what your students' future (or current) careers, their success as writers will depend on the responses they are able to elicit through their writing from their readers. But people at work differ from one another, just as they do in the rest of their lives. To write effective on-the-job communications, students will have to learn about their specific readers and use that knowledge to create communications that these particular persons will find helpful and persuasive.

The book's advice and examples apply this reader-centered approach, whether focusing on content, organization, and other large issues or on the smallest details of sentence construction and table design.

You at the center. Unlike the many textbooks that, by implication, put you in the secondary role of teaching what they say, this book places you at the center of your course. It emphasizes the many indispensable ways you contribute to your students' learning—that your knowledge of your students and their career plans enables you to choose what parts of the book to cover, what to emphasize, and even what to disagree with. It also highlights your ability to give something a book cannot: individualized guidance and feedback.

Support for the course you design. The book's broad coverage and simple, three-part design enable you to choose the topics, assignments, and course design that will best prepare your students with sophisticated yet transferable skills they can use wherever they choose to work after graduation. The book's "Writer's Planning Guides" and "Checklists," its "Libraries of Projects and Cases," and many of its other resources can be downloaded in Word so you can tailor them to your course.

In-depth coverage in an easy-to-learn manner. In most chapters, the major points are distilled into easy-to-remember guidelines whose implications and applications are then elaborated. The guidelines themselves reinforce one another because they all flow from a common set of reader-centered principles and processes.

Numerous richly annotated examples and sample documents. To students, guidelines can be mere abstractions accompanied by concrete examples illustrating their application. Throughout, the book includes sample communications with annotations that illustrate the use of its advice. Moreover, these annotations focus on the writer's purpose, thereby drawing students' attention to the writer's reader-centered decisions and strategies.

New to This Edition

Of course, this new edition offered me a welcome opportunity to refine, update, and respond to new developments in technical communication research and practice. It is also an opportunity to act on recent suggestions from instructors and students, Consequently, this edition includes the following new features.

- **New chapter on "Writing Effectively on Social Media at Work."** Many organizations have learned that social media can increase efficiency—if used in a business-oriented way. This new chapter builds on what students already know about social media and explains how workplace uses of social media differ from their social uses. Several other chapters elaborate on special uses of social media at work.

- **Integration of transfer of learning.** Research over the past two decades demonstrates how difficult it can be for students to adapt and apply what they learned in college to the writing they do on the job. However, more recent research has also uncovered strategies for increasing the transfer of learning to novel situations. Building on this research, Chapter 1 introduces the importance of learning in ways that promote transfer. "Reflect for Transfer" exercises are included at the end of many chapters. A new Appendix B presents reflection assignments you can ask students to complete when they turn in their finished projects.

- **New section of creating professional portfolios in Chapter 2, "Overview of the Reader-Centered Approach: Writing for a Job."** Professional portfolios have become an important, widely used supplement to résumés and job application letters. Chapter 2 also includes a new discussion of the importance of creating a positive presence in social media, which, research shows, most employers check before making job offers.

- **Simplified organization.** To increase flexibility for your teaching and enhance student learning, the book's overall organization has been modified from nine parts in the previous edition to three. Also, the order and content of some chapters are revised. Details are provided below, in the section titled "Organization of this Edition."

- **Streamlined presentation.** When I asked engineering and science students how I could improve this edition, they told me to retain its central features but also suggested several revisions. Their strongest advice was to "Eliminate repetition, redundancy, and things we already know." (See the acknowledgments.) Looking at my text in light of their advice enabled me to create a book that is significantly shorter and sharper, even though it contains additional material.

- **Increased attention to building on what students already know.** The goal is to help students adapt and build on the writing and speaking skills they bring to the course.

- **Chapter 3, "Defining Your Communication's Goals,"** is shortened to emphasize more emphatically its key advice for learning what readers want and what will influence their attitudes and actions.

- **Chapter 4, "Conducting Reader-Centered Research,"** includes new discussions of "What Counts as Good Research in the Workplace?" and ways to conduct research that meets those criteria.

- **Chapter 5, "Using Six Reader-Centered Research Methods,"** adds a new section on "Using Social Media in Your Research."

- **Chapter 6, "Organizing Reader-Centered Communications,"** is reconstructed to present its advice in a crisper, more engaging manner.

- **Chapter 7, "Drafting Reader-Centered Communications,"** presents more comprehensive advice by supplementing the guidelines on organizing the body of a communication with advice on drafting its beginning and ending, which had previously appeared in separate chapters.

- **Chapter 9, "Persuading Your Readers,"** extracts, revises, and relocates in a more logical place advice from the previous edition. The supporting research is updated.

- **Chapter 11, "Writing Reader-Centered Front and Back Matter,"** includes new examples that provide more helpful guidance to students.

- **Chapter 15, "Revising Your Drafts,"** includes a new discussion on revising social media messages at work.

- **Chapter 17, "Creating Communications with a Team,"** replaces earlier sections with new advice based on the latest research, including a five-year study of 180 teams by Google.

- **Appendix A, "Documenting Your Sources,"** includes the new, substantially different (and improved) MLA style, as well as the current APA and IEEE styles.

- **Appendix B,** which is also entirely new, presents "Reflecting for Transfer" assignments that are designed for use by students when they turn in course projects.

Other Major Features

In addition to the major preserved features already named, I have also kept the following because instructors have told me that they increase the book's breadth and effectiveness for teaching and learning.

- **Writer's Tutorials.** The tutorials demonstrate ways students can achieve some of the outcomes described in the text—for instance, by guiding students step-by-step through certain processes. Some are included in the text (see pages 33–35 and 352–354 for examples), and others are accessible in MindTap.

- **Planning Guides and Revision Checklists.** Integrated at key points in the text, they assist students in applying the book's advice as they work on their course projects. Additional ones are available in MindTap. As mentioned above, all can be downloaded in Word so you can modify them to your specific course and assignments.

- **Careful attention to international and intercultural communication.** Global Guidelines, integrated into the chapters, help students learn the many ways that cultural differences affect communication and provide concrete suggestions for increasing their effectiveness in cross-cultural communications.

■ **Marginal notes to extend learning.** Three categories of marginal notes summarize major principles, provide cross-references among chapters ("Learn More"), and challenge students to apply key strategies to their own experiences ("Try This").

■ **Continuous attention to ethics.** Chapter 1 introduces the topic of ethics in technical communication. Ethics guidelines are integrated into the chapters, so ethics becomes a continuous topic throughout a course rather than the topic for one day's reading. In addition, most chapters include special exercises that focus on ethical issues particular to the topic of those chapters.

■ **"Use What You've Learned" exercises.** At the end of most chapters, four types of exercises promote students' ability to apply the book's advice: "Apply Your Expertise," "Explore Online," "Collaborate with Your Classmates," and "Apply Your Ethics." In addition, "Reflect for Transfer" exercises challenge students to think about ways they could use what they have learned in a chapter when writing on the job.

Organization of this Edition

I have simplified the organization of this edition to make it easier for you to select the order in which you want to use the chapters and to provide you with a clearer way to explain to your students the relationships among the chapters you've chosen.

■ **Introduction.** Part I includes two chapters. Chapter 1 can help you explain the nature of writing in the workplace and the ways it differs from most (or all) the writing students do in other courses. Chapter 1 also introduces the reader-centered approach as well as the book's approach to ethics and the transfer of what students learn in your course to the writing they will do on the job.

I include Chapter 2, which focuses on writing for a job, in the introduction because it helps students see in a very personal way why thinking about their reader at every point in the process is the key to writing workplace communications that achieve the results they want. However, other instructors enjoy equal success assigning writing for a job at another point in their courses or not at all.

■ **The Reader-Centered Communication Process.** Part II helps you provide your students with reader-centered guidance for each of the major activities in the technical communication process, beginning with defining the goals for their communications through researching, organizing, drafting, and revising. Four chapters focus on skills that are especially important in technical communication but not most other writing courses: creating graphics, crafting page designs (two chapters), and user-testing drafts.

Most instructors (including me) assign these chapters out of order. As students are working on each course project, we select the chapters from Part II whose content students learn most effectively by applying it to that project.

■ **Applications of the Reader-Centered Approach.** Part III includes a variety of chapters from which you can choose the ones that, based on your knowledge of your students and their career plans, will most help them develop and learn to apply their reader-centered knowledge and skills. These chapters provide detailed advice for communicating in a variety of workplace writing situations, such as writing with a team, writing correspondence, and writing on social media. They also guide students in preparing five major workplace superstructures (genres): proposals, empirical research reports, feasibility studies, progress reports, and instructions. Downloadable projects involving these applications are available in MindTap as Word files, so that you can modify them to suit your particular course and students.

I've also included two appendices. Appendix A is a reference resource that explains how to use the APA, IEEE, and MLA documentation styles. Appendix B includes Reflecting for Transfer activities you can ask students to complete when they turn in their course assignments.

Supporting Materials for Students and Instructors

On Cengage.com students can save on their course materials through our full spectrum of options. Students have the option to rent their textbooks, purchase print textbooks, e-textbooks, individual e-chapters, and audio books, all for substantial savings over average retail prices. Cengage.com also includes access to Cengage Learning's broad range of homework and study tools.

CENGAGE.com

MindTap® English

MindTap is a digital learning solution that helps instructors engage students and transform them into critical thinkers. Each MindTap course also comes with a selection of apps to encourage interactivity, engagement, personalization, and more.

MindTap for *Technical Communication*, Ninth Edition includes:

MindTap·

- **MindTap® Reader eBook.**
- **Aplia homework.** Aplia significantly improves outcomes and elevates thinking by increasing student effort and engagement.
- **InSite.** InSite from Cengage Learning is a full integrated, productivity-enhancing classroom solution that delivers an all-in-one perspective on your students' work.
- **Questia.** Access a vast library of books and articles that instructors can add to the course learning path or eBook and students can use to research.
- **NetTutor®.** NetTutor is staffed with U.S.-based tutors and facilitated by a proprietary whiteboard created for online collaboration in education.
- **Library of Case Studies.** Referenced in several text chapters, Case Studies help students hone their reader-centered communication skills. Some Case Studies are suitable for homework or class discussion; others are appropriate for course projects.
- **Library of Projects.** The Library of Projects provides instructors with a wider selection from which to choose assignments that are most appropriate for their students.
- **Additional annotated sample documents.** These model reader-centered communication in a realistic format.
- **Downloadable and customizable Planning and Revision Guides.** These help students navigate the process of creating many kinds of communication.

- **Additional Writer's Tutorials.** These guide students step-by-step through certain communication processes.
- **Style Guide.** This provides brief, user-friendly guidance on issues of grammar, punctuation, style, and usage.
- **Web Resources.** These direct students to additional online tools and technical communication sites of interest.
- and more!

Instructor's Manual

Accompanying this edition of *Technical Communication: A Reader-Centered Approach* is an updated instructor's manual that includes a thorough introduction to the course, information on how to integrate supplemental materials into the class, advice on teaching the exercises and cases in the textbook, and more. Instructors may download a PDF version from MindTap or from the Instructor Resource Center.

Author's Acknowledgments

Writing a textbook is truly a collaborative effort to which numerous people make substantial contributions. I take great pleasure in this opportunity to thank the many people who generously furnished advice and assistance while I was working on this ninth edition of *Technical Communication: A Reader-Centered Approach*. I am grateful to the following individuals, who prepared extensive and thoughtful reviews of the eighth edition and my preliminary plans for the ninth edition: Krys Adkins (Drexel University), Alyce Baker (Lock Haven University), Scott Banville (Nicholls State University), Ellen Barker (Nicholls State University), Catherine Bean (Anoka-Ramsey Community College), Magdalena Berry (Missouri State University), Heather Burford (Ivy Tech Community College), and Thomas Chester (Ivy Tech Community College).

I would also like to thank the following persons who reviewed the previous two editions and who assisted in the evolution of this book over the past few years: Craig Baehr (Texas Tech University), Gertrude L. Burge (University of Nebraska), Diljit K. Chatha (Prairie View A&M University), Zana Katherine Combiths (Virginia Polytechnic University), Janice Cooke (University of New Orleans), Nancy Coppola (New Jersey Institute of Technology), Tracy Dalton (Missouri State University), Dr. Geraldine E. Forsberg (Western Washington University), Roger Friedmann (English Department, Kansas State University), Dawnelle A. Jager (Syracuse University), Linda G. Johnson (Southeast Technical Institute), Matthew S. S. Johnson (Southern Illinois University Edwardsville), Carole M. Mablekos, Ph.D. (Department of Engineering Management, School of Engineering, Drexel University), Jodie Marion (Mt. Hood Community College), L. Renee Riess (Hill College), Wayne Schmadeka (University of Houston–Downtown), Barbara Schneider (University of Toledo), and William West (University of Minnesota–Minneapolis).

While developing this edition, I have benefited from the thoughtful and energetic assistance of an extraordinary group at Cengage Learning. Ed Dodd and Michael Lepera skillfully shepherded the book through the many phases of development and production. I am very grateful to Ed Dionne for his exceptional support during production. Mathangi Anantharaman and Aruna Sekar contributed abundant resourcefulness and tenacity while conducting photo research and text research, respectively.

I am indebted to all my students for their keen critiques of the book and for the originality of their work, which provides a continuous supply of new ideas. In particular, I thank Jessica Bayles, Erin Flinn, Allen Hines, Billy O'Brien, Joseph Terbrueggen,

Curtis Walor, and Tricia M. Wellspring, whose thorough, thoughtful work provided examples and helped me develop several of the discussions and examples.

This edition owes a great deal to the positive comments and—especially—incisive suggestions from the following science and engineering students at the University of Cincinnati: Daniel Barr, Theo Charles Brooks, Hayden Dillon, Jacob Dorrance, Sam Fintel, Elliott Ice, Timothy Kelleher, Tim Koch, Cameron Meece, Kolawole Omoyosi, Carolyn Kelley Patterson, Brady Perkins, Eduardo Pocasangre, Chelsea J. Rothschild, Chris Sabetta, Osama Saleh, Ethan H. Slaboden, Joseph Treasure, Lam Tse, and Scott J. Welsh.

For this edition, Betty Marak, who has read this book more than anyone else over the years, assisted once more in preparing the manuscript. Diane Bush contributed exceptionally astute copyediting, for which my readers will be as thankful as I am. Tom Collins and Steve Oberjohn have provided enduring assistance over the years. All my work in technical communication benefits from many conversations with and numerous examples of excellent teaching provided by Jean Lutz and Gil Storms. I owe a special thanks to Jeremy Rosselot-Merritt, University of Minnesota, for helping me gather students' advice for making this edition more reader (student)-centered. I am deeply grateful to all of these individuals.

Finally, I thank my family. Their encouragement, kindness, patience, and good humor have made yet another edition possible.

PAUL V. ANDERSON
Miami University
Oxford, Ohio

PART I

INTRODUCTION

Communication, Your Career, and This Book

LEARNING OBJECTIVES

1. Describe the major ways writing at work differs from writing at school.
2. Name and explain the two qualities that writing at work must have to be effective.
3. Summarize in one sentence the reader-centered approach to writing.
4. Use six reader-centered strategies when writing a brief communication.
5. Describe this book's approach to ethics.
6. Tell how to gain lasting value from your course and this book.

MindTap®

Find additional resources related to this chapter in MindTap.

From the perspective of your professional career, communication is one of the most valuable subjects you will study in college.

Surprised? First, consider what employers will be looking for when reading your application for an internship, co-op, or full-time position. At one time, many employers may have focused primarily on the specialized skills you learned in your major. But the world has changed.

In a survey that asked 225 U.S. employers to identify the top qualifications they seek in new employees, the highest number (98%) identified communication skills (Schawbel, 2012). When 400 employers were asked in another survey to identify the most important outcomes of a college education, they listed *communicating effectively in writing* and *communicating effectively orally* as two of the top three (Hart, 2015). The third? *Working effectively with others in teams*, an ability that also depends heavily on communication skills.

There is every reason to believe that similar results would be obtained in other countries. In India, executives of one of the world's largest software consulting companies told me that when hiring new college graduates their company always ranked communication ability above computer skills.

Although employers value communication skills so highly, they are frustrated by the writing skills that most new college graduates bring to the job. Of the 400 employers who ranked writing and speaking effectively at the top of their list of desired skills, only 1 in 4 said that college graduates are well prepared in writing (Hart, 2015). U.S. corporations spend more than $3 billion annually on writing instruction for their employees (National Commission on Writing, 2004).

The gap between what employers want and the qualifications they see in job applicants means that you can prepare yourself to stand out from other applicants by developing your communication expertise.

Communication Expertise Will Be Critical to Your Success

Your communication abilities will continue to be important after you are hired. Writing alone will consume a major part of your time. Newly hired employees spend an average of 20 percent of their time at work writing (Beer & McMurrey, 2009; Sageev & Romanowski, 2001). That's one day out of every workweek! And it doesn't include the time they spend talking in person, on the phone, or on the Internet, whether in person or in meetings. Writing ability will also be a major consideration when you apply for promotions, according to the U.S. National Commission on Writing (2004).

College graduates typically spend one day a week—or more—writing.

Besides being essential to your career, communication expertise will enable you to make valuable contributions to your campus or community. Volunteer groups, service clubs, and other organizations will welcome your ability to present their goals, proposals, and accomplishments clearly and persuasively.

If you enjoy writing and learning about computers, health, engineering, or similar fields, technical communication could become your profession. Private corporations, nonprofit organizations, and government agencies the world over hire professional technical writers and editors. In the United States, technical writing has been rated as one of the top fifty jobs, based on employee satisfaction and projected number of job openings through at least 2020 (Wolgemuth, 2010).

Learning Objectives for This Chapter

Your instructor and this book share the goal of helping you develop the communication expertise needed to realize your full potential on the job and in your community. This chapter and class discussions of it will build the foundation for the rest of your course. As you read and discuss the chapter, focus on learning how to achieve the learning objectives listed on the chapter-opening page. You might imagine that your instructor will give a quiz asking you to do one or more of them.

Characteristics of Workplace Writing

How can students who wrote well and received high grades in school be perceived as poor writers by their employers? Researchers found that the answer is quite simple. The writing done on the job differs in substantial ways from the writing learned in school—and the transition from one type to the other can be difficult. What's valued most highly in school is not what is required of writers at work. In addition, workplace writing involves many new skills not usually taught in college. Your instructor and this book share the goal of giving you an enormous head start on making that transition.

To begin, let's look at what makes writing at work so different from writing at school. First, we need to acknowledge that it won't be completely different. Good grammar and correct spelling still matter. So do many other communication skills you already know. But you will have to adapt much of what you know, for instance how to use social media, in order to use it in what many employers call a "professional" way. And, of course, there are some completely new things for you to learn, such as how to write types of communications you've never written before.

The following sections describe seven important ways workplace writing differs from most college writing—starting with the most crucial of them all.

- It serves **practical purposes**.
- It must satisfy a **wide variety of readers**, sometimes in a single communication.
- It uses **distinctive types of communication**.
- It is **shaped by context**.
- It must **adhere to organizational expectations**.
- It is frequently **created collaboratively**.
- It **uses social media** for practical purposes.

Serves Practical Purposes

The most important difference between the writing you will do at work and most of the writing you do in school concerns *purpose*. On the job, you will write for practical purposes, such as helping your employer improve a product or increase efficiency. Your readers will be supervisors, coworkers, customers, or other individuals who need information and ideas from you in order to pursue their own practical goals. You may already have prepared this type of communication for instructors who asked you to write to real or imagined readers, people who need your information in order to make a decision or take an action.

Most of your school writing—term papers, essay exams, and similar school assignments—has a much different purpose. It is intended to help you learn and to demonstrate your mastery of course material. Although your instructors will read what you write in order to assess your knowledge and decide what grade to assign, they rarely, if ever, need information and ideas from you in order to guide their decisions and actions as they pursue their own goals.

In contrast, communication is the lifeblood of an organization. It is the flow of ideas and information that delivers what you know or have found out to another person who needs your information to do his or her job.

- Sarah, a recent college graduate in metallurgy, discovered the reason that the pistons in a new, lightweight, fuel-saving automobile engine broke in a test. However, her discovery has no value to her employer unless she communicates her finding in a clear and useful way to the engineers who must redesign the pistons.

- Larry, a hospital nutritionist, developed ideas for improving the efficiency of the kitchen where he works. However, the hospital will reap the benefits of his creativity only if he presents his ideas persuasively to the people with the power to implement these changes.

Different purposes profoundly affect the kind of communication you need to produce. For an essay exam or term paper, your purpose is to show how much you know. You succeed by saying as much as you can about your subject. At work, where your purpose is to help your readers make a decision or perform a task, you succeed by telling your readers only what they need, no matter how much more you know. Sarah doesn't need to tell the design engineers everything she learned about the broken pistons. In fact, she shouldn't. She should communicate only the information that will help the engineers make better ones. Extra information will only clog their paths to what they require in order to do their work.

Learning what your readers need and determining the most *helpful* way to present this information are the most critical skills in workplace writing, though they are not relevant for most writing assigned at school. This book and your instructor will help you learn to make this critical transition from school writing to workplace writing.

Must Satisfy a Wide Variety of Readers, Sometimes in a Single Communication

As a student, you usually write to a single reader, your instructor. In contrast, at work, you will often prepare communications that address two or more people who differ from one another in important ways, such as their familiarity with your specialty, the ways they will use your information, and their professional and personal concerns. For example, in his report recommending changes to the hospital kitchen, Larry's readers may include his supervisor, who will want to learn how operations in her area would have to change if Larry's recommendations were adopted; the vice president for finance, who will want to analyze Larry's cost estimates; the director of personnel, who will want to know how job descriptions would need to change; and members of the labor union, who will want assurances that the new work assignments will treat employees fairly.

On the job, you will often need to construct communications that, like Larry's, must simultaneously satisfy an array of persons who will each read it with his or her own set of concerns and goals in mind.

Also, at work, you may often address readers from other nations and cultural backgrounds. Many organizations have clients, customers, and suppliers in other parts of the world. Thirty-three percent of U.S. corporate profits are generated by international trade (Lustig & Koester, 2012). The economies of many other nations are similarly linked to distant parts of the globe. Corporate and other websites are accessed by people around the planet. Even when communicating with coworkers at your own location, you may address a multicultural audience—persons of diverse national and ethnic origins.

LEARN MORE To learn about addressing international and intercultural audiences, read the Global Guidelines included in many chapters.

Uses Distinctive Kinds of Communication

At work, you will create a wide variety of job-related communications that aren't usually prepared at school. Depending on your career, these may be business letters and emails, memos, project proposals, instructions, and progress reports. Each type has its own conventions. In your course, you will study and gain experience writing some of these types (Chapters 20–27), and you will develop strategies for learning about and successfully writing others.

Also, at work, writing involves more than words. In many communications, graphics such as tables, charts, drawings, and photographs are as important as the written text. Equally important, you can make reading easy for your readers by the ways you arrange text and graphics on the page or screen. Figure 1.1 shows a page from an instruction manual that illustrates the importance of graphics and visual design. Chapters 12, 13, and 14 teach you how to create graphics and design pages that your readers will praise.

Shaped by Context

Every communication situation has social dimensions. In your writing at school, the key social relationship usually is that of a student to a teacher. At work, you will have a much wider variety of relationships with your readers, such as manager to subordinate, customer to supplier, or coworker to coworker. Sometimes these relationships will be characterized by cooperation and goodwill. At other times, they will be fraught with competitiveness as people strive for recognition, power, or money for themselves and their departments. To write effectively, you will need to adjust the style, tone, and overall approach of each communication to these social and political considerations.

The range of situational factors that can affect a reader's response is obviously unlimited. The key point is that in order to predict how your readers might respond to a communication, you must understand thoroughly the context in which they will read it.

Adheres to Organizational Expectations

Each organization has a certain style that reflects the way it perceives and presents itself to outsiders. For example, an organization might be formal and conservative or informal and progressive. Individual departments within organizations may also have their own styles. On the job, you will be expected to understand the style of your organization or department and employ it in your writing.

Another important expectation concerns deadlines. At work, they are much more significant—and changeable—than most deadlines at school. A deadline may be pushed back or advanced several times during a project, but no matter what the deadline is, the work must be completed on time. For example, when a company prepares a proposal or sales document, it must reach the client by the deadline the client has set.

FIGURE 1.1

Page That Illustrates the Way Graphics and Visual Design Work Together with Words in Technical Communications

The visually prominent heading explains what readers will learn from this page.

The drawings show exactly what a reader needs to do to clean the printer; they even show a hand performing these tasks.

The drawings include arrows to indicate the direction of movement.

Each numbered drawing corresponds to the step with the same number.

The numbers for the drawings and steps are visually prominent to help readers match each drawing with its corresponding step.
• In the drawings, the numbers are large and bold.
• In the steps, the numbers are bold and placed in a column of their own.

The cautions are highlighted visually.
• Horizontal lines (called rules) set the cautions off from the text.
• The word Caution is printed in bold and blue.

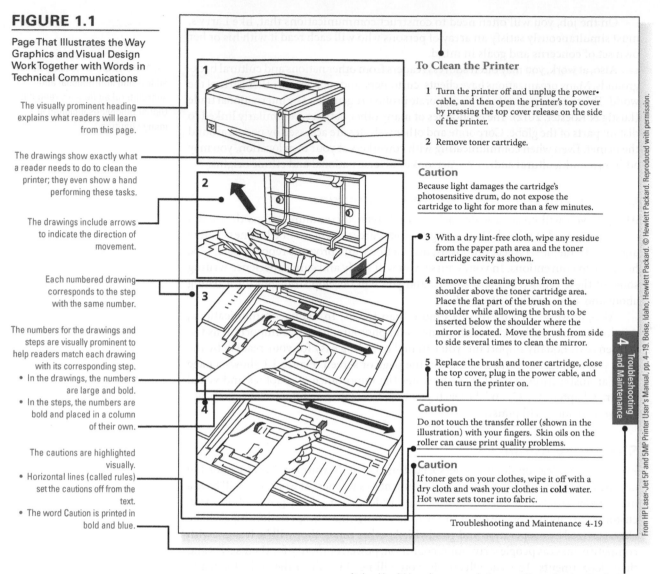

To Clean the Printer

1 Turn the printer off and unplug the power cable, and then open the printer's top cover by pressing the top cover release on the side of the printer.

2 Remove toner cartridge.

Caution

Because light damages the cartridge's photosensitive drum, do not expose the cartridge to light for more than a few minutes.

3 With a dry lint-free cloth, wipe any residue from the paper path area and the toner cartridge cavity as shown.

4 Remove the cleaning brush from the shoulder above the toner cartridge area. Place the flat part of the brush on the shoulder while allowing the brush to be inserted below the shoulder where the mirror is located. Move the brush from side to side several times to clean the mirror.

5 Replace the brush and toner cartridge, close the top cover, plug in the power cable, and then turn the printer on.

Caution

Do not touch the transfer roller (shown in the illustration) with your fingers. Skin oils on the roller can cause print quality problems.

Caution

If toner gets on your clothes, wipe it off with a dry cloth and wash your clothes in **cold** water. Hot water sets toner into fabric.

Troubleshooting and Maintenance 4-19

4 Troubleshooting and Maintenance

From HP Laser-Jet 5P and 5MP Printer User's Manual, pp. 4–19. Boise, Idaho, Hewlett Packard. © Hewlett Packard. Reproduced with permission.

At the side of this and every page in the manual is a colored rectangle that gives the section number and title. These rectangles help readers flip quickly to the specific information they need.

Otherwise, it may not be considered at all—no matter how good it is. Employers sometimes advise that "it's better to be 80 percent complete than 100 percent late."

Finally, under the law, most documents written by employees represent the position and commitments of the organization itself. Company documents can even be subpoenaed as evidence in disputes over contracts and in product liability lawsuits. The documents you write one day could become evidence in a lawsuit five years from now.

Even when the law does not come into play, many communications written at work have moral and ethical dimensions. The decisions and actions they advocate can affect many people for better or worse. Because of the importance of the ethical dimension of workplace writing, this book incorporates in most chapters a discussion of ethical issues that may arise in your on-the-job communications. Ethics are also discussed further at the end of this chapter.

Created Collaboratively

Whatever experience you gain at school in writing collaboratively will benefit you on the job. For long workplace documents, the number of cowriters is sometimes astonishingly large, into the hundreds. Even when you prepare communications alone, you may consult your coworkers, your boss, and even members of your intended audience as part of your writing process.

In one common form of collaboration, you will need to submit drafts of a communication for review by managers and others who have the power to demand changes. The number of reviewers may range from one to a dozen or more. Some drafts go through many cycles of review and revision before obtaining final approval.

At work, increasing numbers of employees are doing their collaborative writing in a globally networked environment. Using the Internet, project team members write reports, proposals, and other documents even though these people may be located in different countries or on different continents. The teams may hold their work sessions entirely online (see Figure 1.2), never meeting in person. Their members may come from different cultures and have different languages as their native languages.

On the job, groups of employees often work together to plan, draft, and revise proposals, reports, and other printed, online, and oral communications.

Uses Social Media for Practical Purposes

What you already know about social media will help you on the job—as long as you can adapt your uses, when at work, for practical purposes. Depending on your job, you may use texts, blogs, and tweets. Various apps are used for research, marketing, branding, and coordinating and managing projects. A key consideration is not to confuse the social and business uses. Both have their own conventions.

Keep in mind also that employers use social media to learn more about job applicants. The next chapter, which is on writing for a job, provides advice about reviewing and possibly revising your social media presence before you begin contacting prospective employers.

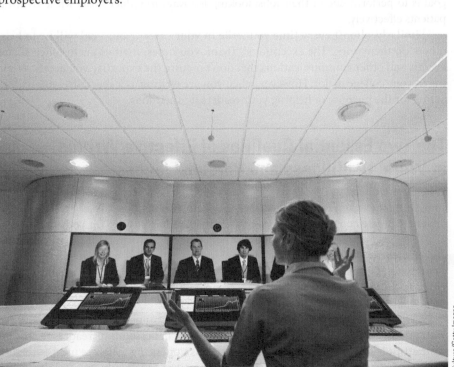

FIGURE 1.2

Writing Team Members in Australia and Germany Conduct a Work Session Online

At Work, Writing Is an Action

As you can infer from the preceding sections, there is tremendous variety among communications written on the job, depending on such variables as their purposes and readers; organizational conventions and cultures; and the political, social, legal, and ethical contexts in which they are prepared. Some people diminish their ability to write effectively in these multifaceted, shifting situations because they mistakenly think of writing as an afterthought, as merely recording or transporting information they developed while acting as specialists in their chosen field.

Nothing could be further from the truth.

When you write at work, you act. You exert your power to make something happen, to change things from the way they are now to the way you want them to be. Consider, again, the examples of Sarah and Larry. Sarah wants to help her team develop a successful engine. Acting as a metallurgical specialist, she has tested the faulty pistons to determine why they failed. To contribute to the success of the engine, she must now perform a writing act. She must now compose sentences, construct tables of data, and perform other writing activities to present her results in a way the engineers will find useful. Similarly, Larry believes that the hospital kitchen is run inefficiently. Acting as a nutrition specialist, he has devised a plan for improving its operation. For his plan to be put into effect, Larry must perform an act of writing. He must write a proposal that persuades the hospital's decision makers to implement his plan.

At Work, Writing Supports the Reader's Action

Just as writing at work is an action, so, too, is reading. The engineers' only reason for reading Sarah's report is to obtain information that helps them do their job, which is to redesign the pistons. When the hospital's administrators read Larry's report, their goal is to perform one of their jobs: looking for ways to reduce costs while treating patients effectively.

Similarly, almost every time you write in your career, your goal will be to help others do their work. The exceptions? Thank-you notes, invitations to office social events, and other communications that don't require the expertise you are developing in your college major. However, when you are writing as a specialist, you will be acting for the purpose of helping your readers do their jobs.

The Two Essential Qualities of Effective Writing at Work: Usefulness and Persuasiveness

When you understand that the purpose of workplace writing is to help readers do something they need to do, you can see why your two major goals as a writer on the job are to create communications that are *useful* and *persuasive* for your readers.

Usefulness defined.

To be *useful*, a communication must help readers perform their tasks effectively and efficiently. Their tasks may be physical, such as installing a new memory card in a computer. Or their tasks may be mental, as when Sarah's readers use her report to redesign the pistons and Larry's readers compare his recommended procedures with those currently used in the hospital kitchen.

Persuasiveness defined.

A communication's *persuasiveness* is its ability to influence its readers' attitudes and actions. For instance, a proposal's goal would be to persuade its readers to change their attitude toward a proposed action (such as upgrading software used at the organization) from neutral—or, possibly, from hostile—to favorable.

In most communications, either usefulness or persuasiveness dominates: usefulness in instructions, for example, and persuasiveness in proposals. However, every workplace communication must possess both to succeed. Instructions are effective only if the intended readers are persuaded to use them. A proposal can persuade only if its readers can easily find, understand, and analyze its content. To see how communications combine usefulness and persuasiveness in a single communication, look at the web page and memo shown in Figures 1.3 and 1.4.

In this home page, writers at Dell met the challenge of creating a site that would be both useful and persuasive to a wide variety of persons, including consumers and business people, novice and expert computer users, and new visitors and longtime customers.

FIGURE 1.3

Web Page That Combines Usefulness and Persuasiveness

Source: Dell, Inc.

Usefulness Strategies

At the top of the page, the writers provide several links that enable people to find specific information quickly.

- A link that takes existing personal and business customers to their accounts.
- A link to a search engine that enables people to locate specific information without navigating through multiple screens.
- A link to support.
- A link that enables people to check the status of orders they've made.

To help people who prefer to buy on the phone rather than online, the writers include the number for Dell's sales agents.

The writers organize their links in ways that help users quickly find the information they want.

- They organize links above the large photo around types of customers.
- They grouped the links across the bottom of the page under headings. Two of the headings feature words users often look for: "Service" and "Deals."

Persuasive Strategies

To highlight the support services Dell provides individual customers, the writers use a photo in which a Dell representative stands very close to a customer, who listens intently.

The writers use a set of icons to represent the many features of their computers.

By providing a list of "headlines" about the company, the writers convey the impression that it is vibrant and innovative while also providing information for which some users may be looking.

To emphasize satisfaction of consumers who purchase Dell computers for home use, the writers include a photo of two relaxed, smiling customers.

FIGURE 1.4

Memo That Combines
Usefulness and
Persuasiveness

In this memo, Frank, a chemist, uses many reader-centered strategies to make his report on test results useful and persuasive to his boss, Alyssa. His report will help her decide whether to use a newly developed plastic to bottle an oven cleaner manufactured by their employer. This product is sold on supermarket shelves.

Usefulness Strategies

In the subject line, Frank tells Alyssa exactly what the memo is about.

Frank uses the first sentence to tell Alyssa how the memo relates to her: It reports on tests she requested.

In the next two sentences, Frank presents the information that Alyssa most needs: The new plastic won't work. She doesn't have to read the entire memo to find this key information.

Using headings and topic sentences, Frank helps Alyssa locate specific details quickly.

He uses bold type to make the headings stand out visually.

By presenting the test results in a list, he enables Alyssa to read through them quickly.

Frank states clearly the significance of each test result.

Persuasive Strategies

To build credibility for his results, Frank describes the test method in detail.

To establish the credibility of the conclusion he draws, Frank presents specific details concerning each of the three major problems discovered in the storage tests.

By using lively active verbs (highlighted in color), he creates a vivid, action-oriented writing style.

By preparing a neat and carefully proofread memo, Frank bolsters confidence in the care he takes with his work.

PIAGETT HOUSEHOLD PRODUCTS
Intracompany Correspondence

October 10, 2017

To   Alyssa Wyatt
From   Frank Thurmond
Subject Test of Salett 321 Bottles for Use with StripIt

We have completed the tests you requested to find out whether we can package StripIt Oven Cleaner in bottles made of the new plastic, Salett 321. We conclude that we cannot, chiefly because StripIt attacks and begins to destroy the plastic at 100°F. We also found other significant problems.

Test Methods
To test Salett 321, we used two procedures that are standard in the container industry. First, we evaluated the storage performance of filled bottles by placing them in a chamber for 28 days at 73°F. We stored other sets of 24 bottles at 100°F and 125° for the same period. Second, we tested the response of filled bottles to environmental stress by exposing 24 of them for 7 days to varying humidities and varying temperatures up to 140°F.

We also subjected glass bottles containing StripIt to the same test conditions.

Results and Discussion
In the 28-day storage tests, we discovered three major problems:

- StripIt attacked the bottles made from Salett 321 at 100°F and 125°F. At 125°F, the damage was particularly serious, causing localized but severe deformation. Most likely, StripIt's ketone solvents weakened the plastic. The deformed bottles leaned enough to fall off shelves in retail stores.

- The sidewalls sagged slightly at all temperatures, making the bottles unattractive.

- StripIt yellowed in plastic bottles stored at 125°F. No discoloration occured in glass bottles at this temperature. We speculate that StripIt interacted with the resin used in Salett 321, absorbing impurities from it.

In the environmental test, StripIt attacked the bottles at 140°F.

Conclusion
Salett 321 is not a suitable container for StripIt. Please call me if you want additional information about these tests.

The Main Advice of This Book: Think Constantly about Your Readers

The key to determining what will make your communications useful and persuasive is to follow the main advice of this book: When writing, think constantly about your readers. Think about what they want from you—and why. Think about how you want to help or influence them and how they will react to what you have to say. Think about them as if they were standing right there in front of you while you talked together.

You may be surprised that this book emphasizes the personal dimension of writing more than such important characteristics as clarity and correctness. Although clarity and correctness are important, they cannot, by themselves, ensure that something you write at work will be successful.

For example, if Larry's proposal for modifying the hospital kitchen is to succeed, he will have to explain the problems created by the present operation in a way that his readers find compelling, address the kinds of objections his readers will raise to his recommendations, and deal sensitively with the possibility that his readers may feel threatened by having a new employee suggest improvements to a system they themselves set up. If his proposal fails to do these things, it will not succeed, no matter how "clear and correct" the writing is. A communication may be perfectly clear, perfectly correct, and yet be utterly unpersuasive, utterly ineffective.

The same is true for all the writing you will do at work: What matters is how your readers respond, as Figure 1.5 illustrates. That's the reason for taking the

FIGURE 1.5

To Write Effectively at Work, You Must Create Communications That Are Useful and Persuasive to Your Readers reader-centered approach described in this book. This approach focuses your attention on the ways you want to help and influence your readers, and teaches specific strategies that you can use to achieve these goals.

The Dynamic Interaction between Your Communication and Your Readers

But what does it mean to think constantly about your readers? The full explanation comes in later chapters. The central point is to be able to imagine your readers in the very act of reading. Not after reading, *but while reading*, so you know what they are looking for and how they will use it, moment by moment, as they read. A solid understanding of your reader in the act of reading your communication will help you make every important decision about it.

An explanation of three basic principles of reading, which have been established by extensive research, will help you understand the importance of thinking about your readers at this level of detail.

- Readers construct meaning.
- Readers' responses are shaped by the situation.
- Readers react moment by moment.

All three research findings stress that writing is not a passive activity for readers but a dynamic interaction between readers and texts.

Readers Construct Meaning

When researchers say that readers construct meaning, they are emphasizing the fact that the meaning of a written message doesn't leap into our minds solely from the words we see. Instead, to derive meaning from the message, we actively interact with it. In this interaction, we employ a great deal of knowledge that's not on the page, but in our heads. Consider, for example, the knowledge we must possess and apply to understand this simple sentence: "It's a dog." To begin, we must know enough about letters and language to decipher the three printed words (*it's*, *a*, and *dog*) and to understand their grammatical relationships. Because young children and people who do not speak English do not possess this knowledge, they cannot read the sentence. Often, we must also bring other kinds of knowledge to a statement in order to understand it. For instance, to understand the meaning of the sentence "It's a dog," we must know whether it was made during a discussion of Jim's new pet or during an evaluation of ABC Corporation's new computer.

In addition to constructing the meanings of individual words and sentences, we build these smaller meanings into larger structures of knowledge. These structures are not merely memories of words we have read.

They are our own creations.

To demonstrate this point to yourself, write a sentence that explains the statement, which you read earlier in this chapter, that workplace writing "serves practical purposes." If you were then to go back and look for a sentence that exactly matches yours, most likely, you wouldn't find one. The sentence you wrote is not one you remembered. Rather, it is the meaning you constructed through your interaction with the text.

The fact that readers construct the meaning they derive from a communication has many implications for you that are explored later in this book. An especially important one is this: You should learn as much as possible about the knowledge your readers will bring to your communication so you can create one that helps them construct the meanings you want them to build.

Readers' Responses Are Shaped by the Situation

A second important fact about reading is that people's reactions to a communication you write will be shaped by the context—including such things as their purpose for reading, their perception of your purpose in writing, their personal stake in the subject discussed, and their past relations with you.

For example, Kate has finished investigating several brands of tablet computers to determine which one she should recommend her company purchase for each of its fifty field engineers. Opening her email, she finds a message from a coworker who asserts that the ABC tablet is "a dog." Kate's response to this statement will depend on many factors. Did the information Kate gathered support this assessment? Was the statement made by a computer specialist, a salesperson for one of ABC's competitors, or the president of Kate's company? Has she already announced publicly her own assessment of the computer, or is she still undecided about it? Depending on the answers to these questions, Kate's response might range anywhere from pleasure that her own judgment has been supported by a well-respected person, to embarrassment that her publicly announced judgment has been called into question.

The range of situational factors that can affect a reader's response is obviously unlimited. The key point is that in order to predict how your readers might respond to a communication you are writing, you must understand thoroughly the context in which they will read it.

Readers React Moment by Moment

The third important fact about reading is that readers react to communications moment by moment. When we read a humorous novel, we chuckle as we read a funny sentence. We don't wait until we finish the entire book. Similarly, people react to each part of a memo, report, or proposal as soon as they come to it. Consider the following scenario.

Imagine that you manage a factory's personnel department. A few days ago, you discussed a problem with Patrick, who manages the data processing department. Recently, the company's computer began issuing some payroll checks for the wrong amount. Your department and Patrick's work together to prepare each week's payroll in a somewhat antiquated way. First, your clerks collect a time sheet for each employee, review the information, and transfer it to time tickets, which they forward to Patrick's department. His clerks enter the information into a computer program that calculates each employee's pay and prints the checks. The whole procedure is summarized in the diagram shown here.

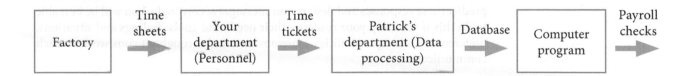

In your discussion with Patrick, you proposed a solution he did not like. Because you two are at the same level in the company, neither of you can tell the other what to do. When you turn on your computer this morning, you find an email message from Patrick. Remember that he is writing to you, in your role as manager of the personnel department.

| To | Your name, Manager, Personnel Department |
| From | Patrick Donaldson, Manager, Data Processing Department |

Subject INCORRECT PAYROLL CHECKS

I have been reviewing the "errors" in the computer files.

Contrary to what you insinuated in our meeting, the majority of these errors were made by your clerks. I do not feel that my people should be blamed for this. They are correctly copying the faulty time tickets that your clerks are preparing.

You and I discussed requiring my computer operators to perform the very time-consuming task of comparing their entries against the time sheets from which your clerks are miscopying.

My people do not have the time to correct the errors made by your people, and I will not hire additional help for such work.

I recommend that you tell your clerks to review their work carefully before giving it to the data processors.

In your role as the person to whom Patrick is writing, how did your feelings toward this memo and its writer evolve as you read each sentence?

Most people who read this memo while playing the role of personnel manager grow defensive the moment they see the quotation marks around the word *errors,* and they become even more so when they read the word *insinuated.* After they read the third paragraph, their defensiveness hardens into a grim determination to resist any recommendation Patrick may make.

A few readers are more even-tempered. Instead of becoming defensive, they become skeptical. They realize Patrick is behaving emotionally, so they evaluate his statements very carefully. When they read his accusation that the personnel department clerks are miscopying the time sheets, they want to see evidence that supports that claim. When the next sentence fails to provide it, they feel disinclined to accept Patrick's recommendations.

Even though different readers react to the first few sentences of this memo in different ways, their early reactions shape their responses to the sentences that follow. Consequently, even though Patrick's recommendation seems sensible enough, not many readers feel inclined to accept it. With a different opening (and some other changes), Patrick could have prompted a much different response to his recommendation.

Patrick's memo underscores the importance of taking a reader-centered approach to writing, one in which you think constantly about your readers. Each reader creates his or her own response to your communications. To write effectively, you must predict these responses and design your messages accordingly. You will be best able to do this if you keep your readers—their needs and goals, feelings and situations, preferences and responsibilities—foremost in mind throughout your work on each communication.

Six Reader-Centered Strategies You Can Begin Using Now

Despite the many ways readers' goals, concerns, feelings, and likely responses can vary from situation to situation, readers approach most on-the-job communications with several widely shared aims and preferences. The following list briefly introduces several reader-centered strategies you can begin using immediately that address these common aims and preferences. All are discussed more fully in later chapters. Many are illustrated in the web page and memo shown in Figures 1.3 and 1.4.

- **Begin by identifying the specific task your readers will perform using your communication.** Develop enough detail to understand how they will proceed. If the task is a proposal, what criteria will they apply? If the task is to perform a procedure, what level of detail will the readers need concerning the steps to be performed?

- **Identify the readers' attitudes that are relevant to the communication.** Imagine how much more effectively Patrick could have written the subject line of his email—as well as its contents—if he had thought about the way the personnel manager's irritation might shape the manager's response to his recommendation.

- **Help your readers quickly find the information needed to perform their tasks.** State your main points at the beginning rather than the middle or end of your communications. Use headings, topic sentences, and lists to guide your readers to the specific information they want to locate. Eliminate irrelevant information that can hide what your readers want.

- **Highlight the points your readers will find to be persuasive.** Present the information your readers will find more persuasive before you present the information they will find to be less persuasive. Show how taking the actions you advocate will enable them to achieve their own goals. When selecting evidence to support your arguments, look specifically for items you know your readers will find to be credible and compelling.

- **Trim away unnecessary words, use the active voice, and put action in your verbs.** The most basic way you can make a communication useful to your readers is to ensure that the readers can readily understand what the communication is saying. In an early and important test of the effectiveness of eliminating unnecessary words, and using the active voice, researchers James Suchan and Robert Colucci (1989) created two versions of the same report. The high-impact version used the techniques. The low-impact one did not. The high-impact version reduced reading time by 22 percent, and tests showed that readers understood it better.

- **Talk with your readers.** Before you begin work on a report or set of instructions, ask your readers, "What do you want in this communication? How will you use the information it presents?" When planning the communication, share your thoughts or outline, asking for their reactions. After you've completed a first draft, ask for their feedback.

If you cannot speak directly with your readers, learn about their likely responses to your communication by creating an imaginary portrait of one of them, and then picture the person reading your communication. Use the responses you see to guide your writing. Does the reader furrow his or her brow in puzzlement? Explain the point more fully. Does the reader twist his or her hands impatiently? Abbreviate your message.

By watching her imaginary reader, Marti, a mechanical engineer, gradually refined one step in a set of instructions so that it met her readers' needs. Here is her first draft.

LEARN MORE For detailed advice about creating an imaginary portrait of your readers, see Chapter 3.

15. Check the reading on Gauge E.

Marti's first draft

While watching her reader's reactions, Marti saw the reader ask, "What should the reading be?" So she answered their question.

Marti's second draft

15. Determine whether the reading on Gauge E matches the appropriate value listed in the Table of Values.

Marti then imagined her reader asking, "Where is the Table of Values?" She revised again.

Marti's third draft

15. Determine whether the reading on Gauge E matches the appropriate value listed in the Table of Values (page 38).

Finally, Marti imagined her reader discovering in the Table of Values that the reading on Gauge E was wrong. Her imaginary reader asked, "What should I do now?" Marti added another sentence.

Marti's addition

If the value does not match, follow the procedure for correcting imbalances (page 27).

Communicating Ethically

So far, this chapter has introduced concepts and strategies that will serve as the springboard from which you can understand and apply the rest of this book's advice for creating effective communications at work. Because effective communications create change—they make things happen—they also have an important ethical dimension. When things happen, people may be affected. Their happiness, health, and well-being may be impacted. You may write a proposal for a new product that could cause physical harm if a certain expensive design feature is not included. You may prepare a report about a new computer service that, without certain safeguards being created, could put people's personal information at risk of being hacked. A manufacturing process you are evaluating may produce waste harmful to the environment. Because technical communications have an important ethical dimension, in addition to providing advice for creating useful, persuasive communications, this book provides advice for creating ethical ones.

At work, you have three major sources for guidance.

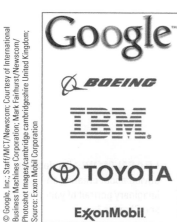

These corporations are among the thousands worldwide that have adopted ethics codes to guide employee and company conduct.

- **Professionals in your specialty have probably developed a code of ethics.** The Center for Business Ethics has collected hundreds of codes of ethics that corporations have adopted as their official policies. For a link, go to MindTap.

- **Your employer may also have developed an ethics code.** Some companies have even hired professional ethics specialists whom employees may consult.

- **Your own sense of values—the ones you developed in your home, community—and studies, is also crucial.**

A basic challenge facing anyone addressing ethical issues is that different people have different values and, consequently, different views of the right actions to take in various situations. This fact should not surprise you. For thousands of years, philosophers have offered various incompatible ethical systems. They have yet to reach agreement. Moreover, people from different cultural backgrounds and different nations adhere to different values. On your own campus, you surely know other students with whom you disagree on ethical issues.

The same thing happens in the workplace. Consequently, this book won't tell you what your values ought to be. Instead, it seeks to help you act in accordance with values you have. Toward that end, it seeks to enhance your sensitivity to often subtle and difficult-to-detect ethical implications so that you don't inadvertently end up preparing a communication that affects people in ways you wish it hadn't. This book also presents some ways of looking at the ethical aspects of various writing decisions—such as the way you use colors in graphs—that you may not have considered before. Ultimately, however, the book's goals are to help you communicate in ways that, after careful consideration, you believe to be ethical and to enable you to build the communication expertise needed to influence others when you want to raise ethical questions.

LEARN MORE To learn strategies for recognizing and addressing the ethical issues that arise on the job, read the Ethics Guidelines included in other chapters.

How to Get Lasting Value from This Book and Your Course

As explained at the beginning of this chapter, new college graduates have difficulty using what they've learned about writing in college once they are on the job. For some, it's almost as if they have to start learning to write all over again.

Your instructor and this book share the goal of helping you carry forward the power and value of what you learn about writing in this course. Based on recent research on the transfer of knowledge and skills from one context to another (Bransford et al., 2000; Yancey et al., 2015), we will help you learn a strategy with two simple parts: *reflection* and a *framework* to guide your reflecting.

Reflection

The first element of the strategy is simply to reflect. Instead of rushing from project to project, pause briefly to use and build your transferable knowledge and skills.

- When you begin work on a writing project or when starting to draft part of a large project, take time to think about what you already know that could help you create this new communication. These include not only strategies and skills you learn in this course but also ones you knew beforehand and learned elsewhere.
- When you finish a project, take time to think about the strategies and skills that you used that you could employ in the future.

Writing down your reflections can help cement them in your memory and may help you think more deeply about them.

Framework

The strategy's second part is your conceptual *framework*—your personal description or "theory"—of writing. It identifies the activities you perform when writing and tells how they fit together. To help you begin to develop your personal framework or theory of workplace writing, here is one (of many) that could be used to describe the reader-centered approach that this book and your instructor teach. As you proceed through the course, you can modify or replace the example framework with your own.

This example framework has two groups of items. As Figure 1.6 shows, you begin by learning about the elements of your writing situation.

- **Your reader**—including his or her specific needs, desires, and expectations.
- **Context**—the practical, personal, political, and other features of the situation that will affect your reader's response—especially initial response—to your message.

FIGURE 1.6

Sample Framework for the
Workplace Writing Process

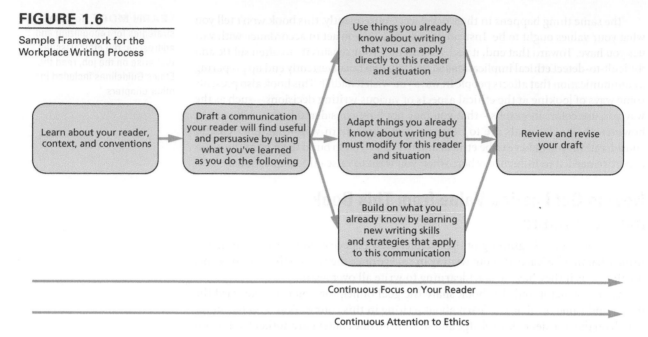

Learn about your reader, context, and conventions

Draft a communication your reader will find useful and persuasive by using what you've learned as you do the following

Use things you already know about writing that you can apply directly to this reader and situation

Adapt things you already know about writing but must modify for this reader and situation

Build on what you already know by learning new writing skills and strategies that apply to this communication

Review and revise your draft

Continuous Focus on Your Reader

Continuous Attention to Ethics

- **Conventions of workplace writing**—the factors that shape your reader's expectations about your communication, such as the tone that is appropriate for it and the conventional features of the superstructure you are using.

The second group includes the writing skills and knowledge you will use to create your communication.

- **Writing skills and strategies you already have that you can use.** These include skills involved at all levels of writing, from overall content and organization to grammar and punctuation, as well as your skill at using or adapting workplace writing conventions.

- **Writing skills and strategies you already know but must adapt** to this specific reader and situation.

- **Writing skills and strategies you need to develop** to create a communication your reader will find useful and persuasive.

In addition, the example framework includes your alertness and response to any ethical issues that may arise as you plan, write, and revise your communication.

In reality, the writing process rarely proceeds in such a straightforward manner as shown in Figure 1.6. Especially for longer communications, there's always back and forth among the activities.

Of course, the more writing knowledge and skills you develop now, the more you can use and adapt in the future. All chapters in this book and all the guidance and assignments your instructor provides will help you extend what you already know about writing.

Developing the Habit of Reflecting

No matter how well you have developed your personal framework or theory of writing by the end of this course, it will be of little value to you unless you use it to reflect after you begin your career. This book is designed to help you develop the habit of reflecting on the framework's four components so that you store away the most valuable lessons from this class for use on the job, and so that you continue to build your knowledge and expertise as you encounter new writing experiences after graduation. Most chapters include reflection exercises. Your instructor may also choose to assign some of the reflection exercises in Appendix B that will help you focus on what you already know as you begin an assignment and on what you have learned when you have completed it.

Also, this book gives (reader-centered) reasons for each piece of advice that it provides. The advice is always presented as guidelines, not rules, leaving it up to your judgment whether it applies, applies with modification, or should be ignored in any particular situation.

In addition, your instructor will give what a book simply cannot: feedback on your writing and reflections that are so important to your learning to write in a new way.

In your effort to learn to write effectively in your career, I wish you good luck and great success.

USE WHAT YOU'VE LEARNED

EXERCISE YOUR EXPERTISE

1. Interview someone who holds the kind of job you might like to have. Ask about the types of communications the person writes, the readers he or she addresses, the writing process and technology the person uses, and the amount of time the person spends writing. Supplement these questions with any others that will help you understand how writing fits into this person's work. According to your instructor's directions, bring either notes or a one-page report to class.

2. Find a communication written by someone who has the kind of job you want, perhaps by asking a friend, family member, or your own employer. Explain the communication's purposes from the point of view of both the writer and readers. Describe some of the writing strategies the writer has used to achieve these purposes.

EXPLORE ONLINE

Explore websites created by two organizations in the same business (airlines, computers, museums, etc.) or two employers for whom you might like to work. Compare the strategies used to make the sites useful and persuasive. Note ways their usefulness and persuasiveness might be increased.

COLLABORATE WITH YOUR CLASSMATES

Working with another student, rewrite the email message by Patrick (see page 14) so that it will be more likely to persuade the personnel manager to follow Patrick's recommendation. Assume Patrick knows the manager's clerks are miscopying because he has examined the time sheets, time tickets, and computer files associated with thirty-seven incorrect payroll checks; in thirty-five cases, the clerks made the errors. Take into account the way you expect the personnel manager to react upon finding an email from Patrick in his or her in-box. Make sure that the first sentence of your revision addresses a person in that frame of mind and that your other sentences lead effectively from there to the last sentence, which you should leave unchanged.

APPLY YOUR ETHICS

As a first step in bringing your personal values to your on-the-job communication, list those values that you think will matter most in your career. Explain situations in which you think it may be especially important for you to be guided by them.

REFLECT FOR TRANSFER

1. Send your instructor a memo briefly describing a writing situation in school or outside. Explain what you knew about your reader and the reader's context that would influence his or her response to your message. Describe one or two decisions you made about how you would write as a result of what you knew about the reader and his or her context. What did you do in this situation that would be helpful to do when you write on the job?

2. Send your instructor a memo briefly describing the writing knowledge and skills you already have that you believe will help you write successfully on the job. What additional things do you feel you will most need to learn in this course?

Case: In MindTap, the case titled "Helping Mickey Cheleni Pick the Right Forklift Truck" is well suited for use with this chapter.

2

Overview of the Reader-Centered Approach: Writing for a Job

Three ways employers read job applications

MindTap®

Find additional resources related to this chapter in MindTap.

This chapter has the twin purposes of providing an overview of the reader-centered writing process—the rest of the book provides details—and helping you write a job application that grabs employers' attention. To fit these two purposes together, the chapter shows how you can take a reader-centered approach to your work on your résumé, application letter, and professional portfolio.

The résumé, letter, and portfolio are ideal subjects for introducing you to the reader-centered process because they illustrate with special clarity the point that the success of your work-related writing depends on your reader's response to them. No matter how good you or anyone else believes your résumé and letter to be, they succeed only if an employer responds by inviting you for an interview or offering you an internship or job. Thus, in addition to helping you use a reader-centered approach to your own on-the-job writing, this chapter will help you see *why* it is important to do so.

Understanding Your Reader

So, let's begin where successful writing on the job always begins: by understanding the reader, the employer you would like to hire you.

From the employer's perspective, the job application process looks much different than it does from your standpoint. You may perceive that the goal of your application is to display the accomplishments and activities that you are most proud of. But the employer isn't focused on what you think are your impressive qualifications. Instead, the employer wants to hire a person who can perform a certain, specific job to help the organization perform at its highest level. Having advertised the opening, the employer will receive your application and many others. The employer's task is to sift and sort through these applicants to find the one person who will best contribute to the organization's success.

The sifting and sorting generally involves the following three steps.

■ **Initial screening.** Many employers receive hundreds, even thousands, of applications a week. One person may be charged with scanning quite quickly through this deluge, hoping to spot promising candidates. Often such initial readers are employees in a personnel office, not specialists in the area where you are seeking a position. They may spend less than a minute on each application. Unless they rapidly find clear, relevant evidence that your qualifications match the employer's needs, they will set your application aside and pick up the next one. To reduce this screening workload, an increasing number of employers first submit all résumés to a computer, often by having applicants apply online. Only those that meet predefined criteria are read by a human.

■ **Detailed examination of the most promising applications.** Applications that pass the initial screening are forwarded to managers and others in the department that wishes to hire a new employee. These readers know exactly what qualifications they desire, and they will examine your materials for those specific skills and abilities.

■ **Preparation for in-depth interviewing.** Applicants whose materials are persuasive to second-stage readers are invited for interviews. Their application materials are usually read with great care at this stage. Often managers and others will use them as the basis for questions to ask during the interview.

The rest of this chapter leads you though the reader-centered approach to creating job applications that lead the employer to single you out as a very promising candidate for its opening, one who should be offered an interview and, ultimately, a job.

Your Three-Part Job Application

In the past, college students have prepared two items for their initial job applications: a résumé and accompanying letter. But those documents can only *describe* your abilities and accomplishments. The Internet and social media make it easy for you to add a third element—a professional online portfolio—that *shows* actual examples of your work and evidence of your accomplishments.

This chapter provides advice about all three elements of your job application materials, beginning with the résumé.

But First: Check Your Online Presence

An essential early step in your job search is reviewing and, if necessary, revising the portrait you have created of yourself through social media. It might seem that only your application materials will provide all the information that employers will have about you. Not so. Ninety-three percent of hiring managers review applicants' profiles on social media, especially Facebook, before making an offer (Davidson, 2014).

For many applicants this means disaster. Fifty-five percent of hiring managers have reconsidered a potential hire as a result of what they saw. In 61 percent of these cases, the reconsiderations were negative. Perhaps surprisingly, evidence of illegal activities and offensive content are not the only sources of disqualification. Sixty-six percent reported that poor spelling and grammar cause major concern.

This chapter and your instructor will help you create a clear, coherent, and distinctive image of what you can bring to an employer. Before you send out your first application for any kind of job, even an internship or volunteer position, review and revise your social media presence. For many employers, it is part of your application.

How to Write an Effective Résumé

If you've prepared a résumé before, you have probably seen much general advice about topics to include (e.g., education), ways to write about them (e.g., give dates), and so on. This general advice is good, but it leads many to create applications that look like everyone else's. You can do better than that.

GUIDELINE 1 **Address your résumé to specific employers**

To know what an employer wants, you must first know who the employer is. If you haven't identified the jobs, internships, or co-op positions you'd like, that's your

TRY THIS Haven't decided on your career yet? Go to a website that offers a free service or test for people trying to find careers they would like. See if it suggests any careers you hadn't thought of.

first step. The traditional, and still indispensable, method of doing so is to talk with other people, such as instructors in your major, staff at the college placement office, friends, and family members. As you speak with each person, ask him or her to suggest others who can assist you. Read newspaper ads and ads in the publications and websites of professional organizations in your field. You can also consult such sites as CareerOneStop (www.careeronestop.org), commercial job boards such as Monster (www.monster.com), and state and local job listings.

Social media have created other sources for leads. LinkedIn enables you to specify various features of the job you'd like (field, location, etc.), and it shows you current openings that match your criteria. See Figure 2.1. LinkedIn also enables employers to search through the descriptions posted by job seekers. At Indeed you can upload your résumé. At LinkedIn, you can create your profile to include résumé information and fashion it as a professional portfolio, as discussed later in this chapter. Many employers announce job openings at their websites. While many people associate Twitter, Facebook, and other social media as recreational or personal resources, a growing number of employers have accounts where they announce job openings. Find companies and professional organizations that do and follow them, asking for instant updates, if that service is available.

GUIDELINE 2 Define your résumé's objectives by learning exactly what your reader wants

Having identified employers to whom you will apply, determine as precisely as you can the specific qualifications they are seeking in the person they wish to hire for the kind of position you want. It can be helpful to begin (but not end) by considering the three general categories of qualifications desired by all employers of college graduates:

FIGURE 2.1

Online Resource for Finding Job Openings

Search for jobs by title, keywords or company, and location

Define criteria for new jobs you'd like to be notified about: location, size, and industry

Sign up for alerts here

Companies with openings

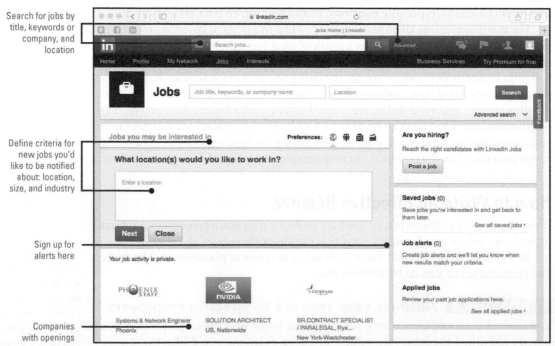

Source: LinkedIn Corporation

What Employers Seek in New Employees

- **Technical expertise.** Employers want to hire people who can perform their jobs adeptly with a minimum of on-the-job training.
- **Supporting abilities.** Most jobs require a wide range of abilities beyond the purely technical ones of your degree or specialty. These often include interpersonal, time management, and project management skills, among others. They almost always include communication skills.
- **Favorable personal qualities.** Employers want to hire motivated, self-directed, and responsible individuals. They also want people who will work well with their other employees as well as with customers, vendors, and others outside the organization.

Of course, for any particular job, employers seek specific technical know-how, supporting abilities, and personal attributes. Instead of asking, "Does this applicant have technical expertise?" an employer will ask, "Can this person analyze water samples for hazardous substances?" "Write computer programs in Linux?" "Design a bridge?" Your task is to find out and list the precise set of technical know-how, supporting abilities, and personal attributes sought by the employers to whom you are applying.

To identify these specific qualifications, learn as much as you can about the daily work of people in the position you would like to hold. Interview someone who holds such a position, talk with instructors in your major, and read materials about careers at your library or campus employment center. If you have an advertisement or job description, study it carefully. As you identify qualifications the employer wants, put them in a list. The longer and more detailed your list, the better prepared you will be to write a résumé that achieves its persuasive objective, which is to demonstrate that you have the knowledge, skills, and personal qualities that the employer seeks.

LEARN MORE For detailed advice about conducting interview and library research, turn to Chapter 5.

GUIDELINE 3 Think creatively about the ways your knowledge and experience match the qualifications the employer wants

Your reader-centered understanding of your résumé's objectives—of who, exactly, your reader is and how, specifically, it must help and persuade this person—will provide a dependable guide throughout the rest of your work on your résumé.

To begin, your knowledge of the particular qualifications the employer is looking for will help you determine what to say specifically about your education, experience, and activities. Of course, you should say where you go to school and what your major is. Your understanding of the employer's focus will, in addition, help you identify the courses and projects to highlight: the ones that most closely match the employer's needs. So that you don't miss any, you might begin with the list you created when detailing the qualifications desired by the employer. Next to each one, write the course, activity, previous job, or other experience through which you gained or demonstrated the desired knowledge or ability. You might list your accomplishments and areas of knowledge, noting the connection each item may have with a qualification for the job you want. You may later decide not to include some of the ideas on your list, but don't hold back on exploring possibilities now.

When planning what to say in their résumés, people often ask, "How long should my résumé be?" The answer is that it should be as short as possible while still presenting the facts employers will find most persuasive. For most undergraduates, this is one page. However, some undergraduates have extensive qualifications that justify

iStockphoto.com/Yuri_Arcurs

Effective résumés are targeted to specific employers who are seeking employees with specific qualifications.

How long should your résumé be?

a second page. Many experienced workers do as well. If your résumé runs past one page by a small amount, redesign your layout, edit your content, or delete something to make it fit on a single page.

GUIDELINE 4 Choose the type of résumé that will display your qualifications most effectively

Although people often think of résumés as being a single genre, there are two common variations, each with its own generic conventions. Choose the one best suited to highlighting your qualifications.

- **Experiential résumé.** In an experiential résumé, you organize information about yourself around your experiences, using headings such as "Education," "Employment," and "Activities." Under these headings, you describe your experiences in ways that demonstrate that you possess the qualifications the employer wants. An experiential résumé is the best choice for most college students and persons new to their careers. Figures 2.2 and 2.3 (pages 27 and 28) show examples of an experiential résumé.

- **Skills résumé.** In a skills résumé, you create key sections around your abilities and accomplishments, using such headings as "Technical Abilities," "Management Experience," and "Communication Skills." Later, you list the college you attended and jobs you've held. Skills résumés work best for people with enough professional experience to be able to list several on-the-job responsibilities and accomplishments in each of several categories. Figure 2.4 (page 31) shows a skills résumé.

GUIDELINE 5 Draft your résumé's text to highlight the qualifications that will most appeal to the employer

It's one thing to identify the things you should say in your résumé. It is quite another to figure out how to say them persuasively. On the one hand, your résumé needs to make your qualifications clear by supplying specific, relevant details. On the other hand, employers want to find information about you quickly. No extra words. No long explanations.

The following sections provide advice about the writing style to use and ways to write persuasively about the major topics you are likely to include. The goal in each case is for you to write your résumé in a way that makes it easiest for your particular reader to see how well your knowledge and experience match the specific qualification he or she is looking for, the ones you identified when defining your résumé's objectives.

Writing Style for Résumés

The best résumés, the ones to which employers respond most favorably, use a crisp, energetic style that conveys many facts with few words.

- **Write concisely.** Typically, résumés contain lists of sentence fragments and phrases, not full sentences. Eliminate every word that you can without losing essential information.

Wordy | Conducted an analysis of the strength of the load-bearing parts

Concise | Analyzed strength of load-bearing parts

■ **Be specific.** Employers find generalities to be ambiguous and unimpressive. Give precise details.

Proficient on the computer	General
Program in Linux, SQL, CAD, and php	Specific
Reduced breakage on the assembly line	General
Reduced breakage on the assembly line by 17%	Specific

■ **Choose the correct verb tenses.** Use the past tense for activities completed. Use the present tense for those you continue to perform.

Designed automobile suspensions	Completed activity
Design aircraft landing gear	Continuing activity

By convention, the word "*I*" is used in only one place in résumés: the professional objective.

Name and Contact Information

Help your readers locate your résumé in a stack of applications by placing your name prominently at the top of the page. Enable them to contact you quickly for an interview or with a job offer by including your postal address, email address, and phone number. If you'll be living at another address during the summer or other part of the year, give the relevant information.

Objective

By naming the job you want, you help yourself focus the rest of your résumé on providing evidence that you possess the knowledge and experience needed to do excellent work in the job you've named. You help your reader identify the criteria to apply when reading your résumé: Do you have the qualifications required for the position you identified?

By taking a reader-centered approach to writing your objective, you will gain a decided advantage over most other applicants for the same position. Many job seekers describe the position they want by telling what they will gain if they have it. They will learn. They will have a chance to advance. And so on. But employers aren't looking for people to give things to. They want employees who will help them achieve their goals. When writing your objective, take the reader-centered approach of telling the employer what you will give, not what you would like to get. Here's a simple procedure for doing that:

1. **Identify the results sought by the department or unit in which you wish to work.** If you don't know what they are, ask a professor, talk with someone in the kind of job you want, or contact your campus employment center for assistance.

2. **State that your objective is to help an employer achieve those results.**

Howard, a senior in health studies, and Jena, a computer science major, used this process to improve the first drafts of their objectives. Their initial drafts read as follows.

Howard: A position where I can extend my knowledge of nutrition and sports studies

Jena: A position as a computer systems analyst in the field of software development with an innovative and growing company

Writer-centered vague objectives

When Howard and Jena considered their statements from their readers' perspectives, they realized that they said nothing about how they would benefit the employers. They revised their objectives to read as follows:

Reader-centered, results-oriented revisions

> **Howard:** To develop nutrition and exercise programs that help athletes achieve peak performance

> **Jena:** To create software systems that control inventory, ordering, and billing

If you are applying for an internship, co-op, or summer job, tell what you will "help" or "assist" the employer to do.

Objective for internship

> An internship helping design high-quality components for the electronics industry

Some students ask whether they really need to include an objective statement. In fact, an objective may be unnecessary for people advanced enough in their careers to be able to begin with a summary of their qualifications and accomplishments, thereby clearly indicating the specific kind of job they are seeking. For people looking for an entry-level job, the objective is essential. Employers are looking for applicants whose qualifications meet their needs. They are not trying to find jobs for applicants.

Education

Place your education section immediately after your identifying information and objective, unless you have had enough work experience to make the knowledge you gained at work more impressive to employers than the knowledge you gained in classes. Name your college, degree, and graduation date. If your grade point average is good, include it. If your average in your major is higher, give it. But don't stop there. Provide additional details about your education that employers will see as relevant to the job you want. Here are some examples.

Facts to Highlight about Your Education

You can highlight any of these credentials by giving it a separate heading (for example, "honors" or "internship").

- Advanced courses directly relevant to the job you want (give titles, not course numbers)
- Courses outside your major that broaden the range of abilities you would bring to an employer
- Internships, co-op assignments, or other on-the-job academic experiences
- Special projects, such as a thesis or a design project in an advanced course
- Academic honors and scholarships
- Study abroad
- Training programs provided by employers

The résumés of Jeannie Ryan (Figure 2.2) and Ramón Perez (Figure 2.3) show how two very different college seniors elaborated on the ways their educations qualify them for the jobs they are seeking.

Work Experience

When drafting your work experience section, list the employers, their cities, your job titles, and your employment dates. Then present facts about your work that will impress employers.

FIGURE 2.2

Experiential Résumé

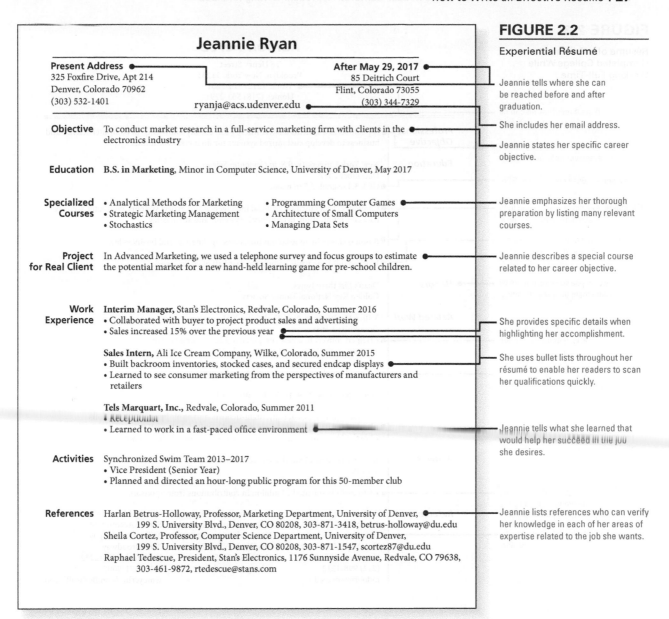

Jeannie Ryan

Present Address	**After May 29, 2017**
325 Foxfire Drive, Apt 214	85 Deitrich Court
Denver, Colorado 70962	Flint, Colorado 73055
(303) 532-1401	(303) 344-7329

ryanja@acs.udenver.edu

Objective — To conduct market research in a full-service marketing firm with clients in the electronics industry

Education — B.S. in Marketing, Minor in Computer Science, University of Denver, May 2017

Specialized Courses
- Analytical Methods for Marketing
- Strategic Marketing Management
- Stochastics
- Programming Computer Games
- Architecture of Small Computers
- Managing Data Sets

Project for Real Client — In Advanced Marketing, we used a telephone survey and focus groups to estimate the potential market for a new hand-held learning game for pre-school children.

Work Experience

Interim Manager, Stan's Electronics, Redvale, Colorado, Summer 2016
- Collaborated with buyer to project product sales and advertising
- Sales increased 15% over the previous year

Sales Intern, Ali Ice Cream Company, Wilke, Colorado, Summer 2015
- Built backroom inventories, stocked cases, and secured endcap displays
- Learned to see consumer marketing from the perspectives of manufacturers and retailers

Tels Marquart, Inc., Redvale, Colorado, Summer 2011
- Receptionist
- Learned to work in a fast-paced office environment

Activities — Synchronized Swim Team 2013–2017
- Vice President (Senior Year)
- Planned and directed an hour-long public program for this 50-member club

References — Harlan Betrus-Holloway, Professor, Marketing Department, University of Denver, 199 S. University Blvd., Denver, CO 80208, 303-871-3418, betrus-holloway@du.edu
Sheila Cortez, Professor, Computer Science Department, University of Denver, 199 S. University Blvd., Denver, CO 80208, 303-871-1547, scortez87@du.edu
Raphael Tedescue, President, Stan's Electronics, 1176 Sunnyside Avenue, Redvale, CO 79638, 303-461-9872, rtedescue@stans.com

Annotations:

Jeannie tells where she can be reached before and after graduation.

She includes her email address.

Jeannie states her specific career objective.

Jeannie emphasizes her thorough preparation by listing many relevant courses.

Jeannie describes a special course related to her career objective.

She provides specific details when highlighting her accomplishment.

She uses bullet lists throughout her résumé to enable her readers to scan her qualifications quickly.

Jeannie tells what she learned that would help her succeed in the job she desires.

Jeannie lists references who can verify her knowledge in each of her areas of expertise related to the job she wants.

Facts to Highlight about Your Work Experience

- **Your accomplishments.** Describe projects you worked on, problems you addressed, goals you pursued, products you designed, and reports you helped write. Where possible, emphasize specific results—number of dollars saved, additional units produced, or extra customers served.
- **Knowledge gained.** Be resourceful in highlighting things you learned that increased your ability to contribute to your future employer. Realizing that her duties of stacking ice cream packages neatly in supermarket freezers might not seem relevant to a job in marketing, Jeannie describes the insight she gained in that job: "Learned to see consumer marketing from the perspectives of manufacturers and retailers" (Figure 2.2).
- **Responsibilities given.** If you supervised others, say how many. If you controlled a budget, say how large it was. Employers will be impressed that others have entrusted you with significant responsibility.

FIGURE 2.3

Résumé of a Person Who
Completed College While
Working Full-Time

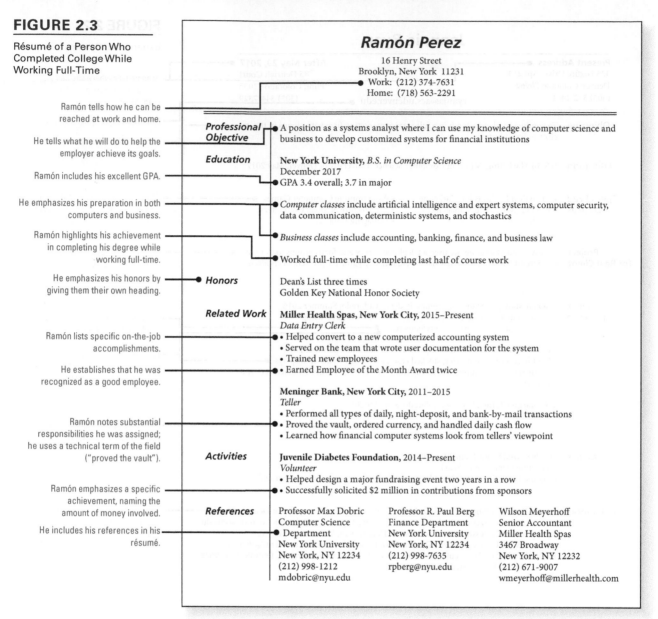

Ramón tells how he can be
reached at work and home.

He tells what he will do to help the
employer achieve its goals.

Ramón includes his excellent GPA.

He emphasizes his preparation in both
computers and business.

Ramón highlights his achievement
in completing his degree while
working full-time.

He emphasizes his honors by
giving them their own heading.

Ramón lists specific on-the-job
accomplishments.

He establishes that he was
recognized as a good employee.

Ramón notes substantial
responsibilities he was assigned;
he uses a technical term of the field
("proved the vault").

Ramón emphasizes a specific
achievement, naming the
amount of money involved.

He includes his references in his
résumé.

Ramón Perez

16 Henry Street
Brooklyn, New York 11231
Work: (212) 374-7631
Home: (718) 563-2291

Professional Objective
A position as a systems analyst where I can use my knowledge of computer science and business to develop customized systems for financial institutions

Education
New York University, *B.S. in Computer Science*
December 2017
GPA 3.4 overall; 3.7 in major

Computer classes include artificial intelligence and expert systems, computer security, data communication, deterministic systems, and stochastics

Business classes include accounting, banking, finance, and business law

Worked full-time while completing last half of course work

Honors
Dean's List three times
Golden Key National Honor Society

Related Work
Miller Health Spas, New York City, 2015–Present
Data Entry Clerk
• Helped convert to a new computerized accounting system
• Served on the team that wrote user documentation for the system
• Trained new employees
• Earned Employee of the Month Award twice

Meninger Bank, New York City, 2011–2015
Teller
• Performed all types of daily, night-deposit, and bank-by-mail transactions
• Proved the vault, ordered currency, and handled daily cash flow
• Learned how financial computer systems look from tellers' viewpoint

Activities
Juvenile Diabetes Foundation, 2014–Present
Volunteer
• Helped design a major fundraising event two years in a row
• Successfully solicited $2 million in contributions from sponsors

References

Professor Max Dobric	Professor R. Paul Berg	Wilson Meyerhoff
Computer Science	Finance Department	Senior Accountant
Department	New York University	Miller Health Spas
New York University	New York, NY 12234	3467 Broadway
New York, NY 12234	(212) 998-7635	New York, NY 12232
(212) 998-1212	rpberg@nyu.edu	(212) 671-9007
mdobric@nyu.edu		wmeyerhoff@millerhealth.com

When organizing and describing your work experience, follow these guidelines for achieving high impact.

Organizing and Describing Your Work Experience

■ **List your most impressive job first**. If your most recent job is your most impressive, list your jobs in reverse chronological order. If an older job, will be more impressive to the employer, create a special heading for it, such as "Related Experience." Then describe less relevant jobs in a later section titled "Other Experience."

- **Put your actions in verbs, not nouns.** Verbs portray you in action. Don't say you were "responsible for the analysis of test data" but that you "analyzed test data."
- **Use strong verbs.** When choosing your verbs, choose specific, lively verbs, not vague, lifeless ones. Avoid saying simply that you "made conceptual engineering models." Say that you "designed" or "created" the models. Don't say that you "interacted with clients" but that you "responded to client concerns."
- **Use parallel constructions.** When making parallel statements, use a grammatically correct parallel construction. Nonparallel constructions slow reading and indicate a lack of writing skill.

LEARN MORE For more information about putting actions in verbs, see Chapter 10, "Put the Action in Your Verbs."

NOT PARALLEL

- Trained new employees
- Correspondence with customers
- Prepared specifications

PARALLEL

- Trained new employees
- Corresponded with customers
- Prepared specifications

Changing *correspondence* to *corresponded* makes it parallel with *trained* and *prepared*.

Activities

At the very least, participation in group activities indicates that you are a pleasant person who gets along with others. Beyond that, it may show that you have acquired certain abilities that are important in the job you want. Notice, for instance, how Jeannie Ryan describes one of her extracurricular activities in a way that emphasizes the managerial responsibilities she held (Figure 2.2).

Synchronized Swim Team 2013–2017

- Vice President (Senior Year)
- Planned and directed an hour-long public program for this 50-member club

Emphasis on management responsibilities

Special Abilities

Let employers know about exceptional achievements and abilities of any sort, using such headings as "Foreign Languages" or "Certifications."

Interests

If you have interests such as golf or skiing that could help you build relationships with co-workers and clients, you may wish to mention them, although a separate section for interests is unnecessary if the information is provided in your activities section.

Personal Data

In the United States, federal law prohibits employers from discriminating on the basis of sex, religion, color, age, or national origin. It also prohibits employers from inquiring about matters unrelated to the job for which a person has applied. For instance, employers cannot ask if you are married or plan to marry. Many job applicants welcome these restrictions because they consider such questions to be personal or irrelevant. On the other hand, federal law does not prohibit you from giving employers information of this sort if you think it will help to persuade them to hire you. If you include such information, place it at the end of your résumé, just

TRY THIS At a website that offers advice to job seekers, find a "model" résumé. Evaluate it in light of the reader-centered advice given in this chapter; what are its strengths and weaknesses?

before your references. It is almost certainly less impressive than what you say in the other sections.

References

Your choice of references can be one of the most important decisions you make about your résumé. The names and titles of the accomplished people who will speak favorably about you can create a very positive impression on an employer.

Who would make a good reference? Select persons who can speak about your ability to excel in the job for which you are applying. Professors are good choices, especially ones who taught you skills needed in the job for which you are applying. These may include professors from outside your major who teach career-related subjects. Your technical communication instructor is an example. Former employers can make excellent references, especially if the work was related to the position for which you have applied. Consider also advisers of campus organizations.

Avoid listing your parents and their friends, who (an employer might feel) are going to say nice things about you no matter what.

Be sure to provide titles, business addresses, phone numbers, and email addresses. If one of your references has changed jobs so that his or her business address doesn't indicate how the person knows you, give the needed information in your cover letter: "My supervisor while at Sondid Corporation."

Obtain permission from the people you want to list as references so they aren't taken by surprise by a phone call. Give your résumé to your references so they can quickly review your qualifications when they receive an inquiry from an employer.

Let's be clear: The process of choosing, getting permission from, and effectively listing each reference requires some work. What about just saying, "References available upon request"? As the first paragraph of this section indicates, doing so prevents you from taking advantage of a great opportunity to impress the employer. The one situation in which it makes sense to omit the names of references is when you are already employed and don't want your current employer to know that you are looking for another job.

Drafting the Text for a Skills Résumé

A skills résumé has the same aims as an experiential one. The chief difference is that a skills résumé consolidates the presentation of your accomplishments and experience in a special section located near the beginning rather than weaving this information into your sections on education, work experience, and activities. Figure 2.4 shows an example.

For this special section, use a title that emphasizes its contents, such as "Skills" or "Skills and Achievements." Within the section, use subheadings that identify the major areas of ability and experience you would bring to an employer. Typical headings include "Technical," "Management," "Financial," and "Communication." However, the specific headings that will work best for you depend on what employers seek when recruiting people for the kind of job you want. For example, the headings in George Shriver's résumé (Figure 2.4) focus specifically on skills required of managers of technical communication departments.

Because you aggregate your skills and accomplishments in a special "Skills" section, the other sections of your skills résumé should be brief in order to avoid redundancy.

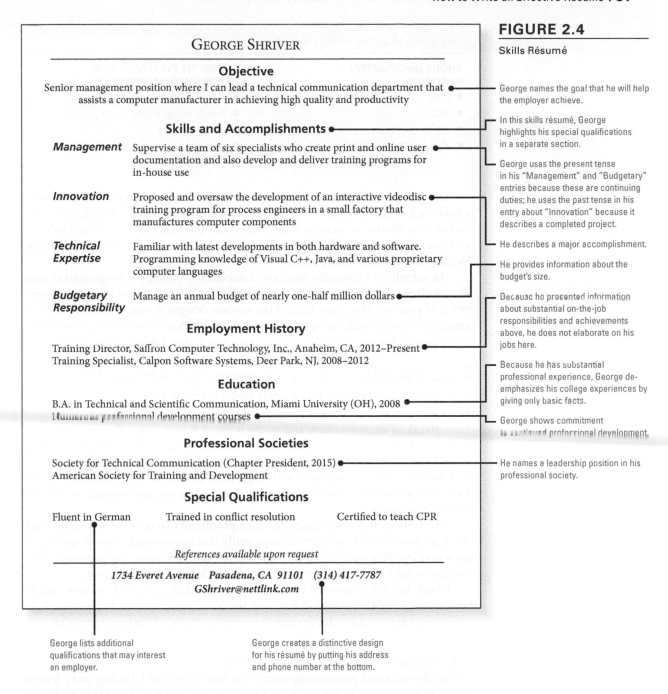

FIGURE 2.4

Skills Résumé

GEORGE SHRIVER

Objective

Senior management position where I can lead a technical communication department that assists a computer manufacturer in achieving high quality and productivity

— George names the goal that he will help the employer achieve.

Skills and Accomplishments

— In this skills résumé, George highlights his special qualifications in a separate section.

Management Supervise a team of six specialists who create print and online user documentation and also develop and deliver training programs for in-house use

— George uses the present tense in his "Management" and "Budgetary" entries because these are continuing duties; he uses the past tense in his entry about "Innovation" because it describes a completed project.

Innovation Proposed and oversaw the development of an interactive videodisc training program for process engineers in a small factory that manufactures computer components

— He describes a major accomplishment.

Technical Expertise Familiar with latest developments in both hardware and software. Programming knowledge of Visual C++, Java, and various proprietary computer languages

— He provides information about the budget's size.

Budgetary Responsibility Manage an annual budget of nearly one-half million dollars

— Because he presented information about substantial on-the-job responsibilities and achievements above, he does not elaborate on his jobs here.

Employment History

Training Director, Saffron Computer Technology, Inc., Anaheim, CA, 2012–Present
Training Specialist, Calpon Software Systems, Deer Park, NJ, 2008–2012

Education

B.A. in Technical and Scientific Communication, Miami University (OH), 2008
Numerous professional development courses

— Because he has substantial professional experience, George de-emphasizes his college experiences by giving only basic facts.

— George shows commitment to continued professional development.

Professional Societies

Society for Technical Communication (Chapter President, 2015)
American Society for Training and Development

— He names a leadership position in his professional society.

Special Qualifications

Fluent in German   Trained in conflict resolution   Certified to teach CPR

References available upon request

1734 Everet Avenue   Pasadena, CA 91101   (314) 417-7787
GShriver@nettlink.com

George lists additional qualifications that may interest an employer.

George creates a distinctive design for his résumé by putting his address and phone number at the bottom.

GUIDELINE 6 Design your résumé's appearance to support rapid reading, emphasize your qualifications, and look attractive

At work, good visual design is crucial to achieving communication goals. Nowhere is that more true than with a résumé, where design must support rapid reading, emphasize your most impressive qualifications, and look attractive. The résumés shown in this chapter achieve these objectives through a variety of methods you can use.

Designing Your Résumé's Visual Appearance

SHORT, INFORMATIVE HEADINGS	DIFFERENT TYPEFACES FOR HEADINGS THAN FOR TEXT
▪ Lists	▪ White space to separate sections
▪ Bullets	▪ Ample margins (3/4" to 1")
▪ Italics	▪ Visual balance
▪ Variety of type sizes	▪ Bold type for headings and key information

You will probably need to experiment with the design of your résumé to create a balanced, attractive page—one that visually emphasizes your qualifications and makes your résumé seem neither too packed nor too thin. To check visual balance, fold your résumé vertically. Both sides should have a substantial amount of type. Neither should be primarily blank.

Do not rely on a computer program's templates or wizards to design your résumé for you. They are generic packages that don't let you create the most favorable presentation of your qualifications. Each of the various designs shown in this chapter's examples are easy to create using the tables in Microsoft Word. For a tutorial on creating your résumé this way, see the Writer's Tutorial on Using Tables to Design a Résumé.

GUIDELINE 7 Revise your résumé to increase its impact and to eliminate errors and inconsistencies

No communication you write needs a higher degree of polish than your résumé. Take plenty of time to review and revise it. Read it over carefully, trying to see it as the employer would. Show your draft to other people. Describe the job you want so they can read from your readers' perspective, imagining what the employer will look for and how the employer will respond to each feature. Good reviewers for your résumé include classmates, your writing instructor, and instructors in your major department.

Run your spell checker again, and then proofread carefully for errors that spell checkers don't catch. Employers say repeatedly that even a single error in spelling or grammar can eliminate a résumé from further consideration. The programs some employers use for initial screening of résumés kick out those with errors.

Check, too, for consistency in the use of italics, boldface, periods, bullets, and abbreviations as well as the format for dates and other parallel items. Finally, check for consistency in the vertical alignment of information that isn't flush against the margin.

Adapting Your Résumé for Different Employers

The reader-centered process emphasizes the importance of tailoring every feature of a communication to the needs and goals of your readers. Does this mean you need to write a new résumé for every job you apply for? Not necessarily. If you are seeking the same type of position with two employers and if both are seeking the same qualifications, the same résumé will probably work equally well with both. However, the greater the differences between the two positions, the more adjustments you should make. In addition to rewriting your objective, you may need to reshape the descriptions of your education, experience, and activities, as well as reorder your résumé's contents.

Figure 2.5 (pages 35–36) shows a Writer's Guide you can use as you prepare your résumé.

USING TABLES TO DESIGN A RÉSUMÉ

This tutorial tells how to design a multicolumn résumé with Word for Mac 2016, which is virtually identical to the latest versions of Word for Windows. If you use a different word processor or get stuck, click on your program's **Help** menu for assistance. After learning the following basic strategies, use your creativity to create a résumé that employers will find useful and persuasive.

Make a Table to Serve as the Visual Framework for Your Résumé

1. Create a new **Blank Document**.
2. Under the **Layout** tab, click on **Margins**.
3. From the dropdown menu, choose **Normal**.
4. Under the **Insert** tab, click on **Table**.
5. In the dropdown window, highlight 3 squares across and 7 down.
   - You can add or delete rows as necessary later.
6. Left click.

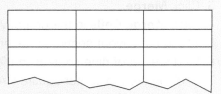

Adjust the Widths of the Columns

1. Under the **View** menu, click on **Ruler**.
2. Move the cursor over the vertical line that separates the left and center columns so that the cursor changes to this: ◄║►
3. Hold down the mouse button.
4. Slide the vertical line to the left until it is at 1.5 inches on the ruler.
   - This narrow left column is for your topic headings. ─────
5. Release the mouse button.
6. Using the same procedure, move the other vertical as far left as you can.
   - This thin column is a gutter between the heading and text columns. ─────

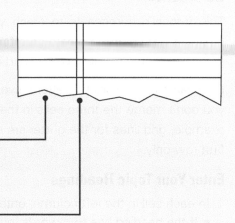

Hide the Table's Borders so They Won't Show When You Print Your Résumé

1. Highlight the entire table (but not anything else).
2. Under the **Table Design** tab, click **Borders**.
3. Click on the arrow next to **Borders**.

4. In the dropdown menu, click **No Border**.
5. Click again on the arrow next to **Border**.
6. At the bottom of the dropdown menu, click on
   View Guidelines.
   • The borders show as thin lines on-screen but will not print.

Create a Heading for Your Résumé

1. Highlight all three columns in the table's top row.
2. Click on **Layout**.
3. Click **Merge**.
4. Click **Merge Cells** in the dropdown menu.
5. Still in the Layout section, click on **Alignment**.
6. Find the set of nine squares in the dropdown menu.
7. Click on the top center square ("Align Top Center").
8. In this wide cell, enter your name and contact information.
9. Click on **Home** to change your name to a larger type and bold.

Be Creative

This tutorial's purpose is to teach you to how to use a word-processing program to create an effective résumé, not to suggest that your résumé should look one particular way. Use your creativity.

For example, here are two alternative headings you can make if you don't merge the three cells in the first row. In the bottom example, grid lines for the gutter are moved to the right in the first row only.

Enter Your Topic Headings

1. In each cell in the left column, enter a topic heading.
   • If the heading has two words, put a **Return** between them.
   • If the column isn't wide enough for the longest word, enlarge the column slightly.
2. Change all the topic headings to bold type.
3. Align the headings on the right side of the column and the tops of their cells.
   • Highlight all the headings.
   • Click **Layout**.
   • Click **Alignment**.
   • In menu, click on the top right square ("Align Top Right").

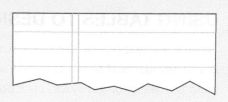

Enter Text

1. In the right-hand column of each row, enter the appropriate text.
2. At the end of each entry, type one **Return**.
   - The **Return** will create a blank line between this entry and the next.
3. If you want to enter multicolumn text in the right-hand column, do the following:
   - Put the cursor in the cell that you want to have columns.
   - Follow the procedure for making a table that is given on the first page of this tutorial.
4. Use bold type, italics, bullets, indentation, and other design elements to create an attractive, easy-to-read design that emphasizes your qualifications.

Experiment with Your Design

- Try using a different typeface for your name and the headings.
- Try aligning the headings on the left side of their column.
- Try adding a vertical rule (line).
   1. Highlight the text cells in the right-hand column.
   2. Under the **Table Design** tab, select **Borders**.
   3. Click the arrow next to **Borders** in the dropdown menu.
   4. In the dropdown menu, click **Left Border**.
- Try adding one or more horizontal rules (lines).
   1. Highlight the cell that has your text.
   2. Under the **Table Design** tab, select **Borders**.
   3. Click on the arrow next to **Borders**.
   4. In the dropdown menu, click **Bottom Border**.

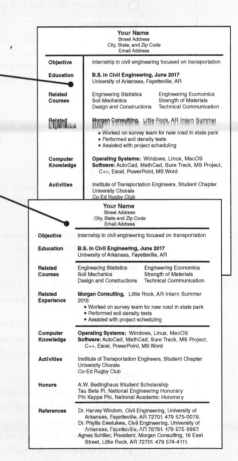

FIGURE 2.5

Writer's Guide for Résumés

Writer's Guide
RÉSUMÉS

This Writer's Guide describes the basic elements of a résumé. Some of the elements would be organized differently in a chronological résumé than in a skills résumé; see pages 30–31.

PRELIMINARY RESEARCH
- ☐ Determine as exactly as possible what the employer wants.
- ☐ Learn enough about the job and employer to tailor your résumé to them.
- ☐ Create a keyword list.

NAME AND CONTACT INFORMATION
- ☐ Enable employers to reach you by mail, phone, and email.

OBJECTIVE
- ☐ Tailor to the specific job you want.
- ☐ Emphasize what you will give rather than what you would like to get.

EDUCATION
- ☐ Tell your school, major, and date of graduation.
- ☐ Provide additional information that shows you are well-qualified for the job you want: academic honors and scholarships, specialized courses and projects, etc.
- ☐ Use headings such as "Honors" and "Related Courses" to highlight your qualifications.

WORK EXPERIENCE
- ☐ Identify each employer's name and city, plus your employment dates.
- ☐ Provide specific details about your previous jobs that highlight your qualifications: accomplishments, knowledge gained, equipment and programs used, responsibilities, etc.

ACTIVITIES
- ☐ Describe your extracurricular and community activities in a way that shows you are qualified, responsible, and pleasant.

INTERESTS
- ☐ Mention personal interests that will help the reader see you as a well-rounded and interesting person.

REFERENCES
- ☐ List people who will be impressive to your readers.
- ☐ Include a mix of references who can speak about your performance in different contexts.
- ☐ Include title, business address, phone, and email address for each reference.
- ☐ Include only people who've given permission to be listed.
- ☐ Omit personal references (family, friends, etc.).

PROSE
- ☐ Present the most impressive information first.
- ☐ Express the action in verbs, not nouns.
- ☐ Use strong verbs.
- ☐ Use parallel constructions.
- ☐ Omit irrelevant information.
- ☐ Use correct spelling, grammar, and punctuation.

FIGURE 2.5

(Continued)

Writer's Guide
RÉSUMÉS

(continued)

VISUAL DESIGN

☐ Look neat and attractive.

☐ Highlight the facts that will be most impressive to employers.

☐ Use headings, layout, and other design features to help readers find specific facts quickly.

ETHICS

☐ List only experiences, accomplishments, degrees, and job titles you've actually had.

☐ Avoid taking sole credit for things you did with a team.

☐ Avoid statements intended to mislead.

Adapting Your Print Résumé or Online Submission for Computer Evaluation

Increasingly, job applicants and employers are requesting, even requiring, that job applications be submitted digitally. Typically, employers who want applications submitted online use a website that has boxes or fields into which you enter your information—a field for education, another for employment, and so on. Typically, these sites ask for the same information that you would put in a résumé but without some of the flexibility you have when creating a résumé yourself.

The most important advice about completing an online application form is to draft your text in a word-processing program, then review and proofread it carefully before pasting it into the employer's form.

Your online application will almost surely be read by a computer before it reaches a human being, so follow the advice given below for scannable résumés. To find applicants who might be invited for a job interview, employers ask the computer to search its database for résumés that have words—*keywords*—that the employers believe would appear in the résumés of good candidates for the opening they want to fill. The computer displays a list of the résumés with the most matches, called *hits*. These are the only résumés a person would read.

To increase the number of hits your résumé will receive, make a list of keywords that employers are likely to ask their computers to look for. Here are some suggestions.

Keywords for Online Job Applications and Résumés

- Words in the employer's ad or job description
- All degrees, certifications, and licenses you've earned: B.A., B.S., R.N. (registered nurse), P.E. (professional engineer license), C.P.A. (certified public accountant)
- Advanced topics you've studied
- Computer programs and operating systems you've mastered: Excel, AutoCAD II, C11, Linux
- Specialized equipment and techniques you've used in school or at work: X-ray machine, PK/PD analysis, ladder-logic
- Job titles and the specialized tasks you performed
- Buzzwords in your field: client server, LAN, low-impact aerobics, TQM (total quality management)
- Names of professional societies to which you belong (including student chapters)
- Other qualifications an employer would desire: leadership, writing ability, interpersonal skills, and such

Increasing the Number of Hits Your Application or Résumé Receives

- Write your keywords as nouns, even if your scannable résumé becomes wordy as a result.
- Be redundant. For instance, write "Used Excel spreadsheet program" rather than "Used spreadsheet program" or "Used Excel." You don't know which term an employer will use, and if both terms are used, you will have two hits.
- Proofread carefully. Computer programs don't catch words that are correctly spelled but are the wrong words (e.g., *in, on, two, too, to*).

Scannable Résumés

A scannable résumé is a printed résumé that will be read into a computer by an optical character reader (OCR). Employers who ask for scannable résumés are usually giving computers responsibility for the first reading. Therefore, follow all of the advice above concerning online applications. An additional challenge is that if the OCR misreads your résumé, some of your information will not be considered. To eliminate this risk, prepare and submit two versions of your résumé, including one that is suited for machine reading.

Ramón Perez prepared the scannable résumé, shown in Figure 2.6. This résumé is longer than Ramón's print résumé (Figure 2.3 on page 28), but length isn't an issue with scannable résumés.

Preparing a Scannable Résumé

- Use a standard typeface (e.g., Times, Arial, or Helvetica).
- Avoid italics, underlining, and decorative elements such as vertical lines, borders, and shading. Boldface, all caps, and bullets are okay.
- Use blank lines and boldface headings to separate sections.
- Use a single column of text.
- Put your name at the top of every page, on a line of its own.
- Use laser printing or high-quality photocopying on white paper.
- Mail your résumé flat and without staples.

FIGURE 2.6

Scannable Résumé

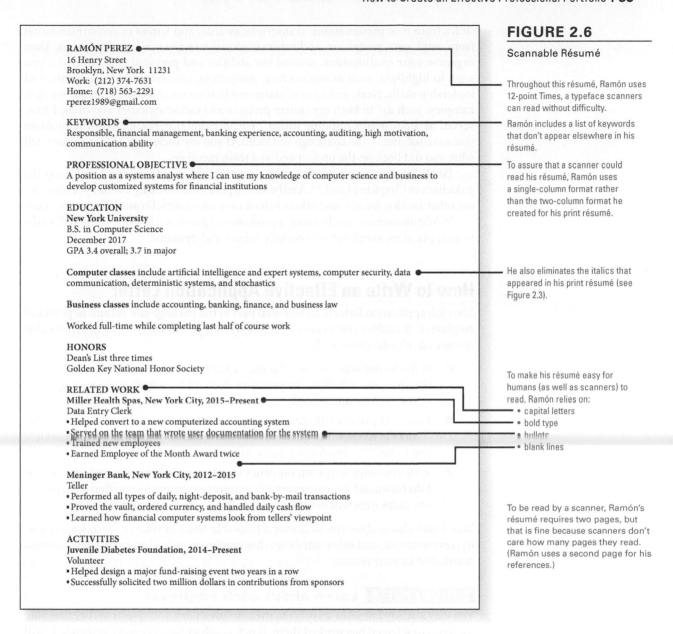

RAMÓN PEREZ
16 Henry Street
Brooklyn, New York 11231
Work: (212) 374-7631
Home: (718) 563-2291
rperez1989@gmail.com

KEYWORDS
Responsible, financial management, banking experience, accounting, auditing, high motivation, communication ability

PROFESSIONAL OBJECTIVE
A position as a systems analyst where I can use my knowledge of computer science and business to develop customized systems for financial institutions

EDUCATION
New York University
B.S. in Computer Science
December 2017
GPA 3.4 overall; 3.7 in major

Computer classes include artificial intelligence and expert systems, computer security, data communication, deterministic systems, and stochastics

Business classes include accounting, banking, finance, and business law

Worked full-time while completing last half of course work

HONORS
Dean's List three times
Golden Key National Honor Society

RELATED WORK
Miller Health Spas, New York City, 2015–Present
Data Entry Clerk
▪ Helped convert to a new computerized accounting system
▪ Served on the team that wrote user documentation for the system
▪ Trained new employees
▪ Earned Employee of the Month Award twice

Meninger Bank, New York City, 2012–2015
Teller
▪ Performed all types of daily, night-deposit, and bank-by-mail transactions
▪ Proved the vault, ordered currency, and handled daily cash flow
▪ Learned how financial computer systems look from tellers' viewpoint

ACTIVITIES
Juvenile Diabetes Foundation, 2014–Present
Volunteer
▪ Helped design a major fund-raising event two years in a row
▪ Successfully solicited two million dollars in contributions from sponsors

Throughout this résumé, Ramón uses 12-point Times, a typeface scanners can read without difficulty.

Ramón includes a list of keywords that don't appear elsewhere in his résumé.

To assure that a scanner could read his résumé, Ramón uses a single-column format rather than the two-column format he created for his print résumé.

He also eliminates the italics that appeared in his print résumé (see Figure 2.3).

To make his résumé easy for humans (as well as scanners) to read, Ramón relies on:
• capital letters
• bold type
• bullets
• blank lines

To be read by a scanner, Ramón's résumé requires two pages, but that is fine because scanners don't care how many pages they read. (Ramón uses a second page for his references.)

How to Create an Effective Professional Portfolio

Nothing will show your abilities more vividly than actual samples of your work and evidence of your accomplishments gathered into an online portfolio. Your school may have software designed to help you create a professional portfolio. Ask your career office. You can use LinkedIn, which allows you to create a full, résumé-like profile and link to external items, such as YouTube videos. And you can create your own portfolio website, a process so simple that you can do it using Microsoft Word. Chapter 20 tells how. Of course, include the online address of your portfolio in your printed résumé and online applications.

Start creating your portfolio by gathering the course, extracurricular, internship, co-op, and volunteer projects that best represent the qualifications desired by the employers to whom you will apply. Along with reports and videos you can include

TRY THIS: Make a list of projects you could include in your professional portfolio. For which ones could you add a photo or create a video?

slides from oral presentations, transcripts, awards, and letters of recommendation from employers, professors, and others who can attest to your qualifications. Then organize your qualifications around the abilities and personal characteristics you want to highlight, such as engineering, computing, communication, diversity, and leadership skills. Next, write a brief statement that serves as a topic sentence for each category, such as "In both my course projects and extracurricular activities, I have served as a consensus-building leader." For individual items, explain the problem you were solving or the challenge you faced. If you are including a team project, tell what you did both on the project and as a team member.

Finally, design your portfolio to be attractive and easy to navigate, following the guidelines in Chapters 14 and 20. And be sure to proofread carefully and ask your instructor, other faculty, friends, and others to look over your portfolio and offer suggestions.

While anyone can easily create a professional portfolio, few do. Yours will make your application stand out as especially robust and dynamic.

How to Write an Effective Application Letter

Your job application letter is an essential part of the package you submit to potential employers. It enables you to answer explicitly and persuasively three questions that résumés don't address directly.

- **Why do you want to work for me instead of someone else?** Answering this question, which most applicants don't address, will make your application stand out immediately.

- **How will you contribute to my organization's success?** Your letter is your chance to tell how the qualifications identified in your résumé will enable you to help the employer achieve its goals.

- **Will you work well with my other employees and the persons with whom I do business?** Because every job requires extensive interaction with others, employers will use your letter to gauge your interpersonal skills.

Your letter also enables you to convey a favorable sense of your enthusiasm, creativity, commitment, and other attributes that employers value but can't easily be communicated in your résumé.

GUIDELINE 1 Learn about each employer

You may already know the answers to the employer's first two questions because you, a parent, or a friend has worked there. If not, conduct the necessary research. It will take time, but ten reader-centered letters will be more productive than many times that number of generic ones.

To answer an employer's first question ("Why do you want to work for me instead of someone else?"), look for things you can praise. You'll find it helpful to categorize the facts you discover as either writer-centered or reader-centered. Examples of *writer-centered facts* are the benefits the organization gives employees or the appealing features of its co-op program. You won't gain anything by mentioning writer-centered facts in your letter because whenever you talk about them you are saying, in essence, "I want to work for you because of what you'll give me." When employers are hiring, they are not looking for people to give things to.

In contrast, *reader-centered facts* concern things the organization is proud of: a specific innovation it has created, a novel process it uses, or a goal it has achieved. These are facts you can build on to create a reader-centered letter.

In addition, research the goals and activities of people who hold the job you want. The more you know about that job, the more persuasively you will be able to answer the reader's second question, "How will you contribute to my organization's success?"

Here are several ways to obtain specific, reader-centered information about an employer.

Learning about Employers

- **Draw on your own knowledge.** If you've already worked for the company, you may know all you need to know. Remember, however, that you need information related to the specific area in which you'd like to work.
- **Ask an employee, professor, or other knowledgeable person.** People often have information that isn't available in print or online.
- **Contact the company.** It might send publications about itself.
- **Search the web.** Start with the employer's own website. Also use resources such as Business Week Online (www.businessweek.com), LexisNexis (www.lexisnexis.com), and the sites for other major newspapers and magazines.
- **Consult your campus's placement office.**
- **Visit the library.** Business newspapers and magazines, as well as trade and professional journals, are excellent sources of information.

LEARN MORE For more detailed advice about how to conduct research, see Chapters 4 and 5.

To answer an employer's second question ("How will you contribute to my organization's success?"), learn as much as you can about the activities of people who hold the particular job you want. What are their goals? What processes, computer programs, or other tools do they use? The more you know about their job, the more persuasively you will demonstrate that you are qualified to perform it.

GUIDELINE 2 Use or adapt the conventional organization for workplace letters

In the workplace, letters have a simple, conventional, three-part structure: introduction, body, and conclusion. There are many ways to organize your answers to an employer's first two questions within that framework.

Use your creativity to make your letter distinctive. The following advice will help you write an original letter that observes the general conventions of business correspondence. Figures 2.7 and 2.8 show how Jeannie Ryan and Ramón Perez applied this advice in their letters.

Introduction

By identifying the job you want in the introduction of your letter, you accommodate the employer's desire to know why you have written. When naming the position, convey your enthusiasm for it.

> I was delighted to see your ad for a chemical engineer in Sunday's newspaper.

To make your enthusiasm credible, explain the *reason* for it. In the following examples, Harlan shaped his explanations to the positions he applied for. To an environmental consulting company, he wrote:

> As an environmental engineering major who will graduate in May, I am eager to help companies assess and remediate hazards.

To a company that makes industrial lubricants, he wrote:

> In my capstone course as a chemical engineering student, I am studying the synthesis of molecules for the same industrial uses that your company's products serve.

Another effective way to begin your letter is to praise an accomplishment, project, or activity you learned about during your research. Praise is almost always welcomed by a reader, provided that it seems sincere. In the following example, Shawana combines the praise with an explanation of her reason for applying to this employer and with her statement of the job she is applying for.

FIGURE 2.7

Job Application Letter to Accompany the Résumé Shown in Figure 2.2

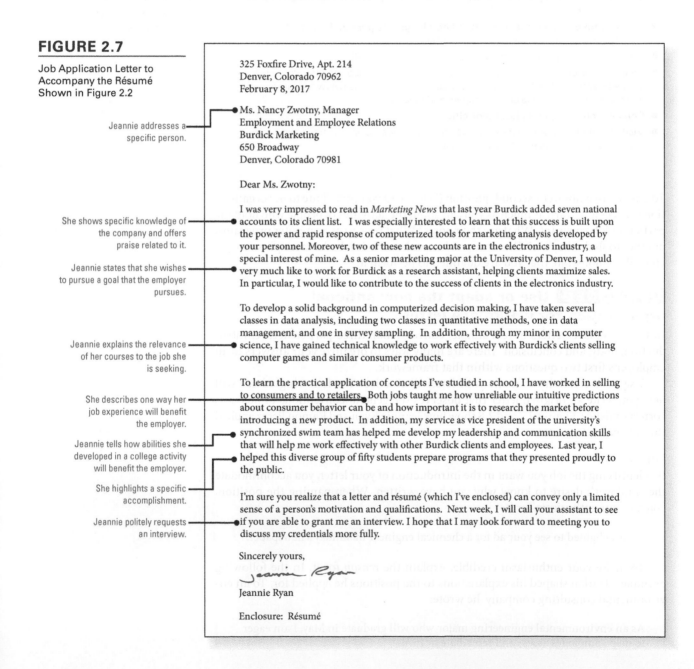

Jeannie addresses a specific person.

She shows specific knowledge of the company and offers praise related to it.

Jeannie states that she wishes to pursue a goal that the employer pursues.

Jeannie explains the relevance of her courses to the job she is seeking.

She describes one way her job experience will benefit the employer.

Jeannie tells how abilities she developed in a college activity will benefit the employer.

She highlights a specific accomplishment.

Jeannie politely requests an interview.

325 Foxfire Drive, Apt. 214
Denver, Colorado 70962
February 8, 2017

Ms. Nancy Zwotny, Manager
Employment and Employee Relations
Burdick Marketing
650 Broadway
Denver, Colorado 70981

Dear Ms. Zwotny:

I was very impressed to read in *Marketing News* that last year Burdick added seven national accounts to its client list. I was especially interested to learn that this success is built upon the power and rapid response of computerized tools for marketing analysis developed by your personnel. Moreover, two of these new accounts are in the electronics industry, a special interest of mine. As a senior marketing major at the University of Denver, I would very much like to work for Burdick as a research assistant, helping clients maximize sales. In particular, I would like to contribute to the success of clients in the electronics industry.

To develop a solid background in computerized decision making, I have taken several classes in data analysis, including two classes in quantitative methods, one in data management, and one in survey sampling. In addition, through my minor in computer science, I have gained technical knowledge to work effectively with Burdick's clients selling computer games and similar consumer products.

To learn the practical application of concepts I've studied in school, I have worked in selling to consumers and to retailers. Both jobs taught me how unreliable our intuitive predictions about consumer behavior can be and how important it is to research the market before introducing a new product. In addition, my service as vice president of the university's synchronized swim team has helped me develop my leadership and communication skills that will help me work effectively with other Burdick clients and employees. Last year, I helped this diverse group of fifty students prepare programs that they presented proudly to the public.

I'm sure you realize that a letter and résumé (which I've enclosed) can convey only a limited sense of a person's motivation and qualifications. Next week, I will call your assistant to see if you are able to grant me an interview. I hope that I may look forward to meeting you to discuss my credentials more fully.

Sincerely yours,

Jeannie Ryan

Jeannie Ryan

Enclosure: Résumé

While reading the August issue of *Automotive Week,* I learned that you needed to shut down your assembly line for only 45 minutes when switching from making last year's car model to this year's model. This 500 percent reduction in shutdown time over last year is remarkable. As a senior in manufacturing engineering at Western University, I would welcome a chance to contribute to further improvements in the production processes at your plant. Please consider me for the opening in the Production Design Group that you advertised through the University's Career Services Center.

Shawana gives specific praise related to the job she wants.

She introduces herself and expresses her desire to contribute.

She identifies the job she wants.

Note that Shawana isn't praising the features of the car or the huge profit the company made. That would be superficial praise anyone could give without having any real understanding of the organization. Instead, Shawana focuses on a specific accomplishment, discovered through her research, that is directly related to her own specialty.

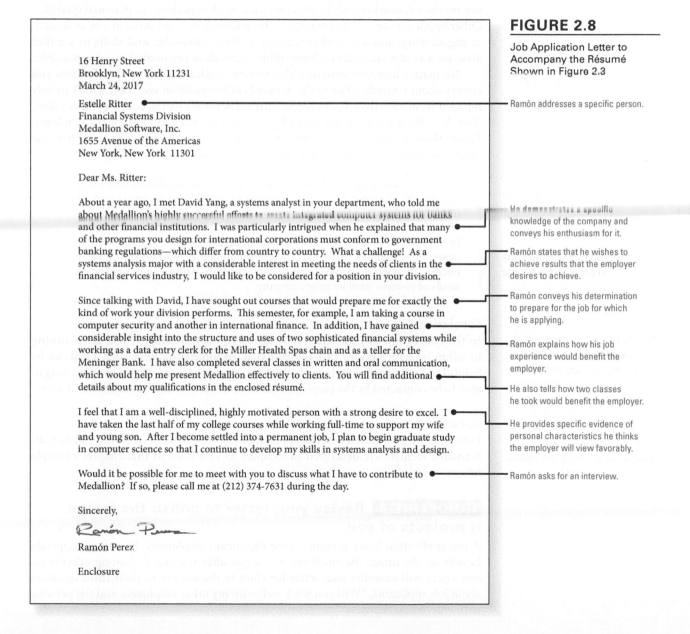

FIGURE 2.8

Job Application Letter to Accompany the Résumé Shown in Figure 2.3

16 Henry Street
Brooklyn, New York 11231
March 24, 2017

Estelle Ritter
Financial Systems Division
Medallion Software, Inc.
1655 Avenue of the Americas
New York, New York 11301

Ramón addresses a specific person.

Dear Ms. Ritter:

About a year ago, I met David Yang, a systems analyst in your department, who told me about Medallion's highly successful efforts to create integrated computer systems for banks and other financial institutions. I was particularly intrigued when he explained that many of the programs you design for international corporations must conform to government banking regulations—which differ from country to country. What a challenge! As a systems analysis major with a considerable interest in meeting the needs of clients in the financial services industry, I would like to be considered for a position in your division.

He demonstrates a specific knowledge of the company and conveys his enthusiasm for it.

Ramón states that he wishes to achieve results that the employer desires to achieve.

Since talking with David, I have sought out courses that would prepare me for exactly the kind of work your division performs. This semester, for example, I am taking a course in computer security and another in international finance. In addition, I have gained considerable insight into the structure and uses of two sophisticated financial systems while working as a data entry clerk for the Miller Health Spas chain and as a teller for the Meninger Bank. I have also completed several classes in written and oral communication, which would help me present Medallion effectively to clients. You will find additional details about my qualifications in the enclosed résumé.

Ramón conveys his determination to prepare for the job for which he is applying.

Ramón explains how his job experience would benefit the employer.

He also tells how two classes he took would benefit the employer.

I feel that I am a well-disciplined, highly motivated person with a strong desire to excel. I have taken the last half of my college courses while working full-time to support my wife and young son. After I become settled into a permanent job, I plan to begin graduate study in computer science so that I continue to develop my skills in systems analysis and design.

He provides specific evidence of personal characteristics he thinks the employer will view favorably.

Would it be possible for me to meet with you to discuss what I have to contribute to Medallion? If so, please call me at (212) 374-7631 during the day.

Ramón asks for an interview.

Sincerely,

Ramón Perez

Ramón Perez

Enclosure

If you have a connection with a current employee or representative of the employer, you might mention it, especially if you can combine this reference with evidence that you are knowledgeable about the organization.

> At last month's Career Fair at the University of Missouri, I spoke with AgraGrow's representative Raphael Ortega, who described your company's novel approaches to no-till agriculture. As an agronomy major, I would welcome the chance to work in your research department.

Body (Qualifications Section)

Sometimes called the *qualifications section,* the body of your letter is the place to explain how your knowledge and experience prepare you to contribute significantly to the employer's organization. Some job applicants divide this section into two parts: one on their education and the other on their work experience or personal qualities. Other applicants devote a paragraph to their knowledge and skills in one area, such as engineering, and a second paragraph to their knowledge and skills in another area, such as communication. Many other methods of organizing are also possible.

No matter how you structure this section, indicate how the specific facts you convey about yourself relate to the demands of the position you want. Don't merely repeat information from your résumé. Instead of simply listing courses you've taken, show how the knowledge you gained will help you do a good job for the employer. Rather than reciting previous job titles and areas of responsibility, indicate how the skills you gained will enable you to succeed in the job you are seeking.

What the student learned

> In my advanced physical chemistry course, I learned to conduct fluoroscopic and gas chromatographic analyses similar to those your laboratory uses to detect contaminants in the materials provided by your vendors.

How this will enable the student to contribute

> In one course, we designed a computer simulation of the transportation between three manufacturing facilities, two warehouses, and seventeen retail outlets. Through this class, I gained substantial experience in designing the kinds of systems used by your company.

Conclusion

In the conclusion of your letter, look ahead to the next step. If you are planning to follow up with a phone call, indicate that. In some situations, such a call can be helpful in focusing an employer's attention on your résumé. If you are planning to wait to be contacted by the employer, indicate where you can be reached and when.

Use a Conventional Format

Finally, when you draft a letter of application, be sure to use a conventional format. Standard formats are described in Chapter 23 and used in this chapter's sample letters.

GUIDELINE 3 Revise your letter to polish the image it projects of you

A job application letter presents some significant challenges. Its success depends heavily on the image the employer has of you after reading it. Among other uses, employers will examine your letter for clues to the answer to their third question about job applicants, "Will you work well with my other employees and the persons with whom I do business?"

TRY THIS At a website that offers advice to job seekers, find a "model" job application letter for someone who has the same amount of work experience that you have. Compare that letter with the ones in Figures 2.7 and 2.8. In what ways could the model letter be improved?

Employers can respond to any aspect of your letter for clues to your interpersonal skills and your personality. When you say what appeals to you about the organization and explain why you are well suited for the job, you are revealing important things about yourself. Notice, for instance, how the first sentence of the second paragraph of Jeannie's letter (Figure 2.7) shows her to be a goal-oriented person who plans her work purposefully. Also, notice how the first paragraph of Ramón's letter (Figure 2.8) shows him to be an enthusiastic person with a firm sense of direction. When reviewing your own draft, pay special attention to the personality you project. Employers will especially value your enthusiasm for their organization and for the work you would do there. These characteristics that Jeannie and Ramón project also suggest that Jeannie and Ramón will get along well with others, approaching team projects and other interactions eagerly and purposefully.

Some people have difficulty indicating an appropriate level of self-confidence. You want the tone of your letter to suggest to employers that you are self-assured, but not brash or overconfident. Avoid statements like this:

> I am sure you will agree that my excellent education qualifies me for a position in your Design Department.

Overconfident tone

The phrase "I am sure you will agree" will offend some readers. And the sentence as a whole seems presumptuous. It asserts that the writer knows as much as the reader about how well the writer's qualifications meet the reader's needs. The following sentence is more likely to generate a favorable response.

> I hope you will find that my education qualifies me for a position in your Design Department.

More effective tone

Achieving just the right tone in the conclusion of a letter is also rather tricky. You should avoid ending your letter like this:

> I would like to meet with you at your earliest convenience. Please let me know when this is possible.

Ineffective, demanding tone

To sound less demanding, the writer might revise the second sentence to read, "Please let me know *whether* this is possible" and add "I *look forward* to hearing from you."

Finally, remember that you must assure that each sentence states your meaning clearly and precisely. And you must eradicate all spelling and grammatical errors.

Adapting Your Application Letter for Different Employers

If you are applying to different employers for the same type of position, you may be able to adapt your application letter by altering only those places where you demonstrate your knowledge of each specific company and tell why you are applying to it. However, if you are applying for different kinds of jobs, revise also your explanations of the match between your qualifications and the job you want.

Figure 2.9 shows a Writer's Guide you can use as you prepare your job application letters.

Ethical Issues in the Job Search

One of the most interesting—and perplexing—features of workplace writing is that the ethical standards differ from one situation to another. The résumé and job application letter illustrate this point because the expectations that apply to them are significantly different from those that apply to other common types of on-the-job writing.

FIGURE 2.9

Writer's Guide for Job Application Letters

Writer's Guide
JOB APPLICATION LETTERS

PRELIMINARY RESEARCH
- ☐ Determine as exactly as possible what the employer wants.
- ☐ Learn enough about the job and employer to tailor your letter to them.

ADDRESS
- ☐ Address a specific individual, if possible.

INTRODUCTION
- ☐ Tell clearly what you want.
- ☐ Persuade that you know specific, relevant things about the reader's organization.
- ☐ Convey that you like the company.

QUALIFICATIONS
- ☐ Explain how the knowledge, abilities, and experiences described in your résumé are relevant to the specific job for which you are applying.

CLOSING
- ☐ Sound cordial, yet clearly set out a plan of action.

PROSE
- ☐ Use clear sentences with varied structures.
- ☐ Use an easy-to-follow organization.
- ☐ Use a confident, but modest tone.
- ☐ Express the action in verbs, not nouns.
- ☐ Use strong verbs.
- ☐ Use correct spelling, grammar, and punctuation.

APPEARANCE
- ☐ Look neat and attractive.
- ☐ Include all the elements of a business letter.

ETHICS
- ☐ Describe your qualifications honestly.
- ☐ Avoid statements intended to mislead.

OVERALL
- ☐ Show that you are aware of your reader's goals and concerns when hiring.
- ☐ Demonstrate that you are a skilled communicator.

For instance, when writing a résumé you are permitted to present facts about yourself in a very selective way that would be considered unethical in most other kinds of workplace writing. For example, imagine that you have suggested that your company reorganize its system for keeping track of inventory. Several managers have asked you to investigate this possibility further, then write a report about it, detailing the new system you'd prefer. In this report, your readers will expect you to include unfavorable information about your system as well as favorable information. If you omit the unfavorable information, your employer will judge that you have behaved unethically in order to win approval of your idea. If you were to submit your résumé to these same readers, however, they would not expect you to include unfavorable information about yourself. In fact, they would be surprised if you did.

Furthermore, these readers would expect you to present the favorable information about yourself in as impressive a language as possible, even though they might feel you were ethically bound to use cooler, more objective writing in your proposal about the inventory system. Here are the ethical guidelines that apply to résumés and job application letters.

Guidelines for an Ethical Résumé and Job Application Letter

- Don't list degrees you haven't earned, offices you haven't served in, or jobs you haven't held.
- Don't list awards or other recognition you haven't actually received.
- Don't take sole credit for things you have done as a team member.
- Don't give yourself a job title you haven't had.
- Don't phrase your statements in a way that is intended to mislead your readers.

If you are unsure whether you are writing ethically in some part of your résumé, check with your instructor or someone else who is familiar with workplace expectations about this type of communication. (You might also ask for advice from more experienced people whenever you are unsure about the ethical expectations that apply to any communication you write at work.)

Writing for Employment in Other Countries

In other countries, résumés and job application letters may look very different from those used in the United States. The traditional Japanese job application, the *rirekisho*, includes a significant amount of personal information to indicate that the applicant comes from an environment that makes it likely the person will enjoy long-term success in the organization (Romney, 2009). Although Japanese employers expect less personal information in applications from persons in other cultures, they still want to know the applicant's nationality, age, and marital status. In France, 75 percent of employers want job application letters written by hand so they can subject the letters to handwriting analysis (White, 2009). French employers believe that this analysis enables them to learn about the applicant's personal traits, which they weigh heavily when making employment decisions.

Because expectations differ so significantly from country to country, conduct reader-centered research about the country to which you will send applications. Among others, Mary Anne Thompson's *Global Résumé and CV Guide* (2000) and http://www.rileyguide.com/intlinfo.html#jsearch are excellent sources.

Interviewing Effectively

Your reward for preparing an effective résumé and application letter is, of course, the opportunity to do some additional reader-centered communicating, this time in a job interview. One way to make yourself stand out in an interview is to show samples of your work. Chapter 22 explains how to create a digital portfolio.

Conclusion

This chapter has demonstrated how the reader-centered approach used by successful workplace communicators can help you create a highly effective résumé and job application letter. The strategies you saw in action here are ones you can use for all your work-related writing.

- Think continuously about your readers.
- Use your knowledge of your readers to guide all your writing decisions.

The rest of this book is devoted to developing your communication expertise by providing detailed reader-centered advice you can use whenever you write at work.

USE WHAT YOU'VE LEARNED

EXERCISE YOUR EXPERTISE

1. Find a sample résumé at your college's Career Services Center or in a book about résumé writing. Evaluate it from the point of view of its intended reader. How could it be improved?

2. Using the web, newspapers, or journal articles, locate four or more openings that appeal to you. Create a unified list of the qualifications they specify. Then identify your skills and experiences that match each item on the employers' list.

3. Complete the Project assignment in MindTap on writing a résumé and a letter of application.

EXPLORE ONLINE

Using the web, find an employer or online job board that asks you to fill out an online résumé form. Evaluate the extent to which the form helps or hinders you from presenting your qualifications in the most persuasive manner.

COLLABORATE WITH YOUR CLASSMATES

Collaborating with another student in your class, work together on developing a keyword list for each of you that you could use when creating a scannable résumé.

APPLY YOUR ETHICS

Using the library or web, read an article that discusses the attitudes of employers toward unethical résumés. Take notes you can share with your class.

REFLECT FOR TRANSFER

1. Think of a kind of communication you expect to prepare on the job. Write a memo to your instructor in which you describe three pieces of advice from this chapter that you could use in that future communication. Explain why each would be helpful.

2. You have undoubtedly heard other advice about writing résumés and application letters. Write a memo to your instructor describing three pieces of advice that differ from advice given in this chapter. For each pair, explain which you feel is better and why.

NOTE: There's no penalty for disagreeing with the book in your memo. However, remember that although you can write the résumés and letters you actually send to employers any way you want, your instructor will probably apply the advice in this chapter, perhaps with some special tweaks suited to what he or she knows about your career plans, when evaluating the materials you turn in. In fact, you may want to ask him or her if this is so.

Case: In MindTap, the case titled "Advising Patricia" is well suited for use with this chapter.

3

Defining Your Communication's Goals

This is the first of fourteen chapters that support your instructor's aim of helping you develop expertise in the many activities people perform as they write at work. This chapter begins where writing begins—by helping you define what you want your communication to accomplish.

"But wait," you may be thinking. "Why do I need to spend time doing that? Won't my goals for writing at work be obvious? I will write the things my employer asks me to write. Shouldn't I just proceed directly to writing?"

Your goals may be obvious, but *your communication's* goals may not be. And you can't achieve your goals unless your communication achieves its goals.

Your résumé and job application letter offer a perfect example. *Your* goal is to obtain a job interview or offer. But you can achieve your goal only if your résumé and letter achieve *their* goal, which is to persuade the employer that you have exactly the qualifications the employer is looking for. It's your definition of *your résumé and letter's* goals that provides you with the basis for making the many decisions, large and small, that you have to make as you write—decisions ranging from what to include and how to organize, to how to phrase a particular sentence. If your résumé and letter succeed in presenting your qualifications in a way that leads the employer to say, "A perfect match with what I'm looking for!" then they have achieved their goal, and you have achieved yours.

When you start your writing process by defining your communication's goals, you are simply following the example of engineers, computer scientists, architects, and professionals in other fields who design products to serve practical purposes. They begin a project by creating a *design specification,* which details what their yet-to-be-designed product needs to include and how it needs to be put together. This document guides the rest of their work on the project, just as your definition of your communication's goals helps you make every decision, large or small, about your communication. Will your reader find your writing to be more useful and persuasive if you present complex or comprehensive data in the body of your text or in an appendix? If you display your data in a bar graph or table? If you use the abbreviation *ppb* or spell out *parts per billion*? Your definition of your communication's goals will enable you to give good answers to these and the many other questions that arise as you strive to help and influence your reader.

Your Goal: To Envision Your Reader's Response to Each Specific Aspect of Your Communication

How can you convert something as abstract as "a definition of your communication's goals" into a dynamic guide that can help you make good, practical writing decisions? *Answer:* You use the information in your definition to make an interactive mental portrait of your reader that enables you to "watch" him or her react to each aspect of its structure, appearance, sentences, word choice, and everything else. In your movie, you can see how your reader responds to your choice. If he or she reacts the way you want, go with it. If not, try something else.

Of course, you can never be absolutely sure how someone will react to anything you write or say. A reader's reactions depend on a complex set of factors. To make your predictions as accurate as possible, your instructor and this book will help you achieve the learning objectives listed on the chapter-opening page as you read this chapter, discuss it in class, and apply its advice while working on the writing assignments in your technical communication class.

Figure 3.1 displays a Writer's Guide for Defining Goals that you can use to apply this chapter's advice. If your instructor asks you to read this chapter when you are beginning work on a course project, you may find it helpful to complete each section of the guide as you read the relevant parts of the chapter.

At the end of this chapter you will find a description of the way Stephanie, a college student, used the guide to prepare a successful recommendation to her boss at a summer job. Stephanie's completed Writer's Guide is included.

How to Determine What Your Communication Must Do to Be Useful

As you learned in Chapter 1, people at work read in order to complete tasks that are part of their job responsibilities. To be effective, communications must help them perform those tasks. That is, they must be *useful*—from the *reader's* perspective.

The following guidelines will help you identify your reader's tasks in ways that enable you to determine how to make your communications most useful to your reader.

Your communication's usefulness is determined by the degree to which its readers can use it to achieve their goals efficiently.

GUIDELINE 1 Describe your reader's goal

Begin creating your mental portrait of your reader by identifying the goal that your communication will help your reader achieve. Stick close to your reader's *immediate* goal—not the overarching goal of creating the best product ever, but the immediate goal of designing one of its parts, deciding which of two companies to buy component parts from, or figuring out how to solve a problem in the manufacturing process for the product.

When naming your reader's goal, use verbs that describe practical actions your reader will perform, such as *choose*, *decide*, *design*, and *solve*, not verbs such as *learn*,

Supervisors managing factory production, executives making decisions, and consumers assembling a product need many different things from the communications they read. To write effectively, you must identify and meet the needs of the particular persons to whom you are writing.

FIGURE 3.1

Writer's Guide for Defining Your Communication's Goals

Writer's Guide
DEFINING YOUR COMMUNICATION'S GOALS

YOUR PURPOSE

1. What are you writing?
2. What outcome do you desire?
3. Who is your reader?

CREATING A USEFUL COMMUNICATION ⟵——— See pages 51–55

1. What task will your communication help your reader perform?
2. What information does your reader want? (What questions will your reader ask?)
3. How will your reader search for the information? (May use more than one strategy.)
____ Sequential reading from beginning to end
____ Reading for key points
____ Reference reading
____ Other (describe)
4. How will your reader use the information?
____ Compare alternatives (What will be the points of comparison?)
____ Determine how the information will affect him or her (or the organization)
____ Perform a procedure (following instructions step by step)
____ Other (describe)

CREATING A PERSUASIVE COMMUNICATION ⟵——— See page 56

1. What is your reader's attitude toward your subject? What do you want it to be?
2. What is your reader's attitude toward you? What do you want it to be?
3. What is your reader's attitude toward your organization? What do you want it to be?

READER'S PROFILE ⟵——— See pages 56–64

1. Job title
2. Familiarity with your topic
3. Familiarity with your specialty
4. Relationship with you
5. Personal characteristics you should take into account
6. Cultural characteristics you should take into account
7. Relevant features of the context in which your reader will read your communication
8. Who else might read your communication?

CONTRAINTS ON THE WAY YOU CAN WRITE ⟵——— See page 64

1. What expectations, regulations, or other factors constrain the way you can write?

ETHICAL TREATMENT OF STAKEHOLDERS ⟵——— See pages 64–65

1. Who, besides your reader, are stakeholders in your communication?
2. How will they view its impact on them?

know, or *understand* that concern mental states. Readers at work read for practical purposes. While they must sometimes learn something in order to achieve their purposes, you should first identify what those purposes are.

By looking at your communication from the perspective of your reader's goals rather than your own, you are acting on the most important fact (already stated) about workplace writing: In order to achieve your goal, you must help your readers achieve theirs.

GUIDELINE 2 Describe the tasks your reader will perform *while reading* your communication

The next step in creating your mental portrait of your reader is to describe the tasks the reader will perform while reading your communication. When writing, your mission is to create a communication that helps your reader perform these tasks quickly and easily. Think of yourself as an assistant to a carpenter or a chef. To know which tool to hand this person—whether a hammer or drill, a knife or whisk—you need to know what your reader is trying to do.

When describing your reader's tasks, be specific. To illustrate the practical value of being specific, notice how the following three reader's tasks can each guide your writing decisions. Notice that the reader's tasks may be physical, mental, or a combination.

Three Reader Tasks

- **To choose.** Usually, readers make choices by comparing the alternatives in terms of specific criteria. Knowing what the criteria are will help you decide what to include in your communication. Knowing that your readers will most likely compare the alternatives criterion by criterion suggests that you organize your information around those criteria.

- **To perform a procedure.** Typically, readers read a step, perform the step, than look back at your instructions to read about the next step. Understanding this process suggests that you present your instructions in lists rather than paragraphs so that each step stands alone, making it easy for the reader to find the next one.

- **To determine how certain changes in policy or procedures will affect them and their unit in the organization.** Usually, readers want to imagine what their work will be like after the change. You can help your readers by including a section that addresses each of your readers' concerns, whether they involve budget, efficiency, prestige, or parking spaces.

GUIDELINE 3 Identify the information and ideas your reader will want your communication to provide

If readers didn't want or need information and ideas from you to perform their tasks, there usually would be little reason for you to write. But you are writing, which means they are counting on you to provide what they desire. A description of what your reader will be looking for is an essential part of the mental portrait of your reader that you are creating.

A particularly powerful way to identify this information is to imagine the questions your reader would ask your communication to answer. For some communications, your reader may have only a few questions—or only one—that are obvious to you. Often, though, you will need to use your creativity and research skills to predict *all* the questions your communication needs to answer. This search may continue throughout your work on your message. When defining your communication's

LEARN MORE For advice about communications in which you offer suggestions or recommendations your reader has not requested, see "Defining the Problem in Unsolicited Communications," page 138.

objectives, your goal is to identify the general nature of the reader's questions so that you know where to focus your energies as you proceed.

As a start, you will often find it helpful to decide whether your reader will be reading in the role of a decision maker, adviser, or implementer. Each role leads to a different set of questions.

Decision Makers

The decision maker's role is to say how the organization will act when it is confronted with a particular choice or problem. Decision makers determine what the company should do in the future—next week, next month, next year. They usually want your communication to provide information that helps them choose between alternative courses of action.

Typical Questions Asked by Decision Makers

- **What are your conclusions?** Decision makers want your conclusions, not the raw data you gathered or the details about your procedures. Conclusions can serve as the basis for decisions. Details cannot.
- **What do you recommend?** Decision makers usually ask you about a topic because you have special knowledge of it. This knowledge makes your recommendation especially valuable to them.
- **What will happen?** Decision makers want to know what will occur if they follow your recommendations—and what will happen if they don't. How much money will be saved? How much will production increase? How will customers react?

Advisers

Advisers provide information and advice for decision makers to consider when deciding what the organization should do. Unlike decision makers, advisers are very interested in details. They need to analyze and evaluate the evidence supporting your general conclusions, recommendations, and projections.

Consequently, advisers ask questions that touch on the thoroughness, reliability, and impact of your work.

Typical Questions Asked by Advisers

- Did you use a reasonable method to obtain your results?
- Do your data really support your conclusions?
- Have you overlooked anything important?
- If your recommendation is followed, what will be the effect on other departments?
- What kinds of problems are likely to arise?

Implementers

Decisions, once made, must be carried out by someone. Implementers are these individuals. Their most important questions are the following.

Typical Questions Asked by Implementers

- **What do you want me to do?** Whether you are writing step-by-step instructions, requests for information, or policies that others must follow, implementers want you to provide clear, exact, easy-to-follow directions.
- **Why do you want me to do it?** To produce satisfactory results, implementers often must know the reason for the policy or directive they are reading. Imagine, for instance, the situation of the managers of a factory who have been directed to cut by 15 percent the amount of energy used in production. They need to know whether they are to make long-term energy savings or compensate for short-term shortage. If the latter, they might take temporary actions, such as altering work hours and curtailing certain operations. However, if the reduction is to be long term, they might purchase new equipment and modify the factory building.
- **How much freedom do I have in deciding how to do this?** People often devise shortcuts or alternative ways of doing things. They need to know whether they have this freedom or whether they must do things exactly as stated.
- **What's the deadline?** To be able to adjust their schedules to include a new task along with their other responsibilities, implementers need to know when the new task must be completed.

Is it always fruitful to determine which of these three roles your reader will play—decision maker, adviser, or implementer? Certainly not. You may have other ways of identifying your reader's questions. The key point is to identify the information you need to include in your communication, by determining what your reader wants to know.

GUIDELINE 4 Describe the way your reader will look for the information

To be able to use the information they want from your communication, readers must first be able to find it. Steve Krug has memorably captured readers' desire to spot the information quickly and easily—in his book on website design, *Don't Make Me Think* (2014). Krug emphasizes that readers of a website—or any other communication—want to think about its content. They don't want to take time figuring out where that content is.

To accelerate your reader's search, identify the ways your reader will look for information and design your communication to match his or her search strategy. Here are four search strategies often used on the job, together with writing strategies that aid readers using them.

Four Ways Readers Search for Information

- **Thorough, sequential reading.** When writing for readers who will read each sentence and paragraph in turn, you can systematically build ideas from one sentence, paragraph, and section to the next. To help your readers understand the larger structure of your topic, you can organize hierarchically and tell readers what the hierarchical structure is before discussing details.
- **Reading instructions.** You can number the steps rather than putting them in bullet lists to help readers quickly find their place as they look back at your communication after completing a step.
- **Reading for key points only.** You can use lists, tables, boldface, headings, and other page-design features to make the key points stand out.
- **Reference reading.** When you write to readers who will seek only specific pieces of information, you can use headings, tables of contents, and indexes to guide your readers rapidly to the information they seek.

LEARN MORE For more information on graphics and page design, see Chapters 12, 13, and 14.

How to Determine What Your Communication Must Do to Be Persuasive

The second essential quality of on-the-job-writing is *persuasiveness*. Your communication's impact on your reader's attitudes is important in *any* on-the-job communication, not just recommendations, proposals, and similar documents people usually think of as persuasive. *Any* communication you write has the potential to affect your reader's attitudes toward you and your employer's organization. Does it lead your reader to view you positively as a competent professional? Does it lead your reader to believe that your employer's organization is capable of addressing his or her needs? Also, persuasion isn't always about changing minds. It can involve reinforcing or shaping as well as reversing your reader's attitudes.

GUIDELINE 1 Describe your reader's current attitudes and what you want them to be after reading your communication

LEARN MORE For a detailed discussion of persuasive strategies, see Chapter 9, "Persuading Your Reader," pages 168–186.

When defining your communication's persuasive objectives, you will often know what attitudes you want your reader to have after reading. In this situation, it may seem that you don't need to do anything other than specify what that end point is. However, the writing strategies that are most likely to achieve the outcome you desire depend also on the starting point—that is, on what the reader's attitude is right now, before reading. For example, if the reader's initial attitude toward your topic is positive, your communication's persuasive purpose will be to reinforce that attitude, making it even stronger than before. Because you can build on the reader's existing attitude, you can probably succeed by presenting only positive points about your position. In contrast, if your goal is to reverse the reader's initial attitude, you can expect resistance, and you would need to use very different writing strategies, such as expressing—and then addressing—the negative points your readers would raise.

GUIDELINE 2 Find out why your reader holds his or her current attitudes

When your communication's goal is to reverse your reader's attitudes, it is important to understand why he or she currently holds them. Those reasons tell you the kinds of rationales the reader will find persuasive, and they indicate the specific assumptions, beliefs, and other evidence you need to counter in order to change the reader's attitudes.

How to Identify Factors that May Influence Your Reader's Responses to Your Communication

As we know from life experience, different people respond differently to the same event or situation. The same goes for writing at work. What one person finds to be highly useful and persuasive, another may not. You can greatly increase your ability to write a useful, persuasive communication if your imaginary portrait of your reader includes the professional, personal, and situational factors that can affect his or her responses to your communication.

GUIDELINE 1 Describe your reader's professional role and characteristics

Start by describing your reader's professional roles and characteristics.

- **Job Title.** People with different titles ask different questions and use the answers differently. When reading a report on the industrial emissions from a factory, an environmental engineer working for the factory might ask, "How are these emissions produced, and what can be done to reduce them?" while the corporate attorney might ask, "Do the emissions exceed Environmental Protection Agency limits; if so, how can we limit our fines for these violations?"

- **Familiarity with Your Topic.** Your readers' familiarity with your topic—company inventory levels, employee morale on the second shift, problems with new software—will determine how much background information you must provide to make your communication understandable and useful to them.

- **Knowledge of Your Specialty.** Readers can use the information you provide only if they understand the terms and concepts you employ. A person unfamiliar with your specialty would want you to explain what *zeroing* and the *Z-axis* are. On the other hand, if you provided those explanations for readers who are familiar with this specialty, they would ask, "Why is this writer making me read about things I already know?" Your goal is to learn enough about your reader to be able to strike the right balance between too little and too much explanation.

- **Relationship with You.** When you are having a conversation, you adjust your speech according to your relationship with the other person. You talk with a friend more informally than you do with a college instructor you don't know well—and both your friend and your instructor might be startled if you didn't make such adjustments. Similarly, at work you should write in a way that reflects the relationships you have with your reader.

> **TRY THIS** Think of a recent letter, email, or instant message you wrote to a friend or family member. How did you adjust your content and style to your purpose, reader, and context? How would you have changed your message if you'd written on the same topic to a different reader? What if your purpose or the context had been different?

GUIDELINE 2 Describe your reader's relevant personal characteristics

A variety of personal characteristics can also influence a reader's responses to your writing. You may be addressing an individual with an especially high or low reading level, weak eyesight, or color blindness. Or your reader may detest the use of certain words, insist on particular ways of phrasing certain statements, or want more details about your topic than most readers would. Your success as a writer can depend on identifying and accommodating these characteristics.

GUIDELINE 3 Describe the context in which your reader will read

At work, people interpret what they read as a chapter in an ongoing story. Consequently, they respond to each message in light of prior events as well as their understanding of the people and groups involved. Fill out your mental portrait of your reader by imagining how the following circumstances might influence his or her response to your communication.

- **Recent events related to your topic.** Maybe you are going to announce the reorganization of a department that has just adjusted to another major organizational change. You'll need to make a special effort to present the newest change in a positive light.

■ **Interpersonal, interdepartmental, and intraorganizational relationships.** Conflicts between individuals and groups in an organization can also create delicate writing situations in which certain ways of expressing your message can appear to support one faction and weaken another, even if you have no intention of doing so.

GUIDELINE 4 Global Guideline: Describe your reader's cultural characteristics

One of the great pleasures for many college graduates is the opportunity to work with persons who live in or have been raised in cultures different from their own. The global reach of business puts many employees at even small companies in touch with employees, suppliers, and customers in many other nations. In many parts of the world, employees in the same building come from many countries and cultural backgrounds. No matter what country you live in, this diversity will enable you to learn about other ways of life and other ways of knowing and being in the world.

Cultural Differences that Affect Communication

Among the many possible differences among people from different cultures are differences in what they expect in writing communications and how they respond to what they read. When writing to someone from a cultural background different from yours, you may need to use writing strategies different from those that would succeed with persons from your own culture. Consequently, you may need to add cultural information to the profile you use to make your mental movie of your reader in the act of reading your communication.

There's no single set of cultural characteristics that applies in every case. To achieve the necessary understanding of your reader, you may need to conduct some research; a later section in this chapter suggests ways to do that. When conducting research, it can be helpful to know some of the kinds of differences you may find. The following sections discuss six. As you read about these differences, remember that each contains a range of possibilities. Between the poles are many gradations.

■ **Amount of detail expected.** Cultures differ in the amount of detail that people expect in written and oral communication (Meyer, 2014). In countries such as Japan, communications provide a small amount of detail because writers and readers both assume that readers can fill in specifics by drawing on their existing knowledge. These are labeled "high-context" cultures because successful communication depends on the large amount of contextual information the readers bring to the message.

In contrast, in "low-context" cultures, such as the United States and much of Northern Europe, writers and readers both assume that readers are responsible for bringing very little contextual knowledge to a communication. Consequently, writers typically provide extensive detail, trying to cover every aspect of their topic thoroughly.

Knowledge of these differences and the ability to tailor communications appropriately is important to you if you are writing from a high- or low-context culture to readers in the other kind of culture. Readers in a high-context culture can be offended if a writer includes more detail than they expect and need. The extra detail would seem to imply that they do not know the things they should know. Similarly, a low-context reader could feel that the writer who has provided a high-context amount of detail was not considering their needs because much of the expected information (even if not truly needed) wasn't provided.

■ **Distance between the top and bottom of organizational hierarchies.** Through research that included the study of managers at IBM facilities in forty countries, Hofstede, Hofstede, and Minkov (2010) distinguished cultures according to the distance they maintain between people at the bottom and the top of an organization's hierarchy. In the United States and in some European cultures, the distance is very small. In other cultures, such as Japanese culture, the distance is much greater. This information is helpful to writers because Hofstede, Hofstede, and Minkov also found that, in general, where the distance is greatest, communication styles are most formal whether the communications are written by people in lower ranks to people above them or vice versa.

■ **Individual versus group orientation.** Hofstede, Hofstede, and Minkov also distinguished cultures that focus on the individual from those that focus on the group. Individualistic cultures honor personal achievement and expect individuals to take care of themselves. The dominant cultures in the United States and Northern Europe provide examples. In group-oriented cultures, success belongs to the group, and people pursue group goals rather than individual ones. Many Asian cultures are group-oriented.

Writers can often increase their effectiveness by adjusting to the individualistic or group orientation of their readers' culture. For example, in persuasive communications in the United States it can be helpful to highlight benefits to the individual, whereas an emphasis on benefits to the readers' organization can be more persuasive to readers in group-oriented cultures.

■ **Preference for direct or indirect statements.** Cultures also vary in the directness with which people typically make requests, decline requests, and express their opinions, particularly negative ones. For example, in the United States and Northern European cultures, writers typically decline a request directly. They may apologize and offer an explanation for their decision, but they will state the denial explicitly. In contrast, Japanese and Korean cultures prefer an indirect style (Meyer, 2014). Instead of declining a request explicitly, they might say that fulfilling it would be difficult or that they need time to think about how to reply. In this way, they save the requester the humiliation of being explicitly denied, while readers in those cultures understand that the request will not be fulfilled.

Using one culture's style when writing to people in another culture can create troublesome misunderstandings. To many U.S. readers, an indirect refusal might be misinterpreted. Because they didn't hear a direct denial, these U.S. readers could believe that the request might be fulfilled later, so they may persist in asking. On the other hand, readers in Japanese and Korean cultures may interpret the direct U.S. style as rude and inconsiderate.

■ **Basis of business decisions.** There are also cultural differences in the ways that business decisions are made. In the United States and many European cultures, organizations typically choose among alternatives on the basis of impersonal evaluations and data analyses. In Arab and other cultures, these same decisions are often made on the basis of relationships. For instance, whereas a U.S. company might choose a company to build a new plant or supply parts for its products by carefully studying detailed proposals from the competitors, an Arab company might select the company

LEARN MORE For more information on communicating between cultures that have different expectations about the appropriate level of formality in written and spoken communications, see Chapter 10, "Global Guideline: Adapt your voice to your readers' cultural background" (page 194).

LEARN MORE For more information on communicating between individualistic and group-oriented cultures, see Chapter 9, Global Guideline: "Adapt to Your Readers' Cultural Background" (page 184).

LEARN MORE For more information on communicating between cultures that have different preferences about direct and indirect statements, see Chapter 10, "Global Guideline: Adapt your voice to your readers' cultural background"(page 194).

represented by a person with whom it would like to do business. These are differences that writers in either kind of culture would need to keep in mind when trying to win business, maintain business relationships, and even respond to complaints from organizations in the other kind of culture.

LEARN MORE Chapter 12, "Global Guideline: Adapt Your Graphics when Writing to Readers in Other Cultures" (pages 236–237) has more information on the interpretation of images, and Chapter 10, "Global Guideline: Consider your readers' cultural background when choosing words" (pages 206–207) has more on the interpretation of words.

■ **Interpretation of images, gestures, and words.** An image, gesture, or word can elicit markedly different responses in different cultures. Even the same words have different meanings in different countries and cultures. Photographs, drawings, and other pictures sometimes depict relationships among people that seem ordinary in one culture but violate the cultural customs of another. Gestures likewise have different meanings in different cultures. When people in the United States signal "Okay" by joining a thumb and forefinger to form a circle, they are making a gesture that is offensive in Germany and obscene in Brazil (Axtel, 2007; Cotton, 2013). To avoid the risk of unintentionally offending others, some experts advise technical communicators to avoid showing hands in graphics that will be viewed by people in other cultures.

Applying Cultural Knowledge when You Write

As the preceding discussion indicates, gaining knowledge of your reader's culture can greatly increase your success in creating a communication that your reader will find useful and persuasive. The discussion is not, however, intended to suggest that gaining general knowledge about a culture can provide you with a recipe for adapting your communication strategies to the needs and expectations of your specific reader. When writing to people in other cultures, your reader-centered goal is the same as when addressing people in your own culture: to understand as fully as possible the relevant facts about the specific reader you are addressing.

LEARN MORE For information about ways that cultural differences can affect writing teams, see Chapter 17, "Global Guideline: Help your team work across cultural differences" (page 312). For information related to oral presentations to people from other cultures, see Chapter 18, "Guideline: Adapt to your listeners' cultural background" (page 327).

How to Gain Knowledge about Your Intercultural Readers

A variety of resources can help you learn about your reader's cultural characteristics. The most helpful are people—including your co-workers—who are familiar with your reader's regional and organizational culture.

You can also consult many helpful print and online sources that present broad descriptions of cultures around the world. Here are some of the most helpful.

■ Cyborlink website: www.cyborlink.com

■ GlobalEDGE website: www.globaledge.msu.edu

■ U.S. Department of State Bilateral Fact Sheets: www.state.gov/r/pa/ei/bgn/

GUIDELINE 5 Learn who all your readers will be

So far, this chapter has assumed that you will know, from the start, just who your readers will be. That may not always be the case. Communications you prepare on the job may find their way to many people in many parts of your organization. Numerous memos and reports prepared at work are routed to one or two dozen people—and sometimes many more. Even a brief communication you write to one person may be copied or shown to others. To write effectively, you must learn who *all* your readers will be so you can keep them all in mind when you write. The following discussion will help you identify readers you might otherwise overlook.

Phantom Readers

The most important readers of a communication may be hidden from you. That's because at work, written communications addressed to one person are often used by others. Those real but unnamed readers are called *phantom readers*.

Phantom readers are likely to be present behind the scenes when you write communications that require some sort of decision. One clue to their presence is that the person you are addressing is not high enough in the organizational hierarchy to make the decision your communication requires. Perhaps the decision will affect more parts of the organization than are managed by the person addressed, or perhaps it involves more money than the person addressed is likely to control.

Much of what you write to your own boss may actually be used by phantom readers. Many managers accomplish their work by assigning it to assistants. Thus, your boss may sometimes check over your communications, then pass them along to his or her superiors.

After working at a job for a while, employees usually learn which communications will be passed up the organizational hierarchy. However, a new employee may be chagrined to discover that a hastily written memo has been read by executives at very high levels. To avoid such embarrassment, identify your phantom readers, then write in a way that meets their needs as well as the needs of the less influential person you are addressing. Because communications are so frequently forwarded to other readers, never put on paper or screen comments that you wouldn't want disseminated, such as private criticisms of person or policy.

Future Readers

Your communications may be put to use weeks, months, or even years after you imagined their useful life was over. Lawyers say that the memos, reports, and other documents that employees write today are evidence for court cases tomorrow. Most company documents can be subpoenaed for lawsuits concerning product liability, patent violation, breach of contract, and other issues. If you are writing a communication that could have such use, remember that lawyers and judges may be your future readers.

Your future readers also may be employees of your company who may retrieve your old communications for information or ideas. By thinking of their needs, you may be able to save them considerable labor. Even if you are asked to write something "just for the record," remember that the only reason to have a record is to provide some future readers with information they will need to use in some practical way that you should understand and support.

Complex Audiences

Writers sometimes overlook important members of their audience because they assume that all their readers have identical needs and concerns. At work, audiences often consist of diverse groups with widely varying backgrounds and responsibilities.

That's partly because decisions and actions at work often impact many people and departments throughout the organization. For instance, a proposal to change a company's computer system will affect persons throughout the organization, and people in different areas will have different concerns: some with recordkeeping, some with data communication, some with security, and so on. People in each area will examine the proposal.

Even when only a few people are affected by a decision, many employers expect widespread consultation and advice on it. Each person consulted will have his or her own professional role and area of expertise, and each will play that role and apply that expertise when studying your communication.

When you address a group of people who will be reading from many perspectives, you are addressing a *complex audience*. To do that effectively, you need to write in a way that will meet each person's needs without reducing the effectiveness of your communication for the others. Sometimes you may have to make a trade-off by focusing on the needs and concerns of the most influential members of your audience. In any case, the first step in writing effectively to a complex audience is to identify each of its members or groups.

Identifying Readers: An Example

To see one way that a writer might identify the members of a complex audience and adjust his or her communication accordingly, consider Thomas McKay's situation. McKay was writing on behalf of his employer, Midlands Research Incorporated, to request compensation from another company, Aerotest Corporation, which had sold Midlands faulty equipment for testing smokestack emissions. McKay addressed his letter to Robert Fulton, Aerotest's Vice President for Sales, but realized that Fulton would distribute copies to many others at Aerotest. To identify these other readers, McKay asked himself who at his own employer's company, Midlands, would be asked to read such a letter if it received one. In this way, McKay identified the following phantom readers.

- Engineers in the department that designed and manufactured the faulty equipment, who would be asked to determine whether Aerotest's difficulties really resulted from flaws in the design
- Aerotest's lawyers, who would be asked to determine the company's legal liability
- Personnel in Aerotest's repair shop, who would be asked to examine the costs Midlands said it had incurred in repairing the equipment

To meet the needs of the diverse readers in his complex audience, McKay created a letter with a modular design, a commonly used workplace strategy in which different, readily distinguishable parts of a communication each address a distinct group of readers. Modular designs are very common at websites, where home pages often have links for different kinds of users. For example, your college's home page may have separate links for current students, prospective students, faculty, and graduates. Figure 3.2 shows the modular design of the home page of a company that

---

FIGURE 3.2

Modular Design of a Website

The National Institute on Drug Abuse designed its website to help each of its major groups of visitors quickly find the information they are seeking.

The site provides a link to a Spanish version for visitors who can read that language more easily than English.

The site has separate areas for five major groups of visitors: researchers, medical and health professionals, patients and their families, parents and teachers, and students and young adults.

The site also provides links for persons who are seeking certain kinds of information, such as details about specific drugs, funding, and news and events.

For persons seeking treatment for themselves or others, the site provides a prominent link to a resource that will help them find a local substance-abuse treatment service.

Scrolling down this page in the section for patients and families, visitors can find links to information on many topics, including prevention, research on treatment, and clinical trials (not shown in this figure).

National Institutes on Health/National Institute on Drug Abuse

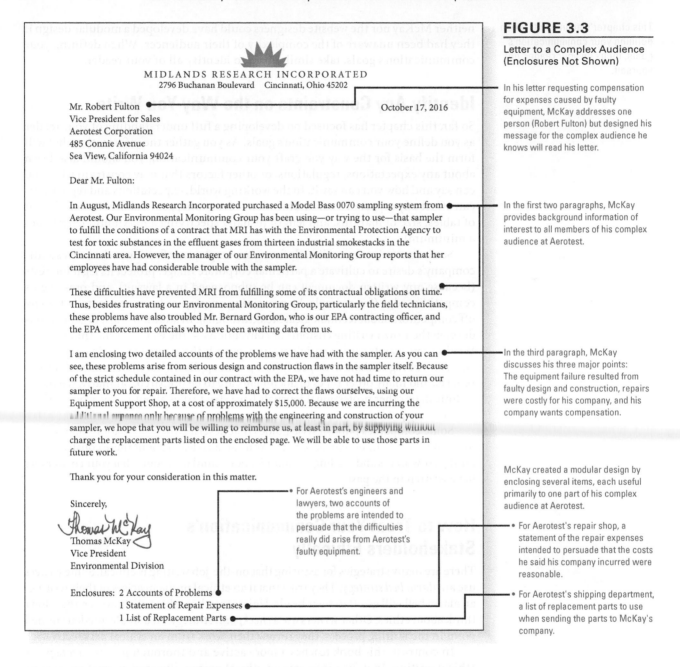

FIGURE 3.3

Letter to a Complex Audience
(Enclosures Not Shown)

In his letter requesting compensation for expenses caused by faulty equipment, McKay addresses one person (Robert Fulton) but designed his message for the complex audience he knows will read his letter.

In the first two paragraphs, McKay provides background information of interest to all members of his complex audience at Aerotest.

In the third paragraph, McKay discusses his three major points: The equipment failure resulted from faulty design and construction, repairs were costly for his company, and his company wants compensation.

McKay created a modular design by enclosing several items, each useful primarily to one part of his complex audience at Aerotest.

The letter contents:

MIDLANDS RESEARCH INCORPORATED
2796 Buchanan Boulevard   Cincinnati, Ohio 45202

Mr. Robert Fulton   October 17, 2016
Vice President for Sales
Aerotest Corporation
485 Connie Avenue
Sea View, California 94024

Dear Mr. Fulton:

In August, Midlands Research Incorporated purchased a Model Bass 0070 sampling system from Aerotest. Our Environmental Monitoring Group has been using—or trying to use—that sampler to fulfill the conditions of a contract that MRI has with the Environmental Protection Agency to test for toxic substances in the effluent gases from thirteen industrial smokestacks in the Cincinnati area. However, the manager of our Environmental Monitoring Group reports that her employees have had considerable trouble with the sampler.

These difficulties have prevented MRI from fulfilling some of its contractual obligations on time. Thus, besides frustrating our Environmental Monitoring Group, particularly the field technicians, these problems have also troubled Mr. Bernard Gordon, who is our EPA contracting officer, and the EPA enforcement officials who have been awaiting data from us.

I am enclosing two detailed accounts of the problems we have had with the sampler. As you can see, these problems arise from serious design and construction flaws in the sampler itself. Because of the strict schedule contained in our contract with the EPA, we have not had time to return our sampler to you for repair. Therefore, we have had to correct the flaws ourselves, using our Equipment Support Shop, at a cost of approximately $15,000. Because we are incurring the additional expense only because of problems with the engineering and construction of your sampler, we hope that you will be willing to reimburse us, at least in part, by supplying without charge the replacement parts listed on the enclosed page. We will be able to use those parts in future work.

Thank you for your consideration in this matter.

Sincerely,

Thomas McKay
Thomas McKay
Vice President
Environmental Division

Enclosures: 2 Accounts of Problems
             1 Statement of Repair Expenses
             1 List of Replacement Parts

• For Aerotest's engineers and lawyers, two accounts of the problems are intended to persuade that the difficulties really did arise from Aerotest's faulty equipment.

• For Aerotest's repair shop, a statement of the repair expenses intended to persuade that the costs he said his company incurred were reasonable.

• For Aerotest's shipping department, a list of replacement parts to use when sending the parts to McKay's company.

makes surgical equipment. To create a modular design for his letter to Aerotest, McKay wrote a one-page letter that provided background information for all his readers (Figure 3.3). He also attached enclosures addressed to specific groups in his complex audience. Two enclosures contained detailed accounts of the problems Midlands encountered with the emissions testing equipment. With these enclosures, McKay provided evidence that the problems encountered by Midlands were, in fact, caused by poor work on Aerotest's part. McKay's third enclosure was a detailed statement of the repair expenses, thereby enabling Aerotest's repair technicians to see that the reimbursement Midlands requested was fully justified. Of course,

This chapter's main point: When defining your communication's goals, focus on your reader, not yourself.

neither McKay nor the website designers could have developed a modular design if they had been unaware of the complexity of their audiences. When defining your communication's goals, take similar care to identify all of your readers.

Identify Any Constraints on the Way You Write

So far, this chapter has focused on developing a full understanding of your reader as you define your communication's goals. As you gather the information that will form the basis for the way you craft your communication, you should also learn about any expectations, regulations, or other factors that may constrain what you can say and how you can say it. In the working world, expectations and regulations can affect any aspect of a communication—even tone, use of abbreviations, layout of tables, size of margins, and length (usually specifying a maximum length, not a minimum).

Some constraints come directly from your employer, reflecting such motives as the company's desire to cultivate a particular corporate image, protect its legal interests (because any written document can be subpoenaed in a lawsuit), and preserve its competitive edge (for example, by preventing employees from accidentally tipping off competitors about technological breakthroughs). In addition, most organizations develop their own writing customs or conventions—"the way we write things here." Writing constraints can also originate from outside the company—for instance, from government regulations that specify how patent applications, environmental impact reports, and many other types of documents are to be prepared. Similarly, scientific, technical, or other professional journals have strict rules about many aspects of the articles they publish.

Some companies publish style guides that describe their regulations about writing. Find out if your employer has one. You can also learn about these constraints by asking co-workers and reading communications similar to yours that your co-workers have written in the past.

How to Treat Your Communication's Stakeholders Ethically

There are many strategies for assuring that on-the-job writing is ethical. Some writers use an *alarm bell strategy*. They trust that if an ethical problem arises in their writing, an alarm bell will go off in their heads. Unless they hear that bell, however, they don't think about ethics. Other writers use a *checkpoint strategy*. At a single, predetermined point in the writing process, they review their work from an ethical perspective.

In contrast, this book teaches a more active and thorough *process strategy* for ethical writing. In it, you integrate an ethical perspective into *every* stage of your work on a communication. It's important to follow a process strategy because at every step of writing you make decisions that shape the way your communication will affect other people. Accordingly, at every step you should consider your decisions from the viewpoint of your personal ethical beliefs about the ways you should treat others.

GUIDELINE 1 Ethics Guideline: Identify your communication's stakeholders

In a process approach to ethical writing, no step is more important than identifying the people you will keep in mind throughout the rest of your writing effort. When you follow the reader-centered approach to writing as explained in this book, you begin by identifying your readers.

To write ethically, you must also identify another group of people: the individuals who will gain or lose because of your message. Collectively, these people are called *stakeholders* because they have a stake in what you are writing. Only by learning who these stakeholders are can you assure that you are treating them in accordance with your own ethical values.

How to Identify Stakeholders

Because communications written at work often have far-reaching effects, it's easy to overlook some stakeholders. If that happens, a writer risks causing accidental harm that could have been avoided if only the writer had thought through all the implications of his or her communications.

To identify the stakeholders in your communications, begin by listing the people or groups who will be directly affected by what you say and how you say it. These individuals may include many other people in addition to your readers. For instance, when Craig was preparing a report for his managers on the development of a new fertilizer, he realized that his stakeholders included not only the managers but also the farmers who would purchase the fertilizer and the factory workers who would handle the chemicals used to manufacture it.

Next, list people who will be affected indirectly. For example, because fertilizers run off the land into lakes and rivers, Craig realized that the stakeholders of his report included people who use these lakes and rivers for drinking water or recreation. Indeed, as is the case with many communications, the list of indirect stakeholders could be extended to include other species (in this case, the aquatic life in the rivers and lakes) and the environment itself.

Finally, think of the people who may be remotely affected. These people may include individuals not yet born. For example, if Craig's fertilizer does not break down into harmless elements, the residue in the soil and water may affect future generations.

GUIDELINE 2 **Determine how your communication's stakeholders will view its impact on them**

A crucial test of the ethical impact of any action is to consider it from the perspective of the persons affected. Sometimes you can do this imaginatively: How would I feel if I were affected this way? But our ability to imagine others' feelings diminishes as we attempt to put ourselves in the position of persons increasingly distant from us in age, experience, and culture. Sometimes, we just need to ask.

To learn how other people will feel about an action or outcome doesn't mean we have to act or write in a way they would prefer. The ethical question for each of us is whether we are treating them fairly and respectfully. What we need to do is be sure that we have settled on proceeding in ways that meet the ethical standards by which we guide our lives.

Putting Your Definition into Action: An Example

To see how defining the goals of your communications provides specific, detailed insights that you can use to write successfully on the job, consider how it helped Stephanie at a summer job with a nonprofit organization that provides Braille translations for books and other reading material requested by people who are blind.

Silatul Rahim Dahman, who is 100 percent blind, writes and reads at his computer at work. He is Information Communication and Technology Manager at the Malaysian Association for the Blind.

© The Nut Graph/www.nutgraph.com

All the Braille translations were prepared by volunteers who worked at home. Stephanie noted that some urgently needed translations, such as those of textbooks that students required for their courses, weren't completed on time because translations were assigned to volunteers on a rotating basis rather than on their ability to complete a translation rapidly. She decided to write a memo recommending to her boss, Ms. Land, that urgently needed translations be assigned to the fastest volunteers instead of to the next volunteer on the list.

Stephanie began by filling out the Writer's Guide shown on page 52. Her completed worksheet is shown in Figure 3.4.

By completing the worksheet, Stephanie gained many useful insights about the most effective way to write her memo. For example, when responding to the questions about her reader's attitudes, Stephanie realized that Ms. Land would probably react defensively to any recommendation about the current system. Ms. Land had created the system and believed that it worked very well. Furthermore, she was the type of person who resists change. By focusing on her reader's characteristics, Stephanie concluded that Ms. Land might even resist the suggestion that a problem existed. To write effectively, Stephanie would have to demonstrate that the current system could be improved—without seeming to criticize Ms. Land.

The worksheet also reminded Stephanie of Ms. Land's belief that student employees like Stephanie didn't understand the complexity of running a nonprofit organization. Therefore, Stephanie predicted, Ms. Land would immediately hunt for holes in Stephanie's knowledge. Consequently, she decided to demonstrate her grasp of the situation by addressing many issues she hadn't previously thought to include: Who would determine which translations were urgent? What criteria would be used? How would the most productive translators be identified? How would the work of the office staff be altered?

Similarly, when filling out the Writer's Guide section on context, Stephanie remembered that recently Ms. Land had successfully resisted pressure from some members of the agency's board of directors to force her to retire so a younger person might take over. Stephanie realized that although a reference to the board members' desire to streamline operations might be persuasive for some readers, it would likely arouse hostility from Ms. Land.

Finally, while completing the section concerning stakeholders, Stephanie realized that, so far, she had thought about her plan only from the perspective of persons requesting rapid Braille translations. She had neglected to consider her plan's impact on the volunteer translators. By talking with a few, she learned that those who might be judged less reliable and less speedy would be deeply offended. To avoid causing them to lose self-esteem, she modified her proposal: The agency would ask all translators how many pages they could translate in a week. Urgent translations would go to those who made the largest commitments. Translators who failed to meet their original commitment would be invited to specify a lower commitment that would better suit their personal schedules.

Because of her expertise at defining objectives, Stephanie was able to write a detailed, diplomatic, four-page memo. After several months of deliberation, Ms. Land accepted Stephanie's proposal. Stephanie's example illustrates the many *immediate* insights you can gain by taking this reader-centered approach. But the benefits will continue. The rest of this book's chapters tell how you can use your reader-centered understanding of a communication's objectives throughout all your work on it.

FIGURE 3.4

Stephanie's Completed Writer's Guide for Defining Your Communication's Goals

Writer's Guide
DEFINING YOUR COMMUNICATION'S GOALS

YOUR PURPOSE

1. What are you writing?

   A proposal for completing more urgent translations on time.

2. What outcome do you desire?

   Adoption of a system that assigns urgent translations to the quickest volunteers.

3. Who is your reader?

   Ms. Land.

CREATING A USEFUL COMMUNICATION

1. What task will your communication help your reader perform?

   Deciding which system for assigning translations is best.

2. What information does your reader want? (What questions will your reader ask?)

   What evidence do you have that there is a problem?

   Who would determine which translations are urgent?

   How would we decide which translators are assigned urgent translations?

   Will your system really work? Are other agencies using it successfully?

3. How will your reader search for the information? (May use more than one strategy)

   __X__ Sequential reading from beginning to end

   _____ Reading for key points

   __X__ Reference reading

   _____ Other (describe)

4. How will your reader use the information?

   __X__ Compare alternatives (what will be the points of comparison?)

   Cost, efficiency, and impact of the change on office staff and volunteers.

   _____ Determine how the information will affect him or her (or the organization)

   _____ Perform a procedure (following instructions step by step)

   _____ Other (describe)

CREATING A PERSUASIVE COMMUNICATION

1. What is your reader's attitude toward your subject? What do you want it to be?

   She believes the current system is the best possible. I want her to see that mine is better.

   (*Continued*)

FIGURE 3.4

(Continued)

2. What is your reader's attitude toward you? What do you want it to be?

Ms. Land thinks I am a good summer employee but that, like all summer employees, I am not knowledgeable enough to make recommendations worth serious consideration.

3. What is your reader's attitude toward your organization? What do you want it to be?

Not relevant since I work in Ms. Land's department.

READER'S PROFILE

1. Job title

Manager of the Translation Department.

2. Familiarity with your topic

Very familiar.

3. Familiarity with your specialty

Very familiar.

4. Relationship to you

Ms. Land is my boss. She likes to maintain a formal superior-subordinate relationship.

5. Personal characteristics you should take into account

She designed the current system and may feel defensive if I suggest it can be improved.

6. Cultural characteristics you should take into account

None.

7. Relevant features of the context in which your reader will read your communication

Ms. Land may have been told that it is time for her to retire.

8. Who else might read your communication?

A few experienced employees. They will have the same perspective as Ms. Land.

CONSTRAINTS ON THE WAY YOU CAN WRITE

1. What expectations, regulations, or other factors constrain the way you can write?

If Ms. Land has been asked to retire, she will be very sensitive to any statement that seems to criticize her. I should avoid statements she might interpret at criticism.

ETHICAL TREATMENT OF STAKEHOLDERS

1. Who, besides your reader, are stakeholders in your communication?

Persons requesting urgent translations. Volunteers.

2. How will they view its impact on them?

Persons making urgent requests will appreciate the better service. Volunteers who aren't selected for the top group may feel that they aren't valued.

Conclusion

This chapter's major lesson is simple: You can greatly increase your ability to write successfully if you define your communication's goals by focusing on your reader, not yourself. Learn enough about your reader and your reader's context to make a mental movie of him or her in the act of reading. Use this movie to predict the way your reader is likely to respond to ways you might write your communication. Using your imagination and creativity in the same ways, ensure that you are writing ethically: Identify your communication's stakeholders and the ways your communication might affect them.

USE WHAT YOU'VE LEARNED

EXERCISE YOUR EXPERTISE

1. Find an example of a communication you might write in your career. Following the guidelines in this chapter, define its objective. Be sure to identify each of the following items.

   - The readers and their characteristics
   - The stakeholders and the ways they might be affected by the communication
   - The final result the writer desires
   - What the communication must do to be useful to its readers
   - What the communication must do to be persuasive to its readers

   Then explain how the communication's features have been tailored to fit its objectives. If you can think of ways the communication might be improved, make recommendations.

2. Using the Writer's Guide shown in Figure 3.1, define the objectives of an assignment you are preparing for your technical communication class.

EXPLORE ONLINE

Find a web page that could be used as a resource for a person in the profession you're preparing to enter. Describe the target readers, what the web page's creators wanted to help these readers do, and the ways they wanted to influence the readers' attitudes and actions. Evaluate their success in achieving these objectives.

COLLABORATE WITH YOUR CLASSMATES

Working with another student, pick a technical or scientific topic that interests you both. Next, one of you should locate an article on the topic in a popular magazine such as *Time* or *Discover*, and the other should locate an article on the topic in a professional or specialized journal. Working individually, you should each study the ways your article has been written so that its target audience will find the article to be useful and persuasive. Consider such things as the way your article opens, the language used, the types of details provided, and the kind of visuals included. Next, meet together to compare the writing strategies used to meet the needs and interests of the two audiences. Present your results in the way your instructor requests.

APPLY YOUR ETHICS

A variety of websites present case studies that describe ethical issues that arise in business, engineering, science, and other fields. Locate, read, and respond to one such case.

REFLECT FOR TRANSFER

1. Write a memo to your instructor describing a communication that you or others received that was either unhelpful or unpersuasive. Identify the ways that the writer could have used the advice in this chapter to improve the communication's effectiveness.

2. Write a memo to your instructor in which you describe a communication you wrote that was useful and persuasive because you had defined its goals along the lines described in this chapter. Explain the features of your communication that were effective because of your thinking about the communication from the perspective of your reader.

   Alternatively, think of a communication you wrote that was not as useful or persuasive as you had hoped because you had not thought about it from your reader's perspective. Tell how you would write it differently now—and why.

Case: In MindTap, the case titled "Announcing the New Insurance Policy" is well suited for use with this chapter.

4

Conducting Reader-Centered Research: Gathering, Analyzing, and Thinking Critically about Information

LEARNING OBJECTIVES

1. Explain the six essential qualities of high-quality workplace research.
2. Gather information and ideas that will be useful and persuasive to your readers.
3. Think critically about the information and ideas you find.
4. Draw conclusions and make recommendations that your reader will find actionable.
5. Observe intellectual property laws and document your sources.

Most chapters in this book focus on the skills needed to present information and ideas usefully and persuasively to your readers. In contrast, this chapter and the next describe strategies for conducting the research that provides the information and ideas your readers need.

Because you have learned research skills in other classes, you may be asking whether you really need to study this chapter. If your instructor asked you to read it, the answer is "Yes." Although the purposes and methods of workplace research and the school-based research taught in more classes overlap, their goals differ in ways that shape your entire research process.

- As described in Chapter 1, school writing is writer-centered. Most of it is about showing how much you know about a subject. In contrast, at work, you write to help and persuade your readers.

- Consequently, when you research on the job, you don't try to locate everything you can discover about your topic. Rather, you concentrate on finding all that *will be useful and persuasive to your reader*.

What to Focus on Learning in this Chapter

As you read this chapter, discuss it in class, and apply its advice to projects your instructor has assigned, focus on the learning objectives on the chapter-opening page. Pay attention, especially, to your instructor's guidance. He or she knows best the specific kinds of research that are most important for you and will be able to emphasize, explain, and supplement this chapter's advice in ways that will be most helpful to you.

What Counts as Good Research in the Workplace?

In the workplace, good writing and good research are so closely related that, in many cases, you can't have one without the other. While we might say that the purpose of good writing is to present research results effectively, we could say with equal validity that the purpose of research is to develop the content you need in order to make your communications useful and persuasive for your readers. Readers at work are acutely aware that when you report the results of good research, you provide them

MindTap®

Find additional resources related to this chapter in MindTap.

with the basis for making excellent decisions and taking effective action. They also know that if they make decisions and take action based on poor research, the result could be disastrous.

What are the qualities of good research on the job? While some on-the-job readers might add others, these six are important to all.

Qualities of Reader-Centered Research

- **FOCUSED on the reader's goal and needs.** All aspects of the research, including the material gathered, analysis of it, conclusions drawn, and recommendations must focus on the reader's goal and tasks. Other material, no matter how interesting to you, must be eliminated.

- **THOROUGH from the reader's perspective.** The reader needs your communication in order to do something practical. To be useful to the reader, the research must be thorough enough to provide a solid foundation for that action or decision.

- **CREDIBLE to the reader.** Every feature of the research needs to be subjected to critical thinking that, in the reader's eyes, forms a solid base for taking action or making a decision.

- **EVIDENCE BASED.** Because of its practical nature, research at work includes not only results ("here's what I found") but also analysis ("here's how it fits together"), conclusions ("the lessons we can take from it"), and recommendations ("here's what I think you should do"). Readers expect each element in this logical chain to be based directly on the results (evidence).

- **UNBIASED.** Readers want reports that depend on unbiased sources that you analyze and present in an unbiased way.

- **ETHICAL.** The research conducted and the actions recommended must be consistent with the ethical principles of the reader. The reader's principles may arise from a professional or institutional code of ethics, laws governing (for instance) intellectual property rights, and the reader's personal values.

The rest of this chapter provides advice for conducting research that has each of these six qualities.

How to Conduct Focused Research

The goal of workplace writing is to provide your reader with the information and ideas needed to perform an action or make a decision. Therefore, the most productive place to begin defining your research goals is to review the goals you identified for your communication when following Chapter 3's guidelines.

Base your research goals on your communication's usefulness and persuasive goals.

- **Identify the information you need in order to write a communication your readers will find to be highly useful.** Review the questions your reader will ask about your topic. The questions you can't answer immediately are the ones you need to research. Also review the kinds of answers your reader will want you to provide. Will he or she want general information or specific details, introductory overviews, or technical explanations? In addition, consider the point of view from which your reader will read your communication. Is the reader going to want it to meet his or her needs as an engineer, accountant, consumer, or producer? The answers to these questions will help you determine not only the kinds of information you must obtain, but also where you should look for them.

Sometimes you will also need to include information your reader did not request because you uncovered facts that are relevant to his or her decision or action. While Toni was evaluating the high-end computer programs her employer might distribute to its engineers, she learned that the company that produces one program might soon be purchased by a larger rival. When such takeovers occur, the smaller company's products often languish—they are still sold but are no longer upgraded, a distinct problem for clients who rely on them. Although her reader did not ask about takeover possibilities, this information is relevant to the decision Toni's reader will make.

■ **Identify the kinds of information and arguments that will make your communication most persuasive in your reader's eyes.** Given your reader's goals, values, and preferences, what kinds of information and arguments should you look for? For example, is your reader primarily interested in efficiency or profitability, safety or consumer acceptance? Consider also the types of evidence and kinds of sources your reader will find most compelling. For instance, is your reader more likely to be swayed by quantitative data or by testimonials from leaders in his or her field?

Although you should define your research goals at the outset, be prepared to revise them as you proceed. Research is all about learning. One thing you may learn along the way is that you need to investigate something you hadn't originally thought important—or even thought about at all.

How to Conduct a Thorough, Focused Search for Information and Ideas

At work, your research goals include gathering all the information your reader will find relevant to the decision or action he or she is considering. Missing information can lead to misguided actions and decisions. Similarly, irrelevant information can distract the reader from what's really crucial. Stay focused on your reader's needs and preferences. The following guidelines suggest five major strategies.

GUIDELINE 1 **Identify the full range of sources and methods that may provide helpful information**

What counts as thorough research depends greatly on the topic you are studying. In the lab of a manufacturing company, some research may involve only a single experiment. When the research involves broad topics that affect people, like air pollution, employment, and health, you may need to use many sources and methods to gather all the information and ideas your readers need.

Primary and secondary research

In your career, you may be involved mostly with *primary research*, which is research where you generate the information yourself through such methods as observation, experimentation, and surveys. Almost certainly, though, you will also conduct *secondary research*, which involves reading the results of other people's research, for instance in professional journals or on the Internet.

Generally, begin by consulting secondary sources—especially ones, such as encyclopedias, that provide an overview of your topic. By gaining a general view of your subject, you increase the ease with which you can locate, comprehend, and interpret the more detailed facts you are seeking. In addition to encyclopedias, general sources include review articles that summarize research on a particular subject and articles in popular magazines.

To provide your readers with complete answers to their questions, take into account the variety of perspectives that people and organizations have on the topic. Consider gathering information from four types of sources.

- Persons affected
- Persons involved
- Other organizations or groups engaged in similar efforts
- Professional publications

Sources that often have helpful information for research affecting people

Imagine, for instance, that your employer has asked you to investigate ways to make some of its processes "greener"—more environmentally friendly. Examine each possibility extensively enough to uncover conflicting evidence, disagreements, and controversies, or to feel certain that none exist. For example, in research on greener processes you might find that one company that proudly uses a manufacturing process to remove harmful chemicals from solid waste projects is opposed by an environmental group because it discharges warm water into a river, damaging the ecological system for miles downstream. Only by looking broadly can you present a full answer to the questions your reader has asked you.

Examine each topic thoroughly enough to uncover conflicting evidence, disagreements, and controversies, or to feel certain that none exist.

GUIDELINE 2 Use secondary sources your readers will find credible and unbiased

When using secondary sources, focus on ones that will be most persuasive to your readers. Your readers will depend on you to provide information they can trust as they make decisions and take action. In addition, they will judge the quality of your research based on the quality of the sources you use.

In some cases, the distinction between sources that are acceptable and those that are unacceptable in your readers' eyes will be very clear. Scientists generally trust research published in peer-reviewed journals because these articles have been judged by other scientists in order to be published. Generally, they do not trust research reported in other places, such as websites.

In some circumstances, you will have a challenging job identifying sources your readers will find credible. The following six questions can help.

In *primary* research, you generate the information or data yourself. In *secondary* research, you work with information or data gathered and reported by someone else.

Questions for Evaluating Research Sources

1. Is it **relevant** to my readers' needs?
2. Will it be **credible** in my readers' eyes?
3. Is it **accurate**?
4. Is it **complete**?
5. Is it **current** and up to date?
6. Is it **unbiased**?

LEARN MORE For suggestions about ways to imagine your readers in the act of reading, see Chapter 3, "Your Goal: To Envision Your Reader's Response to Each Specific Aspect of Your Communication" (page 50).

To answer the questions about relevance and credibility, refer to the imaginary portrait you created of your readers when defining your communication's objectives. Ask yourself, "Can I imagine my readers using this information as they perform the job my research is supposed to help them perform?" If you can't, the source is irrelevant. Your portrait of your readers will also help you learn whether the source will be credible to them. What people in one profession consider credible may lack credibility in another.

The questions about accuracy, completeness, and currency can be difficult to answer when you are researching a subject that is new to you. Follow Guideline 1's advice: Gather information from lots of sources and compare them with one another. Contradictions and inconsistencies may indicate that you should conduct additional research to learn which source to trust.

When your sources disagree, ask whether the person who wrote the information, or that person's employer, may benefit if people believe the information provided. The source's information may still be valid but needs careful scrutiny. Review the reasonableness of its interpretations of evidence. Chapter 5 provides additional advice for evaluating information you find on the Internet.

Note, however, that bias does not necessarily mean that a source should be dismissed. Sometimes your readers will want to know conflicting views.

When evaluating sources, be as cautious about your own biases. Don't dismiss a source simply because it contradicts your views or presents data that fail to support your conclusions. Your readers depend on your thoroughness and integrity.

GUIDELINE 3 Use primary research methods in a credible and unbiased way

To assure that your primary research is credible and unbiased, study the research methods you are going to use. For many kinds of research methods there are procedures and techniques that workplace readers consider to be essential to good studies. Chapter 5 provides detailed advice for skillfully conducting two widely used primary research methods—interviews and surveys. Similarly, when conducting any other kind of primary research in your technical writing course or other classes, learn the procedural details that are crucial to workplace readers.

GUIDELINE 4 Gather information that can be analyzed in subgroups

As explained earlier in this chapter, workplace readers want research that looks for possible differences among various groups of information or data. From a study of the tensile strength of a new alloy it has developed, a manufacturing company will want to test its strength at various temperatures so it can know what applications the alloy might be used for. From research on the appeal of a new cell phone it has developed, an electronics company would want to know about the phone's appeal to specific consumer groups—by age group, economic status, gender, and so on—so it can plan an effective marketing campaign. To be able to provide your reader with this sort of information, you need to gather information that can be analyzed in subgroups.

To identify the subgroups of information that are important to your readers, consider the nature of the decision they are making or the action they are thinking about taking. What would they ask you in an effort to gain a more fine-grained understanding of your results?

GUIDELINE 5 **Take careful notes**

A simple but critical technique for conducting productive, efficient research is to take careful notes every step of the way. Even if you've consulted all relevant and helpful sources, your report to your readers won't be thorough if you've forgotten some of what you found.

When recording the facts and opinions you discover, be sure to distinguish quotations from paraphrases so you can properly identify quoted statements in your communication. Also, clearly differentiate between ideas you obtain from your sources and your own ideas in response to what you find there.

In addition, make careful bibliographic notes about your sources. Include all the details you will need when documenting your sources (see "Ethical Guidelines for Documenting Sources," page 84). For books and articles, record the following details.

Information to Record about Your Sources

BOOKS	ARTICLES
Author's or editor's full name	Author's or editor's full name
Exact title	Exact title
City of publication	Journal title
Year of publication	Volume (and issue, unless pages are
Edition	numbered consecutively
Page numbers	throughout the volume)
	Year of publication
	Page numbers

For interviews, record the person's full name (verify the spelling!), title, and employer, if different from your own. Special considerations apply when your sources are on the Internet; they are described in Appendix A.

Also, record the information you will need if you later have to consult this source again. In a library, jot down the call number of each book; when interviewing someone, get the person's phone number or email address; and when using an online source, copy the universal resource locator (URL).

In addition, keep a list of sources that you checked but found useless. Otherwise, you may find a later reference to the same sources but be unable to remember that you have already examined them.

Intermission

In many projects, all of the research activities are intermingled. Even as you are interpreting some research results, you are gathering others. Even as you are beginning to formulate some recommendations, you discover new information that requires you to revise one of your conclusions. Still, as types of mental work, the activities discussed so far in this chapter, which focus on gathering information, are quite distinct from those described next, which concern analyzing and processing the information you have gathered. Consequently, there is value in pausing here for two reasons.

The first is to provide you with the Writer's Guide for Conducting Reader-Centered Research, shown in Figure 4.1. The other is to give you a chance to think about what the first activities of the reader-centered research process look like in action, using as an example the kind of client project many students (perhaps you) are asked to perform in their technical writing or other courses.

Two engineering students, Anna and Terry, agreed to help their dean find out how effectively engineering majors were being advised about their classes and careers by the faculty in their departments. Having heard both praise and grumbling about the advising, the dean wanted to know what, if anything, needed to be done.

Anna and Terry's research for the dean had all the major characteristics that workplace readers usually want (see "What Counts as Good Research in the Workplace?" pages 70–71). To be certain that their research **focused on the reader's goals and needs,** they listed the major questions they knew the dean would want them to answer: How good is the advising students are receiving? Are they receiving enough advising? How could the advising system be improved? Would some other way of providing advising to engineering students be more effective?

To assure that their study was **thorough**, they surveyed engineering students (persons affected) and faculty (persons involved). To find out whether there might be more effective systems for advising students, they contacted two sources the dean would consider to be **credible**: similar engineering programs (other organizations) and journals on engineering education (professional publications).

FIGURE 4.1

Writer's Guide for Conducting Reader-Centered Research

Writer's Guide
CONDUCTING READER-CENTERED RESEARCH

DEFINING YOUR RESEARCH'S GOALS

1. Identify the information your readers want, need, and will find persuasive.
2. Identify the criteria your readers will use to judge the quality of your research.

GATHERING INFORMATION

1. Use the full range of sources and research methods that might be helpful.
2. Search widely enough to uncover conflicting evidence, disagreements, and controversies or to feel certain none exist.
3. Identify secondary sources your readers will find credible and unbiased.
4. Evaluate each source for its relevance to your readers' needs, credibility, accuracy, completeness, currency, and freedom from bias.
5. Use primary research methods in a credible and unbiased way.
6. Gather information that can be analyzed in subgroups.

TAKING NOTES

1. Record facts and ideas your readers will find useful or persuasive (not everything the source contains).
2. Record details about your sources that will enable you to document them in the communication in which you present your research results.

To assure that their research would be entirely **evidence based**, they decided to ask questions that would enable them to respond with evidence to questions they predicted the Dean might have about subgroups of students. They asked respondents to indicate their department and year in school. Thinking that men and women might feel differently about the advising, they also asked respondents to indicate their sex. To assure that their survey would produce **unbiased** results, Anna and Terry followed carefully the advice given in the next chapter (see "How to Conduct a Survey," pages 103–107) about survey design.

Finally, to treat their respondents **ethically**, they asked students and faculty to respond to the survey forms without giving their names, No individual would need to fear retaliation because of the opinions he or she expressed.

In the rest of this chapter, you will see how Anna and Terry developed evidence-based, reader-focused analyses, conclusions, and recommendations that addressed their reader's needs.

How to Conduct Evidence-Based Analyses

To your readers, the information you gather is of little or no value if you don't make it useful to them. The first step in doing so is to analyze it—to look carefully for the specific items, relationships, and patterns that will be most helpful to your readers as they make their decisions or take their actions. Usually, your hunt for these patterns will be the most creative part of your research, calling on your ability to play with lots of information and data in various ways, seeking to spot correspondences important to your readers.

GUIDELINE 1 Take another look at your research goals

Begin by reviewing your research goals. While gathering information, we all can become enthralled by information that fascinates us but is irrelevant to our readers. We all may be tempted to tell our readers about the findings we worked hardest to obtain, even if these results aren't of any use to our readers. By reviewing your research goals, you remind yourself of the purpose of your analysis.

GUIDELINE 2 Arrange your information in an analyzable form

To analyze research information, you study it, looking for patterns, connections, and contrasts that can help your readers make decisions or take actions that will help them achieve their goals.

Perhaps it is surprising that the most effective way to look for these patterns often doesn't start with mental action, but a physical one: Arrange your data in a table, chart, graph, or other visual display that helps you "see" the relationships. For engineers and scientists in many fields, this is standard practice. To analyze their data, they make a "picture" of them that they can then study. Tables, bar charts, idea trees, and flowcharts can all be helpful in various situations. Some of these displays will yield insights. Others won't. So, make these visual displays quickly, without the refinements you would use if you were preparing the final draft of your report. The key point is that it's often helpful to look, literally, at your information.

Following this guideline, Anna and Terry made many "exploratory" charts of their survey results. Chart A (next page) shows one of those they found to be very useful. The discussion of the next guideline explains.

LEARN MORE For advice on using charts, graphs, idea trees, and similar aids to research, see Chapter 5, "Draw a Picture of Your Topic," (pages 89–90).

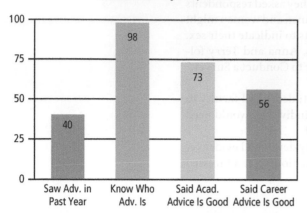

CHART A

Student Survey Responses

Bar chart showing values: Saw Adv. in Past Year = 40, Know Who Adv. Is = 98, Said Acad. Advice Is Good = 73, Said Career Advice Is Good = 56.

The reason for arranging research information in visual displays is to help you discover meaningful relationships, ones that can guide the decisions and actions of your readers. Looking at the two right-hand bars in Chart A, Anna and Terry saw a substantial difference between the percentage of students who said they received "good" academic advice from their faculty advisers (73 percent) and those who said they received "good" career advice (56 percent). This gap was surely an important one to report to the dean because it could support a decision to focus specifically on building professors' skills at providing career advice.

Strategies for Finding Meaningful Relationships

Not every chart or other visual display will disclose a meaningful relationship. The activity of analyzing information often involves many cycles of choosing possible relationships to look for, creating a visual display, finding no meaningful relationships, and choosing another possible relationship to investigate. The process inevitably involves some hits and some misses.

However, your knowledge of both your subject matter and your readers can suggest many likely pieces of information to explore. Also, the following strategies are often fruitful.

Strategies for Finding Relationships Meaningful to Your Readers

- Compare information on related outcomes (for example, academic advising and career advising).
- Compare information about different steps in a process (for example, getting students to know who their advisers are and arrange to meet with them).
- Compare information on the same topic from different sources (for example, surveys, interviews, and books or articles).
- Consider possible cause-and-effect relationships.
- Consider possible correlations.
- Look for relationships where your instinct leads you.

Looking for Meaningful Differences Among Subgroups

As explained in "Guideline 2: Arrange your information in an analyzable form" (page 77), workplace readers desire analysis of subgroups of research results because this analysis enables them to make refined decisions and take targeted actions. Anna and Terry's research illustrates the value of such fine-grained analysis. They predicted that the dean would want a deeper, more detailed understanding of some results shown in Chart A. The two left-hand bars indicated that although almost all students knew who their adviser was, only 40 percent had seen their adviser in the past year. The dean, they thought, would want to know whether there were differences among subgroups of students. If so, the dean could plan targeted actions aimed at increasing the consultation rate for the specific groups that had the lowest rates. Indeed, Anna and Terry found that the percentage of electrical engineering students who saw their advisers in the past year was much higher than that of students in any other department (Chart B). Also, the percentage of nuclear engineering students was much lower than that of any other department. These discoveries formed the basis for one of the actions they recommended to the dean, as described in "Guideline 2: Acknowledge uncertainty" (see below).

Anna and Terry were able to find these and other important differences among subgroups of students only because they had followed the advice given earlier in this chapter to "Gather information that can be analyzed in subgroups." If they had omitted questions about the students' major, year in school, and sex from their survey questionnaire, they would have missed a great deal of information that was important to their reader, the dean.

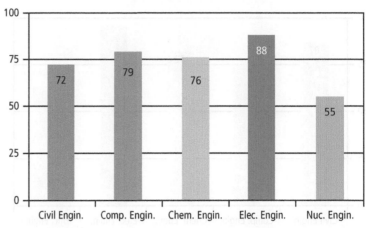

CHART B

Saw Adviser in Past Year

How to Draw Evidence-Based Conclusions

After you've found a meaningful pattern, connection, or other relationship, you need to describe it for your readers. Interpretations may take many forms, including generalizations, explanations, and comparisons, along with exceptions and counterevidence. Figuring out how to interpret a relationship accurately, precisely, and helpfully can sometimes require careful thought. Two of the challenges you may face are choosing among alternative interpretations and dealing with uncertainty.

GUIDELINE 1 Choose conclusions that align with your readers' decisions and actions

There are always alternative ways of understanding a set of data. For example, looking at Chart B, one might say that the students in the Nuclear Engineering Department, who have the worst record of consulting advisers, are less responsible than other engineering majors. They just don't take advantage of the advising that is offered. Alternatively, one might say that the Nuclear Engineering Department is less effective than others at attracting the students to advising. Which interpretation would be most helpful to the dean? Because the dean wants to improve the advising system, it's better to use the second interpretation, which focuses on the department rather than the students. The dean and professors can't control the students' sense of responsibility. They can change the ways they attract students to advising.

GUIDELINE 2 Acknowledge uncertainty

Sometimes, you may need to take a broad and creative look at your information to find a truly helpful interpretation. Trying to understand why students in some departments visit their advisers more often than students in other departments, Anna and Terry compared the data in Chart B with those in Chart C, which shows the percentage of students who said they received good academic advice from their faculty advisers. The overall patterns are very similar, with nuclear engineering the lowest and electrical engineering the highest in both.

Having found this relationship, Anna and Terry faced a dilemma: How should they interpret it? They could say that the more pleased students are with the academic

CHART C

Said Academic Advising "Good"

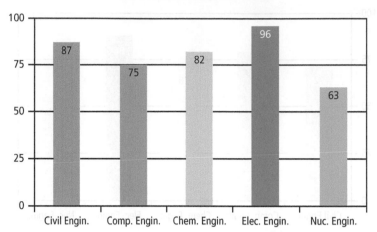

advice they receive, the more likely they are to consult their advisers. Or they could say that the more often students see their advisers, the more pleased they are with the advice they receive.

Which of these two interpretations is correct? From their data, Anna and Terry couldn't tell for sure. Their uncertainty raises an important point about interpretation: When you can't be certain, signal your uncertainty to your readers. One way to do that is simply to tell your readers that alternative interpretations are possible and present them both. Another is to say that your information "suggests" or "appears to indicate" that a relationship exists rather than that the information "demonstrates" or "proves" its existence.

In their report to the dean, Anna and Terry wrote, "The data suggest that students are more likely to visit their advisers if they feel they receive good academic advice when they do." Note that their statement discusses how good the students "feel" the advice is. It doesn't state whether the advice is actually good (or bad) in any department. Anna and Terry didn't have evidence about the quality of the advice itself, only of the students' perception of its quality.

GUIDELINE 3 Explain the significance of your conclusions to your readers

To be sure your readers know what each of your conclusions means to them, you should explain its significance explicitly. In doing this, you move beyond stating the conclusion you draw from your information by describing the decision or action that you believe the conclusion suggests or supports. For instance, as just mentioned, Anna and Terry concluded that "students are more likely to visit their advisers if they feel they receive good academic advice when they do." To state the significance of this conclusion to the dean, they explained the kind of action this conclusion suggested: "These survey results suggest that building professors' skills at career advising would increase student satisfaction with advising and the frequency with which students visit their advisers."

How to Make Evidence-Based Recommendations

In almost all cases, your readers will want you to recommend a decision or action based on your research. The more specific your recommendations, the more helpful they will be.

A weak recommendation that Anna and Terry might have made is that the dean help those professors in the departments with low advising satisfaction become better advisers. It's weak because it merely rephrases the conclusion stated in Guideline 3 on this page. A much more helpful recommendation would be to suggest specific ways in which this assistance might be provided.

The suggestions would be even more helpful if a specific suggestion were also backed by research indicating that the suggested action is likely to achieve its goal. Anna and Terry turned to the information they had gathered from other engineering schools to find specific strategies the dean could use to enhance the advising skills of the professors. They recommended that training sessions be established and that the departments whose students were more satisfied with advising share their strategies with the other departments.

How to Think Critically Throughout Your Research Process

Critical thinking is essential in all aspects of your research but especially when you are analyzing the information you've gathered. At every step, you need to explore alternatives, evaluate possibilities in light of all the relevant evidence, and identify underlying assumptions—including your own. You need to consider the quality of each piece of information by considering the perspectives and biases of the sources. Three ways to do that follow.

GUIDELINE 1 Let go of your anchor

One inhibitor of good analysis is a psychological phenomenon called anchoring (Kahneman, 2013). When people have made an initial commitment of any kind, they tend to view future actions from that vantage point. It's what keeps gamblers at the roulette wheel, thinking that they've got to "win back" the money they've already lost. Thinking rationally, they would know that their odds of winning and losing are the same for each new spin, no matter how much they won or lost on previous spins. For people engaged in research, anchoring can mean that they keep finding evidence to confirm their initial intuitions while ignoring other possibilities. Often, these intuitions are formed even as researchers are gathering information—before they begin their analysis. The way to free yourself from anchoring is to consciously and actively look for alternative ways of seeing and understanding your research information.

GUIDELINE 2 Value counterarguments, counterevidence, and exceptions

In most research, you are likely to find contradictions. Your sources may disagree with one another. A few pieces of information may be inconsistent with everything else you've found. These conflicts can be very valuable. They may signal the need for you to look more deeply for alternative explanations or conclusions—cutting the line to your anchor. They may indicate the need to hedge your conclusions, letting your readers know that there's some uncertainty. They may even be the link to something valuable you can tell your readers. When Anna and Terry discovered that the Electrical Engineering Department received exceptionally high ratings for its academic advising, they saw that this department probably had a lot to teach the others, a point they made to the dean.

GUIDELINE 3 Avoid personal or organizational biases

Without any malicious or selfish intent, it's easy for all of us to see the merits of interpretations and recommendations that would benefit us, our employer's organization, or our department. To serve your readers well, you need to check whether the natural self-interest we all feel is influencing your analysis. The easiest way to do this is to look at your information from your readers' perspective and those of the other stakeholders in the topic you are studying.

How to Observe Intellectual Property Law and Document Your Sources

As you are preparing the communication in which you report the results of your research, you have two important questions to answer.

- Do I need permission to use this material?
- Do I need to document this source in my communication?

The answers to these questions overlap, but they are not identical. To understand their relationship, you need to consider the laws concerning intellectual property as well as the ethical guidelines for acknowledging sources.

Intellectual Property Law

Broadly speaking, intellectual property law includes the following areas.

Three areas of intellectual property law

- **Patent Law.** Governs such things as inventions and novel manufacturing processes.
- **Trademark law.** Pertains to such things as company and product names (Microsoft, Pokémon), slogans ("We bring good things to life"), and symbols (the Nike "swoosh").
- **Copyright law.** Deals with such things as written works, images, performances, and computer software.

When you are writing at work, copyright law will probably be the most important to you. Copyright law was created to encourage creativity while also providing the public with an abundant source of information and ideas. To achieve these goals, copyright law enables the creators of a work to profit from it while also allowing others to use the work in limited ways without cost.

Any communication, such as a report, letter, email, photograph, or diagram, is copyrighted as soon as it is created. If the creator generated the work on his or her own, that individual owns the copyright to it. If the creator made the work while employed by someone else, the copyright probably belongs to the employer. Whether the copyright owner is an individual or an organization, the owner has the legal right to prohibit others from copying the work, distributing it, displaying it in a public forum, or creating a derivative work based on it. When copyright owners grant others permission to do any of these things, they may charge a fee or make other contractual demands. The copyright owner has these rights even if the work does not include the copyright notation or the copyright symbol: ©.

The copyright law does, however, place limits on the copyright owner's rights. First, copyright expires after a certain number of years, which varies depending on the date of publication. Second, the law provides that other people, including you, may legally quote or reproduce parts of someone else's work without their permission if your use is consistent with the legal doctrine of *fair use*. Whether your use is "fair" depends primarily on the following four factors (Stanford University, n.d.).

- **Proportion of the work used.** Your quotation of a few hundred words from a long book is likely to be considered fair use, but quotation of the same number of words from a short pamphlet may not.
- **Publication status.** The law gives greater protection to works that a copyright owner has not published or distributed than to ones the owner has.
- **Economic Impact.** If your use will diminish the creator's profits, it is unlikely the law will consider it to be fair use.

It is also legal for you to use other people's work without their permission if the work is in the *public domain*. Such works include those created by or for the U.S. government and similar entities and works whose copyright has expired. Also, private individuals and organizations sometimes put their work in the public domain. The owners of websites that offer free use of clipart are an example.

Finally, you can generally use work that other people working for your employer created as part of their job responsibilities. In fact, in the workplace it is very common for employees to use substantial parts of communications created by other employees. For instance, when you are creating the final report on a project, you may incorporate portions of the proposal written to obtain the original authorization for the project, as well as parts of progress reports written during the project.

Copyright law permits people and organizations to share their work by relinquishing some of their rights without giving up ownership, an increasingly popular practice. For example, many professional journals allow faculty and students to print copies of their articles as long as this is done for educational, not commercial, purposes. A nonprofit organization called Creative Commons supports these efforts in many ways.

The following guidelines on copyright coverage will help you observe intellectual property law. Note that the principles for text and graphics differ.

Observing Copyright

TEXT

Ask for permission except in the following circumstances:

- You created the source yourself.
- Someone else at your employer's created it.
- The source is in the public domain.
- The copyright owner explicitly includes a statement with the source that it may be used without permission.
- You are using the text for a course project that will not be published on paper or on the web.

  Honor fair use restrictions. Don't use larger amounts of someone else's work than fair use allows.

  When in doubt, ask someone. Intellectual property laws are complicated. If you are uncertain about what to do, consult your instructor or your employer's legal department.

GRAPHICS

Note: Each graphic is separately copyrighted. Consequently, the principle of fair use does not apply. You always need permission to use a graphic, even if it is only a small part of a larger work.

Obtain permission for all graphics unless:

- You created the graphic.
- Your employer is the copyright holder.
- The graphic is in the public domain.
- You are using the graphic for a course project that will not be published on paper or on the web.

Copyright law is different for graphics than for text.

(Continued)

> ### WEBSITES
>
> **Note:** Many people mistakenly believe that anything on the web is in the public domain. Actually, all web content is copyrighted by its creator.
>
> Get permission to use web material unless:
>
> - The site belongs to your employer.
> - The material you are going to use from the site is in the public domain.
> - The site declares that its contents are available for free use.
> - You are using the material for a course project that will not be published on paper or on the web.

Ethical Guidelines for Documenting Sources

On the job, as in college, you have an ethical obligation to credit the sources of your ideas and information by citing them in a reference list, bibliography, or footnotes. Failure to do so is considered plagiarism. However, standards for deciding exactly which sources need to be listed at work differ considerably from the standards that apply at school. By asking the questions that follow, you can usually determine whether you need to credit sources when writing on the job. Note, however, that ethical standards for citing sources differ from culture to culture. The following questions apply in the United States, Canada, and Europe. If you are working in another part of the world, ask your co-workers to help you understand the standards that apply where you are.

LEARN MORE You will find information about how to document your sources in Appendix A.

Determining Whether You Need to Document a Source at Work

- **Did you obtain permission from the copyright owner?** If you obtained the copyright owner's permission, you must document the source.
- **Is the information you obtained from this source common knowledge?** Both in college and at work, you must indicate the source of ideas and information that (1) you have derived from someone else and (2) are not common knowledge.

  However, what's considered common knowledge at work is different from what's considered common knowledge at school. At school, it's knowledge every person possesses without doing any special reading. Thus, you must document any material you find in print.

  In contrast, at work common knowledge is knowledge that is possessed by or readily available to people in your field. Thus, you do not need to acknowledge material you obtained through your college courses, your textbooks, standard reference works in your field, or similar sources.
- **Does my employer own it?** As explained earlier, employers own the writing done at work by their employees. Consequently, it is usually considered perfectly ethical to incorporate information from one proposal or report into another without acknowledging the source.
- **Am I taking credit for someone else's work?** On the other hand, you must be careful to avoid taking credit for ideas that aren't your own. In one case, an engineer was fired for unethical conduct because he pretended that he had devised a solution to a technical problem when he had actually copied the solution from a published article.
- **Am I writing for a research journal?** In articles to be published in scientific or scholarly journals, ethical standards for documentation are far more stringent than they are for on-the-job reports and proposals. In such articles, thorough documentation is required

even for ideas based on a single sentence in another source. Thus, you must document any information you find in print or online. In research labs where employees customarily publish their results in scientific or scholarly journals, even information drawn from internal communications may need to be thoroughly documented.

- **Whom can I ask for advice?** Because expectations about documentation can vary from company to company and from situation to situation, the surest way to identify your ethical obligations is to determine what your readers and employer expect. Consult your boss and co-workers and examine communications similar to the one you are preparing. For clarification about what sources you need to document for your class, ask your instructor.

Conclusion

Conducting reader-centered research is a complex process that calls on your intelligence and creativity in many different ways. This chapter used an example involving survey data to illustrate the process. You would follow the same steps for most other types of primary and secondary research: You define your research goals in a reader-centered way, gather information that will be useful to your readers, and examine the results, looking for relationships and testing them against exceptions and counterevidence in an effort to interpret your findings in ways that can help your readers. In the end, research is a problem-solving process in which, rather than solving the problem yourself, you provide indispensable assistance to your readers so they can solve it.

USE WHAT YOU'VE LEARNED

EXERCISE YOUR EXPERTISE

1. Create a research plan for a project you are preparing for your technical communication course.

2. Choose a concept, process, or procedure that is important in your field. Imagine that one of your instructors has asked you to explain it to first-year students in your major. (See Chapter 5 for guidelines for using each of the following research methods.)

   a. Use brainstorming or freewriting to generate a list of things you might say in your talk.
   b. Use a flowchart, matrix, cluster sketch, or table to generate a list of things you might say.
   c. Compare your two lists. What inferences can you draw about the strengths and limitations of each technique?

3. Imagine that a friend wants to purchase some item about which you are knowledgeable (for example, a motorcycle, MP3 player, or sewing machine). The friend has asked your advice about which brand to buy. Design a matrix in which you list two or three brands and also at least six criteria you recommend that your friend use to compare them. Fill in the matrix

as completely as you can. Each box you can't complete indicates an area you must research. Describe the methods you would use to gather the additional information. (See Chapter 5 for advice about using a matrix as a research tool.)

4. Imagine that you have been asked by the chair of your major department to study student satisfaction with its course offerings. Devise a set of six or more closed questions and four open-ended questions you could use in interviews or in a survey. (For information about interview and survey questions, see Chapter 5, "Conducting the Interview" and "How to Conduct a Survey.")

5. While analyzing their survey data, Anna and Terry compared responses of men with responses of women to this question: "Have you visited your faculty adviser this year?" They displayed their results in the chart shown in Figure 4.2 (page 86). Interpret this result in two or more ways. Which of your interpretations would be the most helpful to the dean of the engineering school? Why? Next, write a sentence that explains the significance of your interpretation to the dean. Finally, make a recommendation to the dean based on your interpretation.

(Continued)

FIGURE 4.2

Chart for Expertise Exercise 5

Comparison of Men and Women Who Reported that They Had Consulted Their Advisers in the Past Year

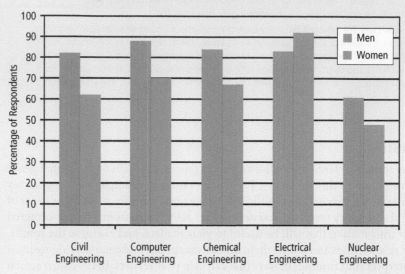

EXPLORE ONLINE

1. Use two search engines and an Internet directory to look for websites on a topic related to your major. How many hits does each produce? Compare the first ten results from each search in terms of the quality of the sites and the amount and kind of information the search engine or Internet directory provides about each one. (For information about using search engines and Internet directories, see "Searching the Internet," pages 91–94 in Chapter 5.)

2. Using a search engine and online library resources, identify three websites, two books, and two articles on a topic related to your field. Which would you find most interesting? Which would be most helpful if you were writing a paper on the topic for a class? Which would be most helpful if you were writing a report on the topic for your employer? (For information about using search engines and online library resources, see "Searching the the Internet," pages 91–94 in Chapter 5.)

3. Using the newspapers available online, find a story that includes a graph. In 200 words, describe the topic of the graph, what the story says it shows, and the additional analysis (such as analysis of subgroups of information) that would provide a fuller understanding of the topic.

COLLABORATE WITH YOUR CLASSMATES

1. Working with another student, choose a topic that interests you both. Find five websites that provide substantial information on your topic. Which sites are most appealing to you initially? Following the advice in "How to Make Evidence-Based Recommendations," pages 80–81, evaluate each site and then compare the results with your initial impression of it.

2. In their survey, Anna and Terry gathered information from the engineering students that let them analyze subgroup data from students according to their major, year in school, and sex. If you were conducting a survey about academic advising at your school, what additional subgroups of students would you want to identify? Why?

3. Imagine that the chair or head of your department has asked you and one or two of your classmates to conduct a survey of students in your major that is exactly eight questions long. As a group, name the survey topic and the chair's reason for wanting the information you would obtain through the survey. List the questions you would ask. Explain how the data obtained from each question would help the chair achieve his or her goal.

APPLY YOUR ETHICS

1. Create a bibliography of sources concerning an ethical issue related to your major or career. Include four websites, one book, and two journal articles that you believe would help you understand various approaches to this issue.

2. One ethical principle for analysis is to use methods that avoid accidentally reaching incorrect conclusions. Even

of thought. It also works well in team projects: An idea expressed by one team member often sparks ideas for others.

The key to brainstorming is to record ideas quickly without evaluating them. If you shift your task from generating ideas to evaluating them, you will disrupt the free flow of associations on which brainstorming thrives.

Start by asking, "What do I know about my subject that will be helpful or persuasive to my reader?" As ideas come, write them down rapidly. When your stream of ideas runs dry, read back through the notes you've made. Doing so may suggest new ideas. When you no longer have new ideas, gather related thoughts into groups to see whether the groups inspire additional ideas.

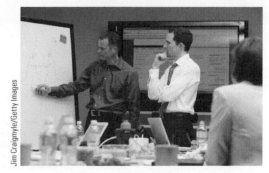

Writing teams at work often brainstorm together to plan their communications.

Freewriting

Freewriting is much like brainstorming. Here, too, you tap your natural creativity, free from the confines of structured thought. You rapidly record your ideas as they pop into your mind. Only this time, you write prose rather than a list.

Freewriting is especially helpful when you are trying to develop your main points, whether in brief communications or in parts of a long communication.

As with brainstorming, start by asking yourself, "What do I know about my subject that will be useful and persuasive to my reader?" As ideas come, write them down as sentences. Follow each line of thought until you come to the end of it, then immediately pick up the next line of thought that suggests itself. Write rapidly without making corrections or refining your prose. If you think of a better way to say something, start the sentence anew. Don't stop for gaps in your knowledge. If you discover that you need information you don't possess, note that fact, then keep writing. You can research for the missing information later. When you finally do run out of ideas, read back through your material to select the ideas worth telling your reader. The rest you throw away. What remains will probably be a little too jumbled to use as a draft, but it could be a gold mine of ideas.

Draw a Picture of Your Topic

Another effective strategy for exploiting your memory and creativity is to explore your topic visually. Examples are shown in Figure 5.1.

LEARN MORE Visualizing is also an important method for analyzing information you have gathered through your research. See Guideline 2: "Arrange your information in an analyzable form" in Chapter 4 (page 77).

Flowchart

When you are writing about a process or procedure, try drawing a flowchart of it. Leave lots of space around each box in the flowchart so you can write notes next to it. Nicole used the flowchart shown in Figure 5.1 to generate ideas for a report in which she recommended improved quality control procedures in the manufacture and delivery of the medical equipment sold by her employer.

Matrix

When you are comparing two or more alternatives in terms of a common set of criteria, drawing a matrix can aid you in systematically identifying the key features of each item being compared. Make a table in which you list the alternatives down the left-hand side and write the topics or issues to be covered across the top. Then, fill in each cell in the resulting table by brainstorming. Blank boxes indicate information you need to obtain. Miguel created the matrix shown in Figure 5.1 while consolidating what he already knew that he could include in a report that would help his employer decide whether to invest in developing new technology for detecting wind shear at airports. Wind shear is a sudden and dangerous atmospheric condition that causes crashes.

A matrix is a table used to generate and organize ideas.

FIGURE 5.1

Four Types of Drawings for Exploring Your Memory

Flowchart

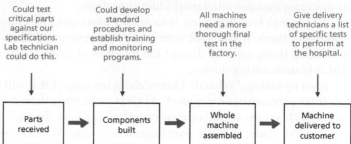

After drawing the flowchart, Nicole wrote above each cell the ideas that occurred to her.

| Could test critical parts against our specifications. Lab technician could do this. | Could develop standard procedures and establish training and monitoring programs. | All machines need a more thorough final test in the factory. | Give delivery technicians a list of specific tests to perform at the hospital. |

Parts received → Components built → Whole machine assembled → Machine delivered to customer

Matrix

Miguel made the bold heading of the matrix, then filled in the thoughts he associated with each cell.

	How It Works	Limitations	Potential Competition
Doppler radar	Detects rapidly rotating air masses, like those found in wind shear	Technology still being researched	General Dynamics Hughes
Infrared detector	Detects slight increases in temperature that often accompany wind shear	Temperature doesn't always rise	None—Federal Aviation Administration suspended testing
Laser sensor	Sudden wind shifts affect reflectivity of air that lasers can detect	Provides only a 20-second warning for jets traveling at a typical speed	Walton Electronic perhaps Sperry

Cluster Sketch

An engineer created this cluster sketch to identify the topics and subtopics he would discuss.

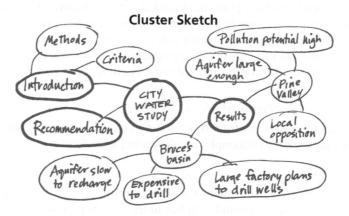

Idea Tree

To identify the report topics, the engineer might also have used an idea tree.

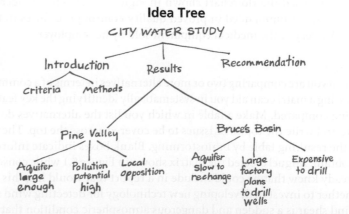

Cluster Sketch

Creating a cluster sketch is a simple, powerful technique for exploring a topic visually. Write your overall topic in a circle at the center of a piece of paper, then add circles around the perimeter that identify the major issues or subtopics, joining them with lines to the main topic. Continue adding satellite notes, expanding outward as far as you find productive. The cluster sketch shown in Figure 5.1 was created by Carol, an engineer who is leading a team assigned to help a small city locate places where it can drill new wells for its municipal water supply.

A variation of the cluster sketch is the idea tree, also shown in Figure 5.1.

TRY THIS Make a cluster sketch or idea tree to generate a list of your abilities that you would like to highlight in an employment interview. Identify general areas of ability and specifics related to each one.

Create and Study a Table or Graphic of Your Data

Often at work you will need to write a communication about data, such as the results of a test you have run, costs you have calculated, or production figures you have gathered. In such cases, many people find it helpful to begin their writing process by making the tables or graphs that they will include in their communication. Then they can begin to interpret the data arrayed before them, making notes about the data's meaning and significance to their readers.

Searching the Internet

The Internet has created a rich and continuously evolving resource for your on-the-job research. Wherever you are, the Internet lets you read technical reports from companies such as IBM, view pictures taken by NASA spacecraft in remote areas of the solar system, or join online discussions on an astonishing array of topics with people around the world. Figure 5.2 lists just a few of the resources the Internet makes available to you.

FIGURE 5.2

Some Major Internet Resources for Research

INTERNET RESOURCES	
Corporate reports and information	IBM posts technical documents, Microsoft offers detailed information on its products, and the World Wildlife Fund reports on its environmental projects. Thousands of other profit and nonprofit organizations do the same.
	Examples
	IBM Research
	http://www.research.ibm.com
	Microsoft TechNet
	http://technet.microsoft.com/
	World Wildlife Fund
	http://www.panda.org
Technical and scientific journals	Many technical and scientific journals are available online, some for free and some for a fee.
	Examples
	Journal of Cell Science
	http://jcs.biologists.org/
	IEEE Transactions on Software Engineering
	http://www.computer.org/tse
Government agencies	Many government agencies have websites at which they provide reports, regulations, forms, and similar resources.
	Examples
	NASA
	http://www.nasa.gov
	National Cancer Institute
	http://www.cancer.gov
	National Park Service
	http://www.nps.gov

Search engines produce a large number of potentially helpful results but don't evaluate them for bias, completeness, or accuracy.

As a resource for research, the web also presents you with significant challenges. Its sheer size can make it difficult to find the exact information your readers need. Search engines are helpful, but not perfect. In addition, no one monitors the web's content to weed out biased, misleading, or inaccurate information.

To conduct web research successfully on the job, you must search through its vast contents efficiently so you don't waste time you should devote to other tasks, and you must evaluate carefully what you find so you don't report wrong information to your readers.

Using Search Engines and Internet Directories Effectively

In addition to search engines like Google and Yahoo, there is another very useful tool for web research: Internet directories. Search engines and Internet directories provide distinctly different kinds of research support, so it is often helpful to use both.

Search engines produce the more comprehensive results. They scour a great deal of the web (though not all of it), looking for resources that include the same words that you enter into the search line. They also determine the order in which they will place their results in the list that appears on your computer screen. For both searching and ranking, they apply a combination of criteria, such as the number of times the words appear on a web page, their location on the page, the number of websites that link to the source, and so on. Because different search engines use different combinations of criteria, they produce different numbers of results with different rankings. On the same day, a Google search for "basking shark" (the second largest of all sharks) returned 286,000 results. Yahoo produced almost four times as many (1,250,000) and placed different results at the top of its list.

Internet directories provide a small number of results that people have judged to be valuable resources, but their criteria for selection may or may not match yours.

Internet directories are created by people, not computers, who search the Internet or evaluate nominations for particularly valuable resources. Consequently, they yield much smaller, more sharply focused results organized in a hierarchical framework that enables you to search systematically for the information you need to obtain for your readers. Using the DMOZ directory (www.dmoz.org), you would first choose the category science, then biology, flora and fauna, Animalia, Chordata, fish, and cartilaginous. You would find twenty results, not hundreds of thousands. Or, you could use DMOZ's search function, which would bring you directly to the page with links to sites judged to have good information about sharks.

Which is better, a search engine or an Internet directory? There's no absolute answer. Search engines can provide an overwhelming number of results, and they don't distinguish between sites created by experts and those created by second graders. Internet directories simplify your search by reducing the number of results, but they miss many useful results, and their selection criteria may not match yours. The DMOZ directory's results for "basking shark" omitted important websites maintained by wildlife and scientific organizations. These and other differences explain why it's often worthwhile to use more than one search engine and more than one directory in the same search.

Also, the web's information on many subjects is incomplete even when all its resources are combined. For some topics, you'll need to look elsewhere, for instance in books and reference journals, to obtain a thorough, balanced understanding of the topic you are researching on your readers' behalf.

TRY THIS Open two browser windows. Using a different search engine in each one, search for information on a topic related to your major or personal interests. Which search engine produces the most hits? Which has the most helpful links at the top of the results?

Using Search Engines Efficiently

A web search is basically a word game. You can significantly increase your skill at it by employing the following strategies.

- **Use as few words as possible.** With the exception of conjunctions, prepositions, and a few other types of words, each word you enter becomes part of the search. For most searches, fewer words are helpful.

- **Use the words most likely to be used on the site you want.** Instead of "stomach hurts," use "stomachache" because this term is more likely to appear on the sites you want.

- **Use precise words.** Especially when researching technical or scientific topics, use the words specialists would use. While "heart attack" is the common term, medical specialists use "myocardial infarction."

- **Try different words.** Keep your vocabulary flexible. If your search isn't producing the results you want, use synonyms. "Gene splicing" will produce different results than "genetic engineering."

- **Reorder words.** Word order influences the number and ranking of search results. A search for "Mars life" produces results focused on space science. One for "life Mars" yields many kinds of results, including those about an old television comedy.

- **Narrow results by adding words.** To sharpen an initial search, add another word you would expect to appear on the websites you are targeting. Adding "science" or "comedy" significantly focuses the results of a search on "life Mars."

- **Narrow results by using advanced and specific search features.** Advanced searches enable you to focus your search in many ways. Among other techniques, you can specify words that should not appear at any site you want and the type of file you want (.doc, .xls, .jpg). Another useful option is to limit the search to certain domains, such as only those for educational institutions (.edu) and governments (.gov) but not businesses (.com and .biz). To use Google's advanced search, enter your search terms and click Enter; on the first page of results, click on the gear icon in the upper right-hand corner.

Google, Yahoo, and other search engines use thousands of PCs to store and search information about websites its computers have visited.

Ways to increase the efficiency of web searches

Evaluating Your Search Results

As you learned in Chapter 4 (page 73), you should carefully evaluate the information you obtain from any source. However, research on the web requires special scrutiny because people can post anything there, whether it's true or false.

Begin by examining the URL (web address) of each search. The most helpful part of a URL is often the site's domain. Different kinds of organizations are assigned to different domains on the web. For instance, in the following URL, the three-letters "edu" identify the "education" domain, meaning that the site belongs to an educational institution, in this case the University of Florida ("ufl").

http://www.flmnh.ufl.edu/fish/Gallery/Descript/baskingshark/baskingshark.html

Location of a site's domain in its address

Other domains often used by researchers include

.com	Commercial (sites for businesses)
.gov	Government (sites for local, state, and federal governments)
.org	Not-for-profit organizations

Depending on what you are looking for, sites in any of these domains may be good sources or bad ones. For example, a .com site may provide useful details about the features of its products but biased information about product quality or lawsuits against the corporation.

Finally, if you decide to visit a site, evaluate its contents critically. In addition to applying the evaluative criteria described in Chapter 4 (pages 73–74), determine whether the site identifies the person or organization that created it, whether you can contact the creator, and when the site was last modified (how up to date it is).

Keeping Records

When conducting Internet research, keep careful records of the sites you find valuable. It's easy to lose your way when searching the Internet, which can make it difficult to relocate a site you need to visit again. Most browsers provide a bookmark feature that lets you add any page you are visiting to a personalized menu of sites you can return to with a single click. Even so, it's best to write down the URL of any site whose information you believe you will provide to your readers.

Be sure to record the date you visit each site. Sites can change and even disappear suddenly, so this date is a crucial part of your bibliographic citation, as Appendix A explains.

Bookmark valuable sites and write down their URLs.

Using Social Media

The feature that makes social media so popular for personal use is the same one that makes them so powerful as research tools in the workplace: social media are interactive. They allow groups of people to ask and answer questions, participate in group discussions on topics of professional interest, and share and comment on ideas and information. Some social media resources allow anyone in the world to participate, while others are open only to members of an organization or company.

Examples of Social Media Used for Research

The three social media most widely used by professionals for research are *discussion boards*, *blogs*, and *wikis*.

Discussion Boards

Discussion boards are websites (sometimes sections of websites) where people can post messages and expect that others may respond. You may have used a discussion board in a college course. When conducting research on the job, you can use a search engine to locate discussion boards on the topic you are studying, read through old messages, and post questions of your own, hoping for responses from others. Many discussion boards have separate areas or threads for specific topics related to their overall subject. For instance, engineeringexchange.com has a discussion board (called a "forum") with such categories as *fluid power*, *motion control*, and *3D CAD*.

Blogs

Blogs (short for web logs) are websites where individuals can post messages that are usually displayed with the most recent posts at the top. Like discussion boards, they often have separate threads or topics. Microsoft, Apple, and other companies maintain them to enable users to ask and answer questions about difficulties they are encountering, shortcuts they have discovered, and similar topics. But there are blogs for many topics in many fields, and probably several on topics you will research in your career.

Wikis

Wikis are websites on which groups of people can collaborate on building resources related to their common interests. The best known is Wikipedia (wikipedia.org).

Contributors can add material and edit existing content, a feature that is ideal for subjects where knowledge advances rapidly or information changes quickly.

Using Social Media for Your Research

Research conducted through social media faces three challenges.

- **Locating social media resources related to your topic.** First, see if your employer has social media resources that are available to the organization's employees. When looking for resources elsewhere, ask your fellow employees for leads. Also, check the websites of professional organizations in the area of your research. For instance, the Institute of Electrical and Electronics Engineers has many communities of specialists in various fields that have discussion boards and blogs. If you are looking for blogs, try a browser search that begins with "blog" followed by your topic, such as "blog energy storage." Also, check the websites of companies that offer products or services related to your research.

- **Receiving responses to your questions.** Professionals who contribute to social media communities often expect that questions will involve advanced concerns. They sometimes decide not to answer or they respond rudely to people who ask elementary questions or who haven't read through earlier entries or used the search function that is available on many social media sites. Before asking for help, find out what you can on your own.

- **Determining the validity of the information provided.** Because many social media are open to large numbers of people, you need to carefully evaluate the quality of the information you find through them. Be alert to potential bias, as well as contributions by people who don't really have the expertise to provide accurate information. Consider the sponsor of the resource. Discussion boards sponsored by professional organizations may be more authoritative than ones sponsored by companies. Also, where possible, check the dates of the contributions. Are they recent enough to provide current information? Whenever possible, check information obtained through social media with other sources.

Using the Library

For research projects, the library will be your best source of information and ideas. Library resources fall into two broad categories.

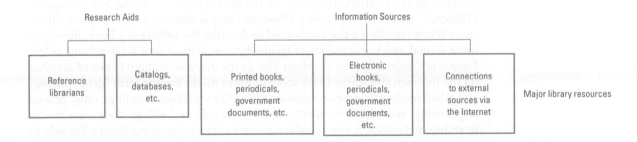

Major library resources

Generally, your excursions in library research will begin with one of the research aids, which can guide you to the most productive information sources. The following

Many employers have their own libraries that include resources directly related to their specialties.

Tell the reference librarian your communication's objectives.

sections will help you use research aids productively and introduce other sources with which you may not be familiar.

Obtaining Assistance from Reference Librarians

You will rarely find any research aid more helpful than reference librarians. They can tell you about specialized resources that you may not be aware of, and they can explain how to use the time-saving features of these resources.

Reference librarians will be able to give you the best help if you indicate very specifically what you want. In addition to stating your topic, describe what the purpose of your communication is, who your readers are, and how your readers will use your communication.

Using the Library Catalog

The library catalog lists the complete holdings of a library, including books, periodicals, pamphlets, recordings, videotapes, and other materials. In most libraries, the catalog is computerized so you can search for items in several ways. If you are looking for a particular book whose title you know or a work written or edited by a person whose name you know, library catalogs are very simple to use. However, when you begin looking for information about a particular topic, your success may depend on your ingenuity and knowledge of how to use the computerized catalog that most libraries have.

To search for a specific topic, you have two choices.

- **Subject search.** To aid researchers, librarians include subject headings in the record for each library item. When you indicate that you want to do a subject search, the computer will prompt you to enter the words that identify the subject you are looking for. The computer will search through all items that have been tagged with the exact words you entered.

- **Word search.** When you indicate that you wish to conduct a word search, you will also be prompted to enter the words that identify your subject. This time, however, the computer will search the entire contents of all its records, including the title, author, and subject lines, as well as tables of contents and other information that particular records might have.

Conducting Subject and Word Searches

Subject and word searches in computerized library catalogs are very similar to keyword searches on the Internet. Therefore, all the advice given in "Using Search Engines Efficiently" (pages 92–93) applies. However, there is also one very important difference: When identifying the words used to describe the subject of a book, librarians use a formal and restricted set of terms that are defined in a large volume titled, *Library of Congress Subject Headings List.* In the discussion about keyword searches on the Internet, you learned that some websites might use the term "gene splicing" and others might use "genetic engineering." In library subject headings, only "genetic engineering" is used. Consequently, while "gene splicing" will produce some results in an Internet search, it won't produce any in a subject search in a library. For help in determining the correct terms for subject searches, you have three resources.

- Many computerized library catalogs will tell you the correct term if you use an incorrect one that it recognizes as a synonym.

- The *Library of Congress Subject Headings List* is available at any library.
- Reference librarians are most willing to assist you.

Many computerized library catalogs will let you choose between abbreviated and extended displays of your search results. Extended displays are usually more helpful because they give you more information to consider as you decide whether to look at the entry for a particular item.

Refining and Extending Your Search

If your initial effort produces too few results, an overwhelming number of them, or an inadequate quality or range of them, there are several ways to refine and extend your search.

- **Look in the catalog entries of books you find for leads to other books.** Catalog entries not only name the subject headings under which a book is cataloged, but they also provide links to lists of other works that also have those subject headings.
- **Narrow your search.** If you receive too many results, you may refine the search in many of the ways described in the discussion of Internet searches (see "Using Search Engines Efficiently," pages 92–93).
- **Use other resources.** Don't limit yourself to resources you locate through the library catalog. Your best source of information may be a corporate publication or other item not listed there. A reference librarian can help you identify other aids to use.
- **When you go to the library shelves, browse.** Sometimes books that will assist you are located right next to books you found through the library catalog. Don't miss the opportunity to discover them. Browse the shelves.

Using Databases

Databases are online research aids that catalog, describe, and often provide access to publications and other resources on particular topics. Your college library's website will include a list of available databases.

Many databases focus on science, technology, engineering, health, and other specialized topics. They respond to your search by giving the title, author, and location of relevant journal articles. Most provide an abstract, or summary, of each item (see Figure 5.3). By scanning through an abstract, you can usually tell whether reading the entire article would be worthwhile. Some databases go farther by enabling you to download copies of the items you would like to read in full.

The first step in locating resources through databases is to identify appropriate ones. Each resource has its own topics and selection criteria. Their topics can be very broad or very narrow. Many overlap in their coverage. An important criterion for choosing which databases to use is the *audience* for the items it indexes. Some databases provide pathways to items for general readers. Others guide users to resources for specialists in a field. You will want to use databases that index items for users like the readers to whom you are writing. A reference librarian can assist you in finding the most promising databases.

To locate items in most databases, you would use a keyword search similar to the one you would use in the library's online catalog. Some databases also support additional, powerful search methods. *SciFinder*, which includes information from

TRY THIS Using the online resources in your college's library, find an article related to your major or a personal interest in a journal you haven't heard of before.

FIGURE 5.3

Abstract from an
Abstracting Index

This abstract illustrates the
research help you can receive when
using online databases
and indexes.

Full citation
In addition to using this information
to locate this item, you can copy and
paste the citation into your notes for
possible inclusion in your references.

**Links to other works
by the author**
These links can help you find other
publications that may also relate to
your research.

Abstract
This summary helps you decide
whether it would be worthwhile to read
the entire article.

Subject codes
This list identifies the keywords used to
index this article. You can use it to link
to other publications that were coded
with these words.

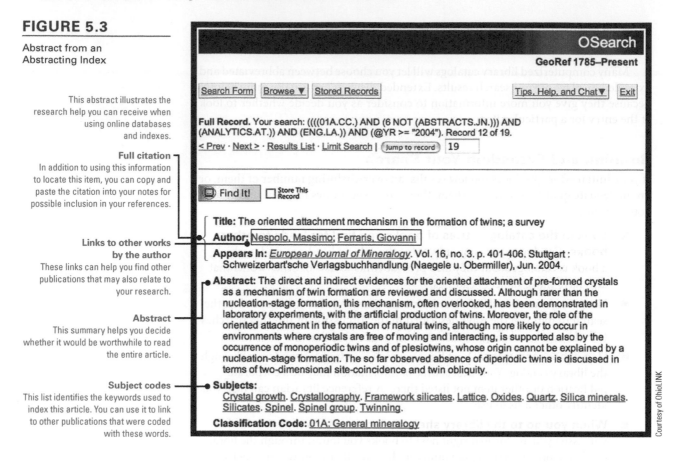

Courtesy of OhioLINK

more than 10,000 chemistry journals worldwide and has information about 49 million chemical substances, enables users to search for items related to a chemical by entering its formula or structure.

Reference Works

When you hear the term *reference works,* you probably think immediately (and quite correctly) of encyclopedias, dictionaries, and similar storehouses of knowledge, thousands of pages long. What you may not realize is that many of these resources, such as the *Encyclopedia Americana,* are now available online or on CD-ROM, so that finding information in them can be very quick and easy.

In addition to such familiar reference works as the *Encyclopædia Britannica,* thousands of specialized reference works exist. Some surely relate to your specialty. For example, there are the *McGraw-Hill Encyclopedia of Science and Technology* (20 volumes), *Elsevier's Dictionary of Medicine and Biology, The Harper Dictionary of Music,* and the *Dictionary of Petroleum Exploration, Drilling & Production.*

Government Documents

Every year, the U.S. Government Printing Office distributes millions of copies of its publications, ranging from pamphlets and brochures to periodicals, reports, and books. Some are addressed to the general public, while others are addressed to specialists in various fields. Sample titles include *Acid Rain, Chinese Herbal Medicine, Poisonous Snakes of the World,* and *A Report on the U.S. Semiconductor Industry.*

Government publications that may be especially useful to you are reports on research projects undertaken by government agencies or supported by government grants and contracts. Annually, the National Technical Information Service acquires more than 150,000 new reports on topics ranging from nuclear physics to the sociology of Peruvian squatter settlements. Chances are great that some relate to your subject.

The following indexes are especially helpful. A reference librarian can help you find many others.

Indexes to U.S. Government Publications

- Monthly Catalog of U.S. Government Publications
  Publications handled by the Government Printing Office
  http://catalog.gpo.gov:80/F?RN=94090418
- Lists of Publications by Specific Agencies
  EPA   http://www.epa.gov/ncepihom/
  NASA   http://www.nasa.gov/news/reports/index.html

Interviewing

At work, your best source of information will often be another person. In fact, people will sometimes be your only source of information because you'll be researching situations unique to your organization or its clients and customers. Or you may be asking an expert for information that is not yet available in print or from an online source.

The following advice focuses on face-to-face interviews, but it applies also to telephone interviews, which are quite common in the workplace.

Preparing for an Interview

Preparing for an interview involves three major activities.

- **Choose the right person to interview.** Approach this selection from your readers' perspective. Pick someone who can answer the questions your readers are likely to ask in a way that your readers will find useful and credible. If you are seeking someone to interview who is outside your own organization, the directories of professional societies may help you identify an appropriate person.

- **Make arrangements.** Contact the person in advance to make an appointment. Let the person know the purpose of the interview. This will enable him or her to start thinking about how to assist you before you arrive. Be sure to say how long you think the interview will take. This will enable your interviewee to carve out time for you. If you would like to record the interview, ask permission in advance.

- **Plan the agenda.** As the interviewer, you will be the person who must identify the topics that need to be discussed. Often, it's best simply to generate a list of topics to inquire about. But if there are specific facts you need to obtain, identify them as well. To protect against forgetting something during the interview, bring a written list of your topics and specific questions. For advice on phrasing questions, see "How to Conduct a Survey" (pages 103–107).

Take a reader-centered approach to selecting your interviewee.

Rob Marmion/Shutterstock.com

A well-planned interview can be productive and enjoyable for both interviewer and interviewee.

CONDUCTING EFFICIENT LIBRARY RESEARCH

By wisely using various features of online library catalogs and indexes, you can increase your research efficiency.

Finding Books

Online library catalogs typically show an opening page that invites you to search by one of four topics: keywords, author, title, and subject.

However, you can often research more efficiently by using an advanced search to sharpen your focus.

1. Limit the search to sources identified by *all* of several keywords.

2. Limit the search to sources that are
   - Written in a certain language.
   - A certain type of resource (e.g., book, periodical).
   - Published before or after a certain date.

3. Examine the search results to determine which items are worth clicking on for more details.

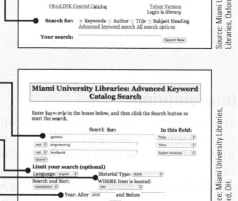

Following Leads to Other Sources

Use the detailed description of a source to find other helpful items.

1. Click on the author's name to see other items written by this person.

2. Note whether the item has a bibliography you can review for leads to other sources.

3. Click on the relevant subject headings to see other items tagged the same way ("Conducting Subject and Word Searches," pages 96–97).

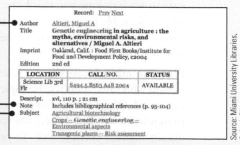

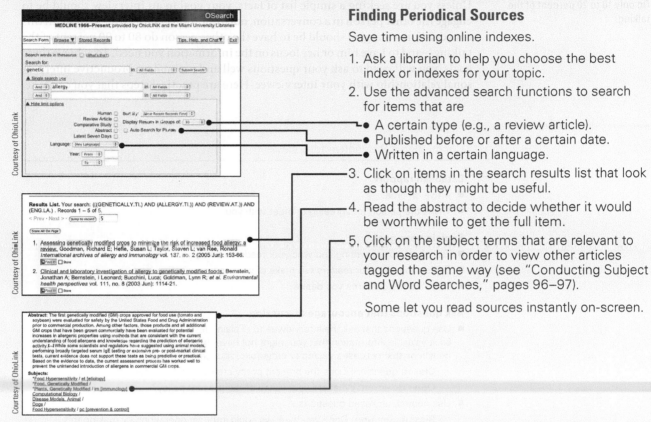

Courtesy of OhioLink

Source: Assessing Genetically Modified Crops to Minimize the Risk of Increased Food Allergy: A Review. International Archives of Allergy and Immunology. Richard E Goodman, Susan L Hefle, Steven L Taylor, Ronald van Ree, Food Allergy Research and Resource Program, University of Nebraska

Finding Periodical Sources

Save time using online indexes.

1. Ask a librarian to help you choose the best index or indexes for your topic.

2. Use the advanced search functions to search for items that are
   - A certain type (e.g., a review article).
   - Published before or after a certain date.
   - Written in a certain language.

3. Click on items in the search results list that look as though they might be useful.

4. Read the abstract to decide whether it would be worthwhile to get the full item.

5. Click on the subject terms that are relevant to your research in order to view other articles tagged the same way (see "Conducting Subject and Word Searches," pages 96–97).

Some let you read sources instantly on-screen.

Conducting the Interview

Do only 10 to 20 percent of the talking.

Unless you are seeking a simple list of facts, your goal in an interview should be to engage the other person in a conversation, not a question-and-answer session. In this conversation, your goal should be to have the other person do 80 to 90 percent of the talking—and to have him or her focus on the information you need. To achieve these goals, you will need to ask your questions well and maintain a productive interpersonal relationship with your interviewee. Here are practical steps that you can take.

Conducting a Productive Interview

Establish rapport.
- Arrive on time.
- Thank the person for agreeing to meet with you.

Explain your goal.
- Tell what you are writing and who your readers will be.
- Explain the use your readers will make of your communication.
- Describe the outcome you desire.

Ask questions that encourage discussion.
- Use questions that ask the interviewee to explain, describe, and discuss. They can elicit valuable information that you might not have thought to ask for. Avoid closed questions that request a yes/no or either/or response.
    Closed question: Does the present policy create any problems?
    Open question: What are your views of the present policy?
- Use neutral, unbiased questions.
    Biased question: Don't you think we could improve operations by making this change?
    Neutral question: If we made this change, what effect would it have on operations?
- Begin with general questions, supplemented by more specific follow-up questions that seek additional details important to you.
    General question: Please tell me the history of this policy.
    Follow-up question: What role did the labor union play in formulating the policy?

Show that you are attentive and appreciative.
- Maintain eye contact and lean forward.
- Respond with an occasional "uh-huh" or "I see."
- Comment favorably on the interviewee's statements.
    Examples: "That's helpful." "I hadn't thought of that." "This will be useful to my readers."

Give your interviewee room to help you.
- If the interviewee pauses, be patient. Don't jump in with another question. Assume that he or she is thinking of some additional point. Look at him or her in order to convey that you are waiting to hear whatever he or she will add.
- If the interviewee begins to offer information out of the order you anticipated, adjust your expectations.

Keep the conversation on track.
- If the interviewee strays seriously from the topic, find a moment to interrupt politely in order to ask another question. You might preface the question by saying something like this: "My readers will be very interested to know . . ."

TRY THIS Write two biased questions intended to get students to give a favorable evaluation of a cafeteria, bookstore, or other facility at your school. Next, write two parallel questions designed to elicit unfavorable evaluations. Finally, write versions of the same questions that are not biased.

Be sure you understand and remember.

- If anything is unclear, ask for further explanation.
- On complicated points, paraphrase what your interviewee has said and then ask, "Have I understood correctly?"
- Take notes. Jot down key points. Don't try to write down everything because that would be distracting and would slow down the conversation.
- Double-check the spelling of names, people's titles, and specific figures.

It's especially important that you assume leadership for guiding the interview. You are the person who knows what information you need to obtain on your readers' behalf. Consequently, you may need to courteously redirect the conversation to your topics.

Concluding the Interview

During the interview, keep your eye on the clock so that you don't take more of your interviewee's time than you requested. As the time limit approaches, do the following.

- **Check your list.** Make sure that all your key questions have been answered.
- **Invite a final thought.** One of the most productive questions that you can ask near the end of an interview is, "Can you think of anything else I should know?"
- **Open the door for follow-up.** Ask something like this: "If I find that I need to know a little more about something we've discussed, would it be okay if I called you?"
- **Thank your interviewee.** If appropriate, send a brief thank-you note by letter, memo, or email.

Conducting a Survey

While an interview enables you to gather information from one person, a survey enables you to gather information from *groups* of people.

At work, surveys are used as the basis for practical decision-making. Manufacturers survey consumers when deciding how to market a new product, and employers survey their employees when deciding how to modify personnel policies or benefit packages. While some surveys require the use of specialized statistical techniques that are beyond the scope of this book, you will usually be able to construct surveys that provide a solid basis for on-the-job decision making simply by following the advice provided in the following sections. This advice is equally valid for paper and online surveys.

At work, surveys support practical decision making.

Deciding What to Ask About

The first step in writing survey questions is to decide exactly what you want to learn.

- **Review your research objectives by focusing on the decisions your readers must make.** Roger worked for a small restaurant chain that asked him to study the feasibility of opening a premium pastry and coffee shop next to a college campus. His readers' questions, Roger knew, would relate primarily to whether there would be enough business to make the shop profitable.
- **Identify the full range of information your readers will find helpful.** Thinking about the information his readers would want, Roger realized

that his survey should ask about the full range of variables that could influence the shop's profitability, such as location, hours of operation, products offered, and pricing.

- **Gather the information needed for analyzing information from subgroups.** Because different groups answer survey questions differently, Roger asked about the respondents' sex, age, income, relationship to the college (student, employee, or not affiliated), and other characteristics. By analyzing responses from various demographic groups, Roger could help his readers understand more precisely the potential market for the shop.

Writing the Questions

More than anything else, the success of your survey depends on the skill with which you write your questions. The following suggestions will help you create an effective questionnaire that provides useful information and elicits the cooperation of the people you ask to fill it out.

Ambiguity is the greatest threat to a survey's value.

- **Avoid ambiguity.** The greatest threat to a survey's value is ambiguity in the questions. If different people interpret a question differently, they will, in fact, be answering different questions, making your data worthless. If they all interpret your question differently than you do, then your interpretation of the results will be erroneous. Take the reader-centered approach of asking yourself how the people responding to the survey might misunderstand each question. Pilot test your questions by asking a few people to tell you what they think each question is asking. In survey questionnaires, as in all communications, what matters isn't what you mean but what your readers think you mean.

TRY THIS If one of your professors asked you to help him or her improve a course by writing a survey for students to take, what three closed questions and what three open questions would you include? Why?

- **Mix closed and open questions.** *Closed questions* allow only a limited number of possible responses. They provide answers that are easy to tabulate. *Open questions* allow the respondent freedom in devising the answer. They provide respondents an opportunity to react to your subject matter in their own terms. See Figure 5.4.

You may want to follow each of your closed questions with an open one that simply asks respondents to comment. A good way to conclude a survey is to invite additional comments.

- **Ask reliable questions.** A *reliable* question is one that every respondent will understand and interpret in the same way. For instance, if Roger asked, "Do you like high-quality pastries?" different readers might interpret the term *high quality* in different ways. Roger might instead ask how much the respondents would be willing to pay for pastries or what kinds of snacks they like to eat with their coffee.

- **Ask valid questions.** A *valid* question is one that produces the information you are seeking. For example, to determine how much business the pastry shop might attract, Roger could ask either of these two questions.

Invalid
- How much do you like pastries?

Valid
- How many times a month would you visit a pastry shop located within three blocks of campus?

The first question is invalid because the fact that students like pastries does not necessarily mean that they would patronize a pastry shop. The second question is valid because it can help Roger estimate how many customers the shop would have.

- **Avoid biased questions.** Don't phrase your questions in ways that seem to guide your respondents to give a particular response.

  - Wouldn't it be good to have a coffee shop near campus?   Biased
  - How would you feel about having a coffee shop near campus?   Unbiased

- **Place your most interesting questions first.** Save questions about the respondent's age or similar characteristics until the end.

- **Limit the number of questions.** If your questionnaire is lengthy, people may not complete it. Decide what you really need to know and ask only about that.

- **Test your questionnaire.** Even small changes in wording may have a substantial effect on the way people respond. Questions that seem perfectly clear to you may appear puzzling or ambiguous to others. Before completing your survey, try out your questions with a few people from your target group.

Selecting Your Respondents

At work, writers sometimes present their survey questions to every person who belongs to the group whose attitudes or practices they want to learn about. For example, an

CLOSED QUESTIONS	
Forced Choice	■ Respondents must select one of two choices (yes/no, either/or).
	Example Would you buy pastries at a shop near campus, yes or no?
Multiple Choice	■ Respondents select from several predefined alternatives. How many times a month would you visit the shop? _____ 1 to 2 _____ 3 to 4 _____ 5 or more
Ranking	■ Respondents indicate an order of preference.
	Example Please rank the following types of pastry, using a 1 for your favorite, and so on.
Rating	■ Respondents pick a number on a scale.
	Example Please circle the number on the following scale that best describes the importance of the following features of a pastry shop: Music Unimportant 1 2 3 4 5 Important Tables Unimportant 1 2 3 4 5 Important
OPEN QUESTIONS	
Fill in the Blank	■ Respondents complete a statement.
	Example When deciding where to eat a late-night snack, I usually base my choice on _____.
Written Response	■ Respondents can frame responses in any way they choose.
	Example Please suggest ways we could make a pastry shop that would be appealing to you.

FIGURE 5.4

Closed and Open Questions

employee assigned to learn what others in her company feel about a proposed change in health care benefits or a switch to flextime scheduling might send a survey questionnaire to every employee.

However, surveys are often designed to permit the writers to generalize about a large group of people (called a *population*) by surveying only a small portion of individuals in the group (called a *sample*). To ensure that the sample is truly representative of the population, you must select the sample carefully. Here are four types of samples you can use.

Your sample should reflect the composition of the overall group.

- **Simple random sample.** Here, every member of the population has an equal chance of being chosen for the sample. If the population is small, you could put the name of every person into a hat, then draw out the names to be included in your sample. If the population is large—all the students at a major university, for example—the creation of a simple random sample can be difficult.

- **Systematic random sample.** To create a systematic random sample, you start with a list that includes every person in the population—perhaps by using a phone book or student directory. Then you devise some pattern or rule for choosing the people who will make up your sample. For instance, you might choose the fourteenth name on each page of the list.

Convenience samples can give unreliable results.

- **Convenience sample.** To set up a convenience sample, you select people who are handy and who resemble in some way the population you want to survey. For example, if your population is the student body, you might knock on every fifth door in your dormitory or stop every fifth student who walks into the library. The weakness of such samples is obvious: From the point of view of the attitudes or behaviors you want to learn about, the students who live with their parents or in apartments may be significantly different from those who live in dorms, just as those who don't go to the library may differ in substantial ways from those who do.

- **Stratified sample.** Creating a stratified sample is one way to partially overcome the shortcomings of a convenience sample. For instance, if you know that 15 percent of the students in your population live at home, 25 percent live in apartments, and 60 percent live in dormitories, you would find enough representatives of each group so that they constituted 15 percent, 25 percent, and 60 percent, respectively, of your sample. Even if you can't choose the people in each group randomly, you would have made some progress toward creating a sample that accurately represents your population.

When creating your sample, you must determine how many people to include. On the one hand, you want a manageable number; on the other hand, you also want enough people to form the basis for valid generalizations. Statisticians use formulas to decide on the appropriate sample size, but in many on-the-job situations, writers rely on their common sense. One good way to decide is to ask what number of people your readers would consider to be sufficient.

Use enough respondents to persuade your readers.

Contacting Respondents

There are three methods for presenting your survey to your respondents.

- **Face to face.** In this method, you read your questions aloud to each respondent and record his or her answers on a form. It's an effective method of contacting respondents because people are more willing to cooperate when someone asks for their help in person than they are when asked to fill out a printed questionnaire. The only risk is that your intonation, facial expressions, or body language may signal that you are hoping for a certain answer. Research shows that respondents tend to give answers that will please the questioner.

- **Telephone.** Telephone surveys are convenient for the writer. However, it can sometimes be difficult to use a phone book to identify people who represent the group of people being studied.

- **Mail or handout.** Mailing or handing your survey forms to people you hope will respond is less time-consuming than conducting a survey face to face or by telephone. Generally, however, only a small portion of the people who receive survey forms in these ways actually fill them out and return them. Even professional survey specialists typically receive responses from only about 20 percent of the people they contact.

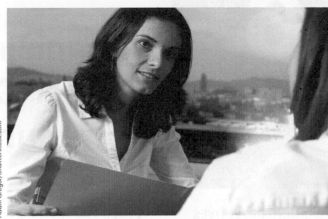

In face-to-face interviews, the interviewer carefully avoids facial expressions, comments, or other indications that he or she wants a particular response to a question.

LEARN MORE See Chapter 4 for detailed suggestions for interpreting survey results.

Interpreting Your Results

Of course, survey results don't speak for themselves. You need to analyze and interpret them in order to make the results meaningful and useful to your readers. Chapter 4 guides you through this process using an extended example that involves the analysis, interpretation, and presentation of survey data.

6

Organizing Reader-Centered Communications

LEARNING OBJECTIVES

1. Identify the exact content your readers need and want.
2. Organize in a way that helps your readers perform their tasks.
3. Organize in a way that makes your main points stand out.
4. Organize in a way that helps your readers quickly find and understand the information they want.
5. Treat your communication's stakeholders ethically.
6. Check the usefulness and persuasiveness of your organization.

MindTap®

Find additional resources related to this chapter in MindTap.

This chapter addresses questions that we all face when taking a reader-centered approach to writing at work. Let's let Toni answer them for us. A recent college graduate, Toni has been asked by her employer, a large construction company, to prepare a report that will help management decide which of several high-end computer programs to distribute to all of its civil engineers. Now that Toni has completed her research, she asks, "What's the best way for me to move from knowing what I know about my topic to creating an effective report about it?"

All features of a communication impact its effectiveness. However, writers often ignore one very important factor: organization. Instead of taking time to plan their communication's organization, they start right in drafting. The amount of time needed to plan a reader-centered organization can vary greatly, depending on such variables as the communication's length, subject matter, readers, and medium (print, website, social media). But even when the time needed is less than a minute, it is never wasted.

When we are organizing communications at work, there are three general—very general—goals for Toni and all of us to pursue. They are hallmarks of effective blueprints for almost any workplace communication: an organization that supports the reader's tasks, makes the main points stand out for the reader, and enables the reader to find and understand quickly the information he or she wants.

This chapter and your instructor will discuss strategies for achieving these three goals and examples of ways the strategies can be applied in some common workplace situations.

GUIDELINE 1 Include everything your reader needs—and nothing else

The first step in organizing is determining what your communication will include. Whether you are writing a short or long communication, whether you have conducted extensive research or just drawn on your own knowledge, you will almost certainly know more about your topic than your reader needs or wants to know.

To write a reader-centered communication, you need to select from everything you know the items required to make your communication useful and persuasive to your reader. Begin by going back to your mental portrait of your reader.

- What task will your reader use your communication to perform? What, exactly, will he or she need from your communication to be able to complete that task?

- What are your reader's current attitudes on your subject, what do you want them to be, and exactly what information, ideas, and arguments do you need to include to persuade your reader to adopt these desired attitudes?

You can determine much of this information, as Chapter 3 explained, by identifying the questions your readers will ask while reading your communication. But remember that you may know some information that will be important to your reader even though your reader won't know to ask about it. For example, in her research, Toni learned that a larger company is about to acquire the company that makes one of the programs her readers are considering. In these situations, the programs made by the company being acquired sometimes receive less attention, less support, and fewer updates. Even though her readers would not have asked about acquisitions, this information will be important to them.

As you identify the information your reader needs, you might list either mentally or in writing the relevant information and then determine how much detail your reader will need and want about each topic. Does your reader prefer lots of detail or messages that convey their contents very succinctly? Is your reader familiar with your topic, or will he or she require background explanations? What else do you know about your reader that will help you determine what he or she needs from you?

When selecting the pieces of your knowledge to include, you are also deciding what to exclude. The decision to leave out things you know about your subject can be difficult. For example, Toni might feel tempted to include information because she found it interesting or because she wants to show how much she knows and how hard she worked. However, such extra information only makes it more difficult for readers to locate and use the information they need.

LEARN MORE For detailed advice about identifying the information your readers want and need, see "How to Determine What Your Communication Must Do to Be Useful" and "How to Determine What Your Communication Must Do to Be Persuasive" (pages 51–56).

GUIDELINE 2 Group together the items your reader will use together

Once you've decided what to include—and not include—you must decide the most effective way of grouping your communication's contents. This can be a simple matter or a complex one. In some cases, a single, simple organizational strategy will do the job. Especially in communications longer than a few hundred words, you may need to combine several strategies.

In every case, however, the best place to begin is your understanding of the way your reader will perform his

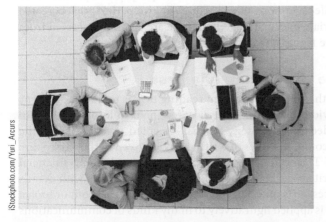

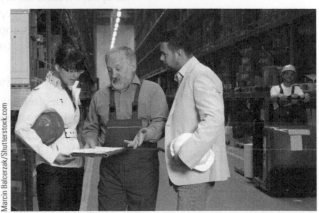

People use communications written at work to perform a wide variety of tasks.

LEARN MORE For advice about imagining your readers' tasks, see Chapter 1, "Six Reader-Centered Strategies You Can Begin Using Now" (page 14) and Chapter 3, "Your Goal: To Envision Your Reader's Response to Each Specific Aspect of Your Communication" (page 50).

or her tasks. As Toni thought about her readers' actions while reading her report, she realized that the managers would want to compare the various computer programs in terms of each specific criterion: cost, reliability, performance, ease of use, and so on. Because she knew that one of her readers' tasks would be to compare the costs of all programs, she put the purchase and other costs of all the programs in the same section. And she did the same for each of the other criteria.

Daniel followed this strategy where it was less obvious which information his readers would use together. He prepared a report summarizing hundreds of research studies on the effects of sulfur dioxide (SO_2) emissions from automobiles and factories. Daniel could have grouped his information in several ways, all of which seemed quite logical. For instance, he could have grouped together all the research published in a given year or span of years. Or he could have discussed all the studies conducted in Europe in one place, all conducted in North America in another place, and so on. However, Daniel is employed by a federal task force. His report would be read by members of the U.S. Congress as they decided how to structure legislation on SO_2 emissions. Daniel knew that, as these readers considered the adequacy of the current emission standards, they would want to read about the impacts of SO_2 emissions on human health separately from effects on the environment. Therefore, he organized his report around kinds of impacts, not around the dates or locations of the studies.

Like Daniel, you can organize your communications in a useful, task-oriented manner by understanding how your reader will use the information you provide.

If your communication has a complex audience, one that includes readers whose responsibilities and interests differ, you can also increase its ease of use by planning a modular design. In a modular communication, different parts are addressed to different readers or groups of readers.

LEARN MORE For more on modular designs, see "Identifying Readers: An Example" and Figure 3.2 (page 62).

Long reports and proposals often use a modular design. Usually such reports have two parts: (1) a very brief summary—called an *executive summary* or *abstract*—at the beginning of the report, designed for decision makers who want only the key information, and (2) the body of the report or proposal, designed for advisers and implementers, who need the details. Typically, the executive summary is only a page or two long, whereas the body may exceed a hundred pages. The body of the report might be divided into still other modules, one addressed to technical experts, one to accountants, and so on.

Website creators often use modular designs. For example, the Epilepsy Project knew that its website would be visited by several distinct groups of people: adults with epilepsy, their family members, kids and teens with epilepsy, and professionals who treat persons with epilepsy, among others. For each group, the Project created a special area with appropriate information. All of these areas can be accessed from links on the home page (see Figure 6.1).

LEARN MORE For advice on creating modular websites, see Chapter 20.

GUIDELINE 3 Give the bottom line first

Readers at work often say that their most urgent reading task is to find the writer's main point. The most obvious way to make it easy for them to locate your main point is to put it first. Indeed, this strategy for task-oriented organization is the subject of one of the most common pieces of writing advice given in the workplace: "Put the bottom line first." The bottom line, of course, is the last line of a financial statement. Literally, the writers are being told, "Before you swamp me with details on expenditures and sources of income, tell me whether we made a profit or took a loss." However, this advice is applied figuratively to many kinds of communications prepared at work.

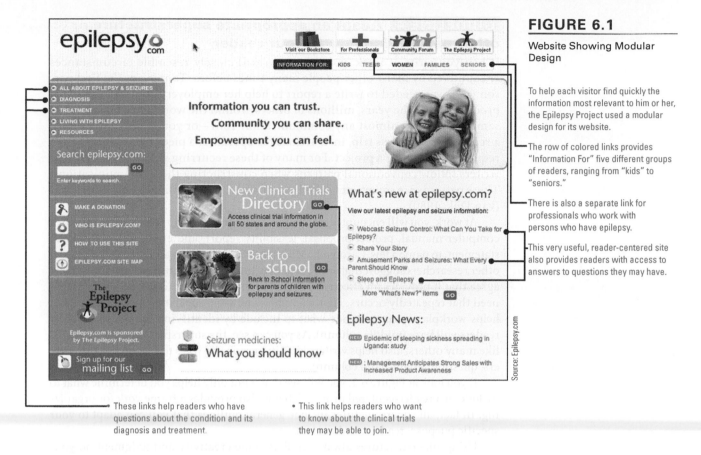

FIGURE 6.1

Website Showing Modular Design

To help each visitor find quickly the information most relevant to him or her, the Epilepsy Project used a modular design for its website.

The row of colored links provides "Information For" five different groups of readers, ranging from "kids" to "seniors."

There is also a separate link for professionals who work with persons who have epilepsy.

This very useful, reader-centered site also provides readers with access to answers to questions they may have.

Source: Epilepsy.com

These links help readers who have questions about the condition and its diagnosis and treatment.

This link helps readers who want to know about the clinical trials they may be able to join.

Don't misunderstand. You don't always need to place your main point in the very first sentence, although this is sometimes the most helpful place for it. When appropriate, you can always provide relevant background information before stating your main point. But don't keep your readers in suspense. As soon as possible, get it out: Is the project on schedule, or must we take special action to meet the deadline? Will the proposed design for our product work, or must it be modified?

To some writers, it seems illogical to put the most important information first. They reason that the most important information is generally some conclusion they reached fairly late in their research and thinking about their subject. Consequently, they think, it is logical to describe the process by which they arrived at the conclusion before presenting the conclusion itself. However, such a view is writer-centered. It assumes that information should be presented in the order in which the writer acquired it. In most workplace situations, such an organization runs counter to the sequence that readers will find most helpful.

When you want to influence your reader's decision or action, putting the bottom line first also increases your communication's persuasiveness. In these communications, the bottom line is the decision or action you advocate. After stating your position, you explain the reader-centered reasons for accepting it. Chapter 9, "Persuading Your Readers," provides a full discussion and examples of this persuasive strategy. Chapter 9 also describes the special circumstances in which it is better to withhold your recommendation. As a general rule, however, you can increase your reader's satisfaction and reading efficiency, as well as your communication's persuasiveness, by creating a task-oriented organization in which you give the bottom line first.

Focus on use, not logic, when you organize.

GUIDELINE 4 **Adapt an appropriate superstructure or other pattern familiar to your reader**

When you write at work, your situation will closely resemble circumstances encountered by many other people many times before. Toni is not the first person who ever needed to write a report to help her employer compare competing products. Over the years, millions of others around the world have done so. The same is true with almost any communication Toni—or you—create. Maybe it's a report on a business trip, instructions for operating a piece of equipment, or a request for funds for a project. For many of these recurring situations, workplace writers follow conventional patterns when constructing their communications. These patterns are sometimes called *genres*. Here, they are named *superstructures* (van Dijk, 1980).

At work, you will encounter many superstructures: the trip report, budget report, computer manual, project proposal, feasibility report, and environmental impact statement, to name a few. Carolyn R. Miller (1984), Herrington and Moran (2005), and other researchers suggest that each superstructure exists because writers and readers agree that it provides an effective pattern for meeting a particular communication need that repeatedly occurs. Figure 6.2 shows how the superstructure for proposals helps workplace writers decide what to include by identifying the questions their readers will have (middle column). As you can see, the superstructure for proposals, like many others, also helps writers identify persuasive ways to respond to the readers' questions (right-hand column).

If you look at Figure 6.2, you will see that it not only helps you determine what to include in a reader-centered proposal, but it also provides a framework for organizing. In fact, all superstructures include general outlines that you can adapt to your specific purpose, readers, and context.

Using superstructures always requires some creativity and judgment on your part. They are not surefire recipes for success. Each represents a general framework

FIGURE 6.2

Superstructure for Proposals

Each topic in the superstructure for proposals answers a particular question by the reader in a way that readers will find useful and in a place where the reader is accustomed to finding the answer.

SUPERSTRUCTURE FOR PROPOSALS		
TOPIC	**READERS' QUESTION**	**HOW THE WRITER WILL ANSWER**
Introduction	What is this communication about?	Provide an overview of the proposed project.
Problem	Why is the proposed project needed?	Describe the problem, need, or goal the project will address.
Objectives	What features will a solution to this problem need in order to be successful?	Identify the features a solution must possess if it is to succeed.
Product or Outcome	How do you propose to do those things?	Describe in detail what the writer will create.
Method	Are you going to be able to deliver what you describe here?	Describe the writer's plan of action; the facilities, equipment and other resources the writer will use; the project schedule; the writer's qualifications to conduct the project; and the writer's project management plan.
Costs	What will it cost?	Provide details about the cost.

for constructing messages in a typical situation. But no two situations are exactly alike. The elements can be combined and reordered in many ways—some can even be dropped—to serve the needs of specific readers and situations. Moreover, for many situations no superstructure exists. However, by looking at communications similar to yours that your co-workers have written you can often find a pattern that is familiar to your readers that you can use.

Chapters 21 through 27 describe the general superstructures for correspondence, social media, reports, proposals, and instructions. They also describe ways of adapting the superstructures to specific readers and purposes. At work you may encounter more specialized superstructures developed within your profession or industry—or even within your own organization.

GUIDELINE 5 Organize hierarchically

When we read, we encounter small bits of information one at a time: First, what we find in this sentence, then what we find in the next sentence, and so on. One of our major tasks as a reader is to build these small bits of information into larger structures of meaning that we can store and work with in our own minds. Represented on paper, these mental structures are hierarchical. They look like outlines (Figure 6.3) or

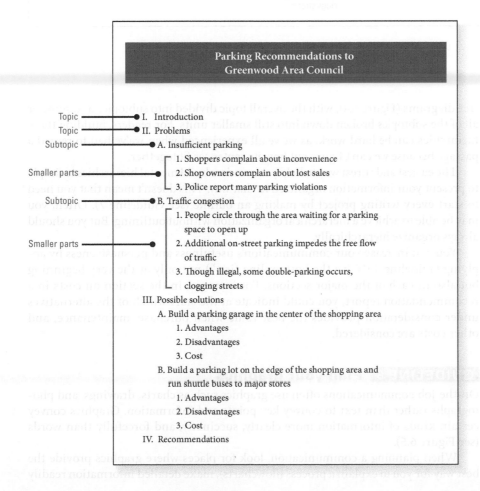

FIGURE 6.3

Outline Showing the Hierarchical Organization of a Report

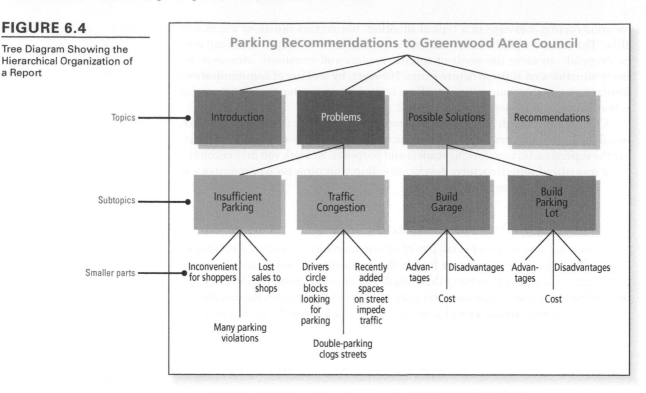

tree diagrams (Figure 6.4), with the overall topic divided into subtopics and some or all of the subtopics broken down into still smaller units. For readers, building these hierarchies can be hard work, as we've all experienced when we've had to reread a passage because we can't figure out how its sentences fit together.

The easiest and surest way to help your readers build mental hierarchies is simply to present your information organized in that way. This doesn't mean that you need to start every writing project by making an outline (see Guideline 7). Often, you may be able to achieve a hierarchical organization without outlining. But you should always organize hierarchically.

You can increase your communication's usefulness and persuasiveness by applying Guideline 3 ("Give the bottom line first") not only at the very beginning but also in each of the major sections. For instance, in the section on costs in a recommendation report, you could indicate at the start which of the alternatives under consideration provides the best value when purchase, maintenance, and other costs are considered.

GUIDELINE 6 **Plan your graphics**

On-the-job communications often use graphics such as charts, drawings, and photographs rather than text to convey key points and information. Graphics convey certain kinds of information more clearly, succinctly, and forcefully than words (see Figure 6.5).

When planning a communication, look for places where graphics provide the best way for you to explain a process (flowcharts), make detailed information readily

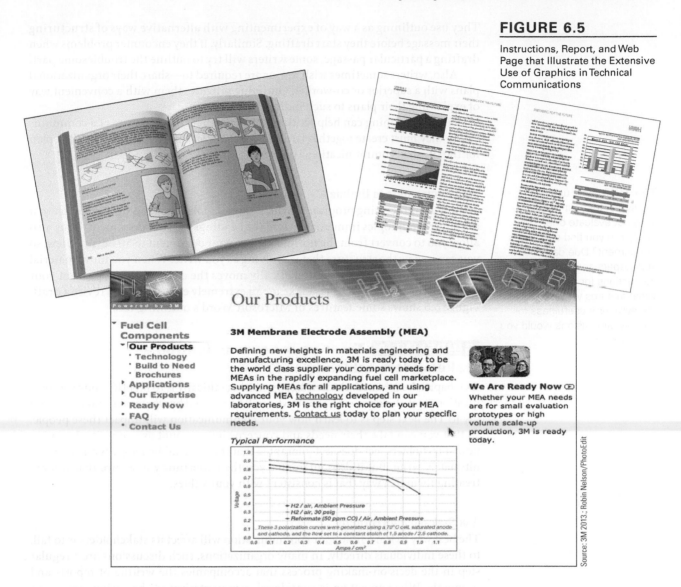

FIGURE 6.5

Instructions, Report, and Web Page that Illustrate the Extensive Use of Graphics in Technical Communications

Source: 3M 2013.; Robin Nelson/PhotoEdit

accessible (tables), clarify the relationship among groups of data (graphs), or show how something looks (drawings or photographs).

Chapter 12 provides detailed advice about where to use graphics and how to construct them effectively. However, don't wait until you read that chapter to begin planning ways to increase your communication's usefulness with graphics.

GUIDELINE 7 Outline, if this will be helpful

When they talk about organizing a communication, many people mention outlining and ask, "Is outlining worth the work it requires?" No single answer to this question is valid for all writers and all situations.

On the job, outlining is rarely used for short or routine messages. However, many writers find outlining helpful for longer, more complex communications.

They use outlining as a way of experimenting with alternative ways of structuring their message before they start drafting. Similarly, if they encounter problems when drafting a particular passage, some writers will try to outline the troublesome part.

Also, writers sometimes wish to—or are required to—share their organizational plans with a superior or co-worker. Outlining provides them with a convenient way of explaining their plans to such individuals.

Finally, outlining can help writing teams negotiate the structure of a communication they must create together (see Chapter 17, "Develop and share a detailed plan for the finished communication," page 309).

Using Computers to Outline

Some word-processing programs include special tools for outlining that enable you to convert your topics immediately into the headings for your document. They also allow you to convert from the normal view of your document to an outline view, so you can review the structure that is evolving as you write. When you move material in the outline, the program automatically moves the corresponding parts of your full text, which can sometimes make for an extremely efficient way to revise a draft. Figure 6.6 shows some features of Microsoft Word's outlining tool.

GUIDELINE 8 Treat your communication's stakeholders ethically

As Chapter 3 explained, the first step in writing ethically is to identify your communication's stakeholders—the people who will be affected by what you say and how you say it. The next step is to learn how your communication will impact these people, so that you can take their needs and concerns into account as you write. Learning how stakeholders feel does not guarantee that they will all be happy with what you ultimately write. But it does guarantee you the opportunity to assure that you are treating them in a way that is consistent with your values.

Ask Stakeholders Directly

The best way to learn how your communication will affect its stakeholders is to talk to these individuals directly. In many organizations, such discussions are a regular step in the decision-making process that accompanies the writing of reports and proposals. When an action is considered, representatives of the various groups or divisions that might be affected meet together to discuss the action's potential impact on each of them.

Similarly, government agencies often solicit the views of stakeholders. For example, the federal agencies that write environmental impact statements are required to share drafts of these documents with the public so that concerned citizens can express their reactions. The final draft must respond to the public's comments.

Even in situations that are traditionally viewed as one-way communications, many managers seek stakeholder input. For instance, when they conduct annual employee evaluations, some managers draft an evaluation, then discuss it with the employee before preparing the final version.

If you are writing a communication for which there is no established process for soliciting stakeholders' views, you can initiate such a process on your own. To hear from stakeholders in your own organization, you can probably just visit or call. To contact stakeholders outside your organization, you may need to use more creativity and also consult with your co-workers.

TRY THIS How useful are the syllabi for your courses? Why not evaluate one? How easily can you find each day's assignment? Due dates for your papers and projects? Instructions for assignments? What features would increase the syllabus's usefulness for you? What revisions would you recommend?

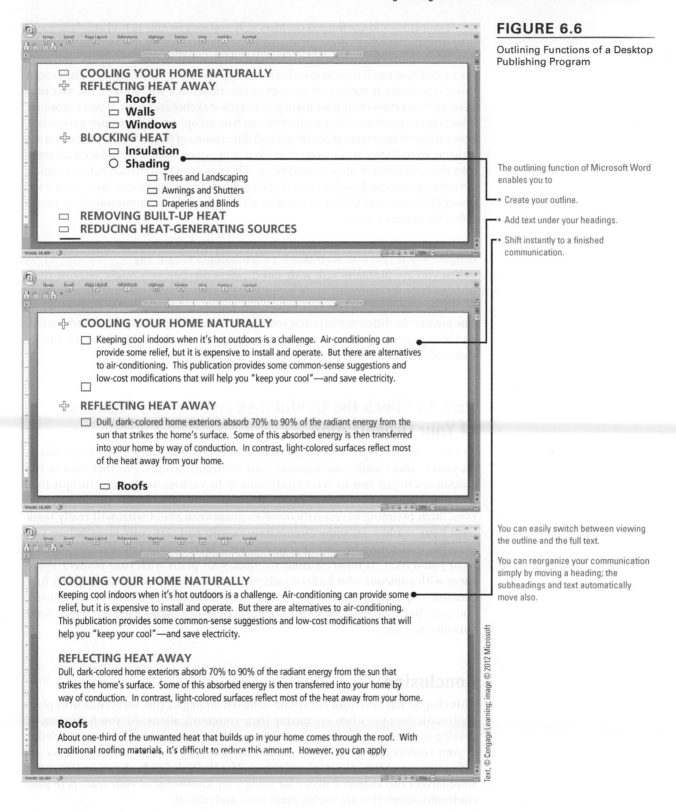

FIGURE 6.6

Outlining Functions of a Desktop Publishing Program

The outlining function of Microsoft Word enables you to

- Create your outline.
- Add text under your headings.
- Shift instantly to a finished communication.

You can easily switch between viewing the outline and the full text.

You can reorganize your communication simply by moving a heading; the subheadings and text automatically move also.

Text, © Cengage Learning; image © 2012 Microsoft

If you already know the stakeholders well, you may be able to find out their views very quickly. At other times, an almost impossibly large amount of time would be needed to thoroughly investigate stakeholders' views. When that happens, the decision about how much time to spend can itself become an ethical decision. The crucial thing is to make as serious an attempt as circumstances will allow. Also, you may need authorization from your managers to seek stakeholder views. If you encounter reluctance to grant you this permission, you have an opportunity to open a conversation with your managers about the ethical dimensions of the communication you are writing. Even if they aren't swayed this first time, you will have introduced the issue into the conversation at your workplace. There is a vast difference between doing everything you can do to investigate stakeholders' views, no matter how limited the possibilities are, and failing to consider all the people your communication might affect for better or worse.

Imagine What Your Communication's Stakeholders Would Say

When you are unable to talk with stakeholders directly, you might *imagine* their response to your communication in much the same way that you imagine your readers' likely responses. Of course, we can never know exactly what others are thinking. The greater the difference between you and the stakeholders—in job title, education, and background—the less likely you are to guess correctly. So it's always best to let stakeholders speak for themselves, whenever possible.

How to Check the Usefulness and Persuasiveness of Your Organization

At the heart of the guidelines you have just read is one common strategy: Focus on your readers while you organize your communications. For that reason, the guidelines urged you to refer continuously to various sources of insight into what will make your communication useful and persuasive in your readers' eyes. Such planning necessarily involves guesswork about what will really work with your readers. In some circumstances, you can test your ideas before you even begin drafting by asking your readers, "Here's what I'm planning. Is this what you'd like?" If it isn't feasible to check your plans with your readers, share them with someone who understands your readers well enough to help you find possible improvements. For projects in your technical communication course, you can check with your instructor. The better your plans, the better your final communication.

Conclusion

This chapter has described the reader-centered strategies that successful workplace communicators use when organizing their communications. As you have seen, to develop expertise in using these strategies you must possess the detailed knowledge of your readers that you gained by following Chapter 3's reader-centered advice for defining your communication's objectives. The Writer's Guide shown in Figure 6.7 summarizes this chapter's advice for using your knowledge of your readers to plan communications that are useful, persuasive, and ethical.

FIGURE 6.7

Writer's Guide for Organizing

Writer's Guide
ORGANIZING

PLAN CONTENT

1. What does your reader need to complete his or her task?

2. What will help to persuade your reader to adopt the attitudes you want him or her to have?

3. What background information does your reader need?

4. How much detail does your reader want and ask for?

5. Have you excluded all information that won't help your communication achieve its objectives?

MAKE CONTENT EASY TO FIND, UNDERSTAND, AND USE

1. What will be your reader's reading task?

2. Which of the following strategies will help you create a useful, persuasive organization? Check all that apply.

☐ Group together items your reader will want to use together
    Which items?

☐ Give the bottom line first (or very early)
    What is your main point overall? For each major section?

☐ Adapt a superstructure or other organizational pattern familiar to your reader
    Which pattern?

☐ Organize hierarchically
    What are the main points of each major part of your communication?

☐ Create a modular design
    What are the two or more major groups of readers?
    What part of your communication will you devote to each group?

☐ Include tables, drawings, or other graphics

☐ Outline your organizational plan

☐ Other

TREAT YOUR COMMUNICATION'S STAKEHOLDERS ETHICALLY

1. Who, besides your readers, are the stakeholders in your report?

2. How might they be affected by it?

3. What do you need to do to treat your readers and other stakeholders ethically?

USE WHAT YOU'VE LEARNED

EXERCISE YOUR EXPERTISE

Imagine that you are employed full-time and have decided to take a course at a local college. First, name a course you might like to take. Next, by following Guideline 1 on page 108, list the questions your employer would want you to answer in a memo in which you ask the employer to pay your tuition and permit you to leave work early two days a week to attend class.

(*continued*)

EXPLORE ONLINE

Develop a list of questions that high school students might ask about studying in your major at your college. Then try to answer these questions by going to your college's website. How accessible are the answers? How complete are they from the viewpoint of high school students? How could the website be made more useful for these students?

COLLABORATE WITH YOUR CLASSMATES

Imagine that you have been hired to create a brochure or website that presents your college department in a favorable light to entering first-year students and students who are thinking of changing their majors. Working with another student and following the guidelines in this chapter, generate a list of things that you and your partner would want to say. How would the two of you group and order this information?

APPLY YOUR ETHICS

Think of a policy of your college or employer that you would like to change. Imagine that you are going to write a report recommending this change. After identifying all the stakeholders, tell how you could gain a complete understanding of the relevant concerns and values of each stakeholder or stakeholder group. Which stakeholders' concerns and values would be most difficult for you to learn about? Why? Would you have more difficulty listening sympathetically to some stakeholders than others? If so, who and why? How would you ensure that you treat these stakeholders ethically?

REFLECT FOR TRANSFER

1. Write a memo to your classmates describing the ways you used two of the organizational strategies described in this chapter. Both strategies may be from the same communication or each from a different one. Explain why you chose each strategy and how well you think it worked for your reader—and why.

2. Write a memo to your instructor describing the two strategies described in this chapter that you predict will be most useful to you in your career. Explain why you chose these two.

Case: In MindTap, the case titled "Filling the Distance Learning Classroom" is well suited for use with this chapter.

Drafting Reader-Centered Communications

This chapter marks a major transition in your study of on-the-job writing. Its focus differs significantly from that of Chapters 3 through 6. They provide advice about the important writing activities you perform before you put fingers to keyboard or pen to paper: defining your communication's goals, researching your topic, and deciding how to organize your message. Beginning with this chapter, you will learn reader-centered strategies for using the results of that early work to create the actual communication you will give to your reader.

The Similarities among Paragraphs, Sections, Chapters, and Short Communications

This chapter's strategies apply to drafting all parts of a communication that are longer than a few sentences, including paragraphs, groups of paragraphs that make up the sections and chapters of longer communications, and even whole communications. They also apply to most short communications, a brief memo or email, for instance. For convenience's sake, this chapter uses the word *segments* to designate these variously sized prose units.

Can the same strategies really apply with equal validity to segments that range in size from a few sentences to an entire communication that may be tens or hundreds of pages long? In fact, it is very possible—for two reasons, one related to usefulness and the other to persuasiveness.

- **Usefulness.** You may have heard a paragraph described as a group of sentences on the same subject. With slight variation, this definition applies to larger segments. A *section* is a group of paragraphs on the same subject, a *chapter* is a group of sections on the same subject, and *an entire communication* is a group of chapters on the same subject.

   As people read any segment, small or large, they try to perform the same mental work, which writers can make easier or more difficult for them. First, readers must determine what the segment is about, and then they must figure out how its parts (whether sentences, paragraphs, sections, or chapters) fit together.

   When writers make it easy for readers to accomplish these two mental tasks, we say that the writing is clear and coherent. Readers can use it efficiently and effectively.

- **Persuasiveness.** Regardless of a segment's size, readers process its arguments the same way: They try to determine exactly what the persuasive

LEARNING OBJECTIVES

1. Draft your communications' segments to create clarity, coherence, and persuasiveness.
2. Draft to help your readers see the organization of your communications.
3. Adapt to your reader's cultural background.
4. Write beginnings that motivate your readers to pay attention.
5. Write endings that support your communications' goals.
6. Examine the human consequences of what you are drafting.

MindTap

Find additional resources related to this chapter in MindTap.

claim is, and they look for the evidence that supports the claim. Writers who make it easy for readers to accomplish these two mental tasks have taken the first—and necessary—steps in writing persuasively.

Because segments of all sizes make the same demands on readers, the same strategies enable you to write segments of all sizes usefully and persuasively. This chapter and your instructor will help you learn these strategies, which focus on three topics: how to start a segment, how to arrange the rest of each segment, and how to help your readers *see* the way you organized each segment. Because most or all of these strategies will be familiar to you, this chapter will review them quickly. In addition, this chapter focuses attention on the segments that begin and end a communication. In the workplace, they serve different functions than in most school writing.

Starting Segments

Some ways of beginning a segment are more effective than others. Guidelines 1 and 2 describe highly effective ways to begin most segments. For some, however, different strategies are superior. The discussion of Guideline 2 also explains when and how to use them.

GUIDELINE 1 Begin by announcing your topic

By far, the most valuable advice for starting segments is something you probably first heard long ago: Begin with a topic sentence. Although we often think of topic sentences as associated with paragraphs, they usually increase the clarity and coherence of segments of any size because they answer the first question readers ask, "What is this communication, section, or paragraph about?"

How important is it to put your topic sentences at the beginning of your paragraphs? In a classic experiment, Bransford and Johnson (1972) asked people to listen to the following passage being read aloud.

Topic statements are more helpful to readers when placed at the beginning of a segment.

Passage used in an experiment that demonstrated the value of placing topic statements at the beginning of a segment

> The procedure is actually quite simple. First you arrange things into different groups. Of course, one pile may be sufficient depending on how much there is to do. If you have to go somewhere else due to lack of facilities, that is the next step; otherwise you are pretty well set. It is important not to overdo things. That is, it is better to do too few things at once than too many. In the short run this may not seem important, but complications can easily arise. A mistake can be expensive as well. At first the whole procedure will seem complicated. Soon, however, it will become just another facet of life. . . . After the procedure is completed, one arranges the materials into different groups again. Then they can be put into their appropriate places. Eventually they will be used once more and the whole cycle will then have to be repeated. However, that is part of life.

The researchers told one group the topic of this passage in advance; they told the other group afterward. Then they asked both groups to write down everything they remembered from what they had heard. People who had been told the topic (washing clothes) before hearing the passage remembered many more details than those who were told afterward. One goal of your writing at work will always be to provide information your readers will remember.

Here are three of the most common and effective ways to provide topic statements at the beginning of your segments.

In addition to writing out a full statement, you can sometimes indicate your topic in other ways. For instance you might start a segment by asking a question, like the one you read a moment ago, "How important is it to put your topic sentences at the beginning of your paragraphs?" The implication is that the segment that follows will answer the questions. Often, a single word will suffice. "Then" indicates that the following segment will tell what happened after the action described in the preceding sentence. "First" indicates the beginning a segment that will include at least one more item in a list.

You can provide your readers with even more assistance in understanding your communications by creating an interlocking set of easy-to-spot topic statements that reveal the organizational hierarchy of your communication. Chandra, a large-animal veterinarian, did so in a report she wrote about the financial situation of the small zoo that employed her. Here is the outline for one part of the report.

I. The Metropolitan Zoo faces a severe budget crisis.
  A. The crisis first surfaced last August.
  B. What is causing the crisis?
    1. The crisis is not caused by rising costs.
    2. The crisis is caused by declining revenues.

In Figure 7.1, you can see how Chandra used these interlocking topic sentences to signal the organization of the flow of thought and reasoning in her report.

LEARN MORE For advice on organizing your communications hierarchically, see "Organize hierarchically" in Chapter 6 (page 113).

Outline corresponding to the page shown in Figure 7.1

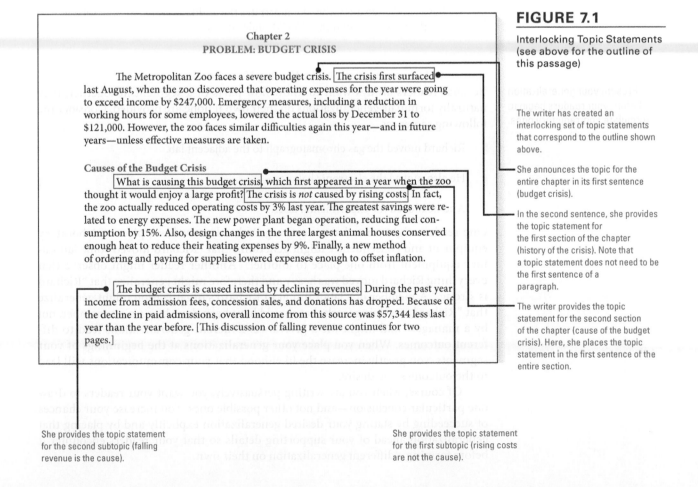

FIGURE 7.1

Interlocking Topic Statements (see above for the outline of this passage)

Chapter 2
PROBLEM: BUDGET CRISIS

The Metropolitan Zoo faces a severe budget crisis. The crisis first surfaced last August, when the zoo discovered that operating expenses for the year were going to exceed income by $247,000. Emergency measures, including a reduction in working hours for some employees, lowered the actual loss by December 31 to $121,000. However, the zoo faces similar difficulties again this year—and in future years—unless effective measures are taken.

Causes of the Budget Crisis

What is causing this budget crisis, which first appeared in a year when the zoo thought it would enjoy a large profit? The crisis is *not* caused by rising costs. In fact, the zoo actually reduced operating costs by 3% last year. The greatest savings were related to energy expenses. The new power plant began operation, reducing fuel consumption by 15%. Also, design changes in the three largest animal houses conserved enough heat to reduce their heating expenses by 9%. Finally, a new method of ordering and paying for supplies lowered expenses enough to offset inflation.

The budget crisis is caused instead by declining revenues. During the past year income from admission fees, concession sales, and donations has dropped. Because of the decline in paid admissions, overall income from this source was $57,344 less last year than the year before. [This discussion of falling revenue continues for two pages.]

The writer has created an interlocking set of topic statements that correspond to the outline shown above.

She announces the topic for the entire chapter in its first sentence (budget crisis).

In the second sentence, she provides the topic statement for the first section of the chapter (history of the crisis). Note that a topic statement does not need to be the first sentence of a paragraph.

The writer provides the topic statement for the second section of the chapter (cause of the budget crisis). Here, she places the topic statement in the first sentence of the entire section.

She provides the topic statement for the second subtopic (falling revenue is the cause).

She provides the topic statement for the first subtopic (rising costs are not the cause).

GUIDELINE 2 Present your generalizations before your details

In many of the segments you write at work, you will present detailed facts about your topic in order to explain or support a general point. You can usually increase the usefulness and persuasiveness of these segments by not only stating the *topic* but also stating the *main point* you want to make at the beginning.

How Initial Generalizations Make Writing Easier to Understand

When you present your generalizations first, you save your readers the work of trying to figure out what your general point is. Imagine, for instance, that you are a manager who finds the following sentences in a report.

Details without an initial generalization

> In a chamber set at 25°C, we passed a gas sample containing 500 micrograms of vinyl chloride monomer through a sampling tube with 5 grams of charcoal. We then divided the tube in half. The front half, which contained 2.3 grams of charcoal, had the entire 500 micrograms. The back half had none.

As you read these details, you probably find yourself asking, "What does the writer want me to get from this?" You would have been saved that labor if the writer had begun the segment with the following statement.

> We have conducted a test that demonstrates the ability of our sampling tube to absorb the necessary amount of VCM under the conditions specified.

How Initial Generalizations Make Writing More Persuasive

Present your generalization before your readers begin to formulate contradictory ones.

Because readers always want to know how parts of a segment fit together, they naturally formulate generalizations on their own if none are provided. Consider the following sentences.

> Richard moved the gas chromatograph to the adjacent lab.

> He also moved the electronic balance.

> And he moved the experimental laser.

One reader might note that everything Richard moved is a piece of laboratory equipment and consequently might generalize that "Richard moved some laboratory equipment from one place to another." Another reader might observe that everything Richard moved was heavy and therefore might generalize that "Richard is strong." A member of a labor union in Richard's organization might generalize that "Richard was doing work that should have been done by a union member, not by a manager" and might file a grievance. Different generalizations lead to different outcomes. When you place your generalizations at the beginnings of your segments, you greatly increase the likelihood that your communications will lead to the outcomes you desire.

Of course, when you are writing persuasively, you want your readers to draw one particular conclusion—and not other possible ones. You increase your chances of succeeding by stating your desired generalization explicitly and by placing that generalization ahead of your supporting details so that your readers encounter it before forming a different generalization on their own.

Sometimes You Shouldn't Present Your Generalizations First

Although you usually strengthen your segments by stating your generalizations before your details, sometimes delaying the generalizations will be more effective. If you begin a segment with a generalization that is likely to provoke a negative reaction from your readers, you may decrease your communication's persuasiveness. In such cases, you usually increase your communication's persuasiveness by postponing your general points until *after* you've laid the relevant groundwork with your details. Chapter 9 provides detailed advice about these situations.

How Guideline 2 Relates to Guideline 1

Taken together, Guidelines 1 and 2 suggest that in most cases you announce your topic and state your main point about it at the beginning of each segment. You can devote a separate sentence to each purpose.

> We conducted tests to determine whether the plastic resins can be used to replace metal in the manufacture of the CV-200 housing. The tests showed that there are three shortcomings in plastic resins that make them unsuitable as a replacement for the metal.

Separate sentences state the topic and the writer's generalization.

Often, however, you can make your writing more concise and forceful by stating both your topic and your main point in a single sentence.

> Our tests showed three shortcomings in plastic resins that make them unsuitable as a replacement for metal in the manufacture of the CV-200 housing.

Topic and generalization are combined in one sentence.

Draft Clear, Coherent, and Persuasive Segments

The guidelines in the preceding section discussed strategies for beginning your segments. The next three guidelines focus on ways to organize the material that follows your opening sentence or sentences.

GUIDELINE 1 Move from most to least important or impressive

In some segments, you will present parallel pieces of information, such as a list of five recommendations or an explanation of three causes of a problem. Whether you devote a single sentence or several paragraphs to each item, you can usually increase your segment's usefulness by presenting the most important or impressive item first and proceeding in descending order from there. This organization helps readers who scan to locate your main points. It also increases your communication's persuasiveness by presenting the strongest support for your arguments in the most prominent spot.

To identify the most important or impressive information, consider your communication from your readers' viewpoint. What information will they be most interested in or find most persuasive? For example, in the segment on the three shortcomings of plastic resins, readers will certainly be more interested in the major shortcoming than the minor ones.

As you think about your segments from your reader's perspective, you may occasionally find one that will be clearer if you ignore this guideline. For instance, to explain clearly the multiple causes of a flooding along a river, you may need to describe events chronologically even though the event that occurred first was not the one with the greatest impact. The key, as always, is to keep your reader in mind while drafting a communication.

GUIDELINE 2 Write segments using patterns familiar to your readers

Reading is always easiest when the segment we are reading follows a familiar pattern. The familiarity enables us to quickly understand how the parts fit together. Some of the patterns most widely used in the workplace are so common in the world in general that we learn them when we're very young. Examples include the chronological, cause-and-effect, and problem-and-solution patterns found so often in children's books and fairy tales. In the workplace, these patterns look a little different, of course, but not so different that they don't help readers. Chapter 8 describes eight of the most helpful ones in the workplace and shows examples of their use in workplace prose and graphics.

GUIDELINE 3 Smooth the flow of thought from sentence to sentence

Whether readers are reading a short segment, such as a paragraph, or a long segment, such as a chapter, they always face the challenge of determining how each sentence they read relates to the sentence they just completed. This connection-making process occurs so rapidly that readers aren't even aware of it until they hit a sentence that doesn't seem to fit. Then, they either stop to figure out the relationship or they push ahead, missing some of the writer's meaning. To avoid these undesirable results, smooth the flow of thought from sentence to sentence, using the following strategies.

Use Transitional Words and Phrases

You surely have a large vocabulary of transitional words and phrases: before, after, during, until, above, below, similarly, because, in contrast, however, nevertheless, and so on. The important point is to remember to use them when they will help your readers follow the flow of thought through your communication. When using them, put them where they will help your readers most: at the beginning of sentences. In that position, they immediately signal the relationship between that sentence and the preceding one.

Use Echo Words

Echo words are another easy and unobtrusive way to guide your readers from one sentence to the next. An echo word is a word or phrase that links a new sentence to a word in the one the reader just completed. For example:

The word echoed is *developed*.
The echo word is *development*.

> We can develop a vaccine for this new variety of the flu. Development will take several weeks, however.

In this example, the noun *development* at the beginning of the second sentence echoes the verb in the first.

There are many other kinds of echo words.

- **Pronouns**

In the second sentence, *It*
echoes *gas chromatograph* in
the first sentence.

> We returned the gas chromatograph. Its frequent breakdowns were disrupting work.

- **Another word from the same "word family" as the word being echoed**

In the second sentence,
oscilloscope echoes
lab equipment in the
first sentence.

> I went to my locker to get my lab equipment. My oscilloscope was missing.

- **A word or phrase that recalls some idea or theme expressed but is not explicitly stated in the preceding sentence**

The company purchased <u>15 tons</u> of cobalt. This <u>supply</u> should meet our needs for six months.

In the second sentence, the words *this supply* echo the *15 tons* in the first sentence.

Like transitional words and phrases, echo words help readers most when they appear at the beginning of a sentence.

Note that if you use *this* or *that* as an echo word at the beginning of a sentence, you should follow it with a noun. If used alone at the beginning of a sentence, they can leave your readers uncertain about what *this* is.

Our client rejected the R37 compound because it softened at temperatures about 500°C. This is what our engineers feared.

Original

In this example, the reader would be unsure whether *This* refers to the client's dissatisfaction or the softening of the R37. The addition of a noun after *This* clears up the ambiguity.

Our client rejected the R37 compound because it softened at temperatures about 500°C. This softening is what our engineers feared.

Revised

GUIDELINE 4 Present background information where it will most help your readers

During your research, organizing, and drafting you may sometimes discover that your readers will need background information in order to fully appreciate the importance of your topic or to understand your message. Ultimately, you must determine whether and what kind of background might be needed in any particular communication, based on your understanding of the specific readers you are addressing. However, you might be especially alert to your readers' possible needs in the following circumstances:

- Your readers are unfamiliar with the situation you are discussing.
- Your readers need to grasp a general principle in order to understand your specific points.
- Your readers are unfamiliar with technical terms you will be using.

When you are organizing, consider where each piece of background information will most help your readers. Not all background information belongs at the beginning of your communication. Information that pertains only to a certain segment should appear at the beginning of that segment or at the place where your communication first introduces the new concept or term. In the beginning of your communication, include only background information that will help your readers understand your overall message.

Help Your Readers See the Organization of Your Communication

You increase your communication's effectiveness by helping your readers see how your communication is organized—how the various segments are related to one another. When readers can see a communication's organization, they can understand the communication more easily and accurately. They can also quickly find a specific piece of information without having to read through the entire communication.

However, even the most carefully constructed organization may not be evident to the reader. That's because information about a communication's subject can be quite distinct from information about its organization. Consider the following diagram.

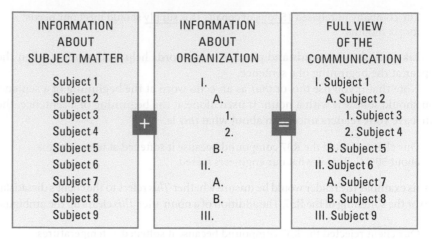

As the writer, you have the full view indicated in the right-hand column. You know how every part fits in with the rest. But your readers' view can be more like the one in the left-hand column: a sequential list of the topics discussed, without any sense of their places in the organizational hierarchy.

The next four guidelines describe ways to integrate organizational and subject matter information so your readers can fully understand the organization you constructed.

GUIDELINE 1 Use headings

Headings are a kind of signpost for mapping your communication's organization. At work, writers use headings not only in long documents, such as reports and manuals, but also in short ones, such as letters and memos. Figure 7.2 shows versions of the same memo, one without and one with headings. To see how effectively headings can signal organization, look at Figure 7.3, which highlights the use of headings on a website.

Headings help readers wherever there is a major shift in topic. In much on-the-job writing, such shifts occur every few paragraphs. Avoid giving every paragraph its own heading, which would give your prose a disjointed appearance rather than helping readers see how things fit together. An exception occurs in communications designed to provide readers quick access to specific pieces of information, as in a warranty, troubleshooting guide, reference manual, or fact sheet. In these documents, headings may even label sentence fragments or brief bits of data, turning the communication into something very much like a table of facts.

To be helpful, each heading must unambiguously indicate the kind of information that is included in the passage it labels.

Creating Text for Headings

- **Ask the question that the segment will answer for your readers.** Headings that ask questions such as "What happens if I miss a payment on my loan?" or "Can I pay off my loan early?" are especially useful in communications designed to help readers decide what to do.
- **State the main idea of the segment.** This strategy is often used in documents that offer advice or guidance. A brochure on bicycling safety uses headings such as "Ride with the Traffic," "Use Hand Signals," and "Ride Single File."
- **Use a keyword or phrase.** This type of heading is especially effective when a full question or statement would be unnecessarily wordy. For instance, in a request for a high-end multimedia production system, the section that discusses prices might have a heading that reads "How Much Will the System Cost?" However, the single word "Cost" would serve the same purpose.

FIGURE 7.2

The Same Memo without and with Headings

Garibaldi Corporation
INTEROFFICE MEMORANDUM

MEMO   June 14, 2017

TO   Vice Presidents and Department Managers
FROM   Davis M. Pritchard, President
RE   PURCHASES

Three months ago, I appo
the purchase of computer
I am establishing the foll

The task force was to bala
each department purchas
(2) to ensure compatibilit
create an efficient electro

I am designating one "pre
vendors.

The preferred vendor, YY
unless there is a compelli
purchases from the prefer
purchase price so that ind

Two other vendors, AAA
Garibaldi; both computer
network. Therefore, the s
price of these machines.

We will select one preferr
The task force will choose
when the choice is made

David Pritchard's first version lacked visual cues to the memo's organization.

By adding headings, he helped readers see how his memo is organized. The bold headings tell readers that these are the main parts of the memo.

By indenting, he indicated that these are the two parts of the computer policy.

Garibaldi Corporation
INTEROFFICE MEMORANDUM

                                               June 14, 2017

TO   Vice Presidents and Department Managers
FROM   Davis M. Pritchard, President
RE   PURCHASES OF TABLET COMPUTERS AND SMARTPHONES

Three months ago, I appointed a task force to develop corporate-wide policies for the purchase of tablet computers and smartphones. We face substantial risks to the security of our data and communications through vulnerabilities of the various operating systems to hacking. Based on the advice of the task force, I am establishing the following policies.

Objectives of Policies
The task force was to balance two possibly conflicting objectives: (1) to ensure that each department purchases the equipment that best serves its special needs and (2) to ensure compatibility among the equipment purchased so the company can create an efficient and, above all, secure network for our communications and data.

Computer Tablet Purchases
I am designating one "preferred" vendor of tablets and two "secondary" vendors.

Preferred Vendor: The preferred vendor, YYY, is the vendor from which all purchases should be made unless there is a compelling reason for selecting other equipment. To encourage purchases from the preferred vendor, a special corporate fund will cover 30% of the purchase price so that individual departments need fund only 70%.

Secondary Vendor: Two other vendors, AAA and MMM, provide tablets very rapidly in some South American countries and may be used when the need is urgent. However, no subsidy will be provided from the corporate fund.

Smartphone Purchases
We will select one preferred vendor and no secondary vendor for smartphones. The task force will choose between two candidates: FFF and TTT. I will notify you when the choice is made early next month.

FIGURE 7.3

Headings Used to Indicate Organization of a Website

Headings are as helpful to readers on web pages as they are in printed communications.

Headings in the navigation bar organize the groups of links.

The page's title gets the largest type.

The title at the top of this column is bold and large.

The headings for these sections, which are also bold, are smaller than the column heading but larger than the text.

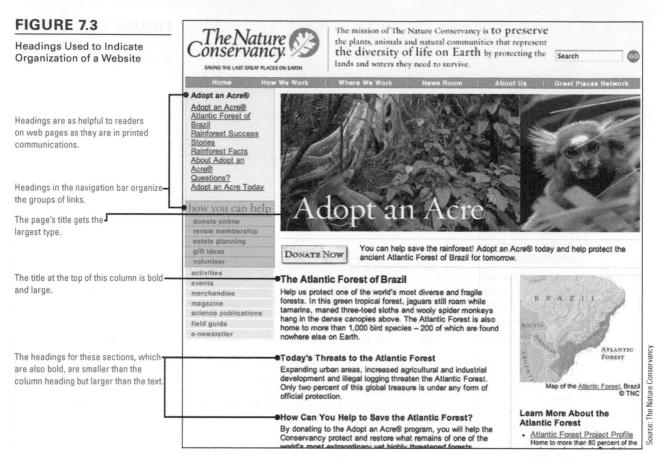

Source: The Nature Conservancy

Often, parallel headings make a communication's content easier to access and understand. For instance, when you are describing a series of steps in a process, parallel phrasing cues readers that the segments are logically parallel: "Opening the Computer Program," "Entering Data," and so on. However, in some situations, a mix of heading types will tell readers more directly what each section is about. Figure 7.4 shows the table of contents from a booklet titled *Getting the Bugs Out* that uses all three types of headings. Notice where they are parallel—and where they are not.

The visual appearance of headings is important to readers. To be useful, headings must stand out visually.

Designing Headings Visually

1. **Make headings stand out from the text, perhaps using these strategies.**
   - ☐ Use bold.
   - ☐ Use a different color than is used for the text.
   - ☐ Place headings at the left-hand margin or center them.
2. **Make major headings more prominent than minor ones. Here are some strategies to consider.**
   - ☐ Make major headings larger.
   - ☐ Center the major headings and position the others against the left-hand margin.
   - ☐ Use all capital letters for the major headings and initial capital letters for the others.
   - ☐ Give the major headings a line of their own and put the others on the same line as the text that follows them.
3. **Give the same visual treatment to headings at the same level in the hierarchy.**

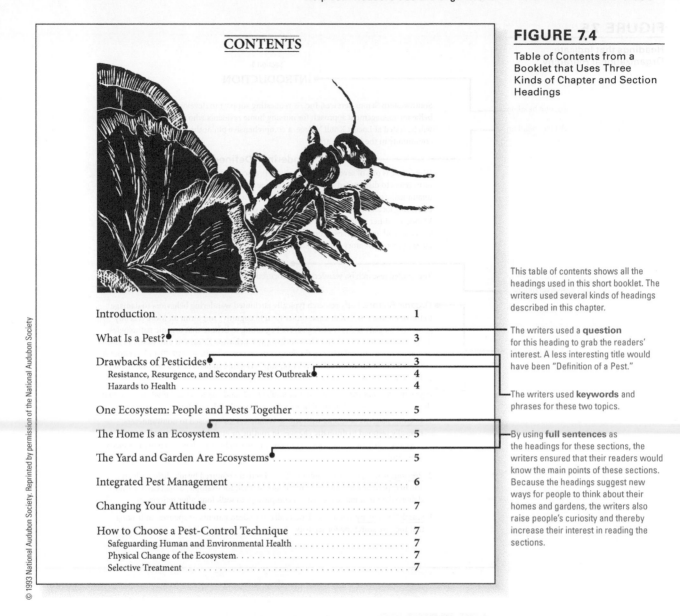

FIGURE 7.4

Table of Contents from a Booklet that Uses Three Kinds of Chapter and Section Headings

CONTENTS

This table of contents shows all the headings used in this short booklet. The writers used several kinds of headings described in this chapter.

The writers used a **question** for this heading to grab the readers' interest. A less interesting title would have been "Definition of a Pest."

The writers used **keywords** and phrases for these two topics.

By using **full sentences** as the headings for these sections, the writers ensured that their readers would know the main points of these sections. Because the headings suggest new ways for people to think about their homes and gardens, the writers also raise people's curiosity and thereby increase their interest in reading the sections.

Usually, you can use two or more of these techniques together so they reinforce one another. By doing so, you can readily create two or three easily distinguishable levels of headings that instantly convey the organizational hierarchy of your communication. Section or chapter titles may also function as a type of heading, as in Figure 7.5.

Sometimes, headings include the numbers and letters that are used in an outline. In most circumstances, however, the numbers and letters of an outline diminish the effectiveness of headings by distracting the readers' eyes from the headings' keywords.

By convention, headings and the topic statements that follow them reinforce each other, with the topic statement repeating one or more keywords from the heading. Topic statements do not contain pronouns that refer to the headings. For example, the heading "Research Method" would not be followed by a sentence that says, "Designing this was a great challenge." Instead, the sentence would read, "Designing the <u>research method</u> was a great challenge."

FIGURE 7.5

Headings that Indicate
Organizational Hierarchy

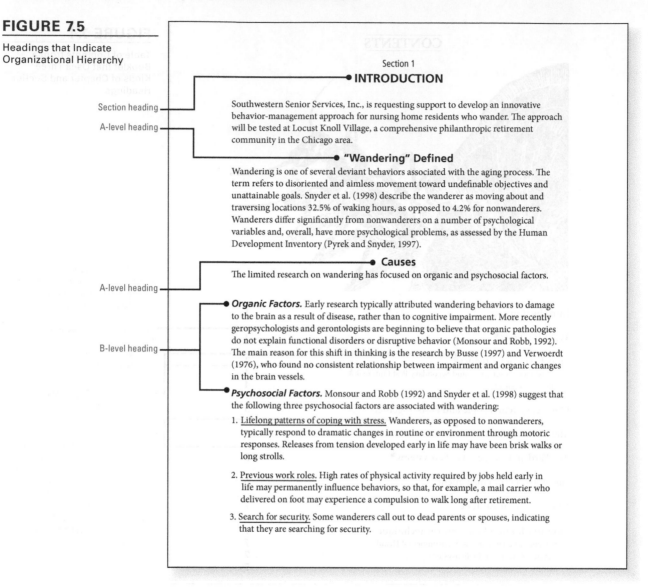

Section 1

INTRODUCTION

— Section heading

— A-level heading

Southwestern Senior Services, Inc., is requesting support to develop an innovative behavior-management approach for nursing home residents who wander. The approach will be tested at Locust Knoll Village, a comprehensive philanthropic retirement community in the Chicago area.

"Wandering" Defined

Wandering is one of several deviant behaviors associated with the aging process. The term refers to disoriented and aimless movement toward undefinable objectives and unattainable goals. Snyder et al. (1998) describe the wanderer as moving about and traversing locations 32.5% of waking hours, as opposed to 4.2% for nonwanderers. Wanderers differ significantly from nonwanderers on a number of psychological variables and, overall, have more psychological problems, as assessed by the Human Development Inventory (Pyrek and Snyder, 1997).

Causes

— A-level heading

The limited research on wandering has focused on organic and psychosocial factors.

— B-level heading

Organic Factors. Early research typically attributed wandering behaviors to damage to the brain as a result of disease, rather than to cognitive impairment. More recently geropsychologists and gerontologists are beginning to believe that organic pathologies do not explain functional disorders or disruptive behavior (Monsour and Robb, 1992). The main reason for this shift in thinking is the research by Busse (1997) and Verwoerdt (1976), who found no consistent relationship between impairment and organic changes in the brain vessels.

Psychosocial Factors. Monsour and Robb (1992) and Snyder et al. (1998) suggest that the following three psychosocial factors are associated with wandering:

1. Lifelong patterns of coping with stress. Wanderers, as opposed to nonwanderers, typically respond to dramatic changes in routine or environment through motoric responses. Releases from tension developed early in life may have been brisk walks or long strolls.

2. Previous work roles. High rates of physical activity required by jobs held early in life may permanently influence behaviors, so that, for example, a mail carrier who delivered on foot may experience a compulsion to walk long after retirement.

3. Search for security. Some wanderers call out to dead parents or spouses, indicating that they are searching for security.

GUIDELINE 2 **Use the visual arrangement of your text**

TRY THIS From the web, download a document written in a language you don't know that uses visual cues to indicate the organization of its contents. Describe in detail the techniques used. Could you improve on them? If you can't download a document, find a website in a language you don't know.

You can also signal your communication's organization through the visual arrangement of your text on the page.

Arranging Text Visually to Signal Organization

■ **Adjust the location of your blocks of type.** Here are three adjustments you can make.

☐ Indent paragraphs lower in the organizational hierarchy. (See Figure 7.3.)

☐ Leave extra space between the end of one major section and the beginning of the next.

☐ In long communications, begin each new chapter or major section on its own page, regardless of where the preceding chapter or section ended.

- **Use lists.** By placing items in a list, you signal readers that items hold a parallel place in your organizational hierarchy. Usually, numbers are used for sequential steps. Bullets are often used where a specific order isn't required.

TRY THIS Why not evaluate this book? Identify places where it provides signposts that indicate a chapter's organization effectively. Where could it do a better job? What do you recommend?

NUMBER LIST
The three steps we should take are
1. _____
2. _____
3. _____

BULLET LIST
Three features of the product are
- _____
- _____
- _____

When you are constructing a list, give the entries a parallel grammatical construction: All the items should be nouns, all should be full sentences, or all should be questions, and so on. Mixing grammatical constructions distracts readers and sometimes indicates a shift in point of view that breaks the tight relationship that should exist among the items.

WITHOUT PARALLELISM

Benefits
- Increased sales
- Decreased production costs
- Morale that is higher

Recommendations
1. Sell our Tallahassee plant.
2. Place the two Baton Rouge plants under a single manager.
3. We should also seek a better location for our Columbia warehouse.

WITH PARALLELISM

Benefits
- Increased sales
- Decreased production costs
- Higher morale ———————— Parallel
                            Not parallel

Recommendations
1. Sell our Tallahassee plant.
2. Place the two Baton Rouge plants under a single manager.
3. Seek a better location for our Columbia warehouse. ——— Parallel
                                                           Not parallel

GUIDELINE 3 Use forecasting statements

Forecasting statements help readers see a communication's organization by announcing the topics that are included in the segment that lies ahead. Often, they appear along with a topic sentence, which they supplement. For instance, here are the first two sentences from a section of a brochure published by a large chain of garden nurseries.

> Our first topic is the trees found in the American Southwest. <u>Some of the trees are native, some imported</u>.

Topic statement followed by a forecasting statement

Forecasting and topic statements are often combined in one sentence.

> Our first topic is the trees—<u>both native and imported</u>—found in the American Southwest.

Topic statement combined with a forecasting statement

Forecasting statements may vary greatly in the amount of detail they provide. The sample sentences above provide both the number and the names of the categories to be discussed. A more general preview is given in the next example, which tells its readers to expect a list of actions but not what these actions are or how many will be discussed.

> To solve this problem, the department must take <u>the following actions</u>.

Forecasting statement

When you are deciding how much detail to include in a forecasting statement, there are three main points to consider.

Writing Forecasting Statements

- **Say something about the segment's arrangement that readers will find helpful.** Usually, the more complex the relationship among the parts, the greater the amount of detail that is needed.

- **Say only as much as readers can easily remember.** A forecasting statement should help readers, not test their memories. When forecasting a segment that will discuss a three-step process, you could name all the steps. If the process has eight steps, state the number without naming them.

- **Forecast only one level at a time.** Don't list all the contents of a communication at its outset. That will only confuse your readers. Tick off only the major divisions of a particular section. If those divisions are themselves divided, provide each of them with its own forecasting statement.

GUIDELINE 4 Use transitions

Transitions help readers see the relationship among segments of a communication as effectively as they signal the relationship between adjacent sentences (see page 126).

Transitions between segments include three elements: a reference to the preceding topic, the topic (or topic statement) that is beginning, and the link between the two.

Transition based on cause and effect

> This large increase in contributions will enable us to expand our free health care program in several ways.

This sentence begins with the preceding topic (the increase in contributions) and ends with the next topic (the ways the free health care program might be expanded). The link between the two is that the contributions have created the opportunity for the expansion. Here's another example.

Transition based on sequence of events

> Having finally reached Dingham Point, the expedition spent the next three days building rafts to cross the river.

In this case, the link between the last topic (about the journey to Dingham Point) and the next one (building the rafts) is that one followed the other.

Sometimes you can signal transitions without using any words at all. For example, in a report that presents three brief recommendations, you might arrange the recommendations in a numbered list. The numbers provide the transition from one recommendation to the next. Similarly, in a memo covering several separate topics, the transition from one to the next might be provided by giving each topic a heading (see "Guideline 1, Use Headings," p. 128).

The numbers or bullets in a list signal transitions from one point to the next.

Global Guideline: Adapt to Your Reader's Cultural Background

The advice you have read so far in this chapter is based on the customs of readers in the United States and other Western countries where readers expect and value what might be called a *linear* organization. In this organizational pattern, writers express their main ideas explicitly and develop each one separately, carefully leading readers from one to another. As international communication experts Myron W. Lustig and Jolene Koester explain, this pattern can be visualized as "a series of steps or progressions that move in a straight line toward a particular goal or idea" (2012, p. 218).

In other cultures, writers and readers are accustomed to different patterns. For example, the Japanese use a nonlinear pattern that many researchers call a *gyre* (Connor & Nagelhout, 2008). The writer approaches a topic by indirection and implication because in Japanese culture it's rude and inappropriate to tell the reader the specific point being conveyed. Communication specialist Kazuo Nishiyama (1999) gives an example: When a Japanese manager says, "I'd like you to reflect on your proposal for a while," the manager can mean "You are dead wrong, and you'd better come up with a better idea very soon. But I don't say this to you directly because you should be able to understand what I'm saying without my being so rude." Similarly, in Hindi (one of the major languages of India), paragraphs do not stick to one unified idea or thought, as they do in the United States and many other Western nations. Linguist Jamuna Kachru (2006) explains that in the preferred Hindi style, the writer may digress and introduce material related to many different ideas.

Because of these cultural differences, serious misunderstandings can arise when your readers are employed by other companies in another country, work for your own company in another country, or even work in your own building but were raised observing the customs of another culture. Such misunderstandings cannot be avoided simply by translating the words of your communication: The whole message must be structured to suit the customs of your readers' culture.

Write a Beginning that Motivates Your Readers to Read

Two segments deserve extra attention because of the special ways they can contribute to a communication's success. They are the first segment, discussed in this section, and the last segment, discussed in the following section.

As you learned in Chapter 1, reading is a dynamic interaction between your readers and your words and graphics. Your readers' response to one sentence or paragraph can influence their reactions to all the sentences, paragraphs, and graph-

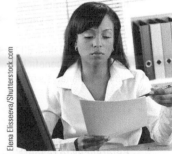

People are more likely to be persuaded by messages they think deeply about.

ics that follow. Consequently, the opening is especially important. It helps to establish the frame of mind readers bring to the rest of your communication. In fact, of the many functions that beginnings perform, the most important is simply to persuade readers to devote their full attention to the message. At work, people complain that they receive too many emails, memos, and reports. As they look at each communication, they ask, "Why should I read this?" Your goal is to convince them not only to pay *some* attention to your message but also to pay *close* attention. Doing so will be especially important when your communication is primarily persuasive, as when you are writing a proposal or recommendation. In a classic study that is still influential today, Petty and Cacioppo (1986) demonstrated that the more deeply people think about a message while reading or listening to it, the more likely they are to hold the attitudes it advocates, resist attempts to reverse those attitudes, and act upon those attitudes.

GUIDELINE 1 Announce your topic and its benefit to your readers

To grab your readers' attention, you must do two things at the outset.

- Announce your topic.
- Tell your readers how they will benefit from the information you are providing.

Be sure to do *both* things. Don't assume that your readers will automatically see the value of your information after you have stated your topic. Benefits that appear obvious to you may not be obvious to them. Compare the following sets of statements:

STATEMENTS OF TOPIC ONLY (AVOID THEM)	STATEMENTS OF TOPIC AND BENEFIT (USE THEM)
This memo tells about the new technology for reducing carbon dioxide emissions.	This memo answers each of the five questions you asked me last week about the new technology for reducing carbon dioxide emissions.
This report discusses step-up pumps.	Step-up pumps can save us money and increase our productivity.
This manual concerns the Cadmore Industrial Robot 2000.	This manual tells how to prepare the Cadmore Industrial Robot 2000 for difficult welding tasks.

LEARN MORE To learn about subject lines, go to Chapter 21, pages 369, 372, and 375–377.

Communications such as memos and emails provide a second opportunity to apply this guideline. You can fill in their subject lines with statements that also name your topic and tell how your communication on that topic will benefit your reader.

Topic only—▪ Subject: Springer Valves

Topic plus benefits (reader will—▪ find recommendations) Subject: Recommendation for Improving Springer Valves

Vague—▪ Subject: Data

Precise (names kind of data and—▪ indicates reader will benefit by finding an analysis of it) Subject: Analysis of Probe Data

GUIDELINE 2 Refer to your readers' request

At work, you will often write because a co-worker, manager, or client has asked you for a recommendation or information. To establish the reader benefit of your reply, simply refer to the request.

References to readers' requests

As you requested, I am enclosing a list of the steps we have taken in the past six months to tighten security in the Information Technology Department.

Thank you for your inquiry about the capabilities of our Model 1770 color 3D printer.

GUIDELINE 3 Offer to help your readers solve a problem

The third strategy for highlighting reader benefits at the beginning of a communication is to tell your readers that your communication will help them solve a problem they are confronting. Most employees think of themselves as problem solvers. Whether the problem involves technical, organizational, or ethical issues, they welcome communications that help them find a solution.

Establishing a Problem-Solving Partnership with Your Readers

To establish a problem-solving partnership, tell your readers these three things.

1. **Tell your readers the problem you will help them solve.** Be sure to identify a problem your readers deem important.
2. **Tell your readers what you have done toward solving the problem.** Review the steps you have taken as a specialist in your own field, such as developing a new

feature for one of your employer's products or investigating products offered by competitors. Focus on activities that will be significant to your readers rather than listing everything you've done.

3. **Tell your readers how your communication will help them as they perform their part of your joint problem-solving effort.** For example, you might say that it will help them compare competing products, understand a new policy for handling purchase orders, or develop a new marketing plan.

Carla, a computer consultant, used this strategy when she wrote the beginning of a report about her trip to Houston, where she studied the billing system at a hotel her employer recently purchased.

Here is how Carla wrote the beginning of her report.

> Last year, our Houston hotel posted a profit of only 4 percent, even though it is almost always 78 percent filled. A preliminary examination of the hotel's operations suggests that its billing system may generate bills too slowly and it may be needlessly ineffective in collecting overdue payments. After examining the hotel's billing cycle and collection procedures, I identified four ways to improve them. To aid in the evaluation of my recommendations, this report discusses the costs and benefits of each one.

Carla names the problem her report will help solve.

Carla describes her work toward solving the problem.

Carla tells how her report will help her readers do their part toward solving the problem.

Must You Always Provide All This Information?

Beginnings that use this strategy may be shorter or longer than Carla's. Just make sure your readers understand all three elements of the problem-solving situation. If one or more of the elements will be obvious to them, there's no benefit in discussing them in detail. For example, if the only persons who would read Carla's report were thoroughly familiar with the details of the problem in Houston and the work she did there, Carla could have written a very brief beginning.

> In this report, I evaluate the billing system in our Houston hotel and recommend ways of improving it.

She included more explanation because she knew her report would also be read by people who were hearing about her trip for the first time.

Here are some situations in which a full description of the problem-solving situation usually is desirable.

- **Your communication will be read by people outside your immediate working group.** The more distant some or all of your readers are, the less likely that all of them will be familiar with your message's context.

- **Your communication will have a binding and a cover.** Bound documents are usually intended for large groups of current readers, and they are often filed for consultation by future readers. Both groups are likely to include at least some readers who will have no idea of the problem-solving situation you are addressing.

- **Your communication will be used to make a decision involving a significant amount of money.** Such decisions are often made by high-level managers who need to be told about the organizational context of the reports they read.

Circumstances in which you may need to describe the problem-solving situation in detail

TRY THIS Think of a way that one of your professors could improve a course you have taken with him or her. Identify ways the professor could better achieve his or her goal by making the change you have in mind. Write the opening sentences of a letter or email you might send to the professor with your suggestion.

Defining the Problem in Unsolicited Communications

At times in your career, you may want to make a request or recommendation without being asked to do so. In these unsolicited communications, you may need to persuade your readers that you can be their partner in solving a problem they didn't realize existed.

Consider the way Roberto accomplished this goal. A computer engineer, he designs computer programs that control manufacturing processes. One program contained two bugs that caused frequent calls from customers wanting assistance. When he and the other engineers received these calls, they guided each caller individually through a work-around. Annoyed by these disruptions, Roberto wanted to stop the calls by fixing the bugs.

However, Roberto knew that his boss wanted him and the other engineers to spend all their time finishing a major new product, not fixing an old one. Consequently, Roberto knew that he wouldn't get assigned to fix the bugs if he gave the writer-centered reason that the customer calls annoyed him. He needed to find a reader-centered problem that he could offer to solve by repairing the bugs. Therefore, he calculated the total time that the engineers spent answering the calls. He then wrote a memo to his boss that opened by demonstrating a problem she hadn't realized: The total amount of time spent responding to customer calls was actually slowing development of the new product. He then offered to help solve this problem—and received permission to repair the bugs.

GUIDELINE 4 Adjust the length of your beginning to the situation

There is no rule of thumb that tells how long the beginning should be. A good, reader-centered beginning may require only a phrase or may take several pages. You need to give your readers only the information they don't already have. Just be sure they know the following:

What your readers need to know

- The reason they should read the communication ("How to Write a Beginning that Motivates Your Reader to Read," page 135)
- The main point of the communication ("Give the bottom line first," page 111)
- The organization of the communication ("Use forecasting statements," page 133)
- The background information they need in order to understand and use the communication ("Present background information where it will most help your readers," page 127)

If you have given your readers all this information, then you have written a good beginning, regardless of how long or short it is.

Here is an opening prepared by a writer who followed all the guidelines given in this chapter.

Brief beginning

> In response to your memo dated November 17, I have called Goodyear, Goodrich, and Firestone for information about the ways they forecast their needs for synthetic rubber. The following paragraphs summarize each of those phone calls.

The following opening, from a two-paragraph memo, is even briefer.

Briefer beginning

> We are instituting a new policy for calculating the amount that employees are paid for overtime work.

At first glance, this single sentence may seem to violate all the guidelines. It does not. It identifies the topic of the memo (overtime pay), and the people to whom the memo is addressed will immediately understand its relevance to them. It also declares the main point of the memo (a new policy is being instituted). Moreover, because the memo itself is only two paragraphs long, its scope is readily apparent. The brevity of the memo also suggests its organization—namely, a brief explanation of the new policy and nothing else. The writer has correctly judged that his readers need no background information.

Figure 7.6 shows a relatively long beginning from a report written by a consulting firm hired to recommend ways to improve the food service at a hospital. Like the brief beginnings given earlier, it is carefully adapted to its readers and to the situation.

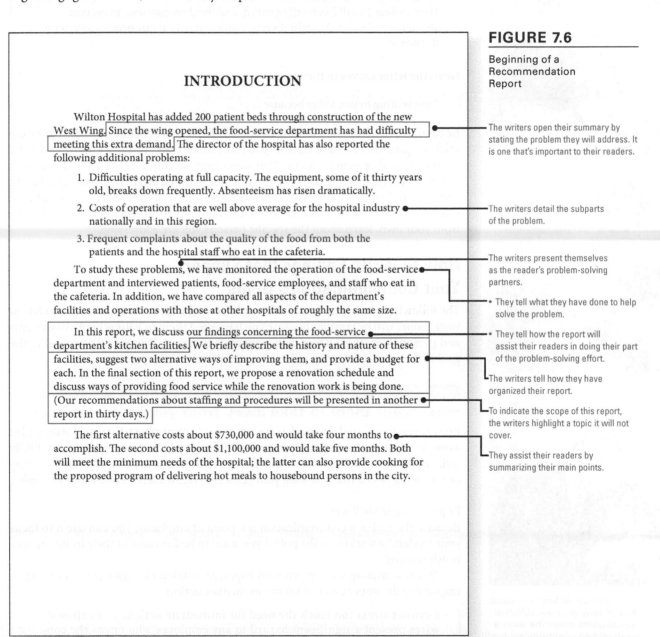

FIGURE 7.6

Beginning of a Recommendation Report

INTRODUCTION

Wilton Hospital has added 200 patient beds through construction of the new West Wing. Since the wing opened, the food-service department has had difficulty meeting this extra demand. The director of the hospital has also reported the following additional problems:

> The writers open their summary by stating the problem they will address. It is one that's important to their readers.

1. Difficulties operating at full capacity. The equipment, some of it thirty years old, breaks down frequently. Absenteeism has risen dramatically.

2. Costs of operation that are well above average for the hospital industry nationally and in this region.

> The writers detail the subparts of the problem.

3. Frequent complaints about the quality of the food from both the patients and the hospital staff who eat in the cafeteria.

To study these problems, we have monitored the operation of the food-service department and interviewed patients, food-service employees, and staff who eat in the cafeteria. In addition, we have compared all aspects of the department's facilities and operations with those at other hospitals of roughly the same size.

> The writers present themselves as the reader's problem-solving partners.

> They tell what they have done to help solve the problem.

In this report, we discuss our findings concerning the food-service department's kitchen facilities. We briefly describe the history and nature of these facilities, suggest two alternative ways of improving them, and provide a budget for each. In the final section of this report, we propose a renovation schedule and discuss ways of providing food service while the renovation work is being done. (Our recommendations about staffing and procedures will be presented in another report in thirty days.)

> They tell how the report will assist their readers in doing their part of the problem-solving effort.

> The writers tell how they have organized their report.

> To indicate the scope of this report, the writers highlight a topic it will not cover.

The first alternative costs about $730,000 and would take four months to accomplish. The second costs about $1,100,000 and would take five months. Both will meet the minimum needs of the hospital; the latter can also provide cooking for the proposed program of delivering hot meals to housebound persons in the city.

> They assist their readers by summarizing their main points.

GUIDELINE 5 Adapt your beginning to your readers' cultural background

Readers' expectations and preferences about the beginning of a communication are shaped by their culture. The suggestions you have just read are suitable for readers in the United States and some other Western countries. However, customs vary widely. For example, the French often open their business correspondence in a more formal way than do people in the United States. In some Spanish-speaking cultures, business letters often begin with statements that seem very elaborate to U.S. readers. In Japan, business letters often begin with a reference to the season. The following example shows one writer's way of adapting to the expectations of a Japanese businessperson (Human Japanese, 2005).

> Here in Seattle, fall has finally arrived, and the days continue to become shorter and shorter. At night it's now cold enough that one needs a sweater to go outside.

Next, the letter moves to the main topic.

> I am writing to you today because . . .

Be cautious, though, when writing to a person from another culture. Avoid operating solely on generalizations or stereotypes. Individuals from the same culture can differ from one another in many ways. Also, don't confuse cultural background with race or national origin. People of a particular race or national origin live in many different cultures, and cultures may have members who are from different races or national origins. When writing to readers from another culture, as when writing to readers from your own, learn about the specific persons you are addressing.

How to Write Endings that Support Your Communication's Goals

The following guidelines introduce you to several ending strategies used widely at work. Some will be familiar because you have used them in your school assignments and personal writing. Others may seem quite different. Pay special attention to the goal that each is used to achieve in the workplace.

GUIDELINE 1 Help your readers remember what you most want them to take away from your communication

Researchers have found that readers are better able to remember what is said at the end of a communication than information presented in the middle. Consequently, a skillfully written ending can increase a communication's usefulness and persuasiveness by stating or reviewing points that you want your readers to recall later. Here are four examples.

Repeat Your Main Point

Because the end of a communication is a point of emphasis, you can use it to focus your readers' attention on the points you want to be foremost in their minds as they finish reading.

In a text message, a construction engineer restated her main point in order to emphasize the importance of taking immediate action.

> I cannot stress too much the need for immediate action. The exposed wires present a significant hazard to any employee who opens the control box to make routine adjustments.

Richard Lord/The Image Works

SuperStock/SuperStock

Stephen Coburn/Shutterstock.com

Readers from different cultures may or may not have different expectations about the ways a work-related communication will begin.

Summarize Your Key Points

This strategy is closely related to the preceding one. In repeating your main point, you emphasize only the information you consider to be of paramount importance. In *summarizing,* you are concerned that your audience has understood the general thrust of your *entire* communication.

Here is the ending of a 115-page book titled *Understanding Radioactive Waste,* whose goal is to help the general public understand the impact of the nuclear power industry's plans to open new plants (Murray, 2003).

> It may be useful to the reader for us to now select some highlights, key ideas, and important conclusions for this discussion of nuclear wastes. The following list is not complete—the reader is encouraged to add items.
>
> 1. Radioactivity is both natural and manmade. The decay process creates alpha particles, beta particles, and gamma rays. Natural background radiation comes mainly from cosmic rays and minerals in the ground.
>
> 2. Radiation can be harmful to the body and to genes, but the low-level radiation effect cannot be proved. Many methods of protection are available.
>
> 3. The fission process gives useful energy in the form of electricity from nuclear plants, but it also produces wastes in the form of highly radioactive fission products.

Key points are summarized.

This list continues, but this sample should give you an idea of how this author ended with a summary of key points.

TRY THIS Find a printed or online communication that ends in a way that is not described in this chapter. How would you describe the strategy? How effective is it, given the communication's purpose and audience?

Refer to a Goal Stated Earlier in Your Communication

Many communications begin by stating a goal and then describe or propose ways to achieve it. If you end by referring to that goal, you sharpen the focus of your communication. The following example comes from a thirty-page proposal prepared by operations analysts in a company that builds customized, computer-controlled equipment used in manufacturing.

> To maintain our competitive edge, we must develop a way of supplying replacement parts more rapidly to our service technicians without increasing our shipping costs or tying up more money in inventory.

Beginning states a goal.

> The proposed reform of our distribution network will help us meet the needs of our service technicians for rapidly delivered spare parts. Furthermore, it does so without raising either our shipping expenses or our investment in inventory.

Ending refers to the goal.

Focus on a Key Feeling

When ending some communications, it may be more important to focus your readers' attention on a feeling rather than a fact. For instance, if you are writing instructions for a product manufactured by your employer, you may want your ending to encourage your readers' goodwill toward the company. Consider this ending of an owner's manual for a clothes dryer. Though the last sentence provides no additional information about the dryer, it seeks to shape the readers' attitude toward the company.

> The GE Answer Center™ consumer information service is open 24 hours a day, seven days a week. Our staff of experts stands ready to assist you anytime.

Ending designed to build goodwill

Ending designed to shape complex attitudes

The following ending is from a booklet published by the National Cancer Institute for people who have been successfully treated for cancer but do not know whether the disease will remain in remission.

> No one expects you to forget you have had cancer or that it might recur. You still might have moments when you feel as if you live perched on the edge of a cliff. They will sneak up unbidden. But they will be fewer and farther between if you have filled your mind with thoughts having nothing to do with cancer.
>
> Cancer might rob you of that blissful ignorance that once led you to believe that tomorrow stretched on forever. In exchange, you are granted the vision to see each today as precious, a gift to be used wisely and richly. No one can take that away.

GUIDELINE 2 Help your readers know what to do next

In much of what you write at work, your readers will already know what to do after reading your communications. With the fact you present or the question you answer, they will have all the information they need to proceed with their task. In other situations, your readers will look to you for advice about what to do next. Whenever you prepare a communication at work, ask yourself whether your readers will want you to suggest their next steps.

Tell Your Readers How to Get Assistance or More Information

At work, a common strategy for ending a communication is to tell your readers how to get assistance or more information. These two examples are from a letter and a memo.

Endings that offer help

> If you have questions about this matter, call me at 523–5221.

> If you want any additional information about the proposed project, let me know. I'll answer your questions as best I can.

By ending in this way, you not only provide your readers with useful information, you also encourage them to see you as a helpful, concerned individual.

Tell Your Readers the Next Steps to Take

In some situations, you can help your readers most by telling them what you think they should do next. If more than one course of action is available, tell your readers how to follow up on each of them.

> To buy this equipment at the reduced price, we must mail the purchase orders by Friday the 11th. If you have any qualms about this purchase, let's discuss them. If not, please forward the attached materials, together with your approval, to the Controller's Office as soon as possible.

Identify Any Further Study that Is Needed

Much of the work that is done on the job is completed in stages. For example, one study might answer preliminary questions and, if the answers look promising, an additional study might then be undertaken. Consequently, one common way of ending is to tell readers what must be found out next.

> This experiment indicates that we can use compound deposition to create microcircuits in the laboratory. We are now ready to explore the feasibility of using this technique to produce microcircuits in commercial quantities.

GUIDELINE 3 **Follow applicable social conventions**

All the strategies mentioned so far focus on the subject matter of your communications. When writing the endings of your communications, it is also important to observe the social conventions that apply in each situation.

Some of those conventions involve customary ways of closing particular kinds of communication. For example, letters usually end with an expression of thanks, a statement that it has been enjoyable working with the reader, or an offer to be of further help if needed. In contrast, formal reports and proposals rarely end with such gestures.

Other conventions are peculiar to specific organizations. For example, in some organizations writers rarely end their memos with the kind of social gesture commonly provided at the end of a letter. In other organizations, memos often end with such a gesture, and people who ignore that convention risk seeming abrupt and cold.

Consider, too, the social conventions that apply to your relationship with your readers. Have they done you a favor? Thank them. Are you going to see them soon? Let them know that you look forward to the meeting.

GUIDELINE 4 **After you've made your last point, stop**

Despite the important ways that endings can enhance the effectiveness of some communications, in other communications your best strategy will be to say what you have to say and stop, without adding any words after your last point. This situation can arise in short communications and when you are using an organizational pattern such as the one for step-by-step instructions (see Chapter 27) that brings you to a natural stopping place.

If your analysis of your purpose, readers, and situation convinces you that you should add something after your last point, review this chapter's other guidelines to select the strategies that will best help your communication achieve its goals.

Ethics Guideline: Examine the Human Consequences of What You're Drafting

When drafting their communications, employees sometimes become so engrossed in the technical aspects of their subject that they forget the human consequences of what they're writing. When this happens, they write communications in which their stakeholders are overlooked. Depending on the situation, the consequences can be quite harmful or relatively mild—but they can always lead to the unethical treatment of other people.

Mining Accidents

An example is provided by Beverly A. Sauer (2003), who has studied the reports written by the federal employees who investigate mining accidents in which miners are killed. In their reports, Sauer points out, the investigators typically focus on technical information about the accidents without paying sufficient attention to the human tragedies caused by the accidents. In one report, for example, the investigators describe the path of an underground explosion as it traveled through an intricate web of mineshafts and flamed out of various mine entrances. At one entrance, investigators write, "Debris blown by the explosion's forces damaged a jeep automobile parked near the drift openings." The investigators don't mention

Miners in the United States would be safer if the persons who prepare mining accident reports paid more attention to the human consequences of their writing, Dr. Beverly Sauer argues.

in this passage that in addition to damaging the jeep, the explosion killed sixteen miners who were in the mineshafts.

In addition to overlooking the victims of the disasters, the investigators' reports often fail to identify the human beings who created the conditions that caused the accidents. One report says, "The accident and resultant fatality occurred when the victim proceeded into an area of known loose roof before the roof was supported or taken down." This suggests that the miner was crushed to death by a falling mine roof because he was careless. Sauer's research showed, however, that in the same mines ten fatalities had occurred in five years—and seven resulted from falling roofs. Managers of the mines were not following safety regulations, and mine safety inspectors were not enforcing the law.

Writing with Awareness of Human Consequences

Of course, there's nothing the inspectors' reports can do on behalf of the deceased miners or their families. However, the stakeholders in the inspectors' reports include other miners who continue to work in what is the most dangerous profession in the United States. As Sauer points out, the investigators' readers include the federal officials responsible for overseeing the nation's mining industry. If the inspectors wrote in ways that made these readers more aware of the human consequences of mining accidents—and of the human failings that often bring them about—the federal officials might be more willing to pass stricter laws and insist that existing regulations be strictly enforced.

In Europe, where government regulation of mining is much stronger, the death and injury of miners are much rarer events than in the United States. The high accident rate that makes mining so dangerous in the United States, she argues, results in part from the way mining investigators write their reports.

Of course, most people aren't in professions where lives are at stake. In any profession, however, it's possible to become so focused on your technical subject matter that you forget the human consequences of your writing.

To avoid accidentally treating others unethically, you can take the following steps.

Avoiding Accidentally Treating Others Unethically

- **When beginning work on a communication, identify its stakeholders** (pages 64–65). Certainly, the stakeholders of the mining-disaster reports include miners whose lives are endangered if government officials don't enact and enforce life-saving safety measures.
- **Determine how the stakeholders will be affected by your communication.** (page 65).
- **Draft your communication in a way that reflects proper care for these individuals.** Be sure that all your decisions about what to say, what *not* to say, and how to present your message are consistent with your personal beliefs about how you should treat other people.

USE WHAT YOU'VE LEARNED

EXERCISE YOUR EXPERTISE

1. Circle all the forecasting statements in Figure 7.5 (page 132). For those segments that lack explicit forecasting statements, explain how readers might figure out the way in which they are organized.

2. Select a communication written to professionals in your field. This might be a letter, memo, manual, report, or article in a professional journal. (Do not choose a textbook.) Identify the guidelines from this chapter that the writer applied when drafting the communication's beginning. Is the beginning effective in achieving the writer's usefulness and persuasiveness goals? If so, why? If not, how could the beginning be improved?

3. Select a communication written to professionals in your field. This might be a letter, memo, manual, technical report, or article in a professional journal. (Do not choose a textbook.) Identify the guidelines from this chapter that the writer applied when drafting that communication's ending. Is the ending effective in achieving the writer's usefulness and persuasiveness goals? If so, why? If not, how could it be improved?

NOTE: Additional exercises are provided at the end of Chapter 8. See page 166.

EXPLORE ONLINE

1. Examine the ways that this chapter's guidelines are applied by a website that explains technical or scientific topics. For example, study the explanation of a medical topic at the National Cancer Institute (www.nci.nih.gov), the National Institute on Drug Abuse (www.nida.nih.gov), or WebMD (www.WebMD.com). Present your analysis in the way your instructor requests.

2. The home page of a website typically looks very different from the opening of a printed document. Nonetheless, the home page has many of the same reader-centered goals as the printed opening. Examine the ways that two websites follow this chapter's guidelines. Consider such things as the communication functions performed by the images, layout, text (if any), and words used for buttons and links. One of the home pages you study should be for a company that sells consumer products such as cars. The other should be for a nonprofit organization or government agency. Present your analysis in the way your instructor requests.

3. For many websites, there is no single ending point. Visitors may leave the site at any point. In fact, links in many sites lead visitors directly to other sites. Visit three different kinds of sites, such as one that provides explanations or instructions, one that enables people to purchase products, and one created by an advocacy group. Within each site, identify the places where one or more of this chapter's guidelines are employed effectively.

COLLABORATE WITH YOUR CLASSMATES

1. Figure 7.7 shows an email whose contents have been scrambled. Each statement has been assigned a number. Working with one or two other students, do the following:

   a. Write the numbers of the statements in the order in which the statements would appear if the email were written in accordance with the guidelines in this chapter. Place an asterisk before each statement that would begin a segment. (When you order the statements, ignore their particular phrasing. Order them according to the information they provide the reader.)
   b. Using the list you just made, rewrite the email by rephrasing the sentences so that the finished message conforms with all the guidelines in this chapter.

2. The following paragraphs are from the beginning of a report in which the manager of a purchasing department asks for better-quality products from the department in the company that provides abrasives. Is this an effective beginning for what is essentially a complaint? Why or why not? Working with another student, analyze this beginning in terms of this chapter's guidelines.

> I am sure you have heard that the new forging process is working well. Our customers have expressed pleasure with our castings. Thanks again for all your help in making this new process possible.
>
> We are having one problem, however, with which I have to ask once more for your assistance. During the seven weeks since we began using the new process, the production line has been idle 28 percent of the time. Also, many castings have had to be remade. Some of the evidence suggests that these problems are caused by the steel abrasive supplies we get from your department. If we can figure out how to improve the abrasive, we may be able to run the line at 100 percent of capacity.

(continued)

FIGURE 7.7

Memo for Collaboration Exercise

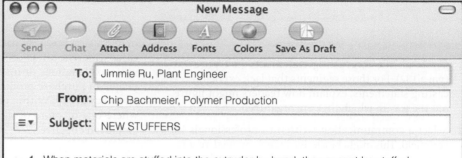

○ ○ ○   **New Message**

Send   Chat   Attach   Address   Fonts   Colors   Save As Draft

To: Jimmie Ru, Plant Engineer

From: Chip Bachmeier, Polymer Production

Subject: NEW STUFFERS

1. When materials are stuffed into the extruder by hand, they cannot be stuffed in exactly the same way each time.
2. We have not been able to find a commercial source for an automatic stuffer.
3. A continuing problem in the utilization of our extruders in Building 10 is our inability to feed materials efficiently into the extruders.
4. An alternative to stuffing the materials by hand is to have them fed by an automatic stuffer.
5. If the materials are not stuffed into the extruder in exactly the same way each time, the filaments produced will vary from one to another.
6. I recommend that you approve the money to have the company shop design, build, and install automatic stuffers.
7. The company shop will charge $4,500 for the stuffers.
8. An automatic stuffer would feed material under constant pressure into the opening of the machine.
9. Currently, we are stuffing materials into the extruders by hand.
10. The shop has estimated the cost of designing, building, and installing automatic stuffers on our ¾-inch, 1-inch and 1½-inch extruders.
11. It takes many working hours to stuff extruders by hand.
12. No automatic stuffers are available commercially because other companies, which use different processes, do not need them.

Image: © Apple Inc.

I would be most grateful for help from you and your people in improving the abrasive. To help you devise ways of improving the abrasive, I have compiled this report, which describes the difficulties we have encountered and some of our thinking about possible remedies.

3. Working with another student, describe the strategies for ending used in the following figures in this text:

FIGURE	PAGE	FIGURE	PAGE
1.4	10	9.6	180
2.7	42	24.1	411
2.8	43	26.3	455
7.2	129	27.7	471

APPLY YOUR ETHICS

1. Using the Internet or a library, find the ethics codes for three corporations. Examine the ways this chapter's guidelines would help these corporations write codes that clearly convey their ethical commitments.

2. It can be particularly challenging to write an effective opening for a communication in which you are advocating for change based on ethical grounds. Think of some organization's practice, procedure, or policy that your values lead you to believe should be changed. Identify the person or group within the organization that has the authority to make the change that you feel is needed. Then, following the advice given in this chapter, draft the opening paragraph of a letter, memo, email, report, or other communication on the topic that you could send to this person or group.

3. Examine the student conduct code or values statement of your school. Which of the guidelines in this chapter does its ending follow? Why? Also, identify some of the specific values that underlie the code's or values statement's provisions.

REFLECT FOR TRANSFER

1. Write a memo to your instructor describing the two pieces of advice in this chapter that you believe will be most useful to you when drafting communications in your career. Explain why.

2. Write a memo to your instructor describing the piece of advice in this chapter that you believe will be easiest for you to apply when drafting communications in your first year on the job. Which one do you expect to be the most difficult. Why?

8

Using Eight Reader-Centered Patterns for Presenting Information and Ideas

You can often increase the helpfulness and persuasiveness of your on-the-job writing by using conventional organizational patterns that are familiar to your readers. This chapter provides detailed, reader-centered advice for using eight patterns—each with its own purpose—that are often used in the workplace. In some brief communications, you might employ only one of these patterns, but in most cases, you will weave two or more together, as described in the section "Combinations of Patterns" (page 165). Each pattern can be clarified for the reader by a drawing, photo, table, or other graphic.

Because each pattern is distinct from the others, this is not a chapter to read straight through. Your instructor will let you know which patterns are most useful as you work on the writing projects in your class. The others are here for your reference during your current course or later in your other classes or on the job.

Grouping Items Formally (Formal Classification)

On the job, you will often need to organize a set of facts, ideas, or other items that could be arranged in many different ways. To create a clear, coherent structure your readers will find useful and persuasive, you can use a strategy called *classification*. There are two kinds of classification, formal (described here) and informal (described in "Grouping Items Informally" on page 150).

In both forms of classification, you arrange your material into groups of related items that satisfy the following criteria.

- **Every item has a place.** In one group or another, every item fits.
- **Every item has only one place.** If there are two logical places for an item, you might describe it in both locations, thereby creating redundancy. Or

LEARNING OBJECTIVES

1. Identify the pattern that will most help your reader understand and use your information.
2. Employ the pattern according to its logical rules.

MindTap®

Find additional resources related to this chapter in MindTap.

Even when using formal classification, group items in a way that will be helpful to your readers.

you might describe it in only one location, thereby requiring your readers to guess where to find the information. Neither alternative is desirable from your readers' perspective.

- **The groupings are useful to your readers.** Items that readers will use together should be grouped together.

How Formal Classification Works

In formal classification, you group items according to a principle of classification—that is, according to some observable characteristic that every item possesses. Usually, you will have several to choose from. While writing a marketing brochure about sixty adhesives manufactured by her employer, a chemical company, Esther could use any characteristic possessed by all the adhesives, such as price, color, and application (that is, whether it is used to bond wood, metal, or ceramic). To choose among these potential principles of classification, she thought about the way consumers would use her brochure. Realizing that they would look for the adhesive best suited to the particular job they were doing, Esther organized around the type of material each adhesive was designed to bond.

Esther organized according to the kinds of material the adhesives will bond because she thought her readers would want to find the adhesives that would work with the materials they had.

In classification, large groups can be organized into subgroups. Within her sections on adhesives for wood, metal, and ceramic, Esther subdivided the adhesives according to the strength of the bond created by each one.

LEARN MORE The formal classification pattern is often combined with other patterns. See "Combining Organizational Patterns" (page 165).

Guidelines for Formal Classification

1. **Choose a principle of classification that is suited to your readers and your purpose.** Consider the way your readers will use the information you provide.
2. **Use only one principle of classification.** For example, you might classify the cars owned by a large corporation as follows:

Valid classification——

Cars built in the United States
Cars built in other countries

This grouping uses only one principle of classification—the country in which the car was manufactured. Suppose you classified the cars this way:

Faulty classification——

Cars built in the United States
Cars built in other countries
Cars that are expensive

This classification is faulty because two principles are being used simultaneously—country of manufacture and cost. An expensive car built in the United States would fit into two categories. To solve this problem, you could use different principles at different levels in a hierarchy.

Corrected classification——

TRY THIS Try classifying the students in your class in two different ways. Include one level of subcategories for each classification. Try to identify someone who could make practical use of information about the students that is organized in each of these ways.

Cars built in the United States
    Expensive ones
    Inexpensive ones
Cars built in other countries
    Expensive ones
    Inexpensive ones

In this case, an expensive car built in the United States would fit into only one category.

Examples of Formal Classification

To see how you could use formal classification to organize the section of a report, look at Figure 8.1, which describes various methods for detecting coronary heart disease. For her principle of classification, Sonja uses the extent to which each method requires physicians to introduce something into the patient's body. Sonja selected this principle of classification because she wanted to focus on a new method for which nothing needs to be placed in the patient's body.

Formal classification can also help you create tables and graphs your readers will find clear and easy to use. Figure 8.2 shows a graph about the percent of municipal solid waste in the United States that is recovered through recycling. The principle of classification is the material from which the waste is made.

LEARN MORE Exercises for organizing with formal classification are in "Use What You've Learned" (page 166).

Detecting Coronary Artery Disease

Coronary artery disease (CAD) is the leading cause of death in industrialized countries. In fact, one-third of all deaths in the U.S. are attributable to it. For this reason, the early detection of CAD has long been regarded as among the most vital areas of medical research, and several moderately invasive diagnostic methods have been developed.

Non-invasive methods gather diagnostic information without introducing anything into a patient's body. These methods include traditional physical examinations and history taking, electrocardiography, and echocardiography (ultrasonic imaging).

In contrast, invasive methods involve introducing a substance or object into the person's body. A moderately invasive method is the thallium test, in which a compound of radioactive thallium-201 is injected into the patient and then distributes itself throughout the myocardium (heart muscle) in proportion to the myocardial blood flow. The low-flow regions are detectable as cold spots on the image obtained from a radioisotope camera set over the chest. Although quite sensitive and accurate, the thallium test is also costly and time-comsuming.

As far as invasive techniques are concerned, the most reliable way to diagnose CAD is by means of cardiac catheterization, in which a catheter is inserted into a large artery (usually in the upper arm or thigh) and advanced to the heart. Once the catheter is positioned in the heart, a radio opaque dye is released through it, making it possible to observe the condition of the coronary arteries with X-rays. Although it produces excellent images and definitive diagnoses, cardiac catheterization is expensive, painful, and time-consuming—and it carries an element of risk. For these reasons, an equally accurate, non-invasive method for early detection of coronary artery disease is greatly to be preferred.

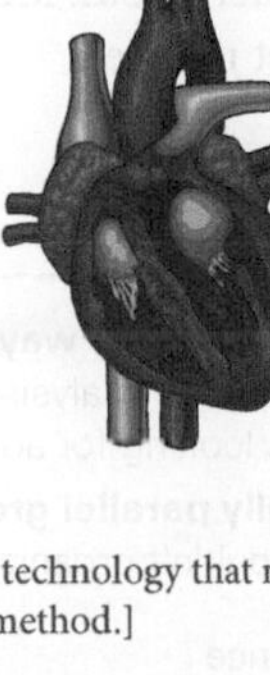

[The chapter continues with a description of a new technology that may provide a highly accurate, non-invasive early detection method.]

From "Wavelet Applications in Medicine," by Metin Akay, *IEEE Spectrum, 34*(5), 50-51 (1997, May). Copyright © 1997 IEEE.

Oguz Ara/Shutterstock.com

FIGURE 8.1

Beginning of a Chapter Organized by Formal Classification

The principle of classification is the extent to which the methods involve placing something in the person's body ("invading" it).

The first group of methods involves no invasiveness.

Several examples are listed.

The second group of methods does involve invasiveness.

Two invasive methods are each described in a separate paragraph: the thallium test and cardiac catheterization.

FIGURE 8.2

Graph Organized by Formal Classification

LEARN MORE For advice about constructing graphics, go to Chapter 12 (page 224) and Chapter 13 (page 242).

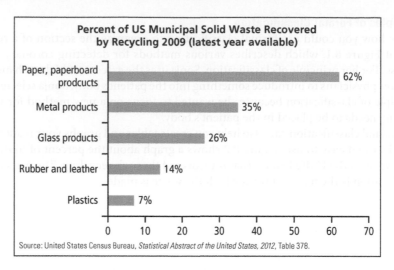

Percent of US Municipal Solid Waste Recovered by Recycling 2009 (latest year available)

- Paper, paperboard prodcuts — 62%
- Metal products — 35%
- Glass products — 26%
- Rubber and leather — 14%
- Plastics — 7%

Source: United States Census Bureau, *Statistical Abstract of the United States, 2012*, Table 378.

Grouping Items Informally (Informal Classification)

Sometimes you will need to organize information about a large number of items but will find it impossible or undesirable to classify them according to the kind of objective characteristic that is necessary for formal classification. Perhaps the items don't share an objective characteristic. Or, grouping according to the objective characteristics that do exist won't serve the readers' needs. Still, to create a clear, coherent paragraph, section, or chapter, you need to arrange the items in a way that achieves the same results as formal classification.

- Every item has a place.
- Every item has only one place.
- The groupings are useful to your readers.

Take a reader-centered approach when using informal classification, just as you do when using formal classification.

The solution is to use informal classification, which replaces the objective principle of classification with a subjective one. For example, Calvin, a mechanical engineer, was asked by his manager to analyze advertisements in three trade journals for the heavy equipment industry. Objective characteristics do exist, such as the size and number of words in each ad. However, these characteristics would not help his employer create effective ads for the new earthmover that Calvin helped design. Instead, Calvin classified the ads according to the type of appeal they made. Obviously, "type of advertising appeal" is not an objective characteristic. Defining an ad's appeal requires subjective interpretation and judgment. Calvin used this informal classification because it best matched his reader's goal: identifying advertising strategies that would appeal to the company's target market.

LEARN MORE The informal classification pattern is often combined with other patterns. See "Combining Organizational Patterns" (page 165).

Guidelines for Informal Classification

1. **Group your items in a way that is suited to your readers and your purpose.** Calvin organized his analysis around "type of advertising appeal" because he knew his employer was looking for advice about the design of ads.

2. **Create logically parallel groups.** For instance, if you were classifying advertisements, you wouldn't organize into groups like this:

   | Focus on price
   | Focus on established reputation |

Focus on advantages over a competitor's product
Focus on one of the product's key features
Focus on several of the product's key features

— Faulty classification

The last two categories are at a lower hierarchical level than the other three. To make the categories parallel, you could combine them in the following way.

Focus on price
Focus on established reputation
Focus on advantages over a competitor's product
Focus on the product's key features
    Focus on one key feature
    Focus on several key features

— Corrected classification

3. **Avoid overlap among groups.** Even when you cannot use strict logic in classifying items, strive to provide one and only one place for each item. To do this, you must avoid overlap among categories. For example, in the following list the last item overlaps the others because photographs can be used in any of the other types of advertisements listed.

Focus on price
Focus on established reputation
Focus on advantages over a competitor's product
Focus on the product's key features
Use of photographs

— Faulty classification

Example of Informal Classification

Figure 8.3 shows how Musa used informal classification to organize advice about preventing identity theft. Thinking from his readers' point of view, he realized that there was no single principle of classification that would make immediate sense to his readers. Therefore, he created categories that he expected his readers to identify with, mixing the kinds of things thieves steal (e.g., credit cards) with the means by

LEARN MORE Exercises for organizing with informal classification are in "Use What You've Learned" (page 166).

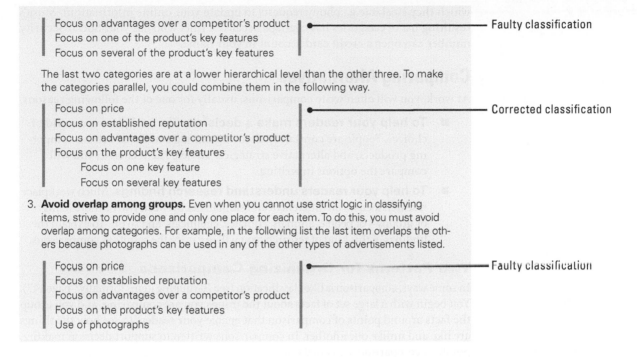

PROTECT YOURSELF FROM IDENTITY THEFT		
Type of Theft	**What Thieves Do**	**Protection**
Social Security Number	• Open credit card accounts in your name • Obtain loans in your name	• Give your number only to businesses you know and trust • Ask to give an alternative identifier when your SS number is requested
Credit Cards	• Steal your wallet or purse • Intercept information when you are buying online • Create fraudulent online businesses • Steal preapproved credit card offers mailed to you	• Check companies with the Better Business Bureau before buying online • Use a secure web browser to buy online
Checks	• Steal your checks or checking account number from your home or office	• Notify your bank immediately if your checks are stolen
Cellular Telephone Service	• Open telephone service in your name • Use your calling card and PIN to make calls	• If this occurs, immediately ask your service provider to close your account, and establish another one with a new PIN.
Internet Account Updates	• Send a phony email requesting your credit card information to update or verify a company's records	• Check with your Internet Service Provider before responding to such a request
Phony Identity Theft Protection Services	• Request personal information in order to "protect" you	• Check out the company with the Better Business Bureau before giving personal information

Based on Chicago Better Business Bureau, http://chicago.bbb.org/identitytheft/

FIGURE 8.3

Table Organized by Informal Classification

The types of identity theft described in this table are organized using informal classification.

• There is not a consistent principle of classification. Most of the categories describe what is stolen (e.g., your Social Security number). The last two describe the means used to steal.

• There is overlap among some of the categories. For example, a thief can use a stolen Social Security number to open a credit card account in your name.

which they steal it (e.g., phony requests to update your online information). Musa's resulting list of categories has overlaps because a thief who has your Social Security number can open a credit card account in your name.

Comparing Alternatives

At work, you will often write comparisons, usually for one of the following reasons.

- **To help your readers make a decision.** The workplace is a world of choices. People are constantly choosing among courses of action, competing products, and alternative strategies. To help them choose, you will compare the options in writing.

- **To help your readers understand research findings.** Much workplace research focuses on differences and similarities between two or more items or groups—of people, animals, climates, chemicals, and so on. To explain the findings of this kind of research, you will organize the results as comparisons.

Two Patterns for Organizing Comparisons

In some ways, comparison is like classification (see "Grouping items formally," page 147). You begin with a large set of facts about the things you are comparing, and you group the facts around points of comparison that enable your readers to see how the things are like and unlike one another. In comparisons written to support decision making, points of comparison are called criteria.

LEARN MORE A point of comparison is very much like a principle of classification. See "Grouping Items Formally (Formal Classification)," page 147.

When writing a comparison, you can choose either of two organizational patterns: the *divided* pattern or the *alternating* pattern. Lorraine's project illustrates the difference. Her employer wants to replace the aging machines it uses to stamp out metal parts for the bodies of large trucks. Lorraine has been assigned to investigate the two machines the company is considering. She can organize the hundreds of facts she has gathered according to the divided pattern or the alternating pattern.

DIVIDED PATTERN	**ALTERNATING PATTERN**
Machine A	**Cost**
Cost	Machine A
Efficiency	Machine B
Construction Time	**Efficiency**
Air Pollution	Machine A
Et cetera	Machine B
Machine B	**Construction Time**
Cost	Machine A
Efficiency	Machine B
Construction Time	**Air Pollution**
Air Pollution	Machine A
Et cetera	Machine B
	Et Cetera
	Machine A
	Machine B

When to Use Each Pattern

To make a reader-centered choice between the divided and alternating patterns, consider the way your readers will use your information. Because the alternating pattern is organized around the criteria, it is ideal when readers want to make point-by-point comparisons among alternatives. Lorraine selected this pattern so her readers can find the information about the costs, capabilities, and other characteristics of both stamping machines in one place. The divided pattern would separate the comparable information about each machine into different sections, requiring readers to flip back and forth to compare the machines in detail.

The divided pattern is well suited to situations where readers want to read all the information about each alternative in one place. Typically, this occurs when both the general nature and the details of each alternative can be described in a short space—say, one page or so. An acoustical engineer used the divided pattern to provide a restaurant manager with information about three sound systems for her business. He described each system in a single page.

Whether you use the alternating or divided pattern, you can usually assist your readers by incorporating two kinds of preliminary information.

TRY THIS Try creating a list of criteria you would recommend that a friend use when comparing smart phones, tablets, videogames, or some other product he or she might be ready to purchase.

- **Description of the criteria.** This information lets your readers know from the start what the relevant points of comparison are.
- **Overview of the alternatives.** This information provides your readers with a general sense of what each alternative entails before they focus on the details you provide.

In both patterns, the statement of criteria would precede the presentation of details. Taking these additional elements into account, the general structure of the two patterns is as follows:

DIVIDED PATTERN	ALTERNATING PATTERN
Statement of Criteria	Statement of Criteria
Overview of Alternatives	Overview of Alternatives
Evaluation of Alternatives	Evaluation of Alternatives
Alternative A	Criterion 1
Criterion 1	Alternative A
Criterion 2	Alternative B
Alternative B	Criterion 2
Criterion 1	Alternative A
Criterion 2	Alternative B
Conclusion	Conclusion

Guidelines for Using Comparisons

1. **Choose points of comparison suited to your readers and purpose.** When helping readers understand something, focus on major points, not minor ones. When helping readers make a decision, choose criteria that truly make a difference.

2. **Discuss each alternative in terms of all your criteria or points of comparison.** If no information is available concerning some aspect of one alternative, tell your readers. Otherwise, they may assume that you didn't try to obtain it.

3. **Arrange the parts in an order your readers will find helpful.** Often, you can help most by discussing the most significant differences first. Sometimes, it's best to briefly mention criteria on which the alternatives are very similar so that readers can then focus on the criteria that make a difference. In comparisons designed to aid understanding, it's usually best to begin with what's familiar to your readers and then lead them to the less familiar.

4. **Include graphics if they will help your readers understand and use your communication.** Tables, graphs, diagrams, and drawings can be especially helpful.

LEARN MORE The comparison pattern is often combined with other patterns. See "Combining Organizational Patterns" (page 165).

LEARN MORE For advice about constructing tables and other graphics, go to Chapter 12 (page 224) and Chapter 13 (page 242).

Examples of Comparisons

Figure 8.4 shows a memo in which the writer uses the alternating pattern of comparison to help a decision maker choose among three alternatives.

In addition to prose, you can use tables and graphs to help readers compare alternatives and make decisions. Figure 8.5 shows a website that helps vacationers pick a campground in New South Wales, Australia.

FIGURE 8.4

Memo Organized to
Help Readers Compare
Alternatives

AdvanceTech
Memorandum

To   Mehash Mehta
From   Kenneth Abney
Date   September 10, 2017

Subject   **Recommendations for Tablet Phones**

Last week, Marisol de Silva asked me to provide you with a comparison of the top tablets that also provide cellphone service. He explained that AdvanceTech might purchase such tablets for all 27 sales representatives and all 140 installation and service technicians.

> Kenneth explains his research method, including the method by which he chose the smartphones to compare.

I have studied product capabilities and published reviews for the three smartphones that received the highest ratings by *PCWorld* magazine: Blacksea Pearl, Kangaroo NZ, and Universalia.

> Kenneth begins his comparison by explaining what all three alternatives have in common.

All three provide high-quality phone service and comparable computing capabilities. The key criteria for selection are ease of use and the ability to meet potential needs created by possible future expansion of our business. Here are my recommendations:

> Kenneth makes two recommendations, one for each of two possible situations.

■ If we continue to do business in North America only, the Universalia is our best choice. Offering both a touchscreen and stylus, along with a large display, it has the longest standby battery life and ample memory.

■ If we will soon open offices outside North America, the Blacksea Pearl would be preferable. It supports the Global System for Mobile Communication (GSM) used elsewhere in the world.

The following table compares the tablets in detail. If you want more information, please let me know.

> He provides his reader with additional details in an easy-to-read table.

	Standby Battery Life	RAM	Memory	Operating System	GSM	Price
Blacksea Pearl	15 Days	8 GB	100 GB	Proprietary	Yes	$800
Kangaroo NZ	7.8 Days	4 GB	50 GB	Proprietary	No	$750
Universalia	20 Days	8 GB	400 GB	Windows Mobile	No	$600

LEARN MORE Exercises
for writing comparisons are in
"Use What You've Learned"
(page 166).

FIGURE 8.5

Website
Designed to
Help Users Make
Comparisons

In this table, the points
of comparison are

• Fees

• Facilities

• Types of camping
available

By using icons rather
than words, the
writers help their
readers read this
table quickly.

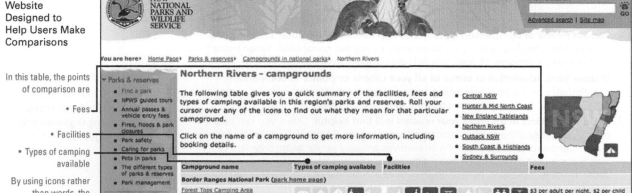

Describing an Object (Partitioning)

In your career, you will have many occasions to describe a physical object for your readers. If you write about an experiment, you may need to describe your equipment. If you write instructions, you may need to describe the machines your readers will be using. If you propose a new purchase, you may need to describe the object you want to buy.

To organize descriptions, use a strategy called partitioning. Partitioning uses the same basic procedure as does classifying (see "How Formal Classification Works," page 148). Think of the object as a collection of parts. Then identify a principle of classification for organizing the parts into groups of related parts. At work, the principle most useful to your readers will usually be location or function.

LEARN MORE For more on principles of classification, see "Grouping Items Formally (Formal Classification)," page 147.

Example: Partitioning a Car

Consider, for instance, how you could use location and function to organize a discussion of the parts of a car.

To organize by location, you might talk about the car's interior (passenger com partment, trunk), exterior (front, back, sides, and top), engine compartment, and underside (wheels, transmission, and muffler).

To organize by function, you might focus on parts that provide power and ones that guide the car. The power-producing parts are in several locations: The gas pedal is in the passenger compartment, the engine is under the hood, and the transmission and axle are on the underside. Nevertheless, you would discuss them together because they are related by function.

Of course, other principles of classification are possible.

Guidelines for Describing an Object

1. **Choose a principle of classification suited to your readers and purpose.** For instance, when describing a car for new owners who want to learn about the vehicle they've purchased, you might organize according to location. Your readers would learn about the conveniences in the passenger compartment, trunk, and so on. In contrast, when describing the car for auto repair specialists, you could organize according to function or the kinds of problems they might encounter.

2. **Use only one basis for partitioning at a time.** To assure that you have one place and only one place for each part you describe, use only one basis for partitioning at a time, just as you use only one principle of classification at a time (see "How Formal Classification Works," page 148).

3. **Arrange the parts of your description in a way your readers will find useful.** When partitioning by location, you might move systematically from left to right, front to back, or outside to inside. When partitioning by function, you might trace the order in which the parts interact during a process that interests your readers.

4. **When describing each part, provide details that your readers will find useful.** For instance, when describing a car tire to a consumer, you might describe the air pressure the tires should have. When describing the same tire to an engineer, you might provide a technical description of the steel belts in the tire's core.

5. **Include graphics if they will help your readers understand and use your information about the object.** Graphics that are especially helpful to readers of an object's description include drawings, diagrams, and photographs.

LEARN MORE The partitioning pattern is often combined with other patterns. See "Combining Organizational Patterns" (page 165).

LEARN MORE For advice about constructing graphics, go to Chapter 12 (page 224) and Chapter 13 (page 242).

LEARN MORE Exercises for organizing with partitioning are in "Use What You've Learned" (page 166).

Description Organized by Partitioning

Figure 8.6 shows a description of the human eye and its ailments, organized by partitioning.

FIGURE 8.6

Passage with Text and Figure that Describe an Object by Partitioning

This description is partitioned according to the steps in the process by which light passes through the eye and is transformed into vision. The organization also corresponds to the location of the parts of the eye.

- Light enters the eye through the cornea.

- The amount of light proceeding farther into the eye is controlled by the iris.

- The light is then focused by the lens.

- Next, the light passes through the vitreous humor to the retina.

- The retina transforms the light energy into electrical impulses.

THE HUMAN EYE AND ITS AILMENTS

Several parts of the eye contribute to creation of sharp, clear vision. Disease or damage to any part can distort the light that passes through it or block the light altogether.

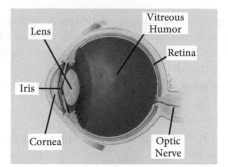

- **Cornea** The cornea protects the inner eye. Disease, infection, or injury can create scars that block or distort light.

- **Iris** The iris is the colored part of the external eye. It expands and contracts to regulate the amount of light entering the eye through the pupil. Skin cancer can attack the iris, limiting its ability to function properly.

- **Lens** A set of muscles attached to the lens controls its shape, thereby focusing the light that passes through it. The lens can develop cloudy areas called cataracts.

- **Vitreous Humor** The vitreous humor is the gel that fills the eyeball. Particles called floaters sometimes drift through it, looking like dust particles or sand in the field of vision.

- **Retina** Nerve cells in the retina transform light energy into electrical impulses that travel to the brain through the optic nerve. The retina can tear or detach from the lining of the eye, causing flawed vision or blindness.

Describing a Process (Segmenting)

At work, you will usually have one of the following reasons for describing a process.

- **To enable your readers to perform the process.** For example, you may be writing instructions that will enable your readers to analyze the chemicals present in a sample of liver tissue, make a photovoltaic cell, apply for a loan, or run a computer program.

- **To enable your readers to understand the process.** For example, you might want your readers to understand the following:

  - **How something is done.** For instance, how coal is transformed into synthetic diamonds.

  - **How something works.** For instance, how the lungs provide oxygen to the bloodstream.

  - **How something happened.** For instance, how the United States developed the space programs that eventually landed astronauts on the moon.

In either case, you need to help your readers understand the overall structure of the process. To do this, you segment the process. Imagine the entire process as a long line or string of steps. Divide the process (or cut the string) into segments at points where one major group of related steps ends and another begins. If some or all of the major groups are large, you may divide them into subgroups, thereby creating an organizational hierarchy. The outline on the right shows how you might segment the steps in the process for building a cabinet.

> **Making a Cabinet**
> Obtaining Materials
>   Preparing the Pieces
>   Cutting the wood
>   Routing the wood
> Assembling the Cabinet
>   Building the base
>   Mounting the doors
> Finishing the Cabinet
>   Sanding
>   Applying the stain
>   Applying the sealant

Principles of Classification for Segmenting

To determine where to segment the process, you need a principle of classification. Commonly used principles include the time when the steps are performed (first day, second day; spring, summer, fall), the purpose of the steps (to prepare the equipment, to examine the results), and the tools used to perform the steps (for example, table saw, drill press, and so on).

Pick the principle that best supports your readers' goals. For instance, if you were writing a history of the process by which the United States placed a person on the moon, you might segment the process according to the passage of various laws and appropriation bills if your readers were members of the U.S. Congress. If your readers are scientists and engineers, however, you might focus on efforts to overcome various technical problems.

LEARN MORE For more on principles of classification, See "Grouping Items Formally (Formal Classification)," page 147.

Guidelines for Segmenting

1. **Choose a principle for segmenting suited to your readers and your purpose.** If you are writing instructions, group the steps in ways that support an efficient or comfortable rhythm of work. If you want to help your readers understand a process, organize around concerns that are of interest or use to them.

2. **Make your smallest groupings manageable.** If your smallest groupings include too many steps or too few, your readers will not see the process as a structured hierarchy of activities or events but rather as a long, unstructured list of steps—first one, then the next, and so on.

3. **Describe clearly the relationships among the steps and groups of steps.** To understand and remember the process, readers need to understand how the parts fit together. Where the relationships among the groups of steps will be obvious to your readers, simply provide informative headings. At other times, you will need to explain the relationships in introductory statements, transitions between groups of steps, and, perhaps, in a summary at the end.

4. **Provide enough detail about each step to meet your readers' needs.** Use your understanding of your readers and the ways they will use your communication to determine the appropriate level of detail.

5. **Include graphics if they will help your readers understand and use your information about the process.** Graphics that are especially helpful to readers of a description of a process are diagrams, drawings, photographs, and flowcharts.

LEARN MORE The segmenting pattern is often combined with other patterns. See "Combining Organizational Patterns" (page 165).

TRY THIS Try segmenting a process you perform regularly, such as going to your first class of the day or doing homework for a course in your major.

LEARN MORE For advice about constructing graphics, go to Chapter 12 (page 224) and Chapter 13 (page 242).

Figure 8.7 shows a passage in which the writer uses segmentation to describe one theory about the way planets are formed. Notice how the writer uses headings to signal the major phases of the process.

In the flowchart shown in Figure 8.8, writers at Lawrence Livermore National Laboratory explained the five-step process by which lasers are used to create nuclear fission.

LEARN MORE Exercises for writing about segmenting are in "Use What You've Learned" (page 166).

FIGURE 8.7

Passage Explaining a
Process That Is Organized
by Segmenting

The topic is introduced. —

First stage is described. —

Alternative outcomes are identified. —

Second stage is described. —

Headings help to distinguish the stages. —

Third stage is described in two parts,
with separate paragraphs devoted to
planetesimals and to planets.

A concluding sentence summarizes the
overall process.

How Planets Are Formed

● Theoretical astrophysicists have developed an elaborate model describing the formation of stars like our Sun. In this theory, planet formation is a natural and almost necessary result of the process of star formation.

A Proto-Star Is Born

● These astrophysicists believe that the process of star formation begins when a dense region of gas and dust in an interstellar cloud becomes gravitationally unstable. The cloud begins a slow, quasistatic contraction as both its internal turbulent and magnetic support are gradually lost through cooling and the outward diffusion of the magnetic field. The core of this cloud condenses to form a proto-star.

● If there is sufficient angular momentum, the collapsing cloud may fragment into two or more smaller pieces, thus leading to the formation of a binary or multiple star.

A Rotating Disc Forms

● The remaining gas and dust from the cloud continue to fall inwards. Because this material must conserve the initial angular momentum of the cloud, it forms into a rotating disc around the proto-star in accordance with Kepler's laws, with the inner portion of the disc rotating more rapidly than the outer region.

● Planets Are Created

● The process of planet formation occurs within this rotating disc. Dust grains collide and stick together, forming larger particles. This collisional growth continues over millions of years and eventually results in the creation of rocky bodies a few kilometers in diameter, known as "planetesimals." At this point the gravitational attraction of the planetesimals begins to dominate their random velocities, increasing their growth rate to the point where planetary cores about a thousand kilometers across can form rapidly.

As the proto-planetary disc continues to evolve, the planetary cores in the inner portion of the disc collide, merge and grow to become terrestrial planets such as Earth and Mars. In the outer disc, the cores grow to larger masses. Once the planetary core reaches about 10 Earth masses, its gravity is sufficient to capture gaseous hydrogen and helium from the disc, forming a gas-giant planet such as Jupiter or Saturn. This rapid-growth phase removes most of the material near the orbit of the growing planet, forming a gap in the disc and effectively shutting off the growth process.

● Thus, scientists think that planets are built from their cores outward by accretion processes in the disc of gas and dust that surrounds a star during its creation.

From Cochran, William (1997, July). "Extrasolar Planets." *Physics World.* Institute of Physics Publishing, Bristol, England.

FIGURE 8.8

Flowchart with Text and Figures
that Describe a Process by
Segmenting

The text explains the steps
in the process.

The drawings illustrate each step.

Three hundred million watts of laser energy are projected into a small container that holds a spherical fusion-fuel capsule.	Reflecting from the container's specially coated walls, the laser beams are converted to x-rays that create a 3-million-degree oven.	The heat energy implodes the fusion-fuel cell to 20 times the density of lead.	The fuel core ignites at 100 million kelvins.	A thermonuclear burn spreads through the fuel, creating gain equivalent to the power of a miniature star lasting less than a billionth of a second.

Science & Technology, June 2005. © Lawrence Livermore National Laboratory. Operated by the University of California for the U.S. Department of Energy. https://www.llnl.gov/str/June05/Aufderheide.html

Describing a Cause-and-Effect Relationship

Questions about cause and effect are very common in the workplace. Profits slumped or skyrocketed. What caused the change? The number of children diagnosed with autism is increasing. How come? If we clear underbrush from forest floors, how will our action affect animal habitats and the severity of forest fires?

At work, you are likely to write about cause and effect for one of two distinct purposes.

- To help your readers understand a cause-and-effect relationship.
- To persuade your readers that a certain cause-and-effect relationship exists.

This section describes strategies for describing cause and effect. The next explains ways to persuade that a certain effect has a particular cause.

Helping Readers Understand a Cause-and-Effect Relationship

Some cause-and-effect relationships are accepted by people knowledgeable about the topic but need to be explained to those who are not.

For instance, geologists agree about what causes the Old Faithful geyser in the U.S. Yellowstone National Park to erupt so regularly. However, many park visitors do not know. Consequently, the park publishes brochures that explain the cause.

Depending on your profession, you may need to explain cause-and-effect relationships to members of the general public; to employees in your own organization, such as co-workers or executives outside your specialty; or to other readers. The following reader-centered guidelines will help you do so effectively in any situation. Figure 8.9, which describes how CDs work, illustrates their application.

Guidelines for Explaining a Cause-and-Effect Relationship

1. **Begin by identifying the cause or effect that you are going to explain.** This advance notice will help them better understand what follows.
2. **Carefully explain the links that join the cause and effect that you are describing.** Remember that you are not simply listing the steps in a process. You want your readers to understand how each step leads to the next one.
3. **If you are dealing with several causes or effects, group them into categories.** Categories help readers to understand a complex set of factors.
4. **Include graphics that will help your readers understand the relationships you are explaining.** Graphics that assist readers in visualizing cause-and-effect relationships include flowcharts, diagrams, and drawings.

LEARN MORE The cause-and-effect pattern is often combined with other patterns. See "Combining Organizational Patterns" (page 165).

LEARN MORE For advice about constructing graphics, go to Chapter 12 (page 224) and Chapter 13 (page 242).

Persuading Readers that a Cause-and-Effect Relationship Exists

About many events, people disagree. They may disagree about past events, such as what caused damage to a jet engine's turbine blade or what caused the disappearance of the Inca civilization in Central America. Alternatively, they may disagree about the future. Examples include disagreements about the effects of growing genetically modified crops or the effect that a proposed price reduction would have on company profits.

FIGURE 8.9

Passage That Uses Text and a Graphic to Explain a Cause-and-Effect Relationship

This passage describes a cause- and-effect relationship to explain how a CD player creates music from a CD. The same technology is used for DVDs.

This paragraph provides background that helps readers understand the cause-and-effect explanation given in the next paragraph.

In this explanation, the effect (variations in the intensity of light) is described first. The explanation follows. In other cause-and-effect passages, the order is reversed.

Many kinds of graphics are used with explanations of cause and effect. In this case, the author uses a diagram to provide a visual description of the process by which pits that "make" the music are created on a CD.

How Light Makes Music

When you listen to your favorite songs on a CD, you are enjoying the music that light makes.

Your CD has three layers. On the bottom is a stiff, protective layer of acrylic. Next is a thin layer of reflective aluminum, which is coated with a clear plastic layer that is 1.2 mm thick. Your CD player projects a shaft of light, in the form of a laser beam, at a tiny spot on the reflective aluminum. As your CD spins, the player reads the variations in the intensity of the light that is reflected from the aluminum.

What causes the intensity of the reflected light to vary? Although a CD seems smooth, its ability to make music depends on millions of pits in the clear plastic layer. Only about 0.5 microns deep and several microns long, they are burned into the aluminum side of the clear plastic layer when the CD is manufactured. Despite being "clear," the plastic layer absorbs some light. It absorbs less where the pits are because the plastic is thinner there. As a CD spins, areas with and without pits pass rapidly under the CD player's laser beam. By reading the variations in the intensity of the reflected light, the CD player makes music.

A Laser Puts Music on a CD by Creating Pits in the Aluminum Side of the Clear Plastic Layer

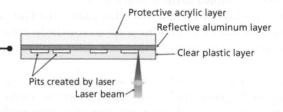

Protective acrylic layer
Reflective aluminum layer
Clear plastic layer
Pits created by laser
Laser beam

Whether you are writing about the causes of a past event or the effects of a future action, your goal is to persuade your readers to accept your explanation. The following guidelines will help you argue persuasively, as will the guidelines in Chapter 9, "Persuading Your Readers." Figure 8.10 illustrates their application.

LEARN MORE For more advice about persuasion, see Chapter 9, "Persuading Your Readers" (page 168).

LEARN MORE For advice about constructing graphics, go to Chapter 12 (page 224) and Chapter 13 (page 242).

Guidelines for Persuading Readers to Accept Your View of Cause and Effect

1. **State your claim at the beginning of your passage.** By stating your claim in advance, you help readers see—and evaluate—the evidence that supports the claim.
2. **Choose evidence your readers will find credible.** As we'll see in Chapter 9, different kinds of evidence are credible in different situations and industries. You will have to understand your readers in order to determine whether data, expert testimony, or some other form of evidence will be most persuasive for them.
3. **Explain your line of reasoning.** The heart of your explanation are the links between the causes and effects you are discussing. Explain them thoroughly.
4. **Avoid faulty logic.** If readers find flaws in your logic they will reject your explanation. See below for descriptions of two logical fallacies that are common in arguments concerning cause and effect.
5. **Address counterarguments.** To persuade readers to accept your account of a cause-and-effect relationship, you must persuade them to reject alternative explanations. Therefore, you need to identify the other explanations they might consider and explain why yours is better.
6. **Include graphics if they will help your readers understand the relationship you are discussing.**

Describing a Cause-and-Effect Relationship | **161**

FIGURE 8.10

Passage Persuading That a
Particular Cause Created a
Specific Effect

WHAT CAUSED THE DEATH OF THE DINOSAURS?

One theory is that a comet, asteroid or other huge extraterrestrial body slammed into the Earth 65 million years ago and ended the 160-million year reign of the dinosaurs. According to this theory, the extraterrestrial body raised a huge dust cloud. Within days the black cloud spread over the Earth, darkening the sun. The air turned cold, and many dinosaurs died. Snow fell. Freezing darkness gripped the Earth for weeks. Plants, cut off from the sunlight that feeds them, couldn't survive. Without plants, the rest of the herbivorous dinosaurs followed, and the carnivores soon afterward. Along with a number of other species, the dinosaurs were gone forever.

— Effect to be explained is announced.

— Possible cause is announced.

— Link between effect and cause is explained: An asteroid created a dust cloud that killed the dinosaurs.

Although many leading paleontologists and evolutionary biologists now accept the asteroid-impact theory, and despite popular accounts implying that the question is settled, it is not. A scattering of critics continue to challenge the whole notion.

Still, the theory is compelling. Every few months a new piece of evidence is added to the list, and most, to the critics' consternation, support the idea of an extraterrestrial impact.

Just recently, for example, scientists at the Scripps Institute of Oceanography, in La Jolla, California, found evidence of organic molecules in the layer of sediments laid down at the time the dinosaurs died; the molecules are exceedingly rare on Earth but relatively common in some meteorites and so, presumably, in some asteroids.

— Evidence of link is presented: Rare molecules in sediment indicate an asteroid may have hit Earth when dinosaurs died.

To put the discovery in perspective, and to appreciate the arguments on both sides of the impact debate, one must first understand the nature of the original finding.

In 1980, Luis Alvarez, a Nobel laureate in physics, his son Walter, a geologist, both at the University of California, Berkeley, and two associates published the theory that a massive impact took place at the end of the Cretaceous Period.

— Additional evidence of link.

The team had found a rare substance in the thin layer of sedimentary clay deposited just on top of the highest, and therefore the most recent, stratum of rock contemporary with those bearing dinosaur fossils. It was the element iridium, which is almost nonexistent in the Earth's crust but 10,000 times more abundant in extraterrestrial rocks such as meteorites and asteroids. Deposits above and below the clay, which is the boundary layer separating the Cretaceous layer from the succeeding Tertiary, have very little iridium.

Because the same iridium anomaly appeared in two other parts of the world, in clay of exactly the same age, the Alvarez team proposed that the element had come from an asteroid that hit the Earth with enough force to vaporize, scattering iridium atoms in the atmosphere worldwide. When the iridium settled to the ground, it was incorporated in sediment laid down at the time.

— Additional evidence continued.

More startling was the team's proposal that the impact blasted so much dust into the atmosphere that it blocked the sunlight and prevented photosynthesis (others suggested that a global freeze would also have resulted). They calculated that the object would have had to be about six miles in diameter.

— Link is restated.

Since 1980, iridium anomalies have been found in more than 80 places around the world, including deep-sea cores, all in layers of sediment that formed at the same time.

— Additional evidence of link.

One of the most serious challenges to the extraterrestrial theory came up very quickly. Critics said that the iridium could have come from volcanic eruptions, which are known to bring up iridium from deep within the Earth and feed it into the atmosphere. Traces of iridium have been detected in gases escaping from Hawaii's Kilauea volcano, for example.

— Challenge to link is explained: molecules perhaps from volcano, not asteroid.

The new finding from Scripps appears to rule out that explanation, though, as a source for iridium in the Cretaceous–Tertiary (K–T) boundary layer. Chemists Jeffrey Bada and Nancy Lee have found that the same layer also contains a form of amino acid that is virtually nonexistent on Earth—certainly entirely absent from volcanoes—but abundant, along with many other organic compounds, in a type of meteor called a carbonaceous chondrite.

— Challenge is refuted by new evidence: Other molecules couldn't have come from volcano.

LEARN MORE Exercises for writing about cause and effect are in "Use What You've Learned" (page 166).

"Death of Dinosaurs: A True Story?" by Boyce Rensberger, from Science Digest, 94, 5, May 1966, pp. 28–32

Logical Fallacies Common in Arguments about Cause and Effect

The following logical fallacies can undermine arguments about cause and effect.

- **Post hoc, ergo propter hoc fallacy.** This fallacy occurs when a writer argues that because an event occurred after another event, it was caused by that event. For example, in an attempt to persuade his employer, a furniture manufacturer, to switch to computerized machinery, Samuel argued that a competitor's profits had risen substantially after making that switch. Samuel's boss pointed out that the competitor's sales increase may have been caused by other changes made at the same time, such as creation of new designs or reconfiguration of sales districts. Samuel's commission of this fallacy didn't mean he was incorrect. However, to persuade his boss that computerization had caused the sales increase, Samuel needed to do more than simply state that it had preceded the increase.

- **Overgeneralization.** Writers overgeneralize when they draw conclusions on insufficient evidence. For instance, writers overgeneralize if they draw a conclusion about the causes of a manufacturing error after examining only 2 percent of the faulty products.

Describing a Problem and Its Solution

Problems and their solutions will be one of the most frequent topics of your on-the-job writing. Sometimes, you will write about a current problem in order to propose a solution to it. At other times, you will write about a past problem, telling how it was solved. The following sections provide advice for writing effectively in both situations.

Proposing the Solution to a Problem

LEARN MORE Chapter 23 on "Writing Reader-Centered Proposals" provides additional information on organizing with the problem-and-solution pattern, which provides the overall structure for proposals.

In proposals to readers outside your organization, you will probably be seeking contracts worth thousands or even millions of dollars. When writing proposals to readers inside your organization, you might be urging support for a large research project or suggesting an improvement in a policy or procedure. Always, your goal will be to persuade your readers to support or approve the problem-solving project you propose.

The following Guidelines will help you achieve that goal. Their application is illustrated in Figure 8.11, where the writer uses the problem-and-solution strategy to recommend that her employer investigate the Kohle Reduktion method of steelmaking. Proposals for further study are common in the workplace.

LEARN MORE The problem-and-solution pattern is often combined with other patterns. See "Combining Organizational Patterns" (page 165).

Guidelines for Persuading Readers to Accept Your Proposed Solution

1. **Describe the problem in a way that makes it seem significant to your readers.** Remember that your aim is to persuade them to take the action you recommend. They will not be very interested in taking action to solve a problem they regard as insignificant.

2. **Describe your method.** Readers want enough detail to feel confident that it is practical and technically sound. Depending on your reader and the situation, this might require a great deal of information or relatively little.

3. **When describing your method, explain how it will solve the problem.** Provide the evidence and reasoning needed to persuade your readers that your method will, in fact, solve their problem.

4. **Anticipate and respond to objections.** As when reading any persuasive segment, your readers may object to your evidence or your line of reasoning. Devote special attention to determining what those objections are so you can respond to them.

5. **Specify the benefit.** With as much specificity as will interest your readers, describe the benefits they and their organization will enjoy if they support or fund your proposed action.

6. **Include graphics if they will help your readers understand and approve your proposed solution.** Review opportunities to use graphics that can help your readers understand the problem and your proposed solution and that can highlight the benefits to your readers of supporting the action you propose. Among the graphics often used in problems and their solutions are tables, flowcharts, drawings, photographs, and diagrams.

LEARN MORE For advice about constructing graphics, go to Chapter 12 (page 224) and Chapter 13 (page 242).

Reporting on a Past Problem-Solving Project

Reports on past problem-solving projects are often used to teach other employees how to approach similar challenges. They are also used to demonstrate a company's capabilities in proposals. The guidelines on page 164 will help you prepare such communications effectively. Figure 8.12 illustrates their use.

FIGURE 8.11

Memo Proposing the Solution to a Problem

MANUFACTURING PROCESSES INSTITUTE
Interoffice Memorandum

June 21, 2017

To   Cliff Leibowitz

From   Candace Olin

RE   Suggestion to Investigate Kohle Reduktion Process for Steelmaking

As we have often discussed, it may be worthwhile to set up a project investigating steelmaking processes that could help the American industry compete more effectively with the more modern foreign mills. I suggest we begin with an investigation of the Kohle Reduktion method, which I learned about in the April 2013 issue of *High Technology*.

A major problem for American steelmakers is the process they use to make the molten iron ("hot metal") that is processed into steel. Relying on a technique developed on a commercial scale over 100 years ago by Sir Henry Bessemer, they make the hot metal by mixing iron ore, limestone, and coke in blast furnaces. To make the coke, they pyrolize coal in huge ovens in plants that cost over $100 million and create enormous amounts of air pollution.

In the Kohle Reduktion method, developed by Korf Engineering in West Germany, the hot metal is made without coke. Coal, limestone, and oxygen are mixed in a gasification unit at 2500°. The gas rises in a shaft furnace above the gasification unit, chemically reducing the iron ore to "sponge iron." The sponge iron then drops into the gasification unit, where it is melted and the contaminants are removed by reaction of the limestone. Finally, the hot metal drains out of the bottom of the gasifier.

The Kohle method, if developed satisfactorily, will have several advantages. It will eliminate the air pollution problem of coke plants, it can be built (according to Korf estimates) for 25% less than conventional furnaces, and it may cut the cost of producing hot metal by 15%.

This technology appears to offer a dramatic solution to the problems with our nation's steel industry: I recommend that we investigate it further. If the method proves feasible and if we develop an expertise in it, we will surely attract many clients for our consulting services.

Problem is identified.
Problem is explained.
Solution is announced.
Solution is explained.
Link between problem and solution is explained.

LEARN MORE Exercises for writing about problems and their solutions are in "Use What You've Learned" (page 166).

Guidelines for Describing Problems and Their Solutions

1. **Begin by identifying the problem.** Make the problem seem significant to your readers. Emphasize the aspects of the problem most directly affected by your solution.
2. **Describe your method.** Provide the details needed by your readers.
3. **Describe the results.** Enumerate the benefits produced. Tell what was learned.
4. **Include graphics that will help your readers understand and use your communication.**

FIGURE 8.12

Passage in which the Writer Describes How a Problem Was Solved in the Past

Rashid describes the general problem: Coal-fired power plants must reduce emissions.

He then focuses on the main issue: Retrofitting with a collection of specialized systems is very expensive.

He uses a photo to enable readers to see what the facility looks like.

Rashid explains that an integrated system can solve the problem.

He highlights the results by presenting them in a table.

Rashid highlights two other benefits of the system.

Pollution Control System Reduces Cost and Makes Fertilizer

Coal-fired power plants produce 36% of electricity worldwide. They also cause air pollution that creates health hazards and damages the environment. To comply with government regulations in the U.S. and elsewhere, thousands of older plants must greatly reduce their emissions or close permanently.

Because pollution control systems have been specialized, power plants need to install several to meet environmental standards: one for sulfur dioxide, another for mercury, and still others for particulates and nitrous oxides. For older plants, retrofitting with several systems can be prohibitively expensive.

At the 50-year-old Burger Plant in Shadyside, Ohio, Powerspan Corporation has successfully demonstrated its patented Electro-Catalytic Oxidation (ECO) system, which combines the functions of four control technologies. Installed as a single unit, it costs much less than the four systems it would replace.

The ECO system not only reduced emissions effectively, but also produced byproducts sold as feedstock to a fertilizer manufacturer, producing a new (though modest) income stream for the plant.

The ECO system can also reduce fuel costs for older plants in areas like Ohio with abundant local deposits of high-sulfur coal. Decades ago these plants had to replace this inexpensive, local resource with more-expensive low-sulfur coal imported from other regions. The more sulfur in the coal, the better the system works, according to Powerspan.

ECO Demonstration Facility

Source: Powerspan Corporation

ECO Pilot Test Restuls	
Pollutant	Removal Efficiency
SO2 (sulfur dioxide)	98%
NOx (nitrogen oxides)	90%*
Hg (mercury)	80–90%
*Inlet NOx is 0.4 lb/mmBtu	

Combining Organizational Patterns

This chapter describes eight organizational patterns in isolation from one another. In practice, you will almost always need to weave two or more together. In a short memo, for instance, you might describe a problem in two sentences, identify its causes in three more, briefly compare alternative solutions, and finally recommend one.

The outline displayed in Figure 8.13 shows how six patterns were integrated in a 25-page feasibility report.

FIGURE 8.13

Outline of a Report that
Weaves Together Six
Organizational Patterns

Like many technical communications,
this report interweaves six
organizational patterns.

Solar Roofs for New Restaurants
A Feasibility Report for the Brendon's Restaurant Chain

I. **Introduction**
   A. The Brendon's Restaurant chain commissioned us to determine the feasibility of
      covering its restaurant roofs with photovoltaic cells (solar cells) that generate
      electricity
   B. Background
      1. Brendon's builds 30 new restaurants a year
      2. Brendon's is committed to environmental responsibility
      3. Federal government has goal of one million solar roofs in the U.S. by 2010
      4. Experiences of other companies ●————————— To organize this part, the writers
         a. Europe   **classified** the companies according
         b. Asia   to their location (partitioning).
         c. United States

II. **Technical Assessment** ●————————————————————————— They organized Chapter II
   A. Problems   around a **problem** and its possible
      1. Solar cells do not generate enough electricity to pay for themselves (though   **solution.**
         this technology is advancing rapidly)
      2. Solar cells collect heat, so additional power would be needed to cool the building
      3. Putting solar cells on top of the regular roof would add 15% to construction ●—— When describing the location of the
         time   solar cells, the writers organized by
   B. Potential Solutions (Used by Applebee's Restaurant in Salisbury, North Carolina)   **cause and effect.**
      1. Use the collected heat to preheat the large amount of water used by the
         restaurant
         a. Collect the heat under the solar cells (instead of venting it)
         b. Use electricity from the solar cells to power fans that blow hot air to the
            preheating system
      2. Use large solar panels that can serve as the roof itself

III. **Components of System** ●——————————————————————— The writers organized each of these
   A. Structure of the Solar Panels   subsections by **partitioning** the
   B. Structure of the Heat Ducts   component into its major subparts.
   C. Structure of the Water Heating System

IV. **Construction Schedule**

V. **Cost Comparison** ●————————————————————————————— They organized their discussion of the
                                                                     construction schedule by gathering
VI. **Conclusion** ●   the steps into related groups of steps
                                                                     (**segmenting** the process).

                                                                     The writers used the divided
                                                                     pattern for **comparisons** in their
                                                                     section on cost.

USE WHAT YOU'VE LEARNED

EXERCISE YOUR EXPERTISE

1. To choose the appropriate principle of classification for organizing a group of items, you need to consider your readers and your purpose. Here are three topics for classification, each with two possible readers. First, identify a purpose that each reader might have for consulting a communication on that topic. Then identify a principle of classification that would be appropriate for each reader and purpose.
   a. *Types of instruments or equipment used in your field*
      Student majoring in your field
      Director of purchasing in your future employer's organization
   b. *Intramural sports*
      Director of intramural sports at your college
      Student
   c. *Flowers*
      Florist
      Owner of a greenhouse that sells garden plants

2. Use a principle of classification to create a hierarchy having at least two levels. Here are some possible topics.

   - Tablet computers
   - Cameras
   - Physicians
   - Exercise machines
   - Skills you will need on the job
   - Tools, instruments, or equipment you will use on the job
   - Items used in your field (rocks if you are a geologist; power sources if you are an electrical engineer)

   After you have selected a topic, identify a reader and a purpose for your classification. Depending on your instructor's request, show your hierarchy in an outline or use it to write a brief discussion of your topic. In either case, state your principle of classification. Does your hierarchy have one and only one place for every item at each level?

3. Partition an object in a way that will be helpful to someone who wants to use it. Some objects are suggested below. Whichever one you choose, describe a specific instance of it. For example, describe a particular brand and model of food processor rather than a generic food processor. Be sure that your hierarchy has at least two levels, and state the basis of partitioning you use at each level. Depending on your instructor's request, show your hierarchy in an outline or use it to write a brief discussion of your topic.

   - Aqualung
   - Microwave oven
   - Graphing calculator
   - Bicycle
   - Some instrument or piece of equipment used in your field that has at least a dozen parts

4. Segment a procedure to create a hierarchy you could use in a set of instructions. Give it at least two levels. Some topics are listed below. Show the resulting hierarchy in an outline. Be sure to identify your readers and purpose. If your instructor requests, use the outline to write a set of instructions.

   - Changing an automobile tire
   - Making homemade yogurt
   - Starting an aquarium
   - Rigging a sailboat
   - Saddling a horse
   - Some procedure used in your field that involves at least a dozen steps
   - Some other procedure of interest to you that includes at least a dozen steps

5. Segment a procedure to create a hierarchy you could use in a general description of a process. Give it at least two levels. Some suggested topics are listed below. Show the resulting hierarchy in an outline. Be sure to identify your readers and purpose. If your instructor requests, use the outline to write a general description of the process addressed to someone unfamiliar with it.

   - How the human body takes oxygen from the air and delivers it to the parts of the body where it is used
   - How television signals from a program originating in New York or Los Angeles reach television sets in other parts of the country
   - How aluminum is made
   - Some process used in your field that involves at least a dozen steps
   - Some other process of interest to you that includes at least a dozen steps

6. One of your friends is thinking about making a major purchase. Some possible items are listed below. Create an outline with at least two levels that compares two or more good alternatives. If your instructor requests, use that outline to write your friend a letter.

   - Stereo
   - Smartphone
   - Binoculars
   - Bicycle
   - DVD player
   - Some other type of product for which you can make a meaningful comparison on at least three important points

7. Think of some way in which things might be done better in a club, business, or some other organization. Imagine that you are going to write a letter to the person who can bring about the change you are recommending. Create an outline with at least two levels in which you compare the way you think

things should be done and the way they are being done now.

8. A friend has asked you to explain the causes of a particular event. Some events are suggested below. Write your friend a brief letter explaining the causes.
   - Static on radios and televisions
   - Immunization from a disease
   - Freezer burn in foods
   - Yellowing of paper

9. Think of a problem you feel should be corrected. The problem might be noise in your college library, shoplifting from a particular store, or the shortage of parking space on campus. Briefly describe the problem and list the actions you would take to solve it. Next, explain how each action will contribute to solving the problem. If your instructor requests, use your outline to write a brief memo explaining the problem and your proposed solution to a person who could take the actions you suggest.

9

Persuading Your Readers

LEARNING OBJECTIVES

1. Focus on your readers' goals and values.
2. Reason soundly.
3. Build an effective relationship with your readers.
4. Organize to create a favorable response.
5. Appeal, when appropriate, to your readers' emotions.
6. Adapt to your readers' cultural background.
7. Persuade ethically.

The purpose of one type of persuasion is to influence other people's actions.

The goal of another type of persuasion is to enable people to collaborate while exploring ideas and possible courses of action.

MindTap®

Find additional resources related to this chapter in MindTap.

Every communication you write at work will have a persuasive dimension, as Chapter 1 explained. Always, for instance, you will want to persuade your readers that you are a careful, intelligent, and knowledgeable professional. Why else will you strive so diligently to eliminate even the occasional misspelling, grammatical mistake, or factual error from your work? But in some communications, persuasion will play a much more central and challenging role: Your primary aim will be to influence your readers' *actions*. This chapter's goal is to work together with your instructor in helping you learn to write these communications effectively.

The Competitive and Collaborative Uses of Persuasion

On the job, you will prepare these primarily persuasive communications in two distinct situations. The first involves *competition*. You may write proposals that enable your company to win contracts that other organizations also want. Perhaps you will work for a nonprofit organization where you will write reports to persuade readers to support better policies for accomplishing some goal, such as protecting the environment or endangered species. You may also write to readers in your own organization to advocate improvements in products or policies that others believe should not be changed. As described in Chapter 3, Stephanie was engaged in this kind of persuasion when she urged her employer to change the way it assigned urgently needed Braille translations.

Less obvious, but equally important, will be situations where you will use your persuasive abilities to contribute to *collaborative* and *cooperative* efforts. They occur when scientists gather to discuss ways to interpret new data, when engineers team up to solve a technical problem, and when medical personnel confer about the best way to respond to the outbreak of an illness. You and your co-workers will work together to explore ideas or possible courses of action. While you won't be competing, you will each use your persuasive skills to "argue" back and forth to assure that the advantages of each alternative are fully understood and the weaknesses of each are fully probed.

To Persuade, Influence Your Readers' Attitudes

More than thirty years of research have established that in both collaborative and cooperative persuasion, the key to influencing readers is to focus on changing their *attitudes*.

What determines a person's attitude toward an idea, object, or action? Researchers have found that it isn't a single thought or argument, but the sum of the various thoughts a person associates with the idea, object, or action under consideration. Imagine the following.

Edward has been asked to recommend a new copy machine for his department. Various thoughts cross his mind: the amount of money in his department's budget, the machine's special features (such as the ability to make copies directly from computer files sent over wifi), its repair record, his experience with another product made by the same company, and his impression of the manufacturer's marketing representative. Each of these thoughts will make Edward's attitude more positive or more negative. He may think the price is reasonable (positive) or too high (negative); he may like the salesperson (positive) or detest the salesperson (negative); and so forth.

Some of these thoughts will probably influence Edward's attitude more than others. Edward may like one copier's special features, but that factor may be outweighed by reports that its repair record is very poor. If the sum of Edward's thoughts associated with the machine is positive, he will have a favorable attitude toward it and may recommend it. If the sum is negative, he will have an unfavorable attitude and probably not recommend it.

> The sum of a reader's positive and negative thoughts about an object, action, or person determines his or her attitude.

In your career, you will have opportunities to influence your readers' attitudes toward a wide variety of subjects, such as products, policies, actions, and other people. Whatever the focus of your persuasive effort, you may use your persuasive powers to change your readers' attitudes in any of the following ways.

> Ways to change readers' attitudes

- **Reverse** an attitude you want your readers to abandon.
- **Reinforce** an attitude you want them to hold even more firmly.
- **Shape** their attitude on a subject about which they currently have no formed opinion.

To help you develop the persuasive skills most useful in the workplace, this chapter draws on studies of persuasion that cover more than 2,500 years, beginning with the Greek philosopher Aristotle and extending to scientific studies conducted very recently. These studies, ancient and contemporary, agree that persuasion requires attention to both thought and feeling, logical and nonlogical (subjective and personal) elements. This chapter provides advice about both. But it begins where all effective workplace writing begins: by looking at all persuasive situations from the perspective of the reader.

Focus on Your Readers' Goals and Values

Easily the most powerful way to prompt your readers to experience favorable or positive thoughts about the idea, action, process, or product you advocate is to tell them how it will help them accomplish some goal they want to achieve.

Aligning your communication's goals with your readers' goals can be easy when you know what your readers' goals are. Suppose you are writing a report on copy machines to Edward. If you know that the quality of the copies produced matters more than the cost to him and that preference would lead him to select the same machine you would choose, you could write a communication that highlights the copier that produces the best copies

However, two challenges can arise. Sometimes, you will learn that your judgment differs from your readers, which creates a persuasive situation discussed later in this chapter. The second challenge is that you can't find out directly from your readers or co-workers what your reader's criteria are. In this case, you may find it helpful to think about the three types of goals that employees often pursue: organizational, values-based, and personal growth goals and achievement.

Organizational Goals

Organizations have goals at many levels.

On the job, you will often write to employees in your organization or other organizations. Because they are hired to advance their employer's interests, employees usually adopt their employer's goals as their own. Consequently, one way to identify your readers' goals is to examine the goals of the organization for which they work.

In any organization, some goals are general and some specific. At a general level, many organizations share the same goals: to increase revenue, operate efficiently, keep employee morale high, and so on. However, each company also has its own unique goals. A company might be seeking to control 50 percent of its market, to have the best safety record in its industry, or to expand into ten states in the next five years. Moreover, each department within a company has specific objectives. Thus, the research department aims to develop new and improved products, the marketing department seeks to identify new markets, and the accounting department strives to manage financial resources prudently. When identifying goals to serve as the basis for persuasion, choose specific goals over general ones.

Also, as you uncover an organization's various goals, focus on the one that are most closely related to the specific idea, action, or process you are advocating. Then craft your communication to show that your recommendation will help them achieve an outcome.

Figure 9.1 shows a marketing brochure that uses organizational objectives to persuade its intended readers (purchasing managers for supermarkets and supermarket chains) to carry a certain product. Notice the boldface statements that proclaim how the product will help supermarkets achieve their goals of building sales volume, stimulating impulse buying, and increasing profits.

Values-Based Goals

Most companies have goals that involve social, ethical, and aesthetic values not directly related to profit and productivity. For instance, Google's *Code of Conduct* famously begins with "Don't be evil" and goes on to detail its obligations to users,

FIGURE 9.1

Brochure That Stresses
Organizational Objectives

In this brochure, Welch's tells food retailers how selling one of its products will help them achieve three organizational goals: greater sales volume, more impulse purchases, and increased profits.

Courtesy of Welch's Foods Inc.

employees, business partners, and even competitors. Johnson & Johnson's *credo* spells out its commitment to treating the company's employees respectfully and fostering their growth and advancement, to supporting the communities in which it has facilities, to protecting natural resources, and (lastly) to providing stockholders with a fair return on their investments. Even in companies that do not have an official credo, broad human and social values may provide an effective foundation for persuasion, especially when you are advocating a course of action that seems contrary to narrow business interests. For example, by pointing to the benefits to be enjoyed by a nearby community, you might be able to persuade your employer to strengthen its water pollution controls beyond what is required by law.

In addition, individual employees bring to work personal values that can influence their decisions and actions. Especially when writing to only one or a few readers, learn whether they have personal values that are related to what you are recommending.

Personal Growth and Achievement Goals

In studies of employee motivation that are still highly respected today, Abraham Maslow (2000) and Frederick Herzberg (2003) highlight another type of benefit you can use to persuade. Both researchers confirmed that, as everyone knew, employees are motivated by such considerations as pay and safe working conditions. However, the studies also found that after most people feel they have an adequate income and safe working conditions, they become less easily motivated by these factors. Consequently, such factors are called *deficiency needs:* They motivate principally when they are absent.

Once their deficiency needs are met—and even before—most people are motivated by so-called *growth needs,* including the desire for recognition, good relationships at work, a sense of achievement, personal development, and the enjoyment of work itself.

Many other studies have confirmed these findings. They demonstrate that one powerful persuasive strategy is to show how the decisions or actions you are advocating will help your readers satisfy their desires for growth and achievement. When you request information or cooperation from a co-worker, mention how much you value his or her assistance. When you evaluate a subordinate's performance, discuss accomplishments as well as shortcomings. When you ask someone to take on additional duties, emphasize the challenge and opportunities for achievement that lie ahead.

Figure 9.2 shows a web page that Chrysler uses to attract new employees. It appeals to college graduates by emphasizing the opportunities they would have with the company to assume responsibility, take on challenges, and grow professionally.

Deficiency needs are needs that disappear when satisfied, as the need for higher pay is satisfied with a pay raise.

Growth needs are needs that are always present, like the need for personal development and enjoyment in one's work.

Reason Soundly

When you are attempting to influence others, it's impossible to overstate the importance of sound reasoning. For Aristotle, who called it *logos,* logical reasoning was one of the three powerful strategies for persuasion. In today's workplace, as in ancient Greece, one of the most favorable thoughts your readers or listeners could possibly have is, "Yeah, that makes sense." One of the most unfavorable is, "Hey, there's a flaw in your reasoning."

Sound reasoning is especially important when you are trying to influence your readers' decisions and actions. In addition to identifying potential benefits that will appeal to your readers, you also must persuade them that the decision or action you advocate will actually bring about these benefits—that the proposed new equipment really will reduce costs enough to pay for itself in just eighteen months or that the product modification you are recommending really will boost sales 10 percent in the first year.

FIGURE 9.2

Page from a Recruiting Website that Focuses on Growth Needs

On this web page from its recruiting website, Chrysler Corporation appeals to potential job applicants by promising to give them substantial challenges and significant opportunities for growth.

Note emphasis on challenge and growth in the following quotations.

"What makes the culture at Chrysler so distinctive? All the things that empower you to bring your best."

"You'll find Chrysler a fast-paced work environment—one that will keep you challenged and growing from day one."

"How far and how fast you grow in your career is yours to own."

"You can apply yourself in ways you never imagined at Chrysler."

"We expect you to bring—and voice—your point of view."

"We respect each other's roles and support each other's growth."

Sound reasoning is also essential when you are describing conclusions you have reached after studying a group of facts, such as the results of a laboratory experiment or consumer survey. In such cases, you must persuade your readers that your conclusions are firmly based on the facts or data and that all sensible readers would reach the same conclusions.

Notice that in each of the situations just mentioned (as well as in any other you might encounter on the job), you must not only use sound reasoning but also convince your readers that your reasoning is sound. The ability to do so is one of the most valuable writing skills you can develop.

Stephen Toulmin developed a way of thinking about reasoning that will help you reason soundly and also show your readers that you do (Hitchcock & Verheij, 2010). You will also find Toulmin's concepts useful when you are analyzing and evaluating the reasoning in other people's communications.

Sound reasoning involves a *claim*, *evidence*, and a *line of reasoning*. The following diagram illustrates the relationship among them.

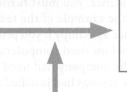

EVIDENCE
The facts, observations, and other evidence that support your claim.

CLAIM
The position you want your readers to accept.

LINE OF REASONING
The connection linking your claim and evidence; the reason your readers should agree that your evidence supports your claim.

Relationship among claim, evidence, and line of reasoning

Imagine that you work for a company that manufactures specialized textiles for astronaut suits, hazardous material suits, and bulletproof clothing. You have found out that a competitor recently increased productivity by using newly developed computer programs to manage its manufacturing processes. If you were to recommend that your employer develop similar programs, your argument could be diagrammed as follows:

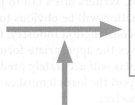

EVIDENCE
Our competitor has increased productivity by using new computer programs.

CLAIM
By using new computer programs, we will increase productivity.

LINE OF REASONING
Experience has shown that actions that increase profits for one company in an industry will usually increase profits of other companies in the same industry.

The line of reasoning explains why the evidence provides support for the claim.

To accept your claim, readers must be willing to place their faith in both your evidence and your line of reasoning.

GUIDELINE 1 Present sufficient and reliable evidence

First, you must convince your readers that your evidence is both sufficient and reliable.

Your evidence is sufficient if it includes all the details your readers want.

To provide *sufficient* evidence, you must furnish all the details your readers will want. For instance, in the example of the textile mills, your readers would probably regard your evidence as skimpy if you produced only a vague report that the other company had somehow used computers and saved some money. They would want to know how the company had used computers, how much money had been saved, whether the savings had justified the cost of the equipment, and so forth.

Your evidence is reliable if your readers will accept it as a valid type of evidence.

To provide *reliable* evidence, you must produce the type of evidence your readers are likely to accept. The type of evidence varies greatly from field to field. For instance, in science and engineering, certain experimental procedures are widely accepted as reliable, whereas common wisdom and unsystematic observations usually are not. In contrast, in many business situations, personal observations and anecdotes provided by knowledgeable people often are accepted as reliable evidence. In general, the following three types of evidence are widely accepted.

The types of evidence readers accept as valid differ from context to context.

- **Data.** Readers typically respond very favorably to claims that are supported by numerical data.
- **Expert testimony.** People with advanced education, firsthand knowledge, or extensive experience related to a topic are often credited with special understanding and insight.
- **Examples.** Specific instances can effectively support general claims.

GUIDELINE 2 Explicitly justify your line of reasoning

To argue persuasively, you must not only present sufficient and reliable evidence, but also convince your readers that you are using a valid line of reasoning to link your evidence to your claim. Writers often fail to justify their line of reasoning in the belief that the explanation will be obvious to their readers. In fact, that is sometimes the case. In the construction industry, for example, people generally agree that if an engineer uses the appropriate formulas to analyze the size and shape of a bridge, the formulas will accurately predict whether or not the bridge will be strong enough to support the loads it must carry. The engineer doesn't need to justify the formulas themselves.

Readers sometimes look aggressively for weaknesses in the writer's line of reasoning.

In many cases, however, readers search aggressively for a weak line of reasoning. In particular, they are wary of arguments based on assumptions that may prove to be false. For example, your readers may agree in principle that if another textile mill like yours saved money by computerizing, then your mill would probably enjoy the same result; however, they may question the assumption that the other mill in actuality is like yours. Maybe it makes a different kind of product or employs a different manufacturing process. If you think your readers will suspect that you are making a false assumption, offer evidence or an explanation to dispel their doubts.

Readers also look for places where writers have overgeneralized by drawing broad conclusions from too few specific instances. If you think your readers will raise such an objection to your argument, mention additional cases. Or, narrow your conclusion to better match the evidence you have gathered. For example, instead of asserting that your claim applies to all textile companies in all situations, argue that it applies to specific companies or specific situations.

GUIDELINE 3 Respond to—and learn from—your readers' concerns and counterarguments

Although it is tempting to focus your attention on the reasons that your readers should agree with you, you can benefit from learning why they may be reluctant to do so.

In fact, learning your readers' reasons for their concerns and counterarguments can help you improve the project you are proposing or refine the position you are advocating. That's because some of these concerns and counterarguments may be valid. By taking time to evaluate the validity of your readers' position, you might identify ways to strengthen your own.

When you are satisfied that your proposed idea or action is as sound as possible, use your insights into your readers' concerns and your understanding of their actual or possible counterarguments to help you increase your communication's persuasiveness. Research shows that when people read, they not only pay attention to the writer's statements but also generate their own thoughts. Whether these thoughts are favorable or unfavorable can have a greater influence on your readers' attitudes than any point you make in the communication itself. Consequently, when you are trying to persuade, avoid saying anything that might prompt negative thoughts in your readers.

One way to avoid arousing negative thoughts is to answer all the important questions your readers are likely to ask while they are reading your communication. If you neglect to address questions that seem obvious to them—"What are the costs of doing as you suggest?" "What do other people who have looked into this matter think?"—they may believe that you have failed to conduct thorough research or that you are hiding facts that would weaken your position.

In addition to asking questions while reading, your readers may generate arguments against your position. For example, if you say there is a serious risk of injury to workers in a building used by your company, your readers may think to themselves that the building has been used for three years without an accident. If you say a new procedure will increase productivity by at least 10 percent, your readers may think to themselves that some of the data on which you are basing your estimate are inaccurate. Readers are especially likely to generate counterarguments when you attempt to reverse their attitudes. Research shows that people resist efforts to persuade them that their attitudes are incorrect (Petty & Cacioppo, 1986).

To deal effectively with counterarguments, you must offer a reason for relying on your position rather than on the opposing one. For example, imagine that you are proposing the purchase of a certain piece of equipment and that you predict your readers will object because it is more expensive than a competitor's product they believe to be its equal. You might explain that the competitor's product, though less expensive to purchase, is more expensive to operate and maintain.

One way to learn about the questions and counterarguments your readers might raise is to ask them about their reasons for holding their present attitudes. Why do they do things the way they do?

Figure 9.3 shows a letter in which the writer skillfully anticipates and then addresses two counterarguments. The writer is the marketing director of a radio station who is attempting to persuade the owner of a web design company to switch some of its ads to her station.

Your readers' counterarguments can help you improve the project or position you are advocating.

TRY THIS Think of a policy or regulation at your school or workplace that you want to see changed. List the good reasons someone supporting the policy or regulation could have for retaining it.

FIGURE 9.3

Letter That Addresses
Counterarguments

In this letter, Ruth Anne attempts to
persuade the owner of a web design
company to switch some
of its radio advertising to her station.

Ruth Anne states the first
counterargument she expects
Mr. McGuffey to raise: a reduction in
the number of times listeners hear
the advertisements.

She addresses the counterargument
by saying the benefit more than offsets
the perceived disadvantage.

Ruth Anne states a second
counterargument, that many of her
station's listeners are "unemployed."

She responds by explaining that these
"unemployed" individuals are part of
Mr. McGuffey's target market.

WGWG

12741 Vienna Boulevard
Philadelphia, PA 19116

August 17, 2017

Mr. Roger L. McGuffey, President
WEB DESIGNERS, Inc.
1200 Langstroth Avenue
Philadelphia, Pennsylvania 19131

Dear Mr. McGuffey:

Thank you for inviting me to discuss the advantages WEB DESIGNERS would
enjoy by switching a portion of your current radio advertising to WGWG.

The switch will increase your audience considerably. According to Arbitron
Market Analytics, the 30 spots per week that you buy from WSER and WFAC
reach 63% of your target market—small business owners who might purchase your
web services. Switching 10 spots to WGWG would increase your reach to 73%.
While this change would reduce from 16 to 13 the number of times the average
listener hears your message, there would still be ample repetition. In exchange, you
would reach 8% more of your target audience. Also, your costs would decrease
significantly because we charge less than the other stations.

You said that you have seen data showing that WGWG has a higher portion of
"unemployed" listeners than either WSER or WFAC. We discovered that in those
statistics, "unemployed" means "unemployed outside the home." The vast majority
are housewives, who, according to a recent survey, represent a large portion of the
entrepreneurs starting at-home companies that do business on the web. Thus, these
"unemployed" persons include a key market for WEB DESIGNERS' services.

I suggest that we meet soon to schedule your advertising on WGWG.

Cordially,

Ruth Anne Peterson
Marketing Director

Build an Effective Relationship
with Your Readers

While reasoning is important in persuasion, it is rarely the only factor that
influences readers' attitudes and actions. Also important is how your readers
feel about you. Even at work, communications are always interpersonal interactions. If your readers feel well disposed toward you, they are likely to consider
your points openly and without bias. If they feel irritated, angry, or otherwise
unfriendly toward you, they may immediately raise counterarguments to every
point you present, making it extremely unlikely that you will elicit a favorable
reaction, even if all your points are clear, valid, and substantiated. Good points
rarely win the day in the face of bad feelings.

The following sections describe two ways you can present yourself to obtain a fair—or even a favorable—hearing from your readers at work: Present yourself as a credible person, and present yourself as a partner, a supporter, not a critic or opponent.

GUIDELINE 1 Establish your credibility

Your credibility rests on whether or not your readers believe you are a good source of information and ideas. If people believe you are credible, they will be relatively open to accepting your judgments and recommendations. If people do not find you credible, they may refuse to give you a fair hearing no matter how soundly you state your case.

Because people's response to a message can depend so greatly on their views of the writer or speaker, Aristotle identified *ethos,* which corresponds roughly to credibility, as the second of the three major elements of persuasion.

Researchers have identified the factors that affect peoples' impression of someone else's credibility. Here are four strategies cited in *Power, Influence and Persuasion* (Harvard Business School Press, 2005).

> The more credible you seem to be, the more open your readers will be to your message.

> Several factors that influence credibility

Strategies for Building Credibility

1. **Technical Expertise.** Expertise is the knowledge and experience relevant to the topic of your communication that your readers believe you possess. *Strategies you can use:*
   - ☐ Mention your credentials.
   - ☐ Demonstrate a command of the facts.
   - ☐ Avoid oversimplifying.
   - ☐ Mention or quote experts so their expertise supports your position.

2. **Trustworthiness.** Trustworthiness depends largely on your readers' perceptions of your motives. If you seem to be acting from self-interest, your credibility is low; if you seem to be acting objectively or from goals shared by your readers, your credibility is high. *Strategies you can use:*
   - ☐ Stress values and objectives that are important to your readers.
   - ☐ Avoid drawing attention to ways you will benefit.
   - ☐ Demonstrate knowledge of the concerns and perspectives of others.

3. **Group membership.** You will gain credibility if you are a member of the readers' own group or a group admired by your readers. *Strategies you can use:*
   - ☐ If you are associated with a group admired by your readers, allude to that relationship.
   - ☐ If you are addressing members of your own organization, affirm that relationship by showing that you share the group's objectives, methods, and values.
   - ☐ Use terms that are commonly employed in your organization.

4. **Power.** For example, simply by virtue of the position, a boss acquires some credibility with subordinates. *Strategies you can use:*
   - ☐ If you are in a position of authority, identify your position if your readers don't know it.
   - ☐ If you are not in a position of authority, associate yourself with a powerful person by quoting the person or saying that you consulted with him or her or were assigned the job by that individual.

In Figure 9.4, the president of a biometric security company uses these strategies when requesting an opportunity to demonstrate a new security system. Figure 9.5 shows how the Land Trust Alliance uses the strategies on its website to persuade people to participate in and support its efforts to preserve farm and natural lands from housing or commercial development.

FIGURE 9.4

Letter that Uses Several
Strategies for Building
Credibility

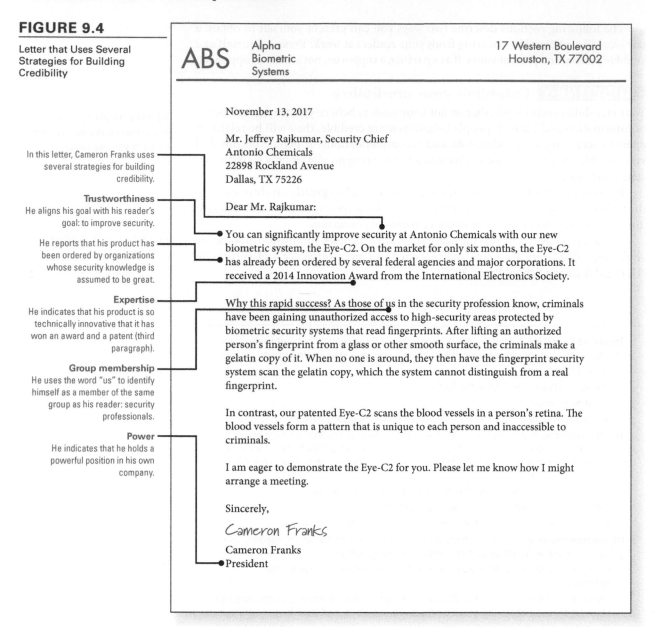

In this letter, Cameron Franks uses several strategies for building credibility.

Trustworthiness
He aligns his goal with his reader's goal: to improve security.

He reports that his product has been ordered by organizations whose security knowledge is assumed to be great.

Expertise
He indicates that his product is so technically innovative that it has won an award and a patent (third paragraph).

Group membership
He uses the word "us" to identify himself as a member of the same group as his reader: security professionals.

Power
He indicates that he holds a powerful position in his own company.

ABS   Alpha
       Biometric
       Systems

17 Western Boulevard
Houston, TX 77002

November 13, 2017

Mr. Jeffrey Rajkumar, Security Chief
Antonio Chemicals
22898 Rockland Avenue
Dallas, TX 75226

Dear Mr. Rajkumar:

You can significantly improve security at Antonio Chemicals with our new biometric system, the Eye-C2. On the market for only six months, the Eye-C2 has already been ordered by several federal agencies and major corporations. It received a 2014 Innovation Award from the International Electronics Society.

Why this rapid success? As those of us in the security profession know, criminals have been gaining unauthorized access to high-security areas protected by biometric security systems that read fingerprints. After lifting an authorized person's fingerprint from a glass or other smooth surface, the criminals make a gelatin copy of it. When no one is around, they then have the fingerprint security system scan the gelatin copy, which the system cannot distinguish from a real fingerprint.

In contrast, our patented Eye-C2 scans the blood vessels in a person's retina. The blood vessels form a pattern that is unique to each person and inaccessible to criminals.

I am eager to demonstrate the Eye-C2 for you. Please let me know how I might arrange a meeting.

Sincerely,

Cameron Franks

Cameron Franks
President

GUIDELINE 2 **Present yourself as a partner, not a critic**

Many people feel threatened and defensive when people make well-intentioned suggestions for improvement to them. They interpret a suggestion as criticism that they are not doing things as well as they should. To circumvent this kind of defensiveness, construct your suggestions in a way that emphasizes your desire to help, to be the reader's partner.

Strategies for Presenting Yourself as a Partner

1. **Praise your readers.**
   □ When writing to an individual, mention one of his or her recent accomplishments.
   □ When writing to another organization, mention something it prides itself on.
   □ When praising, be sure to mention specifics. General praise sounds insincere.

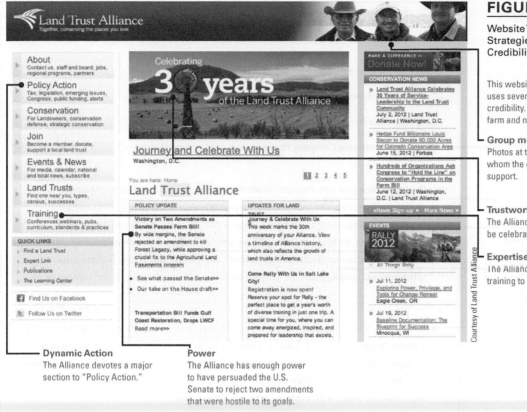

FIGURE 9.5

Website That Uses Several Strategies for Building Credibility

This website by the Land Trust Alliance uses several techniques for building its credibility. The Alliance seeks to preserve farm and natural lands from development.

Group membership
Photos at the top right show people from whom the organization hopes to win support.

Trustworthiness
The Alliance has existed long enough to be celebrating its 30th anniversary.

Expertise
The Alliance has the expertise to provide training to others.

Dynamic Action
The Alliance devotes a major section to "Policy Action."

Power
The Alliance has enough power to have persuaded the U.S. Senate to reject two amendments that were hostile to its goals.

2. **Highlight the goals you share.**
   □ Identify some personal or organizational goal of your readers that you will help them attain.
   □ If you are already your readers' partner, mention that fact and emphasize the goals you share.

3. **Show that you understand your readers.**
   □ Even if you disagree with your readers, state their case fairly.
   □ Focus on areas of agreement.

4. **Maintain a positive and helpful stance.**
   □ Present your suggestions as ways of helping your readers do an even better job.
   □ Avoid criticizing or blaming.

To see how to apply this advice, consider the following situation. Marjorie Lakwurtz is a regional manager for a company that hires nearly 1,000 students in various cities during the summer months to paint houses. The company requires each student to pay a $250 deposit to cover the cost of lost equipment. Noting that the equipment isn't worth even $100, the students believe the company demands the deposits so it can invest the money and keep the interest for itself.

Marjorie agrees with the students and decides to write a memo suggesting that the company change its policy. She could argue for the change in a negative way by telling the company that it should be ashamed of its greedy plot to profit by investing the students' money. Or she could take a positive approach, presenting her suggestion as a way the company can better achieve its own goals. In the memo shown in Figure 9.6, she takes the latter course. Notice how skillfully she presents herself as her reader's partner.

FIGURE 9.6

Memo in Which the Writer Establishes a Partnership with the Reader

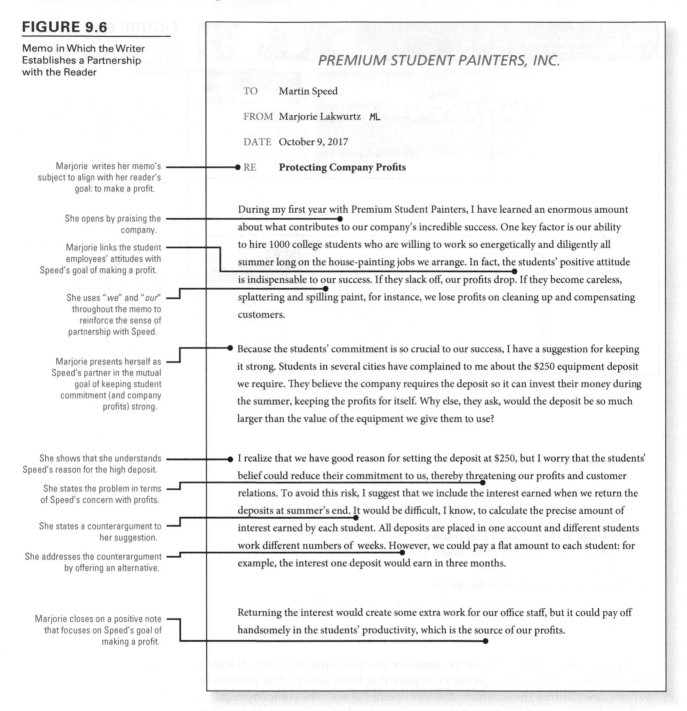

PREMIUM STUDENT PAINTERS, INC.

TO   Martin Speed

FROM Marjorie Lakwurtz ML

DATE October 9, 2017

RE   **Protecting Company Profits**

Marjorie writes her memo's subject to align with her reader's goal: to make a profit.

She opens by praising the company.

Marjorie links the student employees' attitudes with Speed's goal of making a profit.

She uses *"we"* and *"our"* throughout the memo to reinforce the sense of partnership with Speed.

During my first year with Premium Student Painters, I have learned an enormous amount about what contributes to our company's incredible success. One key factor is our ability to hire 1000 college students who are willing to work so energetically and diligently all summer long on the house-painting jobs we arrange. In fact, the students' positive attitude is indispensable to our success. If they slack off, our profits drop. If they become careless, splattering and spilling paint, for instance, we lose profits on cleaning up and compensating customers.

Marjorie presents herself as Speed's partner in the mutual goal of keeping student commitment (and company profits) strong.

Because the students' commitment is so crucial to our success, I have a suggestion for keeping it strong. Students in several cities have complained to me about the $250 equipment deposit we require. They believe the company requires the deposit so it can invest their money during the summer, keeping the profits for itself. Why else, they ask, would the deposit be so much larger than the value of the equipment we give them to use?

She shows that she understands Speed's reason for the high deposit.

She states the problem in terms of Speed's concern with profits.

She states a counterargument to her suggestion.

She addresses the counterargument by offering an alternative.

I realize that we have good reason for setting the deposit at $250, but I worry that the students' belief could reduce their commitment to us, thereby threatening our profits and customer relations. To avoid this risk, I suggest that we include the interest earned when we return the deposits at summer's end. It would be difficult, I know, to calculate the precise amount of interest earned by each student. All deposits are placed in one account and different students work different numbers of weeks. However, we could pay a flat amount to each student: for example, the interest one deposit would earn in three months.

Marjorie closes on a positive note that focuses on Speed's goal of making a profit.

Returning the interest would create some extra work for our office staff, but it could pay off handsomely in the students' productivity, which is the source of our profits.

Organize to Create a Favorable Response

The way you organize a communication may have almost as much effect on its power to persuade as what you say in it. That point was demonstrated long ago in a classic study by researchers Sternthal, Dholakia, and Leavitt (1978). They presented two groups of people with different versions of a talk urging that a federal consumer protection agency be established. One version *began* by saying that the speaker was a highly credible source (a lawyer who graduated from Harvard and had extensive experience with consumer issues); the other version *ended* with that information.

Among people initially opposed to the speaker's recommendation, those who learned about his credentials at the beginning responded more favorably to his arguments than did those who learned about his credentials at the end. Why? This outcome can be explained in terms of two principles you learned in Chapter 1. First, people react to messages moment by moment. Second (and here's the key point), their reactions in one moment will affect their reactions in subsequent moments. In this experiment, those who learned of the speaker's credentials before hearing his arguments were relatively open to what he had to say. But those who learned of the speaker's credentials only at the end worked more vigorously at creating counterarguments as they heard each of his points. After those counterarguments had been recorded in memory, they could not be erased simply with new information about the speaker's credibility.

The order in which you present information can affect the reader's response to your overall message.

Thus, it's not only the array of information that is critical in persuasion but also the way readers process the information. The following guidelines suggest two strategies for organizing to elicit a favorable response.

GUIDELINE 1 Choose between direct and indirect organizational patterns

As you learned in Chapter 6, the most common organizational pattern in the workplace begins with the bottom line—the writer's main point. Communications organized this way are said to use a *direct pattern* of organization because they go directly to the main point and only afterward present the evidence and other information related to it.

In the direct pattern, you state the main point up front, before presenting the supporting evidence.

The alternative is to postpone presenting your main point until you have presented your evidence or other related information. This is called the *indirect pattern* of organization.

In the indirect pattern, you postpone the main point.

To choose between the direct and indirect organizational patterns, focus on your readers' all-important initial response to your message. The direct pattern will start your readers off on the right foot when you have good news to convey: "You're hired," "I've figured out a solution to your problem," or something similar.

The direct pattern also works well when you are offering an analysis or recommending a course of action that you expect your readers to view favorably—or at least objectively—from the start. Leah is about to write such a report, in which she will recommend a new system for managing the warehouses for her employer, a company that manufactures hundreds of parts used to drill oil and gas wells. Leah has chosen to use the direct pattern shown in the left-hand column of Figure 9.7 because her readers (upper management) have expressed dissatisfaction with the present warehousing system. Consequently, she can expect a favorable reaction to her initial announcement that she has designed a better system.

Use the direct pattern of organization when you expect a favorable response from your readers.

The direct pattern is less effective when you are conveying information your readers might view as bad, alarming, or threatening. Imagine, for example, that Leah's readers are the people who set up the present system and believe it is working well. If Leah begins her memo by recommending a new system, she might put her readers immediately on the defensive because they might feel she is criticizing their competence. By using an indirect pattern shown in the

FIGURE 9.7

Comparison of Direct and Indirect Organizational Patterns for Organizing Leah's Memo

The direct pattern presents the recommendation first.

The indirect pattern delays recommendation; it's for use where the reader may react unfavorably.

Direct Pattern	Indirect Pattern
I. Leah presents her recommended strategy.	I. Leah discusses the goals of the present system from the *reader's point of view.*
II. Leah explains why her way of warehousing is superior to the present way.	II. Leah discusses the ways in which the present system does and does not achieve the reader's goals.
III. Leah explains in detail how to implement her system.	III. Leah presents her recommended strategy for achieving those goals more effectively, focusing on the ways her recommendation can overcome the shortcomings of the present system.
	IV. Leah explains in detail how to implement her system.

Consider using the indirect pattern when you believe your readers will respond negatively to your main point unless you prepare them for it beforehand.

right-hand column of Figure 9.7, however, Leah can *prepare* her readers for her recommendation by first getting them to agree that it might be possible to improve on the present system.

You may wonder why you shouldn't use the indirect pattern all the time. It presents the same information as the direct pattern (plus some more), and it avoids the risk of inciting a negative reaction at the outset. The trouble is that this pattern frustrates the readers' desire to learn the main point first.

In sum, the choice between direct and indirect patterns of organization can greatly affect the persuasiveness of your communications. To choose, you need to follow the basic strategy suggested throughout this book: Think about your readers' moment-by-moment reactions to your message.

GUIDELINE 2 Create a tight fit among the parts of your communication

When you organize a communication, you can also strengthen its persuasiveness by ensuring that the parts fit together tightly. This advice applies particularly to longer communications, where the overall argument often consists of two or more subordinate arguments. The way to do this is to review side by side the claims made in the various parts of a communication.

To create a tight fit, be sure all parts of your communication work together to support your overall position.

For example, imagine that you are writing a proposal. In an early section, you describe a problem your proposed project will solve. Here, your persuasive points are that a problem exists and that the readers should view it as serious. In a later section, you describe the project you propose. Check to see whether this description tells how the project will address each aspect of the problem that you described. If it doesn't, either the discussion of the problem or the discussion of the project needs to be revised so the two match up. Similarly, your budget should include expenses that are clearly related to the project you describe, and your schedule should show when you will carry out each activity necessary to complete the project successfully.

The need for a tight fit applies not only to proposals but also to any communication whose various parts work together to affect your readers' attitudes. Whenever you write, make sure the parts work harmoniously together in mutual support of your overall position.

Introduce Emotional Arguments if Relevant

Of Aristotle's three types of persuasive strategies, the appeal to emotions (*pathos*) is used least in workplace communications. In many communications, it makes no appearance at all. In fact, readers would usually consider appeals to emotion in scientific research reports, test reports, and many other types of documents to be highly inappropriate.

In others, appeals to emotion are important. For example, communications on either side of public policy often include them. Organizations that advocate opening new sections of federal land for oil and mineral exploration sometimes include photographs of animals in natural settings to emphasize the measures to be taken to protect them. Organizations opposed to opening the lands sometimes include photographs of damage accompanying such exploration in other areas. Each image is intended to elicit an emotional response.

As they advocate healthy lifestyles, government agencies and private health providers use emotional appeals, as do nonprofit foundations that support research for conditions such as arthritis and breast cancer (see Figure 9.8).

Base your decisions about whether—and how strongly—to appeal to your readers' emotions on your knowledge of your readers and their likely response. Situations in

LEARN MORE For more on having other people review your drafts, turn to "Obtain Truly Helpful Advice from People Who Review Your Drafts–and Give Good Advice When You Are Reviewing Someone Else's Draft" (pages 282–287).

FIGURE 9.8

Websites That Appeal to the Reader's Emotions

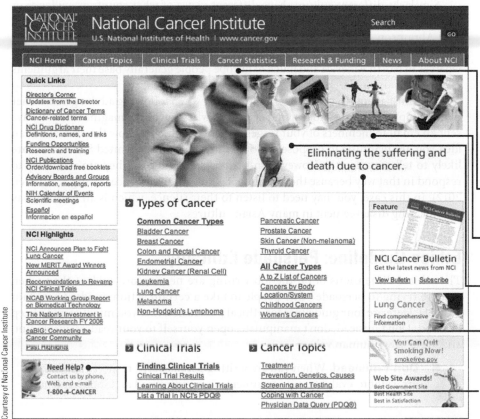

Courtesy of National Cancer Institute

On its home page, the National Cancer Institute (NCI) appeals to the emotions of cancer patients, their families, and friends.

In two ways, the site suggests the NCI understands these emotions and offers hope and relief.

1. The large photo portrays an individual overcome by emotions that many people feel when they visit the site. Another photo farther to the right shows three people on a beach, all apparently healthy and strong. Between these images are the NCI employees who will create this transformation: a researcher and a caregiver who looks directly at the reader.

2. The website identifies in words what many of its readers fear: "suffering and death due to cancer." It also promises to "eliminate" them.

In addition, the site also appeals to the readers' emotions by promising assistance to those who "need help." The NCI employees who will provide this help are represented by a smiling person who seems happy to hear from callers.

which you are using emotional appeals are ones in which it can be especially helpful to ask others to review your drafts.

Global Guideline: Adapt to Your Readers' Cultural Background

Whenever you are writing to readers in a culture other than your own, you will need to learn about the persuasive practices that succeed there. Like all other aspects of communication, every aspect of persuasion can differ from one culture to another.

What people in your culture consider to be persuasive may not be persuasive in another culture.

Consider, for example, the advice conveyed earlier in this chapter: Focus on your readers' goals and values. In different cultures, people's goals and values often differ. The benefits you highlight for readers in one culture will not necessarily appeal to readers in another culture. Marketing specialists are well aware of these differences. To appeal to car buyers, Volvo focuses on safety and durability in the United States, status and leisure in France, and performance in Germany (Ricks, 2006).

Cultural differences in goals and values influence persuasion in the workplace as well as in consumer marketing. For example, in Japan, where group membership is highly valued, the most effective way to influence the attitudes of individual employees is to talk about the success the proposed action will bring to the group. In the United States, where individual achievement is highly valued, appeals to personal goals can be effective in ways they would not be in Japan.

Similarly, different cultures have different views of what counts as evidence or what constitutes credibility. International communication experts Robert Moran, Neil Abramson, and Sarah Moran (2014) report that in some Arab cultures people support their positions through appeals to emotions and through personal relations, whereas in North America and much of Europe people generally base their arguments on evidence and logic.

You must adapt to your readers' culture even when following advice given earlier in this chapter: Listen—and respond flexibly to what you hear. Imagine that you have sent an email to members of your target audience to find out how they feel about an idea you are planning to propose. If your readers are in the United States, they are likely to tell you directly. However, in many Asian cultures, they are not likely to respond in that way because they would not want you to lose face (Lustig & Koester, 2012). In this case, you may need to listen to their silence, which is an indirect way of indicating disagreement in many Asian cultures.

Ethics Guideline: Persuade Ethically

The ethical dimensions of on-the-job writing are never more evident than when you are trying to persuade other people to take a certain action or adopt a certain attitude. Here are four guidelines for ethical persuasion that you might want to keep in mind: Don't mislead, don't manipulate, open yourself to your readers' viewpoint, and argue from human values.

- **Don't mislead.** When you are writing persuasively, respect your readers' right to evaluate your arguments in an informed and independent way. If you mislead your readers by misstating or suppressing facts, using intentionally ambiguous expressions, or arguing from false premises, you deprive your readers of their rights.

- **Don't manipulate.** The philosopher Immanuel Kant originated the enduring ethical principle that we should never use other people merely to get what we want. Whenever we try to influence our readers, the action we advocate should advance their goals as well as our own.

  Under Kant's principle, for instance, it would be unethical to persuade readers to do something that would benefit us but harm them. High-pressure sales techniques are unethical because their purpose is to persuade consumers to purchase something they may not need or even want. Persuasion is ethical only if it will lead your readers to get something they truly desire.

- **Open yourself to your readers' viewpoint.** To keep your readers' goals and interests in mind, you must be open to understanding their viewpoint. Instead of regarding their counterarguments as objections to be overcome, try to understand what lies behind their concerns. Consider ways of modifying your original ideas to take your readers' perspective into account.

  Rather than treat your readers as adversaries, accept them as your partners in a search for a course of action acceptable to all. Management experts call this the search for a "win–win" situation—a situation in which all parties benefit (Covey, 2013).

- **Argue from human values.** Whenever you feel that human values are relevant, don't hesitate to introduce them when you are writing to persuade. Many organizations realize the need to consider these values when making decisions. In other organizations, human values are sometimes overlooked—or even considered to be inappropriate topics—because the employees focus too sharply on business objectives. However, even if your arguments based on human values do not prevail, you will have succeeded in introducing a consideration of these values into your working environment. Your action may even encourage others to follow your lead.

TRY THIS Have you ever felt manipulated by a salesperson or someone urging you to take a certain action? What was it about that individual's arguments that made you feel that you were being manipulated to do something you didn't want to do?

Conclusion

This chapter has focused on writing persuasively in both competitive and cooperative situations. As you can see, nearly every aspect of your communication affects your ability to influence your readers' attitudes and actions. Although the guidelines in this chapter will help you write persuasively, the most important persuasive strategy of all is to keep in mind your readers' needs, concerns, values, and preferences whenever you write.

Figure 9.9 provides a guide you can use when planning your persuasive strategies.

FIGURE 9.9

Writer's Guide for Planning Your Persuasive Strategies

Writer's Guide
PLANNING YOUR PERSUASIVE STRATEGIES

UNDERSTAND YOUR READERS' GOALS ←——— See page 169

1. What organizational goals affect your readers and are related to your topic?

2. What are your readers' values-based goals that are related to your topic?

3. What are your readers' achievement and growth goals?

REASON SOUNDLY ←——— See page 171

1. What kinds of evidence will your readers consider to be reliable and persuasive support for your claims?

2. Where will you need to justify your line of reasoning?

3. Where do you need to be cautious about avoiding false assumptions and overgeneralizations?

RESPOND TO YOUR READERS' CONCERNS AND COUNTERARGUMENTS ←——— See page 175

1. What concerns or counterarguments might your readers raise?

2. How can you respond persuasively to these concerns and counterarguments?

3. Where do you need to be careful about inspiring negative thoughts?

BUILD A RELATIONSHIP WITH YOUR READERS ←——— See page 176

1. What strategies can you use to present yourself as a credible person?

2. What strategies can you use to present yourself as a partner, not a critic?

ORGANIZE TO CREATE A FAVORABLE RESPONSE ←——— See page 180

1. Would a direct or indirect organizational pattern be most effective?

2. What specific strategies can you use to create a tight fit among the parts?

APPEAL TO YOUR READERS' EMOTIONS ←——— See page 183

1. Will your readers think that an appeal to their emotions is appropriate?

2. If so, what kind of emotional appeal would be effective in this situation with your readers?

ADAPT TO YOUR READERS' CULTURAL BACKGROUND ←——— See page 184

1. Do you need to check your draft with people who share your readers' cultural background?

PERSUADE ETHICALLY ←——— See page 184

1. Avoid misleading or manipulative statements.

2. Consider your readers' viewpoint openly.

3. Assure that the relevant human values are considered.

USE WHAT YOU'VE LEARNED

EXERCISE YOUR EXPERTISE

1. Find a persuasive communication that contains at least 25 words of prose. It may be an advertisement, marketing letter, memo from school or work, or a web page. What persuasive strategies are used in the prose and any images that are included? Are the strategies effective for the intended audience? Are all of them ethical? What other strategies might have been used? Present your responses as requested by your instructor.

2. Do the "Unsolicited Recommendation" project given in MindTap.

EXPLORE ONLINE

Study the website of a company or organization related to your major or career. What attitudes toward the organization and its products or services does the site promote? What persuasive strategies are used in the text and images in order to persuade readers to adopt the desired attitudes? How effective are these strategies? Are all of them ethical?

COLLABORATE WITH YOUR CLASSMATES

Working with one or two other students, analyze the letter shown in Figure 9.10, identifying its strengths and weaknesses. Relate your points to the guidelines in this chapter. The following paragraphs describe the situation in which the writer, Scott Houck, is writing.

Before going to college, Scott worked for a few years at Thompson Textiles. In his letter, he addresses Thompson's Executive Vice President, Georgiana Stroh. He is writing because in college he learned many things that made him think Thompson would benefit if its managers were better educated in modern management techniques. Thompson Textiles could enjoy these benefits, Scott believes, if it offered management courses to its employees and filled job openings at the managerial level with college graduates. However, if Thompson were to follow Scott's recommendations, it would have to change its practices considerably. Thompson has never offered courses for its employees and has long sought to keep payroll expenses low by employing people without a college education, even in management positions. (In a rare exception to this practice, the company has guaranteed Scott a position after he graduates.)

To attempt to change the company's policies, Scott decided to write a letter to one of the most influential people on its staff, Ms. Stroh. Unfortunately for Scott, throughout the three decades that Stroh has served as an executive officer at Thompson, she has consistently opposed company-sponsored education and the hiring of college graduates. Consequently, she has an especially strong motive for rejecting Scott's advice: She is likely to feel that, if she agreed that Thompson's educational and hiring policies should be changed, she would be admitting that she had been wrong all along.

APPLY YOUR ETHICS

Find an online or print communication you feel is unethical. Identify the specific elements in the text or images that you feel are untrue, misleading, or manipulative. Present your analysis as your instructor requests.

REFLECT FOR TRANSFER

1. Write a memo to yourself, one you will read one year after you have been on the job. In it, describe the advice given in this chapter that you feel you are most likely to forget between now and then. Explain why this forgotten advice is important to you in your current position or will become important to you in a future position you imagine to be in your future. (Submit your memo to your instructor.)

2. Write a memo to a friend who is not in this class. Explain that you are studying ways to write persuasively in your future career. Describe a piece of advice you think will be especially helpful in your career. Explain why, and give an imaginary on-the-job example. (Submit your memo to your instructor.)

Case: In MindTap, the case entitled "Debating the Drug-Testing Program" is well suited for use with this chapter.

FIGURE 9.10

Letter for "Collaborate with
Your Classmates" Exercise

616 S. College #84
Oxford, Ohio 45056
April 17, 2017

Georgiana Stroh
Executive Vice President
Thompson Textiles Incorporated
1010 Note Ave.
Cincinnati, Ohio 45014

Dear Ms. Stroh:

As my junior year draws to a close, I am more and more eager to return to our company, where I can apply my new knowledge and skills. Since our recent talk about the increasingly stiff competition in the textile industry, I have thought quite a bit about what I can do to help Thompson continue to prosper. I have been going over some notes I have made on the subject, and I am struck by how many of the ideas stemmed directly from the courses I have taken here at Miami University.

Almost all of the notes featured suggestions or thoughts I simply didn't have the knowledge to consider before I went to college! Before I enrolled, I, like many people, presumed that operating a business required only a certain measure of common sense ability—that almost anyone could learn to guide a business down the right path with a little experience. However, I have come to realize that this belief is far from the truth. It is true that many decisions are common sense, but decisions often only appear to be simple because the entire scope of the problem or the full ramifications of a particular alternative are not well understood. A path is always chosen, but is it always the BEST path for the company as a whole?

In retrospect, I appreciate the year I spent supervising the Eaton Avenue Plant because the experience has been an impetus to actually learn from my classes instead of just receiving grades. But I look back in embarrassment upon some of the decisions I made and the methods I used then. I now see that my previous work in our factories and my military experience did not prepare me as well for that position as I thought they did. My mistakes were not so often a poor selection among known alternatives, but were more often sins of omission. For example, you may remember that we were constantly running low on packing cartons and that we sometimes ran completely out, causing the entire line to shut down. Now I know that instead of haphazardly placing orders for a different amount every time, we should have used a forecasting model to determine and establish a reorder point and a reorder quanity. But I was simply unaware of many of the sophisticated techniques available to me as a manager.

I respectfully submit that many of our supervisory personnel are in a similar situation. This is not to downplay the many contributions they have made to the company. Thompson can directly attribute its prominent position in the industry to the devotion and hard work of these people. But very few of them have more than a high school education or have read even a single text on management skills. We have always counted on our supervisors to pick up their management skills on the job without any additional training. Although I recognize that I owe my own opportunities to this approach, this comes too close to the common sense theory I mentioned earlier.

The success of Thompson depends on the abilities of our managers relative to the abilities of our competition. In the past, EVERY company used this common sense approach, and Thompson prospered because of the natural talent of people like you. But in the last decade, many new managerial techniques have been developed that are too complex for the average employee to just "figure out" on his or her own. For example, people had been doing business for several thousand years before developing the Linear Programming Model for transportation and resource allocation problem-solving. It is not reasonable to expect a high school graduate to recognize that his or her particular distribution problem could be solved by a mathematical model and then to develop the LP from scratch. But as our world grows more complex, competition will stiffen as others take advantage of these innovations. I fear that what has worked in the past will not necessarily work in the future: We may find out what our managers DON'T know CAN hurt us. Our managers must be made aware of advances in computer technology, management theory, and operations innovations, and they must be able to use them to transform our business as changing market conditions demand.

I would like to suggest that you consider the value of investing in an in-house training program dealing with relevant topics to augment the practical experience our employees are gaining. In addition, when management or other fast-track administrative positions must be filled, it may be worth the investment to hire college graduates whose coursework has prepared them to use state-of-the-art techniques to help us remain competitive. Of course, these programs will initially show up on the bottom line as increased expenses, but it is reasonable to expect that, in the not-so-long run, profits will be boosted by newfound efficiencies. Most important, we must recognize the danger of adopting a wait-and-see attitude. Our competitors are now making this same decision; hesitation on our part may leave us playing catch-up.

In conclusion, I believe I will be a valuable asset to the company, in large part because of the education I am now receiving. I hope you agree that a higher education level in our employees is a cause worthy of our most sincere efforts. I will contact your office next week to find out if you are interested in meeting to discuss questions you may have or to review possible implementation strategies.

Sincerely,

Scott Houck

10

Developing an Effective, Professional Style

LEARNING OBJECTIVES

1. Create an effective, professional voice that builds your reader's confidence in you.
2. Construct sentences your reader will find easy to understand, easy to remember, and interesting.
3. Choose words that convey your meaning clearly and precisely to your specific reader.

Your instructor's feedback will be especially valuable when you are developing your professional writing style.

MindTap®

Find additional resources related to this chapter in MindTap.

Questions for determining what your readers expect

"You need to use a more professional style, not write like a college student." This is the first response many college students and new graduates hear in their internships or first professional jobs. This advice, always meant to be helpful, can be mystifying. Unless you have been immersed in the workplace, it's not obvious what a professional style is. Furthermore a surprisingly large number of factors shape what counts as a professional writing style in any particular workplace, though not more than you can handle if you know what they are.

Overall, these factors fall into three major groups: the personality or "voice" you project in your writing, the way you construct your sentences, and the words you choose. This chapter provides practical advice about each one.

A special challenge we all face when developing our writing style is that readers don't always interpret our writing in the way we think they will. We might think that we've chosen just the right word; our readers think it is too vague or stuffy. We might think we are sounding formal and smart; they might think we are being long-winded and tedious. For this reason, as you are developing your own professional writing style, pay special attention to your instructor's feedback. He or she can help you learn how your writing is likely to sound to workplace readers.

Create an Effective, Professional Voice

What, exactly, is voice? While reading something you've written, your readers "hear" your voice. Based on what they hear, they draw conclusions about your knowledge, expertise, character, and attitudes. The conclusions they draw can greatly enhance—or detract from—the effectiveness of your communications. The following guidelines suggest six strategies that will help you create an effective, professional voice. Each is associated with choices that affect your readers' impressions of you. These choices and guidelines are summarized in Figure 10.1.

GUIDELINE 1 Find out what's expected

To a large extent, an effective voice is one that matches your readers' sense of what's appropriate. When successful employees are asked to identify the major weaknesses in the writing of new employees, they often cite the inability to use a voice and style that are appropriate to their readers.

Here are three questions that can help you match your voice to your readers' expectations.

- **How formal do your readers think your writing should be?** An informal style sounds like conversation. You use contractions (*can't, won't*),

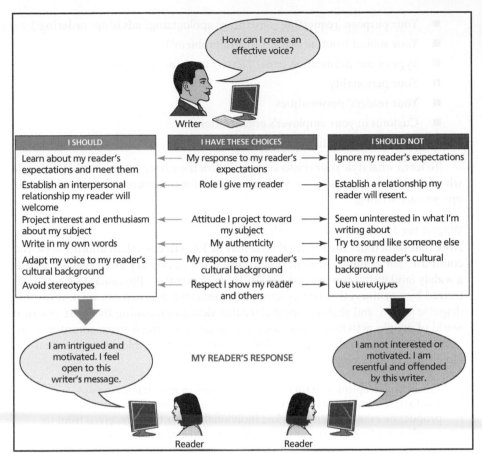

FIGURE 10.1

How to Create an Effective Voice

short words, and colloquial words and phrasing. A formal style sounds more like a lecture or speech, with longer sentences, formal phrasing, and no contractions.

- **How subjective or objective do your readers believe your writing should be?** In a subjective style, you would introduce yourself into your writing by saying such things as "I believe . . ." and "I observed . . . ." In an objective style, you would mask your presence by stating your beliefs as facts ("It is true that . . .") and by reporting about your own actions in the third person ("The researcher observed . . .") or the passive voice ("It was observed that . . .").

- **How much "distance" do your readers expect you to establish between them and you?** In a personal style, you appear very close to your readers because you do such things as use personal pronouns (*I, we*) and address your readers directly. In an impersonal style, you distance yourself from your readers—for instance by avoiding personal pronouns and by talking about yourself and your readers in the third person ("The company agrees to deliver a fully operable model to the customer by October 1").

Here are some major factors that may influence your readers' expectations about style.

- Your professional relationship with your readers (customers? supervisors? subordinates?)

- Your purpose (requesting something? apologizing? advising? ordering?)
- Your subject (routine matter? urgent problem?)
- Type of communication (email? letter? text? formal report?)
- Your personality
- Your readers' personalities
- Customs in your employer's organization
- Customs in your field, profession, or discipline

To learn what style your readers expect, follow the advice in Chapter 3: Ask people who know (including even your readers) and look for communications similar to the one you are writing.

What If the Expected Style Is Ineffective?

Note that sometimes the expected style may be less effective than another style you could use. For example, in some organizations the customary and expected style is a widely (and justly) condemned style called *bureaucratese*. Bureaucratese is characterized by wordiness that buries significant ideas and information, weak verbs that disguise action, and abstract vocabulary that detaches meaning from the practical world of people, activities, and objects. Often, such writing features an inflated vocabulary and a general pomposity that slows or completely blocks comprehension. Here's an example.

Bureaucratese | According to optimal quality-control practices in manufacturing any product, it is important that every component part that is constituent of the product be examined and checked individually after being received from its supplier or other source but before the final, finished product is assembled. (45 words)

The writer simply means this:

Plain English | Effective quality control requires that every component be checked individually before the final product is assembled. (16 words)

Here is another pair of examples.

Bureaucratese | Over the most recent monthly period, there has been a large increase in the number of complaints that customers have made about service that has been slow. (27 words)

Plain English | Last month, many more customers complained about slow service. (9 words)

Bureaucratese is such a serious barrier to understanding that many states in the United States have passed laws *requiring* "Plain English" in government publications and other documents such as insurance policies. This chapter's guidelines will help you avoid bureaucratese. However, some managers and organizations want employees to use that puffed-up style, thinking it sounds impressive. If you are asked to write in bureaucratese, try to explain why a straightforward style is more effective, perhaps sharing this book. If you fail to persuade, be prudent. Use the style that is required. Even within the confines of a generally bureaucratic style, you can probably make improvements. For instance, if your employer expects a wordy, abstract style, you may still be able to use a less inflated vocabulary.

TRY THIS Plain English guidelines exist for almost every profession from architecture to statistics. Find some that apply to your major online. If there aren't any for your field, try a closely related one. For example, if you don't find Plain English for "zoology," try "biology."

GUIDELINE 2 Consider the roles your voice creates for your readers and you

When you choose the voice with which you will address your readers, you define a role for yourself. As manager of a department, for instance, you could adopt the voice of a stern taskmaster or an open-minded leader. The voice you choose also implies a role for your readers. And their response to the role given to them can significantly influence your communication's overall effectiveness. If you choose the voice of a leader who respects your readers, they will probably accept their implied role as valued colleagues. If you choose the voice of a superior, unerring authority, they may resent their implied role as error-prone inferiors—and resist the substance of your message.

By changing your voice in even a single sentence, you can increase your ability to elicit the attitudes and actions you want to inspire. Consider the following statement drafted by a divisional vice president.

> I have scheduled an hour for you to meet with me to discuss your department's failure to meet its production targets last month.

In this draft, the vice president uses a domineering voice.

In this sentence, the vice president has chosen the voice of a powerful person who considers the reader to be someone who can be blamed and bossed around, a role the reader probably does not find agreeable. By revising the sentence, the vice president creates a much different pair of roles for herself and her readers.

> Let's meet tomorrow to see if we can figure out why your department had difficulty meeting last month's production targets.

In this revision, she creates a supportive voice.

The vice president transformed her voice into that of a supportive person. The reader became someone interested in working with the writer to solve a problem that stumps them both. As a result of these changes in voice and roles, the meeting is likely to be much more productive.

GUIDELINE 3 Consider how your attitude toward your subject will affect your readers

In addition to communicating attitudes about yourself and your readers, your voice communicates an attitude toward your subject. Feelings are contagious. If you write about your subject enthusiastically, your readers may catch your enthusiasm. If you seem indifferent, they may adopt the same attitude.

Social media and email present special temptations to be careless about voice because they encourage spontaneity. When you feel a flash of irritation, you may fire off an angry reply you'll soon regret. The danger is increased by the ease with which social media and email can be forwarded to readers you didn't intend to see them. Never include anything you wouldn't be prepared for a large audience to read. Check carefully for statements that your readers might interpret as conveying a different attitude, such as anger or sarcasm, than you intend.

GUIDELINE 4 Say things in your own words

This guideline urges you to avoid a mistake made by many new employees: They assume that a professional voice has no trace of the writer, that it is the same for every person. Trying to create prose that sounds *unlike* them, they end up sounding inauthentic, phony—as if they were trying to persuade their readers that they are someone they aren't.

TRY THIS What style do administrators at your college use? Find a letter or notice from one of the offices at your school. Is the style closer to Plain English or bureaucratese? What adjective would you use to describe the writer's voice? Do different offices use different voices? If so, why?

Too often, they also imagine that in order to sound professional, they need to puff up their prose with big words and long sentences, believing that this style will make them seem sophisticated and impressive. However, writers who write this way end up writing bureaucratese.

Instead, write clear, straightforward sentences containing words you would normally use.

Don't misunderstand this advice. It does *not* mean that you should always use an informal, colloquial style. We all have more than one style. At school, you probably speak differently talking with a professor than when chatting with your friends. Yet, in each case you are able to choose words and express your ideas in ways that feel genuine to you. Similarly, at work choose from among your styles the one that is appropriate to each situation.

To check whether you are using your own voice, try reading your drafts aloud. Where the phrasing seems awkward or the words are difficult for you to speak, you may have adopted someone else's voice—or slipped into bureaucratese, which reflects no one's voice. Reading your drafts aloud can also help you spot other problems with voice—such as sarcasm or condescension.

Despite the advice given in this guideline, it will sometimes be appropriate for you to suppress your own voice. For example, when a report, proposal, or other document is written by several people, the contributors usually strive to achieve a uniform voice so that all the sections fit together stylistically. Similarly, certain kinds of official documents, such as an organization's policy statements, are usually written in the employer's style, not the individual writer's style. Except in such situations, however, let your own voice speak in your writing.

> Whether you are writing in an informal or formal style, choose words and express your ideas in ways that feel genuine to you.

> Sometimes it's appropriate to suppress your own voice.

GUIDELINE 5 Global Guideline: Adapt your voice to your readers' cultural background

From one culture to another, general expectations about voice vary considerably. Understanding the differences between the expectations of your culture and those of your readers in another culture can be especially important because, as Guideline 2 explains, the voice you use tells your readers about the relationship you believe you have with them.

Consider, for instance, the difficulties that may arise if employees in the United States and in Japan write to one another without considering the expectations about voice that are most common in each culture. In the United States and Europe, employees often use an informal voice and address their readers by their first names. In Japan, writers commonly use a formal style and address their readers by their titles and last names. If a U.S. writer used a familiar, informal voice in a communication to Japanese readers, these readers might feel that the writer has not properly respected them. On the other hand, Japanese writers may seem distant and difficult to relate to if they use the formality that is common in their own culture when writing to U.S. readers.

Directness is another aspect of voice. The Japanese write in a more personal voice than do people from the United States, whose direct, blunt style the Japanese find abrupt (De Mente & Botting, 2015). When writing to people in other cultures, try to learn and use the voice that is customary there. Library and Internet research provide helpful information about many cultures. You can also learn about the voice used in your readers' culture by studying communications they have written. If possible, ask for advice from people who are from your readers' culture or who are knowledgeable about it.

> **LEARN MORE** For more advice about adapting communications to your readers' cultural background, go to "Describe your reader's cultural characteristics" (page 58).

GUIDELINE 6 **Ethics Guideline: Avoid stereotypes**

Let's begin with a story. A man and a boy are riding together in a car. As they approach a railroad crossing, the boy shouts, "Father, watch out!" But it is too late. The car is hit by a train. The man dies, and the boy is rushed to a hospital. When the boy is wheeled into the operating room, the surgeon looks down at the child and says, "I can't operate on him. He's my son."

When asked to explain why the boy would call the deceased driver "Father" and why the living surgeon would say, "He's my son," people offer many guesses. Perhaps the driver is a priest or the boy's stepfather or someone who kidnapped the boy as a baby. Few guess that the surgeon must be the boy's mother. Why? Our culture's stereotypes about the roles men and women play are so strong that when people think of a surgeon, many automatically imagine a man.

Stereotypes, Voice, and Ethics

What do stereotypes have to do with voice and ethics? Stereotypes are very deeply embedded in a culture. Most of us use them occasionally, especially when conversing informally. As a result, when we use more colloquial and conversational language to develop our distinctive voice for our workplace writing, we may inadvertently employ stereotypes. Unfortunately, even inadvertent uses of stereotypes have serious consequences for individuals and groups. People who are viewed in terms of stereotypes lose their ability to be treated as individual human beings.

Furthermore, if they belong to a group that is unfavorably stereotyped, they may find it nearly impossible to get others to take their talents, ideas, and feelings seriously. The range of groups disadvantaged by stereotyping is quite extensive. People are stereotyped on the basis of their race, religion, age, gender, sexual orientation, weight, physical handicap, and ethnicity, among other characteristics. In some workplaces, manual laborers, union members, clerical workers, and others are the victims of stereotyping by people in white-collar positions.

The following suggestions will help you avoid stereotypes.

Avoiding Stereotypes

- **Avoid describing people in terms of stereotypes.** In your reports, sales presentations, policy statements, and other communications, avoid giving examples that rely upon or reinforce stereotypes. For example, don't make all the decision makers men and all the clerical workers women.
- **Mention a person's gender, race, or other characteristic only when it is relevant.** To determine whether it's relevant to describe someone as a member of a minority group, ask yourself if you would make a parallel statement about a member of the majority group. If you wouldn't say, "This improvement was suggested by Jane, a person without any physical disability," don't say, "This improvement was suggested by Margaret, a person with a handicap." If you wouldn't say, "The Phoenix office is managed by Brent, a hardworking white person," don't say, "The Phoenix office is managed by Terry, a hardworking Mexican-American."
- **Avoid humor that relies on stereotypes.** Humor that relies on a stereotype reinforces the stereotype. Refrain from such humor not only when members of the stereotyped group are present but at all times.

LEARN MORE For a discussion of stereotypes and word choice, see "Use inclusive language" (page 207).

How to Construct Sentences Your Reader Will Find Easy to Understand, Easy to Remember, and Interesting

Without realizing it, you make many choices with each sentence you write. The choices you make, the way you build your sentences, can have a significant impact on how easy your sentences will be for your reader to understand and remember, as well as how well your communications will hold your reader's interest. The following six guidelines, which are summarized in Figure 10.2, suggest ways to make your six most important choices.

GUIDELINE 1 Simplify your sentences

The easiest way to increase usefulness is to simplify your sentences. Reading is *work*. Psychologists say that much of the work is done by short-term memory. It must figure out how the words in each sentence fit together to create a specific meaning. Fewer words mean less work. In addition, research shows that when you express your message concisely, you make it more forceful, memorable, and persuasive (Smith, 2004).

FIGURE 10.2

How to Write Effective Sentences

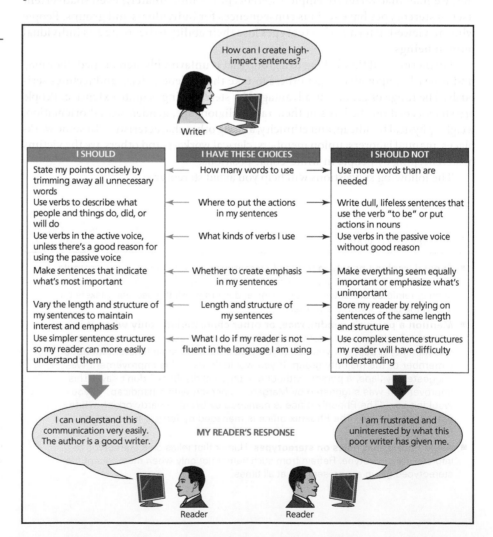

Simplifying Sentences

1. **Eliminate unnecessary words.** Look for places where you can convey your meaning more directly. Consider this sentence:

   | The <u>physical size</u> of the workroom is too small <u>to accommodate</u> this equipment. | Wordy

   With unnecessary words removed in two places, the sentence is just as clear and more emphatic:

   | The workroom is too small for this equipment. | Unnecessary words deleted

2. **Avoid wordy phrases.** Unnecessary words can also be found in many common phrases. "Due to the fact that" can be shortened to "Because." Similarly, "They do not pay attention to our complaints" can be abbreviated to "They ignore our complaints." "At this point in time" is "Now."

3. **Place modifiers next to the words they modify.** Short-term memory relies on word order to indicate meaning. If you don't keep related words together, your sentence may say something different from what you mean.

   | Mandy found many undeposited checks in the file cabinets, which were worth over $41,000.

   Technically, this sentence says that the file cabinets were worth over $41,000. Of course, readers would probably figure out that the writer meant the checks were worth that amount. But readers arrive at the correct meaning only after performing work they would have been saved if the writer had kept related words together by putting *which were worth over $41,000* after *checks* rather than *file cabinets*.

4. **Combine short sentences.** Often, combining two or more short sentences makes reading easier by reducing the total number of words and helping readers see the relationships among the points presented.

   | Water quality in Hawk River declined in March. This decline occurred because of the heavy rainfall that month. All the extra water overloaded Tomlin County's water treatment plant. | Separate

   | Water quality in Hawk River declined in March because heavy rainfalls overloaded Tomlin County's water treatment plant. | Combined

GUIDELINE 2 Put the action in verbs

Most sentences are about action. Sales rise, equipment fails, engineers design, managers approve. Clients praise or complain, and technicians advise. Yet, many people bury the action in nouns, adjectives, and other parts of speech. Consider the following sentence.

| Our department accomplished the conversion to the new machinery in two months. | Original

It could be energized by putting the action (*converting*) into the verb.

| Our department converted to the new machinery in two months. | Revised

Not only is the revised version briefer, but it is also more emphatic and lively. Furthermore, according to researchers E. B. Coleman (1964), when you put action in your verbs, you can make your prose up to 25 percent easier to read.

To create sentences that focus on action, do the following.

Focusing Sentences on Action

■ **Avoid sentences that use the verb _to be_ or its variations (_is, was, will be,_ and so on).** The verb _to be_ often tells what something is, not what it does.

Original	The sterilization procedure <u>is a protection</u> against reinfection.
Revised	The sterilization procedure <u>protects</u> against reinfection.

■ **Avoid sentences that begin with _It is_ or _There are_.**

Original	<u>It is</u> because the cost of raw materials has soared that the price of finished goods is rising.
Revised	Because the cost of raw materials has soared, the price of finished goods is rising.
Original	<u>There are</u> several factors causing the engineers to question the dam's strength.
Revised	Several factors cause the engineers to question the dam's strength.

■ **Avoid sentences where the action is frozen in a word that ends with one of the following suffixes: -tion, -ment, -ing, -ion, -ance.** These words petrify the action that should be in verbs by converting them into nouns.

Original	Consequently, I would like to make a <u>recommendation</u> that the department hire two additional programmers.
Revised	Consequently, I <u>recommend</u> that the department hire two additional programmers.

Although most sentences are about action, some aren't. For example, topic and forecasting statements often introduce lists or describe the organization of the discussion that follows.

Topic sentence for which the verb _to be_ is appropriate	There are three main reasons the company should invest money to improve communication between corporate headquarters and the out-of-state plants.

GUIDELINE 3 **Use the active voice unless you have a good reason to use the passive voice**

Another way to focus your sentences on action and actors is to use the _active voice_ rather than the _passive voice_. To write in the active voice, place the actor—the person or thing performing the action—in the subject position. Your verb will then describe the actor's action.

Active voice

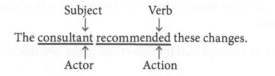

In the passive voice, the subject of the sentence and the actor are different. The subject is _acted upon_ by the actor.

Passive voice

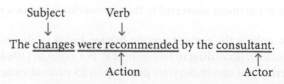

Here are some additional examples.

The Korean ore was purchased by us.	Passive voice
We purchased the Korean ore.	Active voice

Research shows that readers comprehend active sentences more rapidly than passive ones (Young, 2010). Also, the active voice eliminates the vagueness and ambiguity that often characterize the passive voice. In the passive voice, a sentence can describe an action without telling who did it. For example, "The ball was hit" is a grammatically correct sentence but doesn't tell who or what hit the ball. With the active voice, the writer identifies the actor: "Linda hit the ball."

The following sentence illustrates the importance of ensuring that readers understand who the actor is.

The operating temperatures must be checked daily to protect the motor from damage.	Passive voice

Will the supervisor of the third shift know that he is the person responsible for checking temperatures? In the passive voice, this sentence certainly allows him to imagine that someone else, perhaps a supervisor on another shift, is responsible.

Although the passive voice generally reduces readability, it has some good uses. One occurs when you don't want to identify the actor. The following sentence is from a memorandum in which the writer urges all employees to work harder at saving energy but avoids causing embarrassment and resentment by naming the guilty parties.

There are some places where the passive voice is appropriate.

The lights on the third floor have been left on all night for the past week, despite the efforts of most employees to help us reduce our energy bills.	Passive voice

Another good reason for using the passive voice is discussed in Chapter 7, ("Smooth the flow of thought from sentence to sentence," page 126).

GUIDELINE 4 Emphasize what's most important

Another way to write clear, forceful sentences is to direct your readers' attention to the most important information you are conveying.

Emphasizing What's Most Important

1. **Place the key information at the end of the sentence.** As linguists Joseph Williams and Joseph Bizup (2013) point out, you can demonstrate to yourself that the end of the sentence is a place of emphasis by listening to yourself speak. Read the following sentences aloud:

Her powers of concentration are extraordinary.
Last month, he topped his sales quota even though he was sick for an entire week.

   As you read these sentences aloud, notice how you naturally stress the final words *extraordinary* and *entire week*.

   To position the key information at the end of a sentence, you may sometimes need to rearrange your first draft.

The department's performance has been superb in all areas.	Original
In all areas, the department's performance has been superb.	Revised

(Continued)

2. **Place the key information in the main clause.** If your sentence has more than one clause, use the main clause for the information you want to emphasize. Compare the following versions of the same statement.

Original

> Although our productivity was down, our profits were up.

Revised

> Although our profits were up, our productivity was down.

In the first version, the emphasis is on profits because *profits* is the subject of the main clause. The second version emphasizes productivity because *productivity* is the subject of the main clause. (Notice that in each of these sentences, the emphasized information is not only in the main clause but also at the end of the sentence.)

3. **Emphasize key information typographically.** Use boldface and italics. Be careful, however, to use typographical highlighting sparingly. When many things are emphasized, none stand out.

4. **Tell readers explicitly what the key information is.** You can also emphasize key information by announcing its importance to your readers.

> Economists pointed to three causes of the stock market's decline: uncertainty about the outcome of last month's election, a rise in inventories of durable goods, and— *most important*—signs of rising inflation.

GUIDELINE 5 Vary your sentence length and structure

If all the sentences in a sentence group have the same structure, two problems arise: Monotony sets in, and (because all the sentences are basically alike) you lose the ability to emphasize major points and deemphasize minor ones.

You can avoid such monotony and loss of emphasis in two ways.

- **Vary sentence length.** Longer sentences can be used to show the relationships among ideas. Shorter sentences provide emphasis in the context of longer sentences.

This short sentence receives emphasis because it comes after longer ones.

The final sentence is also emphasized because it is much shorter than the preceding one.

> In April, many amateur investors believed that another rally was about to begin. Because exports were increasing rapidly, they predicted that the dollar would strengthen in global monetary markets, bringing foreign investors back to Wall Street. Also, unemployment dropped sharply, which they interpreted as an encouraging sign for the economy. They were wrong on both counts. Wall Street interpreted rising exports to mean that goods would cost more at home, and it predicted that falling unemployment would mean a shortage of workers, hence higher prices for labor. Where amateur investors saw growth, Wall Street saw inflation.

- **Vary sentence structure.** For example, the grammatical subject of the sentence does not have to be the sentence's first word. In fact, if it did, the English language would lose much of its power to emphasize more important information and de-emphasize less important information.

One alternative to beginning a sentence with its grammatical subject is to begin with a clause that indicates a logical relationship.

Introductory clause

> After we complete our archaeological survey, we will know for sure whether the proposed site for our new factory was once a Native American camping ground.

Introductory clause

> Because we have thoroughly investigated all the alternatives, we feel confident that a pneumatic drive will work best and provide the most reliable service.

| GUIDELINE 6 | **Global Guideline: Adapt your sentences for readers who are not fluent in your language** |

The decisions you make about the structure of your sentences can affect the ease with which people who are not fluent in English understand your message. Companies in several industries, including oil and computers, have developed simplified versions of English for use in communications for readers in other cultures. In addition to limited vocabularies, simplified English has special grammar rules that guide writers in using sentences that will be easy for their readers to understand. Because many readers may not need this degree of simplification, learn as much as possible about your specific readers. Also, remember that simplifying your sentence structure should not involve simplifying your thought.

LEARN MORE For more advice on adapting communications to your readers' cultural background, go to "Describe your reader's cultural characteristics" (page 58) and see the other Global Guidelines throughout this book.

Guidelines for Creating Sentences for Readers Who Are Not Fluent in English

- **Use simple sentence structures.** The more complex your sentences, the more difficult they will be for readers to understand.
- **Keep sentences short.** A long sentence can be hard to follow, even if its structure is simple. Set twenty words as a limit.
- **Use the active voice.** Readers who are not fluent in English can understand the active voice much more easily than they can understand the passive voice.

Choose Words that Convey Your Meaning Clearly and Precisely

When selecting words, your first goal should be to enable the reader to grasp your meaning and obtain the information he or she wants quickly and accurately. Your word choices also affect your reader's attitudes toward you and your subject matter. The following six guidelines are summarized in Figure 10.3. The most important point to remember is that words that are understandable and meaningful to one reader may not be to another. Choose words in the same way you make all of your other decisions about writing: with your specific reader in mind.

| GUIDELINE 1 | **Use concrete, specific words** |

Almost anything can be described either in relatively abstract, general words or in relatively concrete, specific ones. You may say that you are writing on a piece of *electronic equipment* or that you are writing on *a laptop computer connected to a 3D printer*. You may say that your employer produces *consumer goods* or that it makes *cell phones*.

When groups of words are ranked according to degree of abstraction, they form *hierarchies*. Figure 10.4 shows such a hierarchy in which the most specific terms identify concrete items that we can perceive with our senses; Figure 10.5 shows a hierarchy in which all the terms are abstract, but some are more specific than others.

Concrete words help your readers understand precisely what you mean. If you say that your company produces television shows for a *younger demographic segment,* they won't know whether you mean *teenagers* or *toddlers.* If you say that you study *natural phenomena,* your readers won't know whether you mean *volcanic eruptions* or the *migration of monarch butterflies.*

TRY THIS Pick a group of words used in your major, a hobby, or a sport. List words in ranked order, moving from more abstract to more concrete.

FIGURE 10.3

How to Decide Which
Words to Use

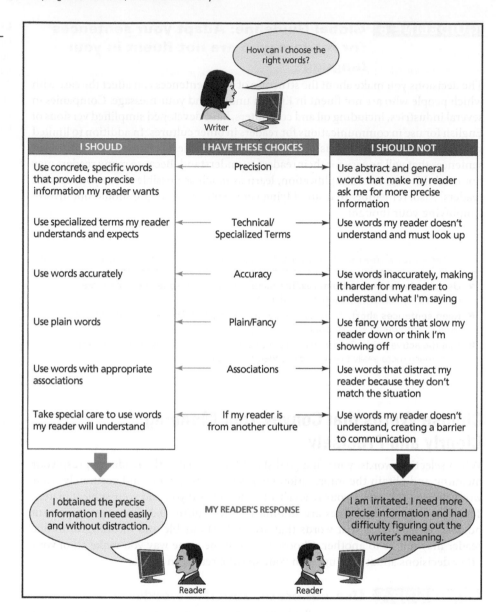

Such vagueness can hinder readers from getting the information they need in order to make decisions and take action. Consider the following sentence from a memo addressed to an upper-level manager who wanted to know why production costs were up.

Original | The cost of one material has risen recently.

This sentence doesn't give the manager the information she needs to take remedial action. In contrast, the following sentence, using specific words, tells precisely what the material is, how much the price has risen, and the period in which the increase took place.

Revised | The cost of the bonding agent has tripled in the past six months.

Of course, abstract and general terms do have important uses. For example, in scientific, technical, and other specialized fields, writers often need to make general points, describe the general features of a situation, or provide general guidance for

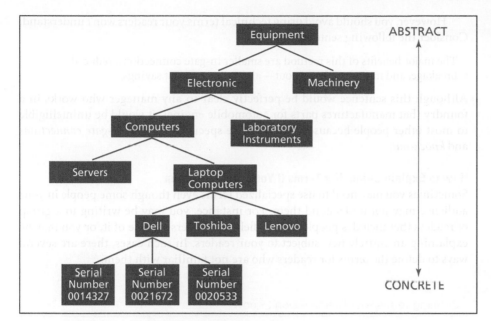

FIGURE 10.4

Hierarchy of Related Words that Move from Abstract to Concrete

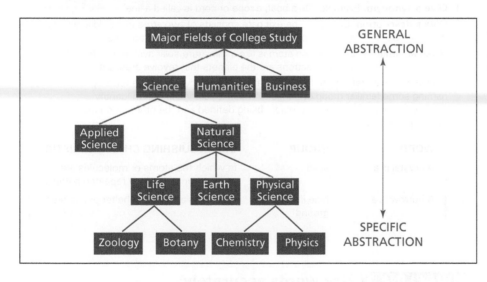

FIGURE 10.5

Hierarchy of Related Words that Move from a General to a Specific Abstraction

action. Your objective when choosing words is not to avoid abstract, general words altogether, but rather to avoid using them when your readers will want more concrete and specific ones.

GUIDELINE 2 Use specialized terms when—and only when—your readers will understand them

You can increase the usefulness and persuasiveness of your writing by using wisely the specialized terms of your own profession.

In some situations, specialized terms help you communicate effectively.

- **They convey precise, technical meanings economically.** Many terms have no exact equivalent in everyday speech.
- **They help you establish credibility.** By using the special terms of your field accurately, you show your fellow specialists that you are adept in it.

However, you should avoid using technical terms your readers won't understand. Consider the following sentence.

> The major benefits of this method are smaller in-gate connections, reduced breakage, and minimum knock-out—all leading to great savings.

Although this sentence would be perfectly clear to any manager who works in a foundry that manufactures parts for automobile engines, it would be unintelligible to most other people because it includes the specialized terms *in-gate connections* and *knock-out*.

How to Explain Unfamiliar Terms If You Must Use Them

Sometimes you may need to use specialized terms even though some people in your audience may not understand them. For instance, you may be writing to a group of readers that includes people in your field and others outside of it, or you may be explaining an entirely new subject to your readers. In such cases, there are several ways to define the terms for readers who are not familiar with them.

Defining Terms Your Readers Don't Know

1. **Give a synonym.** Example: On a boat, a rope or cord is called a line.
2. **Give a description.** Example: The exit gate consists of two arms that hold a jug while it is being painted and then allow it to proceed down the production line.
3. **Make an analogy.** Example: An atom is like a miniature solar system in which the nucleus is the sun and the electrons are the planets that revolve around it.
4. **Give a classical definition.** In a classical definition, you define the term by naming some familiar group of things to which it belongs and then identify the key distinction between the object being defined and the other members of the group. Examples:

WORD	GROUP	DISTINGUISHING CHARACTERISTIC
A crystal is a	solid	in which the atoms or molecules are arranged in a regularly repeated pattern.
A burrow is a	hole in the ground	dug by an animal for shelter or habitation.

GUIDELINE 3 Use words accurately

Whether you use specialized terms or everyday ones and whether you use abstract, general terms or concrete, specific ones, you must use all your words accurately. This point may seem obvious, but inaccurate word choice is all too common in on-the-job writing. For example, people often confuse *imply* (meaning to *suggest* or *hint,* as in "He implied that the operator had been careless") with *infer* (meaning to *draw a conclusion based upon evidence,* as in "We infer from your report that you do not expect to meet the deadline"). It's critical that you avoid such errors. They distract your readers from your message by drawing their attention to your problems with word choice, and they may lead your readers to believe that you are not skillful or precise in other areas—such as laboratory techniques or analytical skills.

How can you ensure that you use words accurately? There's no easy way. Consult a dictionary whenever you are uncertain. Be especially careful when using words that are not yet part of your usual vocabulary. Pay attention as well to the way words are used by other people.

GUIDELINE 4 Choose plain words over fancy ones

You can also make your writing easy to understand by avoiding fancy words where plain ones will do. At work, some writers do just the opposite, perhaps thinking that fancy words sound more official or make them sound more knowledgeable. The following list identifies some commonly used fancy words; it includes only verbs but might have included nouns and adjectives as well.

FANCY VERBS	EQUIVALENT COMMON VERBS
ascertain	find out
commence	begin
compensate	pay
constitute	make up
endeavor	try
expend	spend
fabricate	build
facilitate	make easier
initiate	begin
prioritize	rank
proceed	go
terminate	end
transmit	send
utilize	use

There are two important reasons for preferring plain words over fancy ones.

- **Plain words promote efficient reading.** Research has shown that even if your readers know both the plain word and its fancy synonym, they will still comprehend the plain word more rapidly (Smith, 2004).

- **Plain words reduce your risk of creating a bad impression.** If you use words that make for slow, inefficient reading, you may annoy your readers or cause them to conclude that you are behaving pompously, showing off, or trying to hide a lack of ideas and information behind a fog of fancy terms. Consider, for instance, the effect of the following sentence in a job application letter.

I am transmitting the enclosed résumé to facilitate your efforts to determine the pertinence of my work experience to your opening.

Pompous word choices

Don't misunderstand this guideline. It doesn't suggest that you should use only simple language at work. When addressing people with vocabularies comparable to your own, use all the words at your command, provided that you use them accurately and appropriately. This guideline merely cautions you against using needlessly inflated words that bloat your prose and open you to criticism from your readers.

GUIDELINE 5 Choose words with appropriate associations

The three previous guidelines for choosing words relate to the literal or dictionary meaning of words. At work, you must also consider the associations your words have for your readers. In particular, be especially sensitive to your words' *connotation* and *register*.

Connotation

Connotation is the extended or suggested meaning that a word has beyond its literal meaning. For example, according to the dictionary, *flatfoot* and *police detective* are synonyms, but they connote very different things: *flatfoot* suggests a plodding, perhaps not very bright cop, while *police detective* suggests a trained professional.

Verbs, too, have connotations. For instance, to *suggest* that someone has overlooked a key fact is not the same as to *insinuate* that she has. To *devote* your time to working on a client's project is not the same as to *spend* your time on it.

Research on the impact of connotation

The connotations of your words can shape your audience's perceptions of your subject matter. Researchers Raymond W. Kulhavy and Neil H. Schwartz (1981) demonstrated those effects in a classic experiment for which they created two descriptions of a company that differed in only seven out of 246 words. In one, the seven words suggested stiffness, such as *required* and *must*. In the other, those seven words were replaced by ones that suggested flexibility, such as *asked* and *should*. None of the substitutions changed the facts of the overall passage.

The researchers found that people who read the flexible version believed that the company would actively commit itself to the welfare and concerns of its employees, voluntarily participate in affirmative action programs for women and minorities, receive relatively few labor grievances, and pay its employees well. People who read that version also said they would recommend the company to a friend as a place to work. People who read the stiff version reported opposite impressions of the company. That readers' impressions of the company could be affected so dramatically by just seven nonsubstantive words highlights the great importance of paying attention to the connotations of the words you use.

Register

Linguists use the term *register* to identify a second characteristic exhibited by words: their association with certain kinds of communication situations or context. For example, in an ad for a restaurant we might expect to see the claim that it offers *amazingly* delicious food. However, we would not expect to see a research company boast in a proposal for a government contract that it is capable of conducting *amazingly* good studies. The word *amazingly* is in the register of consumer advertising but not in the register of research proposals.

If you inadvertently choose words with the wrong register, your readers may infer that you don't fully grasp how business is conducted in your field, and your credibility can be lost.

GUIDELINE 6 Global Guideline: Consider your readers' cultural background when choosing words

Take special care in your choice of words when writing to readers in other cultures. Some words whose meaning is obvious in your own culture can be misunderstood or completely mystifying to readers from other cultures. Misunderstanding can even occur when you are writing to readers in other cultures where the native language is English. In the United States, people play football with an oblong object that they try to carry over a goal line or kick through uprights. In England, India, and many other parts of the world, football is played with a round object that people are forbidden to touch with their hands and attempt to kick into a net—a game called soccer in the United States.

Yuri Arcurs/Shutterstock.com

At work, even small departments often include a rich diversity of employees.

The following guidelines will help you choose words your readers will understand in the way you intend. Of course, different readers in other cultures have different levels of facility with English, so follow the guidelines only to the extent that your readers require.

Guidelines for Choosing Words for Intercultural Communications

- **Use simple words.** The more complex your vocabulary, the more difficult it will be for readers not fluent in English to understand you.
- **Use the same word each time you refer to the same thing.** For instance, in instructions, don't use both *dial* and *control* for the same part of a test instrument. Those two terms may be synonyms in your language, but they will each be translated into a different word in the other language, where the translated words may not be synonyms.
- **Avoid acronyms your readers won't understand.** Most acronyms that are familiar to you will be based on words in *your* language: AI for Artificial Intelligence; ACL for Anterior Cruciate Ligament.
- **Avoid slang words and idioms.** Most will have no meaning for people in other cultures. Instead of "We want a level playing field," say, "We want the decision to be made fairly." Instead of saying "We want to run an idea past you," say, "We'd like your opinion of our idea."

LEARN MORE For more advice on adapting communications to your readers' cultural background, go to "Describe your reader's cultural characteristics" (page 58) and see the other Global Guidelines throughout this book.

GUIDELINE 7 Ethics Guideline: Use inclusive language

When constructing your voice, use language that includes all persons instead of excluding some. For example, avoid sexist language because it supports negative stereotypes. Usually these stereotypes are about women, but they can also adversely affect men in certain professions such as nursing. By supporting negative stereotypes, sexist language can blind readers to the abilities, accomplishments, and potential of very capable people. The same is true of language that insensitively describes people with disabilities, illnesses, or other limitations.

LEARN MORE For another discussion of stereotypes and ethics, see "Avoid stereotypes," page 195.

Using Inclusive Language

1. **Use nouns and pronouns that are gender-neutral rather than ones containing the word *man*.**

    Instead of: businessman, workman, mailman, salesman

    Use: businessperson, manager, or executive; worker; mail carrier; salesperson

    Instead of: man-made, man hours, man-sized job

    Use: synthetic, working hours, large job

2. **Use plural pronouns or *he* or *she* instead of sex-linked pronouns when referring to people in general.**

    Instead of: "Our home electronics cater to the affluent shopper. She looks for premium products and appreciates a stylish design."

    Use the plural: "Our home electronics cater to affluent shoppers. They look for premium products and appreciate a stylish design."

    Instead of: "Before the owner of a new business files the first year's tax returns, he might be wise to seek advice from a certified public accountant."

    Use *he or she*: "Before the owner of a new business files the first year's tax returns, he or she might be wise to seek advice from a certified public accountant."

(Continued)

3. **Refer to individual men and women in a parallel manner.**

Instead of: "Mr. Sundquist and Anna represented us at the trade fair."

Use: "Mr. Sundquist and Ms. Tokagawa represented us at the trade fair"
or "Christopher and Anna represented us at the trade fair."

4. **Revise salutations that imply the reader of a letter is a man.**

Instead of: Dear Sir, Gentlemen

Use: The title of the department or company or the job title of the person
you are addressing: Dear Personnel Department, Dear Switzer Plastics
Corporation, Dear Director of Research

5. **When writing about people with disabilities, refer to the person first, then
the disability.**

Instead of: the disabled

Use: people with disabilities

Conclusion

Your writing style can make a great deal of difference to the success of your communications. The voice you use, sentence structures you employ, and words you
select affect your readers' attitudes toward you and your subject matter as well as the
usefulness and persuasiveness of your writing. This chapter has suggested many things
you can do to develop a highly usable, highly persuasive style. Underlying all these
suggestions is the advice that you take the reader-centered approach of considering
all your stylistic choices from your readers' point of view.

USE WHAT YOU'VE LEARNED

EXERCISE YOUR EXPERTISE

1. Imagine that you are the head of the Public Safety
Department at your college. Faculty and staff have been
parking illegally, sometimes where there aren't parking spots. Sometimes individuals without handicaps
are parking in spots reserved for those with handicaps. Write two memos to all college employees announcing that beginning next week, the Public Safety
Department will strictly enforce parking rules—something it hasn't been doing. Write the first memo in a
friendly voice and the second in a stern voice. Then
compare the specific differences in organization, sentence structure, word choice, and other features of writing to create each voice. (Thanks to Don Cunningham,
Auburn University, for the idea for this exercise.)

2. Without altering the meaning of the following sentences, reduce the number of words in them.

   a. After having completed work on the data-entry
   problem, we turned our thinking toward our next
   task, which was the processing problem.

   b. Those who plan federal and state programs for
   the elderly should take into account the changing
   demographic characteristics in terms of size and
   average income of the composition of the elderly
   population.

   c. Would you please figure out what we should do
   and advise us?

   d. The result of this study will be to make total
   white-water recycling an economical strategy for
   meeting federal regulations.

3. Rewrite the following sentences in a way that will keep
the related words together.

   a. This stamping machine, if you fail to clean it twice
   per shift and add oil of the proper weight, will cease
   to operate efficiently.

   b. The plant manager said that he hopes all employees
   would seek ways to cut waste at the supervisory
   meeting yesterday.

   c. About 80 percent of our clients, which include over
   1,500 companies throughout North and South

America and a few from Africa, where we've built alliances with local distributors, find the help provided at our website to be equivalent in most cases to the assistance supplied by telephone calls to our service centers.

d. Once they wilt, most garden sprays are unable to save vegetable plants from complete collapse.

4. Rewrite the following sentences to put the action in the verb.

a. The experience itself will be an inspirational factor leading the participants to a greater dedication to productivity.

b. The system realizes important savings in time for the clerical staff.

c. The implementation of the work plan will be the responsibility of a team of three engineers experienced in these procedures.

d. Both pulp and lumber were in strong demand, even though rising interest rates caused the drying up of funds for housing.

5. Rewrite the following sentences in the active voice.

a. Periodically, the shipping log should be reconciled with the daily billings by the Accounting Department.

b. Fast, accurate data from each operating area in the foundry should be given to us by the new computerized system.

c. Since his own accident, safety regulations have been enforced much more conscientiously by the shop foreman.

d. No one has been designated by the manager to make emergency decisions when she is gone.

6. Create a one-sentence, classical definition for a word used in your field that is not familiar to people in other fields. The word might be one that people in other fields have heard of but cannot define precisely in the way specialists in your field do. Underline the word you are defining. Then circle and label the part of your definition that describes the familiar group of items that the defined word belongs to. Finally, circle and label the part of your definition that identifies the key distinction between the defined word and the other items in the group. (Note that not every word is best defined by means of a classical definition, so it may take you a few minutes to think of an appropriate word for this exercise.)

7. Create an analogy to explain a word used in your field that is unfamiliar to most readers. (Note that not every word is best defined by means of an analogy, so it may take you a few minutes to think of an appropriate word for this exercise.)

EXPLORE ONLINE

Using your desktop publishing program, examine the readability statistics for two communications. These might be two projects you're preparing for courses, or they might be a course project and a letter or email to a friend or family member. What differences, if any, do you notice in the statistics? What accounts for the differences? Are the statistics helpful to you in understanding and constructing an effective writing style in either case? Read the explanations of the scores that are provided with your desktop publishing program; these may be provided in the program's Help feature. Do the interpretations of your scores agree with your own assessment of your communications? If not, which do you think is more valid? (To learn how to obtain the readability statistics with your program, use its Help feature.)

COLLABORATE WITH YOUR CLASSMATES

Working with another student, examine the memo shown in Figure 10.6. Identify places where the writer has ignored the guidelines given in this chapter. You may find it helpful to use a dictionary. Then write an improved version of the memo by following the guidelines in this chapter.

APPLY YOUR ETHICS

1. Find a communication that fails to use inclusive language and revise several of the passages to make them inclusive.

2. The images in advertising often rely on stereotypes. Find one advertisement that perpetuates one or more stereotypes and one that calls attention to itself by using an image that defies a stereotype. Evaluate the ethical impact of each image. Present your results in the way your instructor requests.

REFLECT FOR TRANSFER

1. Write a memo to your classmates explaining a time when—either in writing or when speaking—you or someone else confused his or her audience and lost credibility by not following one or more of the guidelines in this chapter.

2. Write a memo to your classmates identifying two guidelines in this chapter you believe will be the most difficult for you to apply on the job—and why. For each, describe a strategy you might use to overcome this difficulty.

FIGURE 10.6

Memo for Collaboration
Exercise

MEMO

July 5, 2017

TO   Gavin MacIntyre, Vice President, Midwest Region

FROM   Nat Willard, Branch Manager, Milwaukee Area Offices

The ensuing memo is in reference to provisions for the cleaning of the six offices and two workrooms in the High Street building in Milwaukee. This morning, I absolved Thomas's Janitor Company of its responsibility for cleansing the subject premises when I discovered that two of Thomas's employees had surreptitiously been making unauthorized long-distance calls on our telephones.

Because of your concern with the costs of running the Milwaukee area offices, I want your imprimatur before proceeding further in making a determination about procuring cleaning services for this building. One possibility is to assign the janitor from the Greenwood Boulevard building to clean the High Street building also. However, this alternative is judged impractical because it cannot be implemented without circumventing the reality of time constraints. While the Greenwood janitor could perform routine cleaning operations at the High Street establishment in one hour, it would take him another ninety minutes to drive to and fro between the two sites. This is more time than he could spare and still be able to fulfill his responsibilities at the High Street building.

Another alternative would be to hire a full-time or part-time employee precisely for the High Street building. However, that building can be cleaned so expeditiously, it would be irrational to do so.

The third alternative is to search for another janitorial service. I have now released two of these enterprises from our employ in Milwaukee. However, our experiences with such services should be viewed as bad luck and not affect our decision, except to make us more aware that making the optimal selection among companies will require great care. Furthermore, there seems to be no reasonable alternative to hiring another janitorial service.

Accordingly, I recommend that we hire another janitorial service. If you agree, I can commence searching for this service as soon as I receive a missive from you. In the meantime I have asked the employees who work in the High Street building to do some tidying up themselves and to be patient.

11

Writing Reader-Centered Front and Back Matter

In the workplace, many communications longer than a page are held together with a paper clip or staple. Usually because of their importance or length, others have many of the features found in the books you buy for your classes. Some of these features are called *front matter* because they precede the opening chapter or section. Front matter includes the following items.

- Cover
- Title Page
- Executive Summary or Abstract
- Table of Contents
- List of Figures and Tables

The other book-like features are called *back matter*, for an obvious reason.

- Appendixes
- Reference List, Endnotes, or Bibliography
- Glossary or List of Symbols
- Index

Communications with front and back matter are often accompanied by transmittal letters when they are delivered to their readers.

To assist you in creating these book-like elements effectively, this chapter begins with guidelines for planning front and back matter, followed by advice about writing each element.

How to Plan Front and Back Matter

To plan front and back matter, begin in the same way you plan any other feature of your communication: by thinking about how your readers will use your communication and how you want to influence their attitudes—as well as the expectations of your readers and employer.

GUIDELINE 1 Review the ways your readers will use the communication

Front and back matter can make especially important contributions to a communication's usefulness. The way you write your title and design your table of contents can increase the efficiency with which readers can locate the information they want. To help readers in this way, you need to know what your readers will be looking for and

LEARNING OBJECTIVES

1. Plan the front and back matter of a communication you are writing.
2. Write a reader-centered letter of transmittal.
3. Write reader-centered front matter.
4. Write reader-centered back matter.

MindTap®

Find additional resources related to this chapter in MindTap.

how they will use your communication. As you begin work on the front and back matter, review the information you developed while defining your communication's usefulness objectives (see Chapter 3).

GUIDELINE 2 **Review your communication's persuasive goals**

By thoroughly understanding your readers' goals, values, and attitudes, you can determine what should be included and emphasized in the executive summary to lead readers to make the decision or take the actions you advocate. Reviewing the information you developed while defining your communication's persuasive objectives, you can also enhance the effectiveness of your title and other elements of your front and back matter.

GUIDELINE 3 **Find out what's required**

Many organizations distribute instructions that tell employees in detail how to prepare front and back matter. These instructions, sometimes called *style guides*, can describe everything from the maximum number of words permitted in an executive summary to the size and color of the title on a report's cover. If you are writing to one of your employer's clients, the *client's* requirements may be the ones you need to follow. Whatever the source, obtain the relevant style guide and follow its requirements precisely.

GUIDELINE 4 **Find out what's expected**

Even if nothing is explicitly required, your employer and your readers probably have expectations about the way you draft front and back matter. Also, beyond what's required, there are often unspoken expectations. Look at communications similar to the one you are preparing to learn what expectations you should meet.

A Word about Conventions and Local Practice

As mentioned above, requirements and expectations about front and back matter vary from organization to organization. For instance, one employer may want the title of any report to be placed in a certain spot on the cover, and another employer may want the title in another spot. Despite these variations, formal reports written on the job look very much alike. The formats described in this chapter reflect common practices in the workplace. You will have little trouble adapting them to other, slightly different versions when the need arises.

How to Write a Reader-Centered Transmittal Letter

When you prepare formal reports and proposals, you will often send them (rather than hand them) to your readers. In such cases, you will want to accompany them with a letter (or memo) of transmittal, which may be placed on top of your communication or bound into it. Transmittal letters typically contain the following elements.

- **Introduction.** In the introduction to your letter, introduce the accompanying communication and explain or remind readers of its topic. Begin on an upbeat note. Here is an example written to a client.

> We are pleased to submit the enclosed report on evaluating three possible sites for your company's new manufacturing plant.

FIGURE 11.1

Memo of Transmittal Written at Work

This letter of transmittal accompanied the empirical research report shown in Figure 24.1 (page 411).

ELECTRONICS CORPORATION OF AMERICA
MEMORANDUM

To   Myron Bronski, Vice-President, Research
            MCB
From   Margaret C. Barnett, Satellite Products Laboratory

Date   September 29, 2017

Re   REPORT ON TRUCK-TO-SATELLITE TEST

On behalf of the entire research team. I am pleased to submit the attached copy of the operational test of our truck-to-satellite communication system.

The test shows that our system works fine. More than 91% of our data transmissions were successful, and more than 91% of our voice transmissions were of commercial quality. The test helped us identify some sources of bad transmissions, including primarily movement of a truck outside the "footprint" of the satellite's strongest broadcast and the presence of objects (such as trees) in the direct line between a truck and the satellite.

The research team believes that our next steps should be to develop a new antenna for use on the trucks and to develop a configuration of satellites that will place them at least 25° above the horizon for trucks anywhere in our coverage area.

We're ready to begin work on these tasks as soon as we get the okay to do so. Please let me know if you have any questions.

Encl: Report (2 copies)

Margaret opens with an introductory sentence that mentions the accompanying report and tells its topic.

In the body of her brief memo, Margaret summarizes the report's findings that will be of most interest to her reader. To do this, she includes very precise information, using exact percentages in the first sentence, for instance. Note that she has not included the broader range of information included in the report's executive summary (see Fig. 11.6, page 218).

Again tailoring her memo to her reader, an executive decision maker, Margaret describes in specific detail the next step recommended by the research team.

Margaret closes on an upbeat note and indicates that she would welcome questions.

■ **Body.** The body of a transmittal letter usually highlights the communication's major features. For example, researchers often state their main findings and the consequences of what they've learned, repeating information in their communication's abstract or summary.

■ **Closing.** Transmittal letters commonly end with a short paragraph (often one sentence) that states the writer's willingness to work further with the reader or promises to answer any questions the reader may have.

Figure 11.1 shows a memo of transmittal.

How to Write Reader-Centered Front Matter

The following sections will help you prepare each element of a communication's front matter in a reader-centered way.

Cover

A cover typically includes the communication's title, the name or logo of the organization that created it, and the date it was issued. In some organizations, covers also include the writer's name and a document number used for filing. To write a reader-centered title, use precise, specific terms that tell your current readers exactly what the communication is about and help your future readers determine whether the communication will be useful to them.

Figure 11.2 shows the cover of a research report.

FIGURE 11.2

Cover of a Formal Report

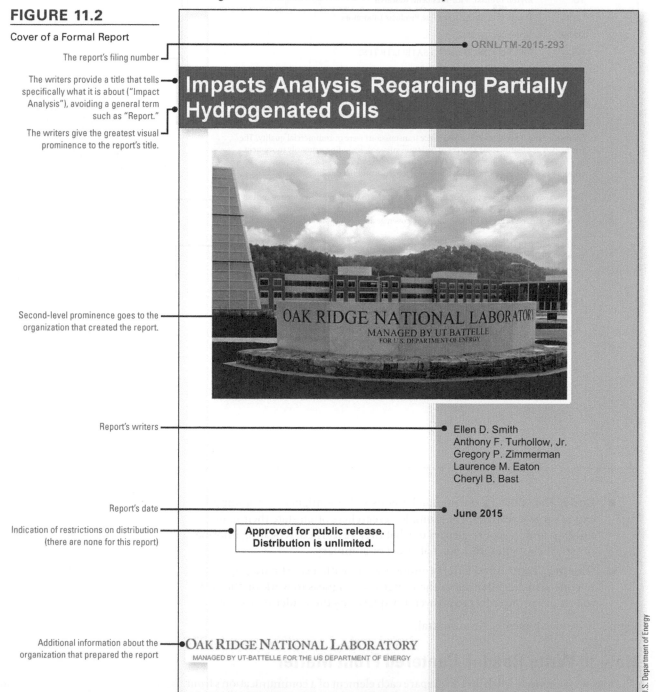

The report's filing number

The writers provide a title that tells specifically what it is about ("Impact Analysis"), avoiding a general term such as "Report."

The writers give the greatest visual prominence to the report's title.

ORNL/TM-2015-293

Impacts Analysis Regarding Partially Hydrogenated Oils

Second-level prominence goes to the organization that created the report.

OAK RIDGE NATIONAL LABORATORY
MANAGED BY UT BATTELLE
FOR U.S. DEPARTMENT OF ENERGY

Report's writers

Ellen D. Smith
Anthony F. Turhollow, Jr.
Gregory P. Zimmerman
Laurence M. Eaton
Cheryl B. Bast

Report's date

June 2015

Indication of restrictions on distribution (there are none for this report)

**Approved for public release.
Distribution is unlimited.**

Additional information about the organization that prepared the report

OAK RIDGE NATIONAL LABORATORY
MANAGED BY UT-BATTELLE FOR THE US DEPARTMENT OF ENERGY

U.S. Department of Energy

Title Page

A title page repeats the information on the cover and usually adds more. For instance, your employer may want you to include contract or project numbers and such cross-referencing information as the name of the contract under which the work was performed. A title page may also provide copyright and trademark notices. Note that some employers require such information on the cover instead.

Figure 11.3 shows the title page of the report whose cover is shown in Figure 11.2.

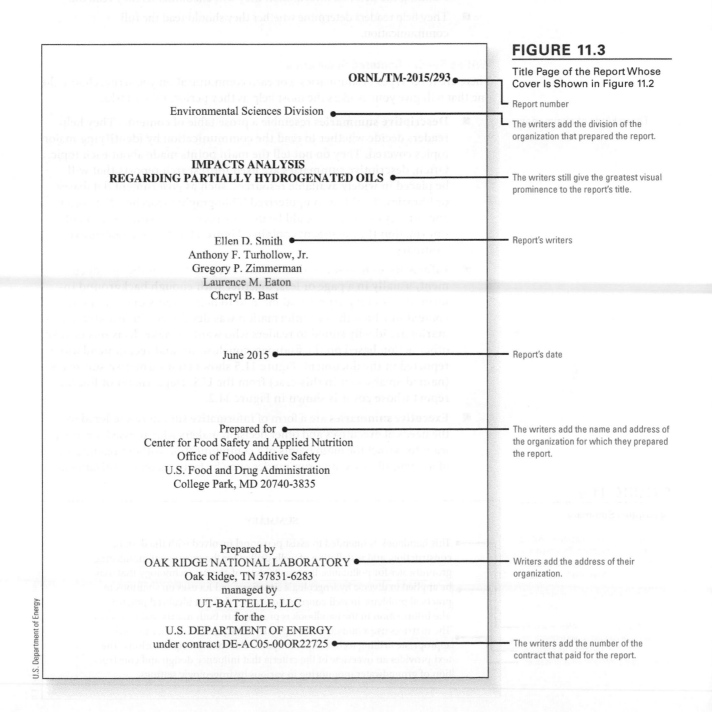

FIGURE 11.3

Title Page of the Report Whose Cover Is Shown in Figure 11.2

ORNL/TM-2015/293 — Report number

Environmental Sciences Division — The writers add the division of the organization that prepared the report.

IMPACTS ANALYSIS
REGARDING PARTIALLY HYDROGENATED OILS — The writers still give the greatest visual prominence to the report's title.

Ellen D. Smith — Report's writers
Anthony F. Turhollow, Jr.
Gregory P. Zimmerman
Laurence M. Eaton
Cheryl B. Bast

June 2015 — Report's date

Prepared for — The writers add the name and address of the organization for which they prepared the report.
Center for Food Safety and Applied Nutrition
Office of Food Additive Safety
U.S. Food and Drug Administration
College Park, MD 20740-3835

Prepared by
OAK RIDGE NATIONAL LABORATORY — Writers add the address of their organization.
Oak Ridge, TN 37831-6283
managed by
UT-BATTELLE, LLC
for the
U.S. DEPARTMENT OF ENERGY
under contract DE-AC05-00OR22725 — The writers add the number of the contract that paid for the report.

U.S. Department of Energy

Summary or Abstract

Most reports and proposals that have front matter include a summary, sometimes called an *abstract*. These summaries serve three purposes.

Purposes of summaries

- They help busy managers learn the main points without reading the entire document.
- They help all readers build a mental framework for organizing and understanding the detailed information they will encounter as they read on.
- They help readers determine whether they should read the full communication.

Writing Reader-Centered Summaries

There are three types of summaries. For each communication you write, choose the one that will give your readers the most help as they perform their tasks.

Three kinds of summaries

- **Descriptive summaries** resemble a prose table of contents. They help readers decide whether to read the communication by identifying major topics covered. They do not tell the main points made about each topic. Often, descriptive summaries are used for research reports that will be placed in widely available resources, such as government databases or libraries. To aid in computerized bibliographic searches, they often contain keywords that would be used by persons seeking the kind of information the document contains. Figure 11.4 shows a descriptive summary.

- **Informative summaries** distill the main points from the full document, usually in a page or less, together with enough background information about purpose and method to help readers understand the context in which the key information was developed. Informative summaries are ideally suited to readers who want to make decisions or take other action based on the findings, conclusions, and recommendations reported in the document. Figure 11.5 shows an informative summary (named an abstract in this case) from the U.S. Department of Energy report whose cover is shown in Figure 11.2.

- **Executive summaries** are a form of informative summaries tailored to the needs of executives and other decision makers who, pressed for time, want to extract the main points of a communication without reading all of it. Typically, executive summaries open by focusing on organizational

FIGURE 11.4

Descriptive Summary

In a descriptive summary, writers tell readers the topics discussed in a communication without telling what it says about each topic.

In this example, the writers do the following:

- Identify the intended audience and the way the handbook will assist them
- Describe its contents.
- Tell how the information is presented.

SUMMARY

This handbook is intended to assist personnel involved with the design, construction, and installation of wells drilled for the purpose of monitoring groundwater for pollutants. It presents state-of-the-art technology that may be applied in diverse hydrogeologic situations and focuses on solutions to practical problems in well construction rather than on idealized practice. The information in the handbook is presented in both matrix and text form. The matrices use a numerical rating scheme to guide the reader toward appropriate drilling technologies for particular monitoring situations. The text provides an overview of the criteria that influence design and construction of groundwater monitoring in various hydrogeologic settings.

U.S. Environmental Protection Agency

FIGURE 11.5

Abstract from the Report Whose Cover Is Shown in Figure 11.2

Impacts Analysis   *Regarding Partially Hydrogenated Oils*

ABSTRACT

The U.S. Food and Drug Administration (FDA) has the responsibility under the Federal Food, Drug, and Cosmetic Act (FD&C Act) [21 *United States Code* (U.S.C) 301 et seq.] for assuring that the U.S. food supply is safe, sanitary, wholesome, and honestly labeled. Toward that end, FDA exercises approval authority over substances permitted for use as food additives. Substances that are generally recognized as safe (GRAS) are not subject to regulation as food additives under the FD&C Act.

Partially hydrogenated oils (PHOs), such as partially hydrogenated soybean and cottonseed oils, have been used in food for many years based on self-determinations by industry that such use is GRAS. However, based on new scientific evidence establishing the health risks associated with the consumption of *trans* fatty acids (also called *trans* fat), FDA has determined that there is no longer a consensus among qualified scientific experts that PHOs—which are the primary dietary source of industrially-produced *trans* fat—are safe for human consumption either directly or as ingredients in other food products. FDA therefore is issuing a declaratory order to revoke the GRAS status of PHOs for use in food, thus making PHOs subject to regulation as food additives.

Under the FDA Order, food manufacturers would no longer be permitted to sell PHOs, either directly or as ingredients in another food product, without prior FDA approval. Therefore, the U.S. food industry would be reasonably expected to use oils and fats from other sources as replacements for PHOs in all U.S. food products. One potential replacement would be palm oil imported from sources outside the United States (most likely from Southeast Asia).

This impacts analysis has been prepared by the staff of the Oak Ridge National Laboratory (ORNL) to assess and document the potential effects of the FDA Order on the environment of the United States. The purpose of this report is to provide input to FDA regarding an FDA determination as to whether the actions reasonably expected to result from the FDA Order may have significant environmental effects that would require preparation of an environmental impact statement or environmental assessment in accordance with FDA regulations for implementing the National Environmental Policy Act of 1969 (42 U.S.C 4321, et seq.).

The analysis considers the effects of the FDA Order revoking the GRAS status of PHOs on land use, water resources, air quality, waste management, transportation, and resources energy. An economic analysis of impacts to agriculture is included. Impacts to other environmental resources were judged to be either inconsequential or too local in nature to be anticipated in this analysis. No projected environmental impacts in the U.S. were judged to be significant.

U.S. Department of Energy

As background, the writers explain that their client, the Food and Drug Administration, does not regulate food additives that have GRAS ("generally recognized as safe") status.

The writers explain that there is no longer scientific consensus that partially hydrogenated oils are safe, so the FDA determined that it should regulate them.

The writers explain that the FDA realizes that the regulation will probably lead food manufacturers to begin using oils and fats from other sources.

The writers identify the purpose of the study that the FDA has paid them to investigate: Will the switch to other fats and oils have an impact on the environment of the United States. If so, an environmental impact study would be required.

In the first two sentences, the writers identify the possible effects they studied.

In the third, they identify possible effects they did not study.

In the final sentence, they give the "bottom line" of their report: No projected impacts were judged to be significant.

questions and issues and briefly describe the investigative or research methods used by the writer. Often half or even more of the summary is devoted to precisely summarizing major conclusions and recommended actions—the type of information most helpful to decision makers. Figure 11.6 shows an executive summary.

Summaries that appear in printed and online bibliographic resources are usually called *abstracts*. In many scientific and engineering fields, the term *abstract* is also used for the summary that appears at the front of long reports and proposals.

FIGURE 11.6

Executive Summary

In this executive summary, the writers include the key information from each of the report's sections, condensing a 28-page report into fewer than 250 words.

Because an executive summary is addressed to decision makers, the writers present the information on which their readers would decide what action to take. (Compare with the abstract in Figure 11.5.)

In their first sentence, the writers state the **main point** they make in the body of their report: The airport should purchase a new system.

Because their readers would not be familiar with the details of the Accounting Department's computers, the writers provide the **background information** these readers need in order to understand the rest of the executive summary.

The writers pinpoint the **problem**.

They briefly describe the three **possible solutions** they investigated. Their language echoes the statements they made in the preceding list of the problem's sources.

The writers conclude their summary with their **recommendation and the reason for it**.

EXECUTIVE SUMMARY

The Accounting Department recommends that Columbus International Airport purchase a new operating system for its InfoMaxx computer. The airport purchased the InfoMaxx in 2013 to replace an obsolete and failing Hutchins computer system. However, the new InfoMaxx computer has never successfully performed one of its key tasks: generating weekly accounting reports based on the expense and revenue data fed to it. When airport personnel attempt to run the computer program that should generate the reports, the computer issues a message stating that it does not have enough internal memory for the job.

Our department's analysis of this problem revealed that the InfoMaxx would have enough internal memory if the software used that space efficiently. Problems with the software are as follows:

1. The operating system, BT/Q-91, uses the computer's internal memory wastefully.
2. The SuperReport program, which is used to generate the accounting reports, is much too cumbersome to create reports this complex with the memory space available on the InfoMaxx computer.

Consequently, we evaluated three possible solutions:

1. Buying a new operating system (BT/Q-101) at a cost of $35,000. It would double the amount of usable space and also speed calculations.
2. Writing a more compact program in LINUX, at a cost of $50,000 in labor.
3. Revising SuperReport to prepare the overall report in small chunks, at a cost of $40,000. SuperReport now successfully runs small

We recommend the first alternative, buying a new operating system, because it will solve the problem for the least cost. The minor advantages of writing a new program in LINUX or of revising SuperReport are not sufficient to justify their cost.

Whether you are writing a descriptive, informative, or executive summary, follow these reader-centered guidelines.

Writing Reader-Centered Summaries

■ **Make it 100 percent redundant with the communication.** This purposeful redundancy provides a complete and understandable message to the reader who reads nothing else in the communication. It also means that the summary can't serve as the introduction, even though the introduction that follows it may seem somewhat repetitive.

- **Mirror the structure of the overall communication.** Include information from each major part of the communication, presented in the order of the parts in the overall communication.
- **Meet the needs of your readers.** For instance, if you know that your readers will be primarily interested in a novel method you used, provide more detail about the method than you would for readers who are primarily interested in your results. Likewise, in deciding what to include from any part of the full document, pick the information your specific readers will find most useful.
- **Be specific.** Replace general terms with precise ones. Instead of writing that it was "hot," write that it was "150°." Rather than writing the less expensive alternative "saved money," write that it saved "$43,000 per month." The more specific your abstract, the more useful it will be to your readers.
- **Keep it short.** Summaries are typically only 2 to 5 percent of the length of the body of the communication (not counting attachments and appendixes). That's between half a page and a whole page for every twenty pages in the body of your communication.
- **Write concisely.** Abstracts need to be lean but highly informative. Keep them short by eliminating unnecessary words, not by leaving out information important to your readers.

LEARN MORE For more strategies for writing concisely, see "Simplify your sentences," in Chapter 10 (page 196).

Table of Contents

By providing a table of contents, you help readers who want to find a specific part of your communication without reading all of it. Your table of contents also assists readers who want an overview of the communication's scope and contents before they begin reading it in its entirety.

Tables of contents are constructed from chapter or section titles and headings. They are most useful to readers if they have two (perhaps three) levels. The highest level consists of the chapters or section titles. Often they are generic: Introduction, Method, Results, and so on. The second level consists of the major headings within each chapter or section. They guide readers more directly to the information they are seeking. Using features of Microsoft Word, you can create a table of contents automatically from your communication's headings.

Figure 11.7 shows the table of contents from the report whose cover is shown in Figure 11.2. Note that the words *Table of* do not appear.

Lists of Figures and Tables

Some readers search for specific figures and tables. If your communication is more than about fifteen pages long, you can assist readers by preparing a list of figures and tables. Figure 11.8 (page 221) shows the list of tables from the U.S. Department of Energy report whose cover is shown in Figure 11.2. That report also contains a list of figures with the same format.

How to Write Reader-Centered Back Matter

Depending on your readers and purpose, you may include one or more of the following kinds of back matter: appendixes; reference list, works cited, or bibliography; glossary; and index.

Appendixes

Appendixes enable you to create a modular organization in which you present information you want to make available to some of your readers even though you know it won't interest all of them. For instance, in a research report you might create an appendix for a two-page account of the calculations you used for data analysis.

FIGURE 11.7

Table of Contents of the Report Whose Cover Is Shown in Figure 11.2

Impacts Analysis *Regarding Partially Hydrogenated Oils*

CONTENTS

Front matter

ABSTRACT i
LIST OF TABLES v
LIST OF FIGURES vi
ACRONYMS AND ABBREVIATIONS vii

Body of the report. In addition to the chapter titles, the writers include the second- and third-level headings.

1. INTRODUCTION AND BACKGROUND 1
 1.1 WHAT IS FDA'S ACTION? 1
 1.2 WHY IS THIS IMPACTS ANALYSIS BEING PREPARED? 2
 1.3 WHAT RESOURCE ISSUES ARE CONSIDERED IN THIS DOCUMENT? 4
 1.4 WHAT RESOURCE ISSUES HAVE BEEN ELIMINATED FROM THIS DETAILED ANALYSIS? 4
2. ENVIRONMENTAL IMPACTS 7
 2.1 LAND USE 8
  2.1.1 Overview 8
  2.1.2 Potential Impacts of the Action 10
 2.2 WATER RESOURCES 12
  2.2.1 Overview 12
  2.2.2 Potential Impacts of the Action 14
 2.3 AIR QUALITY 16
  2.3.1 Overview 16
  2.3.2 Potential Impacts of the Action 16
 2.4 WASTE MANAGEMENT 19
  2.4.1 Overview 19
  2.4.2 Potential Impacts of the Action 20
3. REFERENCES 25
APPENDIX A AGRICULTURAL ANALYSIS A-1

Note that lowercase Roman numerals are used for the front matter's page numbers.

U.S. Department of Energy

By placing this account in an appendix, you help readers who will want to use the same calculations in another experiment, but you accommodate readers who aren't interested in wading through the calculations while reading the body of your report. Appendixes can also assist readers by providing important reference information in a readily accessed place. That is a function of Appendix A in the back of this text.

When you create appendixes, list them in your table of contents (see Figure 11.7) and give each an informative title that indicates clearly what it contains. Also, mention each appendix in the body of your report at the point where your readers might want to refer to it: "Printouts from the electrocardiogram appear in Appendix II."

Begin each appendix on its own page. Arrange and label the appendixes in the same order in which they are mentioned in the body of the communication. If you

U.S. Department of Energy

FIGURE 11.8

List of Tables (Partial) from the Report Whose Cover Is Shown in Figure 11.2

Impacts Analysis   *Regarding Partially Hydrogenated Oils*

LIST OF TABLES

The writers provided a list of tables to assist readers interested in finding and reviewing the data provided in them.

Page numbers of the tables in the appendix begin with the appendix letter.

The numbers of the tables in the appendix begin with the appendix letter.

The writers also included a list of figures (not shown).

have only one appendix, label it simply "Appendix." If you have more than one, you may use Roman numerals, Arabic numerals, or capital letters to distinguish them.

References List, Endnotes, or Bibliography

Place your reference list, endnotes, or bibliography immediately after the body of your communication or after the appendixes. Appendix A explains which sources to cite and how to construct reference lists, endnotes, and bibliographies.

Glossary and List of Symbols

When writing at work, you will sometimes use terms or symbols that are unfamiliar to some of your readers. If you use each term or symbol only once or in only one small segment of your communication, you can explain its meaning in the text. Imagine, however, that you are writing a report in which a special term appears on pages 3, 39, and 72. If you define the term every time it occurs, your report will be repetitious. On the other hand, if you define the term only the first time it appears, your readers may forget the definition by the time they encounter the term again. To solve this problem, create a glossary. It will provide the definition where readers can easily locate it if they need to but won't encumber their reading if they don't.

Figure 11.9 shows the first page of the glossary for the an U.S. Environmental Protection Agency (EPA) report. The author places the terms to be defined in bold-face and color so readers can find them easily. Figure 11.10 shows a list of symbols.

Index

If your communication is too long for your readers to thumb through quickly, give them a quick path to specific pieces of information by creating an index. Identify

LEARN MORE For additional information about documenting your sources, including links to other documentation systems, go to Appendix A.

FIGURE 11.9

First Page of the Glossary of a Report

In this glossary, the writers provide precise definitions for key terms in their report.

They use bold type and color to make the terms being defined stand out.

GLOSSARY OF TERMS

Availability — The probability that equipment will be capable of performing its specified function when called upon at any random point in time. Calculated as the ratio of run time to the sum of run time and downtime.

Bottle — A single plastic container, also referred to as a container.

Cascade — To sequentially increase the belt speed in a series of conveyors.

Contamination — Bottles improperly removed at a station.

Effective Bottle Feed Rate — The average rate at which a specific type of bottle is fed during a mixed composition test. Calculated as the total number of bottles of a specific type fed divided by the total run time for a specific test.

Extended Test — A test series consisting of many replicates at pre-selected conditions.

U.S. Environmental Protection Agency

the kinds of information they might want to locate without reading the rest of the communication. If several index topics can be gathered under a single word, indent them under the main word to create second-level entries. See the example index in Figure 11.11.

Some desktop publishing programs can help you create an index by generating an alphabetized list of the words used in your communication. From this list, you can index those that will help your readers find the information they desire.

FIGURE 11.10

List of Symbols

This list of symbols, from a study concerning computer modeling of music, is several pages long.

Knowing that readers will search for symbols they have seen in the text, the writer has placed the symbols on the left side—the side seen first by people who read from left to right.

To aid the readers' search, the writer has arranged the symbols in alphabetical order.

List of Symbols

a_k	allpass filter coefficients
$\mathbf{a}$	vector of allpass filter coefficients
A	cross-sectional area
$A(z)$	allpass transfer function
B	area ratio
c	speed of sound
$\mathbf{c}$	vector of transfer functions or of cosine functions
C	scaling coefficient
$C(z)$	transfer function of a subfilter in the Farrow structure
d	fractional part of the total delay D
D	total delay to be approximated
$D(z)$	denominator of $A(z)$

Helsinki University of Technology, Laboratory of Acoustics and Audio Signal Processing, Finland. Courtesy of Vesa Valimaki, Helsinki University of Technology

FIGURE 11.11

Index

This index is from the service manual for a diesel engine used in farm machinery.

The writers use second-level headings to organize related topics.

To help readers scan through the index, the writers use white space and capital letters to break the index into sections.

12

Creating Reader-Centered Graphics

MindTap®

Find additional resources related to this chapter in MindTap.

In communications you write at work, you will often convey some of your information and ideas in tables, graphs, photographs, flowcharts, drawings, and other visual representations. In fact, graphics are very common in communications written in engineering, science, and other technical and specialized fields.

Why? They are able to convey some information much more concisely, clearly, memorably, and persuasively than words. Sometimes, they even replace words entirely, as in the wordless instructions shown in Figure 12.1.

Because graphics are such a powerful means of communication, developing your ability to create effective ones is just as important as your ability to write clear, informative, and compelling prose. Each graphic you create will have its own specific objectives, which include being useful to your readers and influencing their attitudes, even if only to persuade them that you are a skilled communicator.

To create graphics that will set you apart at work, you may need to develop several abilities to a higher level than you probably already have. These include spotting places where graphics will be more effective than prose, making those graphics especially useful and persuasive to readers in your culture and others, integrating them with your prose, and using graphics ethically. This chapter and your instructor will help you enhance those abilities.

Identify Places Where Graphics Would Increase Your Communication's Effectiveness

Throughout your work on a communication, search actively for places where graphics can help you achieve your communication objectives. Contemporary society has become increasingly visual. More and more, people gain information through images rather than words. Paradoxically, most of us rely primarily on words when we want to convey information to others. Consequently, unless you devote time to looking for them, you may miss opportunities to increase your communication's effectiveness with graphics.

GUIDELINE 1 Find places where graphics would make your communication easier to use

Begin by making a reader-centered search for places where graphics can increase your communication's usefulness to your readers—and perhaps save space as well. The many possibilities range from creating a table to help your readers find and use data, facts, or advice to helping them understand how to perform an action that would be difficult and time-consuming to grasp if described in prose.

Courtesy of Aero Safety Graphics Inc.

FIGURE 12.1

Wordless Instructions for Leaving a Plane in an Emergency

Wordless instructions are often used when the readers do not share a common language, as is the case with airline passengers.

In these instructions, arrows show passengers how to remove the cover from the latch, pull the latch, remove the door, and throw the door out of the plane.

Think, for instance, of the wordless instructions for exiting an airplane in an emergency (Figure 12.1).

The first page of Chapter 13 (page 242) describes a variety of ways you can use graphics to help your readers and identifies the kind of graphics that are commonly used in the workplace to provide that assistance to readers. The rest of Chapter 13 is devoted to helping you create eleven types of reader-centered graphics.

GUIDELINE 2 Find places where graphics can increase your communication's persuasiveness

Graphics can greatly increase a communication's persuasive impact. Often a graphic enables you to support your persuasive points in an especially dramatic way. For example, a manufacturer of windows used the bar graph shown in Figure 12.2 to persuade homeowners that they could greatly reduce heating and cooling bills by installing double-pane replacement windows that cost more (and yield a greater profit) than single-pane windows.

Many other kinds of visual representations can enhance a communication's persuasiveness. A drawing can help readers envision the desirable outcomes of projects you propose. A photograph of a polluted stream can forcefully portray a problem you want to motivate your readers to address.

TRY THIS In addition to using arrows, what other design strategies help readers of the airline instructions understand what they should do? *Alternative:* Find a set of wordless instructions for something you own. What strategies did the designer use to make them easy to follow?

FIGURE 12.2

Graphic That Makes a Persuasive Point

With this bar graph, a company that installs replacement windows for older homes aims to persuade customers to purchase more expensive double-pane windows.

To emphasize the benefits of double-pane windows, the company used green to color the bars for the money saved by those windows. Green is associated with savings and environmental benefits. For the single-pane windows, the company used red, a color associated with financial loss.

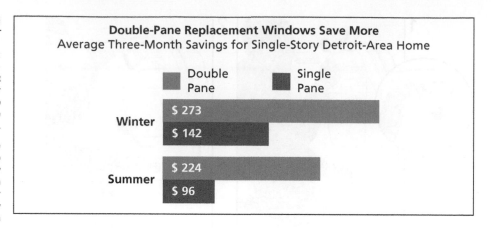

Choose the Type of Graphic Best Matched to Your Communication's Goals

Once you've decided where to use a graphic, pick the type that will achieve your communication's objectives most effectively. Numerical data can be presented in a table, bar graph, line graph, or pie chart. The components of an electronic instrument can be represented in a photograph, sketch, block diagram, or schematic. Most information can be presented in more than one type of graphic.

GUIDELINE 1 Consider your readers' tasks

Different types of graphics support different reading tasks. Consider, for example, Ben's choices.

Ben has surveyed people who graduated over the past three years from three departments in his college. Now he wants to report to the alumni what he has learned about their average starting salaries. He could do this with a table, bar graph, or line graph (see Figure 12.3). Which type of graphic would be best? To decide, Ben must determine the specific task the alumni will want to perform by using the graphic. If they want to learn the average starting salary of people who graduated in their year from their department, a table will help them most. If they want to compare the average starting salaries in their department with the average starting salaries of people who graduated in that same year from other departments, a bar graph will work best. And if the alumni's task will be to determine how the average starting salary in their department changed over the years and to compare that change with the changes experienced by other departments, a line graph will be the most useful.

GUIDELINE 2 Consider your readers' attitudes

When selecting the type of graphic you will use, think about your readers' attitudes as well as their tasks. Pick the type of graphic that most quickly and dramatically communicates the evidence that supports your persuasive point.

Consider, for instance, Akiko's choices. Akiko recommended a change in the design of one of her company's products. To show how the company has benefited from this change, she has tallied the number of phone calls received by the company's

Average Starting Salaries in Three Departments

Department	Year of Graduation		
	2014	2015	2016
A	30,653	32,898	49,519
B	42,289	44,904	52,698
C	46,172	61,538	66,357

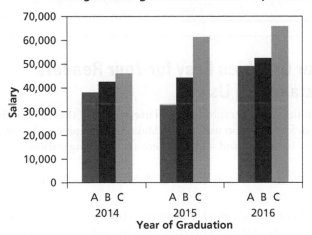

Average Starting Salaries in Three Departments

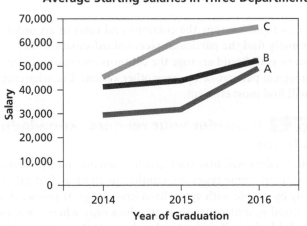

FIGURE 12.3

Three Ways of Showing Average Starting Salaries for Three Departments over Three Years

A table helps a reader who wants to quickly find a specific piece of information, such as the average starting salary for graduates of Department B in 2015.

A bar graph helps a reader who wants to make comparisons, for instance, a comparison among the average starting salaries of three departments in 2015.

A line graph helps a reader who wants to see trends in the average starting salaries for one department or compare trends for two or more departments.

toll-free help line during the months immediately before and after implementation of the new design. As Figure 12.4 shows, if Akiko presents her data in a table, her readers will have to do a lot of subtracting to appreciate the impact of her recommendation. If she presents her data in a line graph, they will be able to recognize her accomplishment at a glance.

FIGURE 12.4

Comparison of the Persuasiveness of Data Presented in a Table and a Line Graph

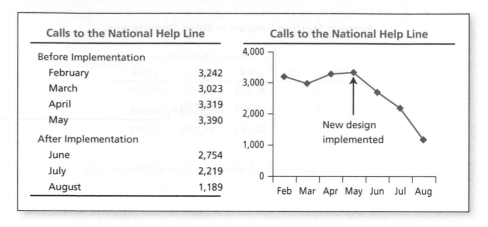

Make Your Graphics Easy for Your Readers to Understand and Use

Having chosen the type of graphic you will use, you must then design the item itself. When doing so, focus first on usefulness. Make your graphics, like your prose, easy for your readers to understand and use. Here are some suggestions.

GUIDELINE 1 Design your graphics to support your readers' tasks

First, follow this familiar, reader-centered strategy: Imagine your readers in the act of using your graphic. Then design it to support your readers' efforts. In drawings or photographs for step-by step instructions, for example, show objects from the same angle that your readers will see them when performing the actions you describe.

Likewise, in a table, arrange the columns and rows in an order that will help your readers rapidly find the particular pieces of information they are looking for. Maybe this means you should arrange the columns and rows in alphabetical order, according to a logical pattern, or by some other system. Use whatever arrangement your readers will find most efficient.

GUIDELINE 2 Consider your readers' knowledge and expectations

Of course, your readers will find your graphics useful and persuasive only if they can understand them. Some types are familiar to all of us, but other types can be interpreted only by people with specialized knowledge. If you work in a field that employs specialized graphics, use these graphics only when communicating with readers in your field who will understand and expect them. However, when writing to other readers, use an alternative type of graphic—or include explanations these readers need in order to interpret the special-use graphic.

GUIDELINE 3 Simplify your graphics

You can also make your graphics easy to understand and use by keeping them simple.

- **Include only a manageable amount of material.** Sometimes, it's better to separate your information into two or more graphics than to cram it all into one.

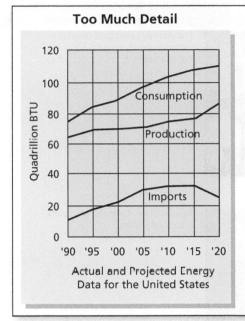
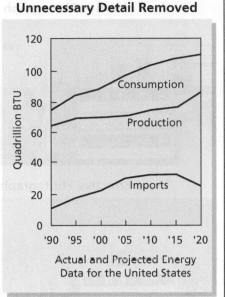

FIGURE 12.5

A Line Graph with and without Grid Lines

Replacing the grid lines with tick marks helps the readers more easily follow the graphed lines and read the labels.

- **Eliminate unnecessary details.** Like unnecessary words in prose, superfluous details in graphics create extra, unproductive work for readers and obscure the really important information. Figure 12.5 shows how the elimination of extraneous detail can simplify a graph. Figure 12.6 shows two ways to simplify a graphic showing complicated equipment by paring it down to the essentials.

GUIDELINE 4 Label the important content clearly

Labels help readers locate the information in a graphic and understand what it shows. In tables, label every row and column. In diagrams, label every part that is important to your readers. But avoid labeling other features. Unnecessary labels clutter a graphic, making it difficult to understand and use.

For each label, carefully choose the appropriate word or words and place them where they are easy to see. If necessary, draw a line from the label to the item. Avoid placing a label on top of an important part in your graphic. Figure 12.7 shows good placement of labels in a photograph. Note that labels placed in a graphic are much easier than a key for readers to use. See Figure 12.8.

GUIDELINE 5 Provide informative titles

Titles help readers to find the graphics they are looking for and also to know what the graphics contain once they locate them. Typically, titles include both a number (for example, "Figure 3" or "Table 6") and a description ("Effects of Temperature on the Strength of M312").

Make your titles as brief—and informative—as possible. Use more words if they are needed to give your readers precise information about your graphic. Don't write, "Information on Computer Programs," but write instead, "Comparison of the Speed and Capabilities of Three Database Programs." Be consistent in the

FIGURE 12.6

Two Photographs and
a Drawing of the Same
Equipment

Less Effective Photograph

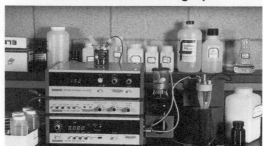

The cluttered background makes it
difficult to identify the parts of the
testing equipment.

More Effective Photograph

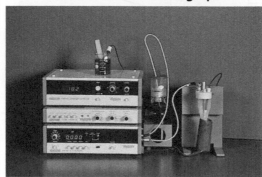

Removal of the clutter allows the parts
to be distinguished.

Drawing

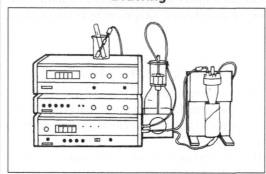

A line drawing shows very clearly some
parts that are not obvious even in the
uncluttered photograph.

placement of your titles. For example, place all titles above your figures or all
below your figures.

Use Color to Support Your Message

In recent years, there's been an explosion of color in workplace communications.
With color, you can clarify your messages, speed your readers' comprehension, and
make your information easy for your readers to use.

The impact of color on readers is determined partly by physiology and partly by
psychology. Based on what researchers know about these responses, the following
suggestions will enable you to use color in a reader-centered way:

How it works

Here's how a blind person would make use of the Robotic Guide:

Braille instructions: The visitor would use a Braille piece of paper, hanging on the handle, to find the number of the item he or she needs.

RFID antenna: The diamond-shaped tube is the antenna for a Radio Frequency Identification (RFID) finder that interacts with RFID tags on the store shelves to find the right location.

Laptop computer: The brains, located in the middle.

Laser range finder: This box on the robotic base helps the robot maneuver around objects — other shoppers or store displays.

Robotic base: The hunk of equipment at the bottom with the wheels.

Keypad: Buyer punches in product number here.

RFID reader: Gray box that sits on the back of the robotic base, behind the laser. It powers the RFID antenna and sends data to the laptop.

Source: Vladimir Kulyukin, Utah State University. Photo by Jeff D. Allred for USA TODAY

© Jeffrey D. Allred

FIGURE 12.7

Good Placement of Labels

This photograph shows a "seeing eye" robot developed at Utah State University to assist shoppers who are blind. It directs them to the products they want to purchase.

The creator of this graphic, Jeff D. Allred, placed the labels around the robot, person, and guide dog.

He printed the labels against white rectangles, an effective way of making the text readable when labels are placed inside a photograph. When creating photographs, people sometimes also place the arrows against a white background to make them easier to see.

In the labels, Jeff has printed each component's name in bold, followed by an explanation of the component and what it does.

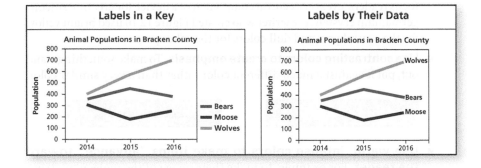

FIGURE 12.8

Labels Placed in a Graphic versus Use of a Key

Difficult to Use
When labels are placed in a key, readers must repeatedly shift their attention from the graphic to the key in order to interpret the information provided.

Easy to Use
When labels are placed next to the items they identify, readers can immediately link each part to its label.

1. **Use color primarily for clarity and emphasis, not decoration.** Color attracts the eye, making it a perfect tool for emphasizing a certain element of a graphic. However, color can also distract if it is used in a way that draws your reader's eye away from important content.

    Deploy colors in a way that promotes rather than hinders comprehension.

2. **Choose color schemes, not just single colors.** Readers see a color in terms of its surroundings. As a result, a color's appearance can change if surrounding colors are changed. To illustrate this point, design

TRY THIS Find a page in a book or magazine where color is used especially well.

Adapted from Color for the Electronic Age by Jan V. White. © 1990 by Jan V. White. Used by permission of Watson-Guptill Publications, a division of Random House, Inc.

FIGURE 12.9

Some Ways a Color's Appearance Is Affected by the Surrounding Colors

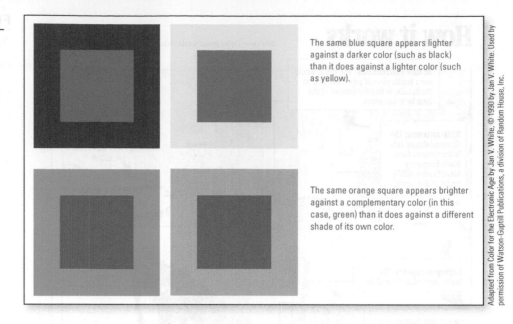

The same blue square appears lighter against a darker color (such as black) than it does against a lighter color (such as yellow).

The same orange square appears brighter against a complementary color (in this case, green) than it does against a different shade of its own color.

expert Jan V. White (1990) uses blocks of color like those shown in Figure 12.9. The top pair of blocks shows that the same shade of blue appears lighter viewed against a dark color than against a pale color. As the bottom pair of blocks demonstrates, the same pure color looks much different when it is surrounded by another shade of the same color than when it is surrounded by a complementary color. The following advice builds on the ways color schemes affect readers.

- **Use bright colors to focus your readers' attention.** For example, to focus readers' eyes on a central component of a device for creating silicon-germanium crystals, the writer who created Figure 12.10 used bright yellow for the central part and dull colors for the other parts.

- **Use contrasting colors to create emphasis.** To make something stand out, put it against a much different color rather than a very similar color.

Greater contrast creates more emphasis.

> More emphasis   Less emphasis

- **Use warm, intense colors to make items "advance" toward the reader.** For example, to make their image of the molecular structure of cytokine "closer" to their readers, the scientists who created Figure 12.11 used warm, intense colors for the molecule and set them against a black background.

3. **To promote easy reading, use a high contrast between text and background.** We traditionally use black print on white paper because this color combination creates a very high contrast that makes letter identification easy. However, backgrounds of colors other than white have become common in on-the-job communications. The chart at the top of the facing page shows the difficulty of reading type when the contrast with the background is reduced.

20%	This type is surprinted in black.	This type is dropped out in white.	20%
40%	This type is surprinted in black.	This type is dropped out in white.	40%
60%	This type is surprinted in black.	This type is dropped out in white.	60%
80%	This type is surprinted in black.	This type is dropped out in white.	80%
100%	This type is surprinted in black.	This type is dropped out in white.	100%

Reducing contrast reduces readability.

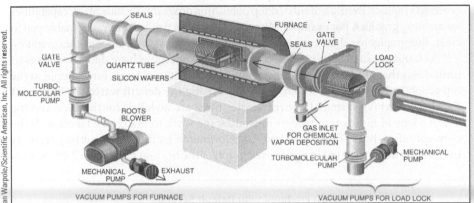

FIGURE 12.10

Bright Color Used to Focus Readers' Attention

The bright color at the center of this figure draws the readers' attention there.

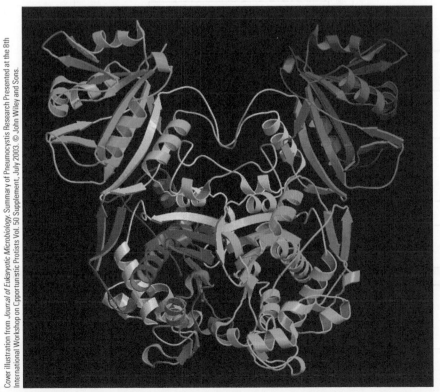

FIGURE 12.11

Warm, Intense Colors Used against a Dark Background to Make Important Material Advance toward the Reader

4. **Use the associations that colors already have for your readers.** In many contexts, colors have specific associations and even symbolic meaning. Driving through a city, we associate red with "Stop" and green with "Go." Use such associations when choosing colors for your graphics. Note the importance, however, of being alert to the fact that different colors have different associations in different contexts (White, 1990). In business, blue is associated with stability, but to a doctor it connotes death. In other contexts, blue suggests sky, water, and cold.

Use Graphics Software and Existing Graphics Effectively

Software companies offer a variety of powerful programs that can help you create reader-centered graphics. Even standard desktop publishing programs include many capabilities for creating graphics. For example, Microsoft Word and similar programs enable you to create line graphs, bar charts, organizational charts, and drawings, among other graphics.

As handy as such software features are, they sometimes have default settings or limitations that produce results that are not reader-centered. Look, for example, at the top graph in Figure 12.12, which was created using the default setting in Microsoft Excel. The text is too small to be read easily, there are no labels for the axes, and the use of a legend means that readers must do some work to match the labels with the corresponding lines. These problems are eliminated using the formatting features in Excel, as demonstrated in the revised graph at the bottom of Figure 12.12. Always

FIGURE 12.12

Comparison of Default and Revised Versions of a Graph Created in Microsoft Excel

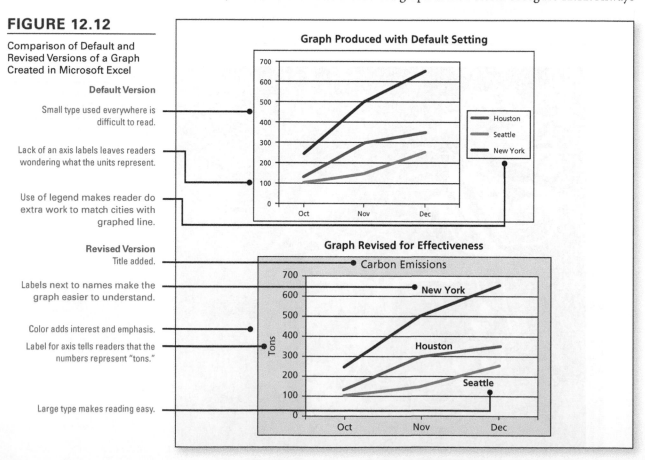

Default Version

Small type used everywhere is difficult to read.

Lack of an axis labels leaves readers wondering what the units represent.

Use of legend makes reader do extra work to match cities with graphed line.

Revised Version

Title added.

Labels next to names make the graph easier to understand.

Color adds interest and emphasis.

Label for axis tells readers that the numbers represent "tons."

Large type makes reading easy.

review graphics you produce with software from your readers' perspective, making sure that it will be as useful and persuasive as possible for them.

The same caution applies when you are incorporating into a communication a graphic previously made by you or by co-workers. A graphic that was extremely well-suited for the audience and purpose for which it was originally created is not necessarily designed in a way that will best address your objectives in a new situation or with different readers.

How to Integrate Your Graphics with Your Text

To enable your graphics to achieve their potential for usefulness. and persuasiveness, carefully integrate them with a communication's prose. Here are four strategies you can use to create a single, unified message in which your graphics and prose work harmoniously together.

GUIDELINE 1 Introduce your graphics in your text

When people read, they read one sentence and then the next, one paragraph and then the next, and so on. When you want the next element they read to be a table or a chart rather than a sentence or a paragraph, you need to direct their attention from your prose to the graphic and tell them how the graphic relates to the statements they just read. There are various ways of doing this.

LEARN MORE For advice on integrating graphics into oral presentations and web pages, see Chapters 18 and 22.

> In a market test, we found that Radex was much more appealing than Talon, especially among rural consumers. See Figure 3.

Two sentences

> As Figure 3 shows, Radex was much more appealing than Talon, especially among rural consumers.

One sentence with an introductory phrase

> Our market test showed that Radex was much more appealing than Talon, especially among rural consumers (Figure 3).

One sentence with the reference in parentheses

Sometimes your reference to a graphic will have to include information your readers need in order to understand or use the graphic. For example, here is the way the writers of an instruction manual explained how to use one of their tables.

> In order to determine the setpoint for the grinder relay, use Table 1. First, find the grade of steel you will be grinding. Then read down column 2, 3, or 4, depending upon the grinding surface you are using.

Writer tells reader how to use the graphic.

Whatever kind of introduction you make, place it at the exact point where you want your readers to focus their attention on the graphic.

GUIDELINE 2 Place your graphics near your references to them

When your readers come to a statement asking them to look at a graphic, they lift their eyes from your prose and search for the graphic. You want to make that search as short and simple as possible. Ideally, you should place the graphic on the same page as your reference to it. If there isn't enough room, put the graphic on the facing page or the page that follows. If you place the figure farther away than that (for instance, in an appendix), give the number of the page on which the figure can be found.

FIGURE 12.13

Explanatory Text Incorporated in a Drawing

To help readers understand the stages of the water cycle, a writer used this drawing.

The drawing illustrates one way to integrate text into a graphic to clarify meaning for readers. Text is placed close to the arrow it labels.

The large arrows Identify the overall process. The small arrows indicate variations in the way water moves through the overall process.

Many versions of this graphic are available with the labels written in different languages.

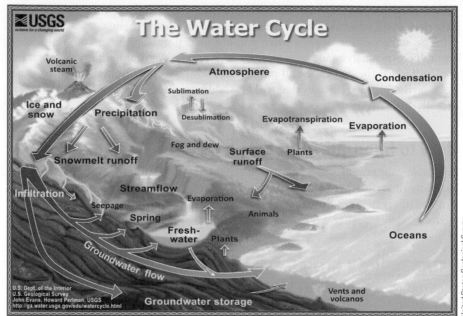

United States Geological Survey

GUIDELINE 3 State the conclusions you want your readers to draw

One way to integrate your graphics into your text is to state explicitly the conclusions you want your readers to draw from them. Otherwise, readers may draw conclusions that are quite different from the ones you have in mind.

For example, one writer included a graph that showed how many orders she thought her company would receive for its rubber hoses over the next six months. The graph showed that a sharp decline would occur in orders from automobile plants, and the writer feared that her readers might focus on that fact and miss the main point. So she referred to the graph in the following way:

Writer tells reader how to use the graphic.

> As Figure 7 indicates, our outlook for the next six months is very good.
> Although we predict fewer orders from automobile plants, we expect the slack
> to be taken up by increased demand among auto parts outlets.

GUIDELINE 4 When appropriate, include explanations in your figures

Sometimes you can help your readers understand your message by incorporating explanatory statements into your figures. Figure 12.13, which uses text and arrows to explain the water cycle, provides an example.

Global Guideline: Adapt Your Graphics When Writing to Readers in Other Cultures

Visual language, like spoken and written language, differs from nation to nation. Whereas people in Western nations typically read graphics from left to right, the Japanese read them from right to left—the same way they read prose. Moreover, though many technical symbols are used worldwide, others are not.

A major U.S. corporation reports that it once encountered a problem with publications intended to market its computer systems abroad because they included photos of telephones used in the United States rather than the much different-looking telephones used in the other countries (Anonymous). The design of many other ordinary objects differs from country to country. If you use a picture of an object that looks odd to people in your target audience, the effect you are striving for may be lost. Customs vary even in such matters as who stands and who sits in various working situations. Certain hand gestures that are quite innocent in the United States are regarded as obscene elsewhere in the world.

Colors too have different connotations in different cultures. While yellow suggests caution or cowardice in the United States, it is associated with prosperity in Egypt, grace in Japan, and femininity in several other parts of the world (Thorell & Smith, 1990).

The point is simple: Whenever you write for readers in another culture, review your draft graphics with people familiar with that culture.

Use Graphics Ethically

Graphics involve their own set of ethical issues for you to be aware of.

GUIDELINE 1 Ethics Guideline: Avoid graphics that mislead

Graphics can mislead as easily as words can. When representing information visually, you have an ethical obligation to avoid leading your readers to wrong conclusions. This means not only that you should refrain from intentional manipulation of your readers but also that you should guard against accidentally misleading readers. Here are some positive steps you can take when creating several common types of graphs.

Ethical Bar Graphs and Line Graphs

To design bar graphs and line graphs that convey an accurate visual impression, you may need to include zero points on your axes. In Figure 12.14, the left-hand graph makes the difference between the two bars appear misleadingly large because the Y-axis begins at 85 percent instead of at zero. The center graph of Figure 12.14, gives a more accurate impression of the data represented because the Y-axis begins at zero.

If you cannot use the entire scale, alert your readers by using hash marks to signal a break in the axis and, if you are creating a bar graph, in the bars. See Figure 12.14.

Note, however, that zero points and hash marks are sometimes unnecessary. For example, in some technical and scientific fields, certain kinds of data are customarily represented without zero points, so readers are not misled by their omission.

Ethical Pictographs

Pictographs can also mislead. The graphs in Figure 12.15 represent the percentage of an apple harvest that a grower can expect to be graded Extra Fancy. The left-hand graph

TRY THIS Using your favorite search engine, search for "misleading graphs." Can you find one that seems to be misleading through carelessness, one that the creator appears to have made to manipulate readers, and one that is called misleading but wouldn't really cause anyone to draw an inaccurate conclusion?

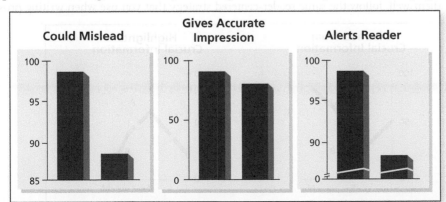

FIGURE 12.14

Creating Ethical Bar Graphs

Because its Y-axis begins above zero, the graph on the left can mislead.

The two graphs on the right illustrate ways to avoid misleading.

FIGURE 12.15

Creating Ethical Pictographs

The left-hand pictograph makes the difference between the two amounts larger than it actually is because the red apple is larger in height *and* width, thereby enlarging its area disproportionately.

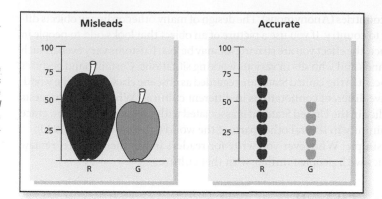

makes the percentage of Red Delicious apples seem much larger than that of Golden Delicious, even though they differ by only 20 percent. That's because the picture of the Red Delicious apple is larger in height *and* width, so its area is much greater. The right-hand graph represents the data accurately: by making columns that differ in height alone.

Ethical Use of Color

Color, too, can be used unethically. For example, because bright colors attract the eye, they can distract the reader's focus from the most important point. The left-hand graph in Figure 12.16 uses a bright red line for unimportant data to draw attention away from the crucial bad news, which shown by a dull green line. The right-hand graph corrects the problem by using bright red for the important news.

GUIDELINE 2 Ethics Guideline: Obtain permission and cite the sources for your graphics

LEARN MORE For more information on copyright and fair use, see "How To Observe Intellectual Property Law and Document Your Sources" (pages 82–84).

Copyright law treats graphics in a way that differs significantly from its treatment of text. Under the "fair use" provision of the law, you may quote a small part of a text without permission. However, you must obtain permission for *every* graphic you wish to use, even if you want to use only one out of hundreds in a particular book. The only exceptions are graphics in the public domain because they belong to a government agency or are owned by your employer. Even for graphics in the public domain, cite the source of your graphics.

You must obtain permission and cite sources for graphics obtained at websites as well as those from print documents.

Conclusion

Graphics can greatly increase the clarity and impact of your communications. To use them well, follow the same reader-centered strategy that you use when writing prose:

FIGURE 12.16

Using Color Ethically

The left-hand graph misleads readers if Trend B represents the crucial information.

The right-hand graph shows how to use color to avoid misleading.

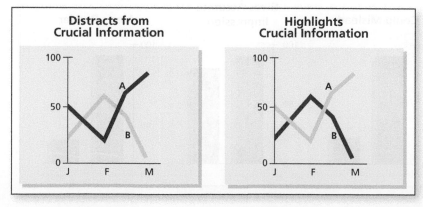

Think about the tasks your readers will perform while reading and think about the ways you want your communication to shape your readers' attitudes. Doing so will enable you to decide where to use graphics, determine the most effective types of graphics to use, make them easy to understand and use, and integrate them successfully with your prose.

Figure 12.17 provides a guide you can use when designing graphics. Chapter 13 supplements this chapter's general advice by providing detailed information for constructing eleven types of graphics often used at work.

FIGURE 12.17

Writer's Guide for Creating Graphics

WRITER'S GUIDE
CREATING GRAPHICS

PLANNING
1. Identify places where graphics will increase your communication's usefulness.
2. Identify places where graphics will increase your communication's persuasiveness.

SELECTING
1. Select the types of graphics that will best support your readers' tasks.
2. Select the types of graphics that will effectively influence your readers' attitudes.

DESIGNING
1. Design graphics that are easy to understand and use.
2. Design the graphics to support your readers' tasks.
3. Design graphics that your readers will find persuasive.
4. Keep graphics simple enough for easy use.
5. Label content clearly.
6. Provide informative titles.

USING COLOR
1. Use colors to support your message.
2. Use color for emphasis, not decoration.
3. Choose a color scheme, not just individual colors.
4. Provide high contrast between text and background.
5. Select colors with appropriate associations.

INTEGRATING WITH THE TEXT
1. Introduce each graphic in the text.
2. Tell your readers the conclusions you want them to draw.
3. Provide all explanations your readers will need in order to understand and use each graphic.
4. Locate each graphic near the references to it.

ADDRESSING AN INTERNATIONAL AUDIENCE
1. Check your graphics with persons from the other nations.

USING GRAPHICS ETHICALLY
1. Avoid elements that might mislead your readers.
2. Obtain permission from the copyright owner of each image that is not in the public domain.

USE WHAT YOU'VE LEARNED

EXERCISE YOUR EXPERTISE

1. Create an outline for an assignment in your course. On the finished outline, list each place where a graphic could make your communication more useful and persuasive in your reader's eyes. For each place, identify the type of graphic that will be most effective in achieving your communication's goals.

2. Using a desktop publishing program, make a table that displays the number of hours in a typical week that you spend in each of several major activities, such as attending class, studying, eating, and visiting with friends or family. The total number of hours should equal 168. Include brief headings for the columns and rows. Next, use your desktop publishing program to convert your table to a graph or chart. Refine your table so that the graph is as easy to read as possible. For assistance in using these features of your program, use the program's Help feature. Alternatively, do this exercise using a different kind of data.

3. Using a desktop publishing program's drawing feature, create a simple illustration of a piece of equipment used in your major or in one of your hobbies or activities. For assistance in using the drawing feature of your program, use the program's Help feature.

EXPLORE ONLINE

Find websites of companies that provide products or services related to your major, including some in your country and some in a country in a very different part of the world. For instance, if you are in Asia, look at sites in the Middle East, Africa, Europe, or North America. Using the guidelines in this chapter, compare the graphics in the two sets of websites. Although you will have looked at too few websites to make any broad generalizations about the use of graphics in the two countries, what hypotheses can you form about their similarities and differences?

COLLABORATE WITH YOUR CLASSMATES

1. Team up with another student. First, each of you is to locate two different types of graphics in journals, textbooks, or other sources in your field (including websites). Working together, evaluate each of the graphics from the perspective of the guidelines in this chapter. What features of the designs are effective? How could each graphic be improved?

2. Working with another student, exchange drafts of a project in which you are both using one or more graphics. Using the guidelines in this chapter, review the graphics in the drafts. What features are most effective in achieving the writer's usefulness and persuasiveness goals? How could the graphics be made more effective?

APPLY YOUR ETHICS

Graphs and various types of charts appear frequently in the newspapers and popular magazines. Occasionally, they are criticized for presenting data in a misleading fashion. Find a graphic that you feel may mislead readers. Explain the reason for your assessment and tell how the graphic could be redesigned. Also, identify persons who might be adversely affected or who might unfairly benefit if some action were taken on the basis of the misinterpretation. Present your results in the way your instructor requests.

REFLECT FOR TRANSFER

1. Write a memo to your instructor describing a writing project you prepared for another course that could have been improved by inclusion of a graphic. Tell what the paper was about, the point the graphic could have helped you make more clearly or persuasively, and the type of graphic you would have used.

2. Ask an instructor in your major or a mentor from your co-op or internship to tell you the kinds of graphics you are likely to use in your career. For one type, ask what would distinguish an excellent from a poor graphic of that kind. Report the results in a memo to your classmates.

Creating Eleven Types of Reader-Centered Graphics

This chapter includes a set of Writer's Tutorials that provide detailed reader-centered advice for creating eleven types of graphics that are widely used on the job. The table below will help you choose the type that is best suited to your purpose and direct you to guidelines and models for constructing effectively the type you've selected.

TO HELP YOUR READERS . . .	BEST GRAPHICS	PAGE
Tables and Graphs		
■ Find and use data, facts, or advice	Table	242–243
■ Understand the relationships among variables	Line graph	244–245
■ Compare quantities	Bar graph, Pictograph	246–247
■ See a trend	Line graphs, Bar graph	244, 246
■ See the relative sizes of the parts of a whole	Pie chart	249
Images		
■ See how something looks	Photograph, Drawing, Screenshot	250–251, 252, 254–255
■ Understand how something is constructed	Photograph, Drawing	250–251, 252–253
■ Understand a process	Flowchart	256–257
■ Understand how to do something	Photograph, Drawing, Screenshot	250–251, 252, 254–255
Management Graphics		
■ Understand the structure of an organization	Organizational chart	258
■ Understand the schedule for completing a project	Schedule chart	259

LEARNING OBJECTIVES

1. Choose the types of graphics that are best matched to your communication's goals.
2. Design the details of each graphic you choose so that will be clear, useful, and persuasive to your readers.

MindTap®

Find additional resources related to this chapter in MindTap.

TABLES

How to Create Formal Tables

Table number

Title
Be precise and specific.

Units of measure
Include the units if they won't be
obvious to your readers.

Row headings
Use precise words your
readers will understand.

Order headings in a way that will
be most useful to your readers.

Footnotes
Place them below the table.
Use letters, not numbers, so your
readers don't confuse a footnote
with data.

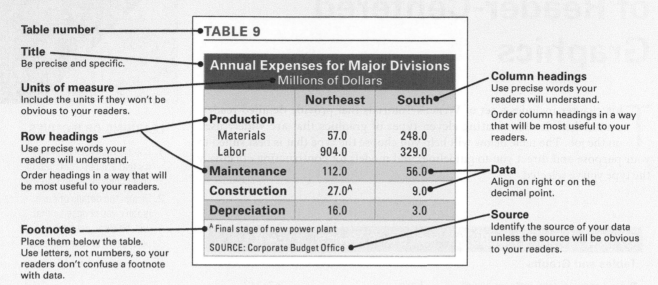

TABLE 9

Annual Expenses for Major Divisions Millions of Dollars		
	Northeast	**South**
Production		
Materials	57.0	248.0
Labor	904.0	329.0
Maintenance	112.0	56.0
Construction	27.0[A]	9.0
Depreciation	16.0	3.0

[A] Final stage of new power plant

SOURCE: Corporate Budget Office

Column headings
Use precise words your
readers will understand.

Order column headings in a way
that will be most useful to your
readers.

Data
Align on right or on the
decimal point.

Source
Identify the source of your data
unless the source will be obvious
to your readers.

How to Create Informal Tables

Informal tables flow right into the text. They are useful when the preceding
sentence tells what the table is about and when the interpretation of the data
is obvious.

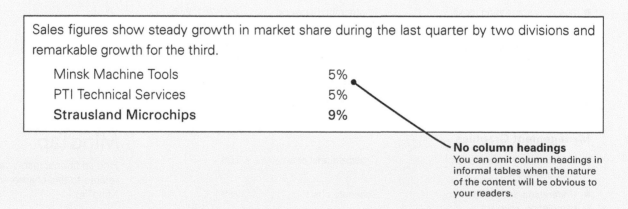

Sales figures show steady growth in market share during the last quarter by two divisions and
remarkable growth for the third.

Minsk Machine Tools	5%
PTI Technical Services	5%
Strausland Microchips	**9%**

No column headings
You can omit column headings in
informal tables when the nature
of the content will be obvious to
your readers.

How to Create Tables that Contain Only Text

Compared with paragraphs, tables can present many kinds of textual information in a manner that is much easier for readers to understand and use. The title is usually centered or aligned left over the entire table. Headings and other content are also usually centered or aligned left in their columns.

Lawn Mower Troubleshooting		
Problem	**Cause**	**Action**
Engine fails to start	A Blade control handle disengaged.	A Engage blade control handle.
Hard starting or loss of power	A Spark plug wire loose. B Carburetor improperly adjusted.	A Connect and tighten spark plug wire. B Adjust carburetor.

In this table, the title and column headings are centered.

The table's content is aligned to the left in all three columns.

Comparisons of Foods		
Food	**Calories**	**Fat**
Apple	101	0
Apple Pop Tart	191	2.9 g

In this table, the title, column headings, and content are all centered.

Tips for Creating Reader-Centered Tables

- Use extra space or draw horizontal lines to guide your readers' eyes across rows or groups of rows.
- Make key information stand out with bold type, color, and highlighting.
- Sort row and column headings to help readers find the information they want.

UNSORTED	**SORTED**
Fruits	Nutritious Foods
Sweets	Fruits
Legumes	Grains
Grains	Legumes
Fried products	Nonnutritious Foods
	Fried products
	Sweets

Group related items under a common heading.

Arrange headings and subheadings in an order that will help the readers find what they want. In this example, alphabetizing is used for subheadings. The heading "Nutritious Foods" is placed before "Nonnutritious Foods" for emphasis.

- Avoid tables that are too large for readers to use easily.
  - ☐ Include only what your readers need.
  - ☐ If the table is still large, divide it into two or more separate tables.

LINE GRAPHS

How to Create Line Graphs

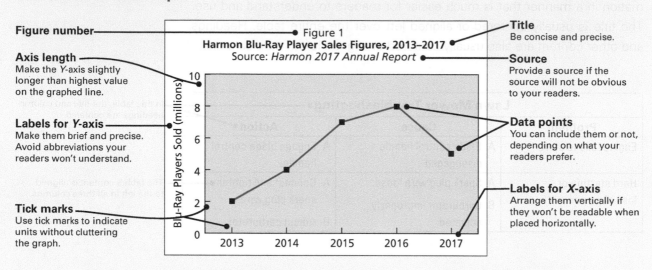

Figure number

Axis length
Make the Y-axis slightly longer than highest value on the graphed line.

Labels for Y-axis
Make them brief and precise. Avoid abbreviations your readers won't understand.

Tick marks
Use tick marks to indicate units without cluttering the graph.

Figure 1
Harmon Blu-Ray Player Sales Figures, 2013–2017
Source: *Harmon 2017 Annual Report*

Title
Be concise and precise.

Source
Provide a source if the source will not be obvious to your readers.

Data points
You can include them or not, depending on what your readers prefer.

Labels for X-axis
Arrange them vertically if they won't be readable when placed horizontally.

(Graph y-axis: Blu-Ray Players Sold (millions), 0 to 10; x-axis: 2013, 2014, 2015, 2016, 2017)

How to Create Line Graphs Comparing Trends

By using two or more lines to represent a quantity over time, you enable your readers to compare trends.

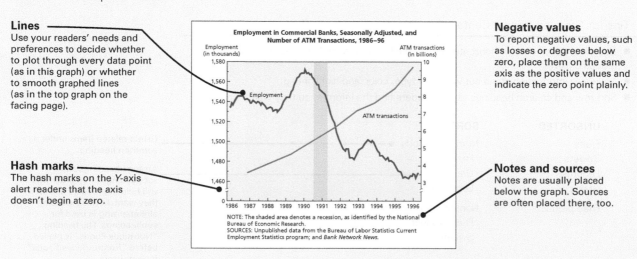

Lines
Use your readers' needs and preferences to decide whether to plot through every data point (as in this graph) or whether to smooth graphed lines (as in the top graph on the facing page).

Hash marks
The hash marks on the Y-axis alert readers that the axis doesn't begin at zero.

Employment in Commercial Banks, Seasonally Adjusted, and Number of ATM Transactions, 1986–96

Employment (in thousands)
ATM transactions (in billions)

Employment

ATM transactions

(x-axis: 1986–1996)

NOTE: The shaded area denotes a recession, as identified by the National Bureau of Economic Research.
SOURCES: Unpublished data from the Bureau of Labor Statistics Current Employment Statistics program; and *Bank Network News*.

Negative values
To report negative values, such as losses or degrees below zero, place them on the same axis as the positive values and indicate the zero point plainly.

Notes and sources
Notes are usually placed below the graph. Sources are often placed there, too.

How to Create Line Graphs Showing Interactions among Variables

In technical and scientific communications, your readers may want to see the interactions of several variables.

Labels for axes
Make the labels precise. In this graph, the writers used highly technical labels appropriate for the scientists who are their readers.

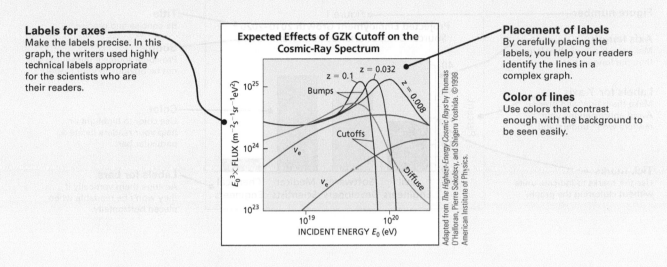

Expected Effects of GZK Cutoff on the Cosmic-Ray Spectrum

Adapted from *The Highest-Energy Cosmic Rays* by Thomas O'Halloran, Pierre Sokolsky, and Shigeru Yoshida. ©1998 American Institute of Physics.

Placement of labels
By carefully placing the labels, you help your readers identify the lines in a complex graph.

Color of lines
Use colors that contrast enough with the background to be seen easily.

Tips for Creating Reader-Centered Line Graphs

- Use different colors to enable readers to distinguish readily among the lines. If you can't use color, use different styles for the lines—dashes for one, dots for another, and so on.
- If possible, begin lines at zero to avoid misleading readers.
- If it isn't practical to start the lines at zero, use hash marks to signal this fact to your readers.

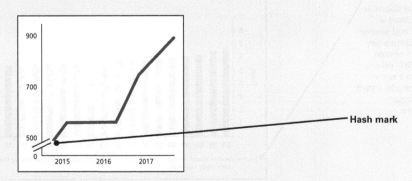

Hash mark

BAR GRAPHS

How to Create Bar Graphs

Figure number —

Axis length —
Make the *Y*-axis slightly longer than the longest bar.

Labels for *Y*-axis —
Make them brief and precise. Avoid abbreviations your readers won't understand.

Tick marks —
Use tick marks to indicate units without cluttering the graph.

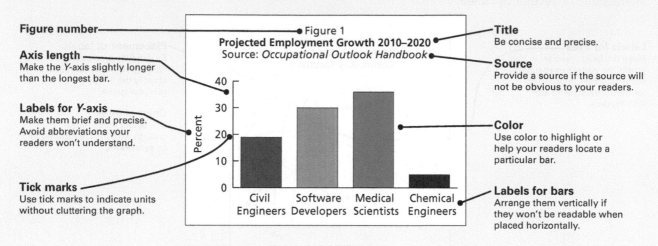

Figure 1
Projected Employment Growth 2010–2020
Source: *Occupational Outlook Handbook*

Title
Be concise and precise.

Source
Provide a source if the source will not be obvious to your readers.

Color
Use color to highlight or help your readers locate a particular bar.

Labels for bars
Arrange them vertically if they won't be readable when placed horizontally.

How to Create Bar Graphs Showing Trends

By using a series of bars to represent a quantity over time, you enable your readers to see an overall trend.

Figure number, title, and source
Some employers and scientific fields tell writers to place a figure number, title, and source below a bar graph. Others say these three elements should be placed above it. Still others place some above and some below, as in this example. Learn the expectations of your employer and readers.

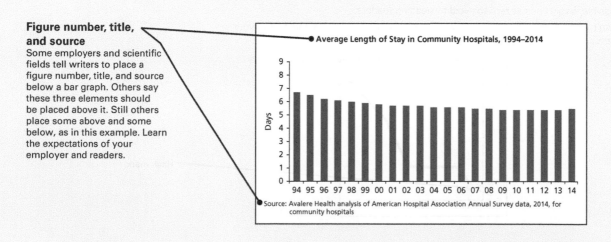

Average Length of Stay in Community Hospitals, 1994–2014

Source: Avalere Health analysis of American Hospital Association Annual Survey data, 2014, for community hospitals

How to Create Multibar Graphs Showing Several Comparisons

With a multibar graph, you help your readers compare the same groups on many topics (e.g., male and female participants in each of several sports activities).

Labels for bars
When you are using horizontal bars, you can line up their labels using the last letter of each word rather than the first letter.

Key to bars
When the bars are repeated, use a key to label them. Otherwise, place labels next to the bars, as in the top graph on the opposite page.

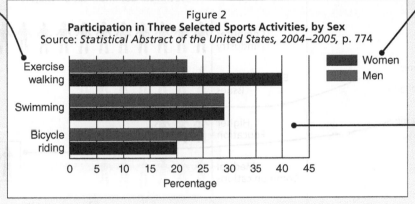

Figure 2
Participation in Three Selected Sports Activities, by Sex
Source: *Statistical Abstract of the United States, 2004–2005,* p. 774

Grid lines
Use grid lines when tick marks aren't enough to enable readers to readily gauge each bar's length.

More Types of Bar Graphs

Bar graphs come in many other varieties.

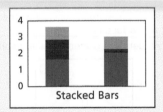

Stacked Bars

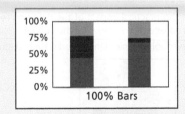

100% Bars

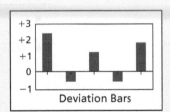

Deviation Bars

Tips for Creating Reader-Centered Bar Graphs

- Arrange bars in the order that your readers will find most helpful. Alternatives include arranging them alphabetically, chronologically, or from longest to shortest.
- If possible, begin bars at zero to avoid misleading your readers.
- If it isn't practical to start the bars at zero, use hash marks to signal this fact to your readers.

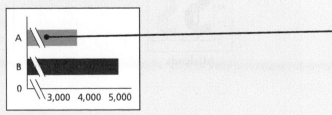

Hash mark

PICTOGRAPHS

How to Create Pictographs

Figure number

Title
Be concise and precise.

Process

1. Create a bar graph (see page 246).

2. Replace bars with symbols.

3. Use tick marks instead of grid lines.

4. Include a key explaining the value of each symbol.

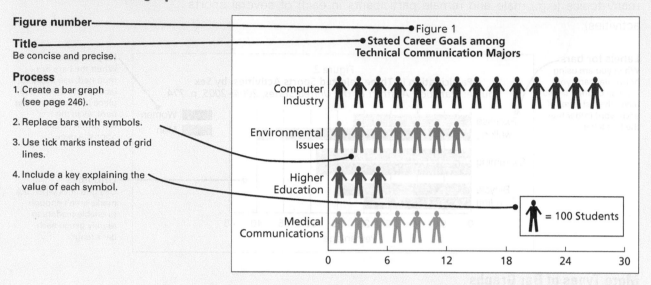

Figure 1
Stated Career Goals among Technical Communication Majors

Computer Industry

Environmental Issues

Higher Education

Medical Communications

= 100 Students

0   6   12   18   24   30

Tips for Creating Reader-Centered Pictographs

- Emphasize the practical impact of the data you represent. For example, use silhouettes of people to represent the workers who will be hired at a new factory.

- Keep the symbols simple. Unneeded details distract readers and clutter the pictograph.

Correct

Too much detail

- To avoid distortion, repeat a symbol several times rather than using a larger one.

Misleading size
When symbols are enlarged, they increase in both height and width (left-hand example), giving them a disproportionately large area.

LEARN MORE You can obtain symbols free from clipart sites. In your Internet browser, search for "Clipart."

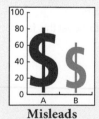

Misleads

Accurate

PIE CHARTS

How to Create Pie Charts

Process
1. Draw a circle.

2. Divide the circle into wedges proportional to each item's percentage of the whole.

3. Label each wedge; include the percentage in each label.

4. Use a different color or shading for each wedge.

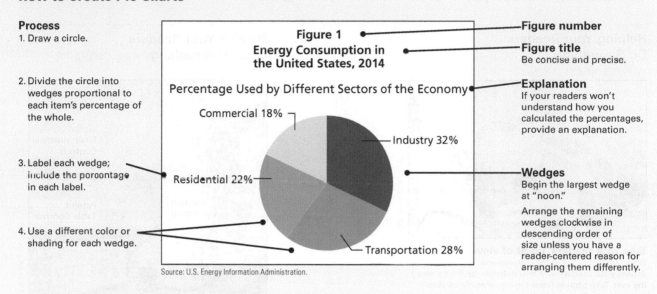

Figure number

Figure title
Be concise and precise.

Explanation
If your readers won't understand how you calculated the percentages, provide an explanation.

Wedges
Begin the largest wedge at "noon."

Arrange the remaining wedges clockwise in descending order of size unless you have a reader-centered reason for arranging them differently.

Figure 1
Energy Consumption in the United States, 2014

Percentage Used by Different Sectors of the Economy

Commercial 18%
Industry 32%
Residential 22%
Transportation 28%

Source: U.S. Energy Information Administration.

How to Avoid Distortion in Pie Charts

Stick with two-dimensional pie charts. Three-dimensional ones can create a visual distortion.

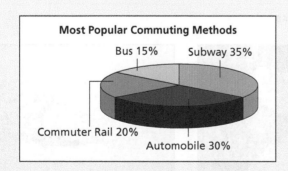

Most Popular Commuting Methods

Bus 15%
Subway 35%
Commuter Rail 20%
Automobile 30%

Visual distortion
The 30 percent wedge looks largest because the reader sees both the top and the side of the slice. However, the 35 percent slice represents a larger portion of the chart.

PHOTOGRAPHS

How to Create Photographs that Are Useful to Your Readers

Helping Your Readers Do Something

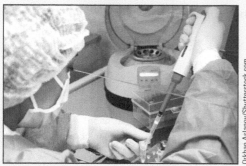

Orkhan Aslanov/Shutterstock.com

Take your readers' point of view
When photographing for instructions, include all the details your readers will find helpful—and crop away the rest. Take photos from the same angle of view that your readers will have.

Helping Your Readers Locate Something

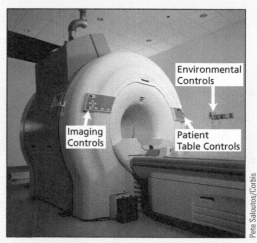

Environmental Controls

Imaging Controls

Patient Table Controls

Pete Saloutos/Corbis

Provide context and pointers
In addition to the item your readers want to find, show enough of the surrounding area so they know the general place in which to search. Arrows and labels are very helpful to readers.

Helping Your Readers See How Something Looks

Display condition
Photographs often surpass words in showing readers the condition of an object.

Show identifying marks
Photographs help readers identify animals, plants, and other objects.

Focus on what matters most
Fill your image with the details most useful or persuasive to your readers.

Sascha Burkard/Shutterstock.com

Red-Eyed Leaf Frog

tkachuk/Shutterstock.com

Condition of *Eastern Pegasus*

How to Use Software to Create Reader-Centered Photographs

Using ordinary desktop publishing software, you can increase the effectiveness of the photographs you have scanned or taken with a digital camera. The following examples use features available in Microsoft Word.

Cropping

By cropping (or trimming) a photograph, you eliminate unnecessary detail and enlarge the items you want your readers to see.

Crop tool

1. Click on the photograph.

2. Click on a handle that appears at the edges of the photo.

3. Drag the handle to the desired position.

Note: If you cannot find the Crop tool, click on Help and type "Crop" into the Search window.

Pichugin Dmitry/Shutterstock.com

Labeling

Label the items that are significant to your readers.

The Text Box tool
To create a label, use the **Text Box** and **Line** tools. They are under the **Insert** tab in Word 2016 and on the **Drawing** toolbar in Word 2016 for Mac.

Image courtesy Jeff Schmaltz, MODIS Land Rapid Response Team at NASA GSFC

Typhoons Mindulle and Tingting spin side by side off the coast of China

Label placement
Labels can be placed inside (left-hand photo) or outside (right-hand photo) a photograph.

Position them so they do not cover important parts of the image.

Image courtesy Jeff Schmaltz, MODIS Land Rapid Response Team at NASA GSFC

Typhoons Mindulle and Tingting spin side by side off the coast of China

Tips for Creating Reader-Centered Photographs

- Plan your photographs carefully in advance, considering the angle of view and removing items that would clutter the picture.
- Include a person or familiar object in a photo if this will help your readers understand the size of the object you are photographing.

DRAWINGS

How to Create Drawings

Process

1. Identify the features that are most important for your readers to see.

2. Choose the angle of view that will help your readers most.

3. Focus your readers' attention on the key features by using color, heavier lines, and labels.

4. Eliminate unnecessary and distracting details.

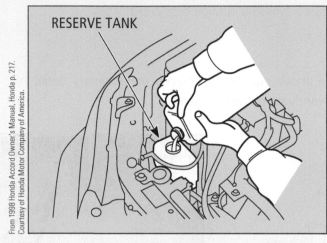

RESERVE TANK

From 1998 Honda Accord Owner's Manual, Honda p. 217. Courtesy of Honda Motor Company of America.

About this drawing

This drawing helps car owners locate the reserve tank for radiator fluid.

● A label identifies the tank.

● The tank, container, and arms are highlighted by the use of heavy lines. Lighter lines are used for everything else.

● The shading of everything else also highlights the tank, container, and arms.

● The angle of view is close to the one that owners would have when filling the tank.

How to Create Drawings that Show How to Do Something

Instructional drawings

This drawing shows how the wires will look when a reader has pigtailed them.

It shows the reader exactly how to hold the switch, using the same angle of vision that the reader would have when performing this step.

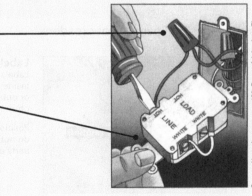

Source: Home Improvement 123: Expert Advice from The Home Depot (Des Moines: Meredith, 1995): 169.

Tips for Creating Reader-Centered Drawings

■ Show three sides of an object (e.g., front, top, side) unless you have a reader-centered reason for using another view. A three-sided perspective usually gives readers the most information about the item.

■ Labels can be placed in a drawing or next to it.

How to Create Drawings Showing Appearance and Structure

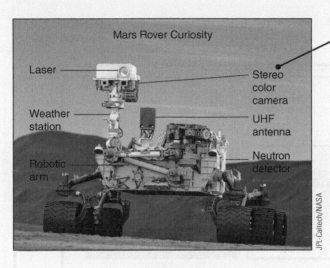

Mars Rover Curiosity

Laser

Weather station

Robotic arm

Stereo color camera

UHF antenna

Neutron detector

Show appearance
Drawings of the external views of complex objects enable significant details to stand out more distinctly than in photographs.

Show structure
Cutaway views show structural relationships that photographs cannot. In this example, the automobile parts include a piston, camshaft, flywheel, transmission, and driveshaft.

Epidermis

Dermis

Hair follicle

Fat

Blood vessel

Sensory nerve

Fibroblasts

Sweat gland

Cross section
Cross section views show parts as they would appear if the object were sliced open. This drawing shows a cross section of human skin.

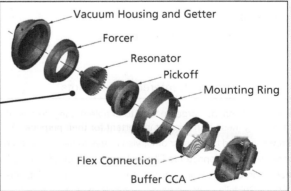

Vacuum Housing and Getter

Forcer

Resonator

Pickoff

Mounting Ring

Flex Connection

Buffer CCA

Exploded drawing
Exploded views show each part separately while also enabling readers to see how the parts fit together.

SCREENSHOTS

How to Create Screenshots

Using programs included on most operating systems, you can easily make the following types of screenshots:

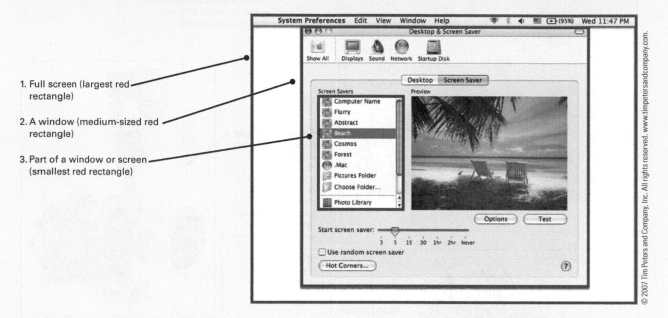

1. Full screen (largest red rectangle)

2. A window (medium-sized red rectangle)

3. Part of a window or screen (smallest red rectangle)

Creating a Screenshot in Windows:

1. To take a full screenshot, press **Print Screen**.
2. To capture a window's image, press **Print Screen** and **Alt** simultaneously.
3. Paste the image into your document.
4. Click on the image.
5. Click on the **Format** tab.
6. Click on **Crop**.
7. Drag the black lines from the corners.

Creating a Screenshot in Mac OS:

1. To take a full screenshot, press these keys simultaneously: **Command**, **Shift**, and **3**.
2. To take a cropped screenshot
   - Press these keys simultaneously: **Command**, **Shift**, and **4**.
   - Drag the cursor around the **desired area**.
3. Place the image in your document by pulling down the **Insert** menu and choosing **Picture>From File**.

Tips for Creating Reader-Centered Screenshots

- Make your screenshots large enough to enable readers to read the words and recognize the icons—unless reading these items is not important for their purpose.
- Use arrows, colored borders, or other visual cues to help readers locate items.
- If your readers are supposed to fill in fields in a window, fill in the fields in your screenshot (instead of leaving the fields blank).
- To present directions compactly, you can sometimes overlap screenshots.
- See "Ethics Guideline: Obtain permission and cite the sources for your graphics" (page 240) for information concerning copyright and screenshots.

Helping Readers Locate an On-screen Item

Show a sufficient area of a screen or window to let the reader know where
to look for the item.

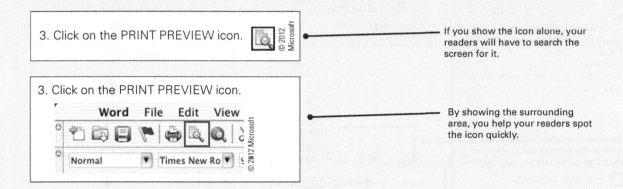

If you show the icon alone, your
readers will have to search the
screen for it.

By showing the surrounding
area, you help your readers spot
the icon quickly.

Guiding Your Readers through a Sequence of Steps

In some cases, you can overlap screenshots to guide your readers through a
series of steps, as this example shows.

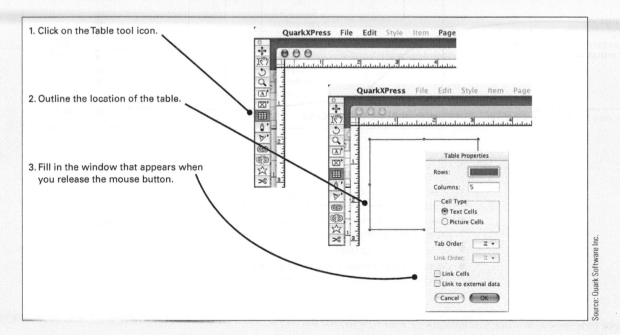

FLOWCHARTS

How to Create Flowcharts

Process

1. List the steps in the process.
2. Create a symbol or drawing for each type of step.
3. Label each step.
4. Arrange symbols from left to right, top to bottom, or in a circle.

Symbols

The symbols in the left-hand chart carry a specific meaning to readers in many fields:

☐ = process or activity

◇ = decision

Arrows

Arrows show the direction of activity.

Drawings

In the right-hand chart, each drawing is an image readers will associate with the activity represented.

Title

The title may be at the bottom or top.

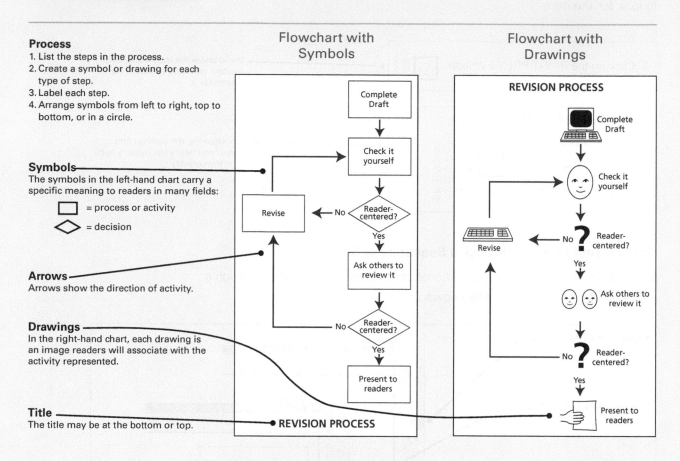

Flowchart with Symbols

Flowchart with Drawings

Tips for Creating Reader-Centered Flowcharts

- Choose between symbols and drawings by considering the knowledge and expectations of your readers.
- Make symbols and labels large enough for your readers to see plainly.
- When writing to readers who work in a field that uses specialized flowchart symbols, use the symbols in your flowchart.

How to Create Flowcharts with Drawings

In this example, the flowchart and caption work together to help readers understand the carbon cycle.

Arrows
The writers used arrows to indicate the flow from one element in the carbon cycle to another.

Labels
They used captions to label processes and states in the carbon cycle. Some labels include explanations, such as the one for photosynthesis.

Caption
They used a caption to summarize the process represented in the drawing.

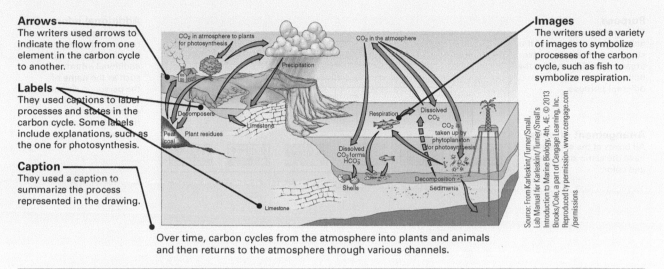

Images
The writers used a variety of images to symbolize processes of the carbon cycle, such as fish to symbolize respiration.

Over time, carbon cycles from the atmosphere into plants and animals and then returns to the atmosphere through various channels.

How to Create Flowcharts with Specialized Symbols

Using this flowchart, a computer consulting company described the general structure of a computer system it proposed to create for a buying club. Members pay a fee to join and then order products online or by phone.

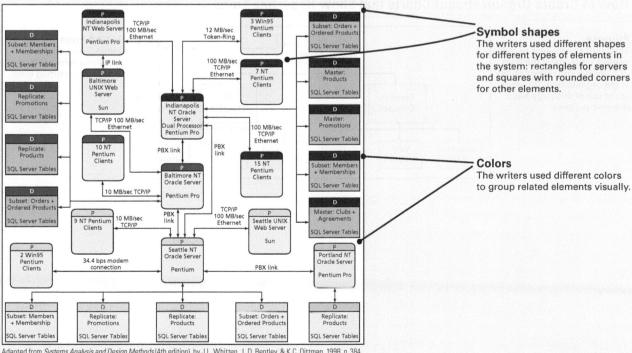

Symbol shapes
The writers used different shapes for different types of elements in the system: rectangles for servers and squares with rounded corners for other elements.

Colors
The writers used different colors to group related elements visually.

Adapted from *Systems Analysis and Design Methods* (4th edition), by J.L. Whitten, L.D. Bentley, & K.C. Dittman, 1998, p. 384. Copyright ©1998 McGraw-Hill, Inc.

ORGANIZATIONAL CHARTS

How to Create Organizational Charts

Purpose
This chart's purpose is to help readers understand the structure of the units in this organization. The chart at the bottom of this page has a different purpose.

Arrangement of boxes
All boxes at the same level have the same size, shape, and color.

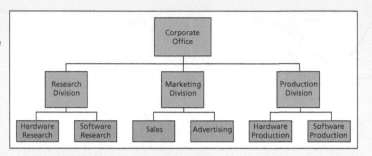

Additional information
Depending on how readers will use the chart, the boxes may include additional information, such as the name of the person in charge of each unit, the number of employees in the unit, or the unit's central phone number.

How to Create Organizational Charts that Show Reporting Lines

Purpose
This chart's purpose is to help readers understand the lines of authority and responsibility in the organization. The chart at the top of this page has a different purpose.

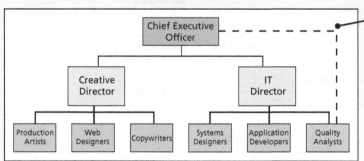

Dotted lines
Dotted lines indicate consulting relationships or secondary reporting lines.

SCHEDULE CHARTS

How to Create Schedule Charts

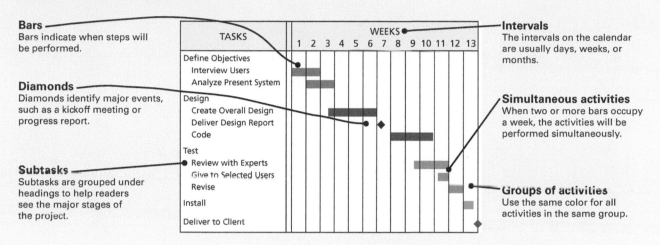

Bars
Bars indicate when steps will be performed.

Diamonds
Diamonds identify major events, such as a kickoff meeting or progress report.

Subtasks
Subtasks are grouped under headings to help readers see the major stages of the project.

Intervals
The intervals on the calendar are usually days, weeks, or months.

Simultaneous activities
When two or more bars occupy a week, the activities will be performed simultaneously.

Groups of activities
Use the same color for all activities in the same group.

How to Create Gantt Charts

Gantt charts are detailed schedule charts used to plan and monitor progress on complex projects. Usually created using special computer programs, they may involve hundreds or thousands of tasks. However, you can adopt some of their conventions in simpler schedule charts. Gantt charts are named for Henry Gantt, the early twentieth-century engineer who invented them.

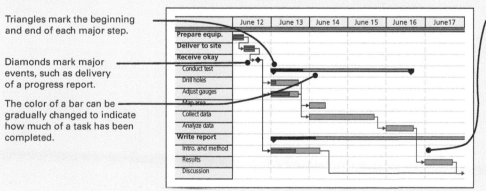

Triangles mark the beginning and end of each major step.

Diamonds mark major events, such as delivery of a progress report.

The color of a bar can be gradually changed to indicate how much of a task has been completed.

Bars linked by red lines indicate the "critical path" in a project, which is the sequence of steps that must be completed in the time specified for the entire project to be finished by its deadline.

Designing Reader-Centered Pages and Documents

You build your communications out of *visual* elements: the dark marks of your words, sentences, and paragraphs against the light background of the page, as well as your drawings and graphs and tables. Your readers *see* the visual design of these elements before they read and understand your message. And what they see has a powerful effect on the success of your communications, on their usefulness and persuasiveness.

Here, for example, are some of the ways good design can enhance your communication's *usefulness*.

- **Good design helps readers understand your information.** For example, visual design can signal the hierarchy of ideas and information in a report, thereby helping your readers quickly understand your meaning. Similarly, when writing instructions, you can place a direction and a graphic next to each other to indicate that they work together to explain a step.

- **Good page design helps readers locate information**. At work, readers often want to find part of a communication without reading all of it. With headings and other design elements, you can help them do that quickly.

- **Good design helps readers notice highly important content.** With a good design, you can emphasize for readers the content that is especially important to them, such as a warning in a set of instructions or a list of actions to take in a recommendation report.

Good design can also affect readers' attitudes, thereby increasing your communication's *persuasiveness*.

- **Good design encourages readers to feel good about the communication itself.** You've surely seen pages—perhaps in a textbook, set of instructions, or website—that struck you as uninviting, even ugly. As a result, you may have been reluctant to read them. And, undoubtedly, you've seen printed or online pages that you found attractive, so that you approached them eagerly and receptively. Good design has the same impact on readers of work-related communications. It increases readers' willingness to read documents and websites carefully and thoroughly.

- **Good design encourages readers to feel good about the communication's subject matter.** The considerable impact of design on readers' attitudes toward subject matter is illustrated in a famous

LEARNING OBJECTIVES

1. Create a page design that helps your reader see how your communication is organized.
2. Use page design to unify a long communication visually.
3. Select a font that's easy for your readers to read.
4. Choose the physical characteristics that support your communication's goals.

Good design increases usefulness.

Good design increases persuasiveness.

MindTap®

Find additional resources related to this chapter in MindTap.

experiment by researcher Karen Schriver (1997). She asked people to comment on two sets of instructions for a microwave oven. The sets differed only in their page design. When people commented on the well-designed version, which they preferred, they also said that the oven—which they had never seen—was easy to use. No matter what you are writing about, good page design can generate a more positive attitude toward your subject.

Because page design can have such a significant impact on your communication's usefulness and persuasiveness, you should approach design in the same reader-centered manner that you use when drafting text and graphics: Think continuously about your readers, including who they are, what they want from your communication, and the context in which they will be reading it.

Design Elements of a Communication

As you take this reader-centered approach, it will be helpful to think about the building blocks of page design in the way that professional graphic designers do. When they look at a page, they see six basic elements.

The six design elements of a page

- **Text.** Paragraphs and sentences.
- **Graphics.** Drawings, tables, photographs, and so on—including their captions.
- **Headings and titles.** Labels for sections of your communication.
- **White space.** Blank areas.
- **Headers and footers.** The items, such as page numbers, that occur at the top or bottom of each page in a multipage document.
- **Physical features.** These include paper, which may take many shapes and sizes, and bindings, which come in many forms.

The Writer's Tutorial on Designing Grid Patterns for Print (pages 263–264) displays a small sample of the numerous ways you can arrange these six elements as you create designs that will help you achieve your communication objectives.

LEARN MORE For advice about applying this chapter's advice to the design of web pages and websites, see Chapter 20.

This chapter's advice for working with the six visual elements of page design is just as valid for pages on the web as for pages on paper. Chapter 20 explains their application to online communications and provides additional advice about designing web pages and websites.

Help Your Reader See How Your Communication Is Organized

The first goal of page design is to enable your readers to see—just by glancing at the page—how its content is organized and how the parts fit together. As explained above, this knowledge will help your readers readily and fully understand the ideas and information you have prepared for them, and it helps them locate particular facts or thoughts. The following guidelines will help you create simple, attractive, and meaningful visual relationships that signal your communication's organization.

GUIDELINE 1 Create a grid to serve as the visual framework for your page

To begin, you need to decide where on your page to place each of the first five visual elements (text, graphics, headings and titles, white space, and headers and footers). For this purpose, graphic designers draw a grid of vertical and horizontal lines that provides the framework for their pages.

The simplest grid is one you know well: the grid used for most college papers, as illustrated here.

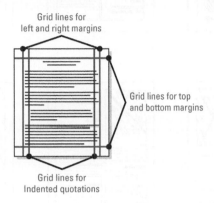

Grid lines for
left and right margins

Grid lines for top
and bottom margins

Grid lines for
Indented quotations

The simplest grid design

Using grid patterns creatively, you can build an unlimited array of functional and persuasive designs. See Figure 14.1 for some samples.

With such variety being possible, how can you explore and assess grids you might use for a particular communication? Graphic designers accomplish these tasks by sketching various possibilities for the same page. These sketches are called *thumbnails* because they are miniature versions of possible end results. Take a look at Figure 14.1 again. The nine sketches shown there are thumbnails used by a graphic artist to try out and evaluate possible designs for the *same* communication. Notice that the different designs place different levels of emphasis on the figure, title, text, and headings. Using them, the designer was able to select the design most likely to achieve her communication objectives.

As you create thumbnails and choose among them, keep three aspects of your communication in mind.

- **Usefulness.** Will your readers want to read straight through your report from start to finish? The best design may be the one used for college reports: lots of paragraphs grouped under headings and subheadings. Are you preparing instructions for a process that requires readers to read one step, look away to perform the step, and then look back at the directions to find the next step? In this case, a page design that relied on paragraphs would make reading difficult. Readers will be able to read a step, do a step, and locate the next step more easily if your page design separates the steps visually and numbers the steps prominently.

- **Persuasiveness.** The ways you want to affect your readers' attitudes should also influence your page design. Is it likely that your readers will be indifferent to the topic of your report unless you attract their attention? You may need to create a design that features appealing or intriguing photos or illustrations. Many technical magazines do this. Do you need to persuade your readers that the process described in your instructions is

Aspects of your communication
to consider as you explore and
assess thumbnails

FIGURE 14.1

Nine Thumbnail Page Designs
for the Same Material

Each of these nine thumbnails
describes a different way to present
the same content.

Notice how different designs give
different levels of emphasis to the title,
text, figure, and headings. The choice
among these designs would depend
on the readers and purpose of this
communication.

easy? Then you may need to use larger photos and more white space than
you otherwise would.

■ **Content.** Finally, you should also consider the amount and kinds of content you have. If you are writing instructions for which there is one graphic
for every step, you may want to use a two-column design that places all the
steps in the left-hand column and the corresponding figures in the right-
hand column. However, if you have a graphic for every eight or ten steps,
devoting one column to graphics would create an unattractive and not
especially useful design.

The rest of this chapter provides advice that will help you construct thumbnails and
also build your communication's actual pages.

DESIGNING GRID PATTERNS FOR PRINT

Basic Designs

Research Journal

This page from *Water Environmental Research* uses a basic two-column design. Both columns have the same width.

- The title and author are centered in an area that goes from the left-hand margin of the left column to the right-hand margin of the right column.
- In the footer, the page number is flush with the left margin. Other information is flush against the right margin.

Page Layout from Water Environment Research, Vol. 7, No. 4, July-August 2005, p. 372. Reprinted from Water Environment Research, Vol. 77, No. 4, July-August 2005, p. 372. Copyright (c) 2005 Water Environment Federation: Alexandria, VA

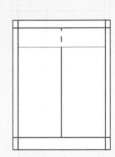

Computer Manual

Apple Computer uses a two-column design for this manual.

- The wide column is for the text and figures. The chapter title and most of the information in the footer align with the left-hand margin of this column.
- The narrow column is used only for the major headings. By extending headings into this column, the writers made them easy for readers to spot.

The page number aligns with the wide column's right margin.

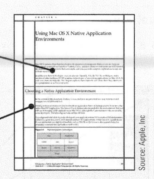

Source: Apple, Inc

Chapter from a Book

This page uses a four-column grid.

- The chapter number, title, and writers' names are centered across all columns.
- The graphic and its caption occupy the two center columns.
- Below the caption for the graphic, the text spans the two left columns and the two right columns.

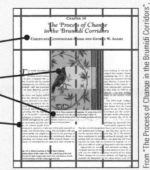

From "The Process of Change in the Brumidi Corridors", by Christiana Cunningham-Adams and George W. Adams. Constantine Brumidi: Artist of the Capitol, published in Featured Senate Publications, Y 11/3:103-27, July 11, 2030.

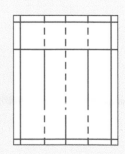

Variations on Basic Designs

Science Journal

To create a dynamic design, this page creates some asymmetries within a three-column grid.

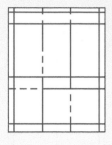

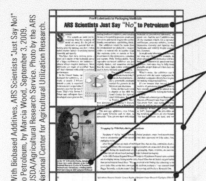

With Biobased Additives, ARS Scientists "Just Say No!" to Petroleum, by Marcia Wood, September 3, 2009. USDA/Agricultural Research Service. Photo by the ARS National Center for Agricultural Utilization Research.

- The chapter title spans all three columns.
- A graphic is centered between the left and center columns, intruding the same amount into each.
- A photo and its caption are placed in the left column.
- A text box spans the center and right columns.
- The tops of the photo and text box are aligned with different horizontal grid lines.
- The page number and footer text are centered vertically in the footer. The page number is flush against the left-hand vertical grid line. The footer text is centered horizontally across the two right-hand columns.

Research Laboratory Annual Report

In this page from one of its annual reports, the Lawrence Livermore National Laboratory uses a three-column grid.

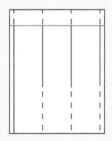

Diagram Organized by Segmentation, from Lawrence Livermore Lab, Science & Technology Review, June 2004, http://www.llnl.gov/str/Juno05/Aufderheide.html.

To create a dynamic design, the laboratory extended some visual elements across the outside margins so that they reach the edges of the page.

- The banner at the top of the page goes to both edges.
- The color behind the left side of the title also goes to the edge.
- The photo at the bottom reaches the right edge of the page.

The title, text, and footer remain strictly within the grid columns.

Introduction to Guidelines 2 through 4

Whether you are making thumbnails or creating a draft of your communication, you need to place your communication's contents into the grid with which you are working. For thumbnails, you use sketchy lines like those in Figure 14.1 to show where you will position your text, headings and titles, graphics, white space, and headers and footers. For your drafts, you place the contents in your desktop publishing pages. Based on design principles described by Robin Williams (2014), the following parts of this chapter provide advice about how to do that.

Design Principles

WILLIAMS'S PRINCIPLES	PARTS OF THIS CHAPTER
Alignment	Guideline 2: Align related elements with one another (page 267)
Grouping	Guideline 3: Group related items visually (page 268)
Contrast	Guideline 4: Use contrast to establish hierarchy and focus (page 270)
Repetition	Use Page Design to Unify a Long Communication Visually (page 271)

GUIDELINE 2 Align related elements with one another

One goal of visual design is to help your readers see how your information is organized. The first principle of visual design—*alignment*—helps you do this. By comparing the following two designs, you can see how readily alignment establishes connections among related items.

Arbitrary Placement	Alignment Establishes Relationships	
Martina Alverez AAA Consultants, Inc. 4357 Evington Street Minneapolis, MN 50517 (416) 232-9999	Martina Alverez AAA Consultants, Inc. 4357 Evington Street Minneapolis, MN 50517 (416) 232-9999	In the right-hand design, the writer has used alignment to establish relationships that are not apparent in the left-hand design.

In this demonstration, the right-hand design connects related items with one another by aligning them along the invisible grid lines at the right and left margins.

All visual elements can be arranged in one of three ways with respect to the vertical grid lines: flush left, flush right, or centered.

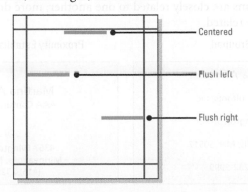

Ways visual elements can align with vertical grid lines

A page's visual elements can be aligned with horizontal grid lines in much the same way: top, bottom, or centered.

Headers and footers usually align flush right or left. A heading, figure, or other item can span two or more columns as long as it aligns within the grid system. Extending items across columns is one way of emphasizing them.

To place a drawing or other item with irregular outlines within a grid, try enclosing it in a rectangle whose sides, top, and bottom run along grid lines.

To coordinate the more complex array of visual elements included in many on-the-job communications, writers add more vertical and horizontal lines, all arranged with the readers' needs and attitudes in mind.

Sample multicolumn grid designs

In the left-hand example, a pair of vertical grid lines is used to create a simple, two-column page. The white space between the columns is called a *gutter*.

In the center page, the writer presents three figures, each accompanied by an associated block of text. To connect each figure to its text, the writer aligns each text-and-figure pair vertically in a column of its own, leaving varying amounts of white space at the bottoms of the columns.

The right-hand example represents a set of instructions in which the writer places all the directions in the left-hand column and all the illustrations in the right-hand column. Horizontal gutters separate the direction-and-illustration pairs from one another.

The Writer's Tutorial on Designing Grid Patterns for Print (pages 263–264) shows sample grid designs for several types of communication.

GUIDELINE 3 Group related items visually

The second design principle is *grouping* (sometimes called *chunking*). It relies on the way that readers judge the relationship between adjacent items by interpreting the distance, or amount of white space, between them. Less distance (less white space) means that two items are closely related to one another; more distance signals that they are not closely related.

In the right-hand design, the writer has used grouping (adjustment of the white space) to establish relationships that are not apparent in the left-hand design.

Not Grouped

Martina Alverez

AAA Consultants, Inc.

4357 Evington Street

Minneapolis, MN 50517

(416) 232-9999

Proximity Establishes Groupings

Martina Alverez
AAA Consultants, Inc.

4357 Evington Street
Minneapolis, MN 50517
(416) 232-9999

Here are some ways in which you can use white space to group elements in your page designs.

- Use less white space below headings than above to place the heading in closer proximity to the text it labels than to the preceding section.
- Use less white space between titles and the figures they label than you use to separate the title and figure from adjacent items.
- In lists that have subgroups, use less white space between items within a subgroup than between one subgroup and another subgroup.

In Figure 14.2, notice how grouping helps to convey the organization of the page.

Ways to use grouping

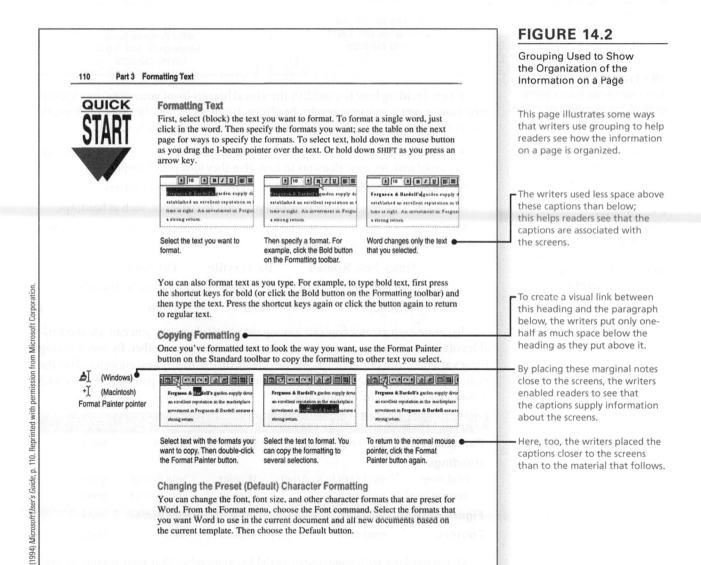

From Microsoft, (1994) *Microsoft User's Guide*, p. 110. Reprinted with permission from Microsoft Corporation.

FIGURE 14.2

Grouping Used to Show the Organization of the Information on a Page

This page illustrates some ways that writers use grouping to help readers see how the information on a page is organized.

The writers used less space above these captions than below; this helps readers see that the captions are associated with the screens.

To create a visual link between this heading and the paragraph below, the writers put only one-half as much space below the heading as they put above it.

By placing these marginal notes close to the screens, the writers enabled readers to see that the captions supply information about the screens.

Here, too, the writers placed the captions closer to the screens than to the material that follows.

GUIDELINE 4 Use contrast to establish hierarchy and focus

Usually, you will want some items to stand out more than others. For instance, you may want to indicate that some items are at a higher level than others within your organizational hierarchy. Or you may want to focus attention on certain items, such as a warning in a set of instructions. However, nothing will stand out if everything on the page looks the same. To make some things stand out, you must use *contrast*, the third of Williams's design principles.

Color, type size, and bold are used in the right-hand design to establish a focus and a visual hierarchy.

No Contrast

> Martina Alverez
> AAA Consultants, Inc.
>
> 4357 Evington Street
> Minneapolis, MN 50517
> (416) 232-9999

Contrast Establishes Emphasis

> **Martina Alverez**
> **AAA Consultants, Inc.**
>
> 4357 Evington Street
> Minneapolis, MN 50517
> (416) 232-9999

TRY THIS Some advertisements ignore some or all of the suggestions made in Guidelines 1 through 4. Find an example and identify the suggestions it follows and the ones it doesn't.
How does ignoring the suggestions improve the ad's effectiveness?
How would ignoring the same suggestions affect a communication you are preparing for your technical communication course?

When deciding how to establish the visual hierarchy of your page, pay special attention to text items: paragraphs, headings, figure titles, and headers and footers. To create distinctions among these items, you can control four variables.

- **Size:** Larger type for more prominent items, such as major headings. However, avoid making some elements so large that they look out of proportion.
- **Type treatment:** Plain (called Roman), bold, and italic letters.
- **Color:** Different color for elements you want to stand out, such as headings.
- **Font:** Serif and sans serif. The letters of serif fonts have lines, called serifs, across the ends of their strokes.

> Times New Roman   Baskerville   Palatino

Sans serif fonts do not have serifs (in French, *sans* means "without").

> Helvetica   Arial   **Folio**

To create even greater contrast among the visual elements, you can use the variables of size, type treatment, font, and color to reinforce one another. By coordinating these variables with one another, you can create distinctive appearances for all of the visual elements on a page. The following table illustrates this principle. Figure 14.3 shows a page created using this table.

Element	Font Category	Size	Treatment	Color
Paragraph	serif (e.g., Times)	12 point	Roman	black
Headings				
First level	sans serif (e.g., Arial)	16 point	bold Roman	green
Second level	sans serif	14 point	bold Roman	green
Figure caption	sans serif	10 point	bold Roman	black
Footers	sans serif	10 point	italic	black

When working with page design variables, remember that your reason for creating visually contrasting elements is to *simplify* the page for your readers so they can readily see the role played by each element on the page. Don't create so many distinctions that the page seems chaotic to readers. In most cases, for example, stick to only two colors and two fonts for the text.

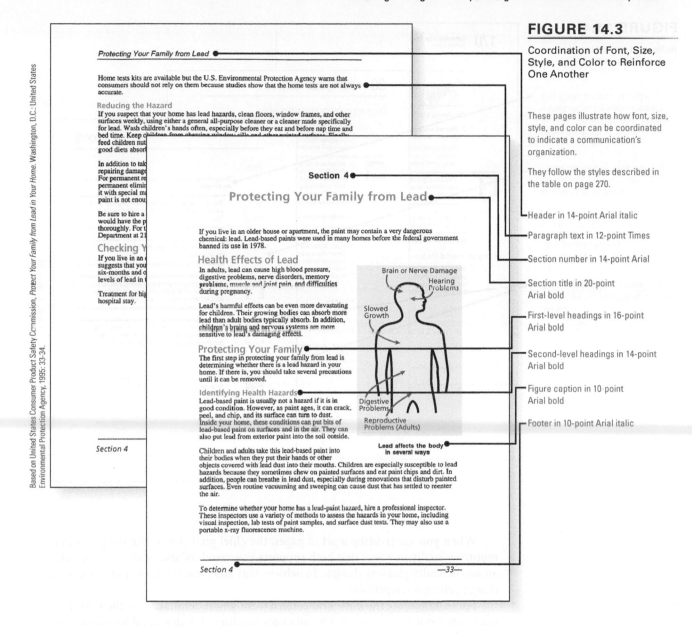

FIGURE 14.3

Coordination of Font, Size, Style, and Color to Reinforce One Another

These pages illustrate how font, size, style, and color can be coordinated to indicate a communication's organization.

They follow the styles described in the table on page 270.

Header in 14-point Arial italic

Paragraph text in 12-point Times

Section number in 14-point Arial

Section title in 20-point Arial bold

First-level headings in 16-point Arial bold

Second-level headings in 14-point Arial bold

Figure caption in 10-point Arial bold

Footer in 10-point Arial italic

Other Visual Organizers

Other tools you can use to help readers see the organization of a page include *rules,* which are lines that mark the boundaries between parts, and *icons,* which highlight information for readers or help them locate a certain kind of content. See Figure 14.4.

Use Page Design to Unify a Long Communication Visually

In communications that are more than a page long, you should think not only about creating well-designed single pages but also about creating a well-designed *set* of pages.

FIGURE 14.4

Rules and Icons Used to Signal Organization

The first sentence on page xxvii explains that the icons' purpose is to make "finding your way around" the guide easier.

The icons help readers distinguish special kinds of information from the rest of a page's contents:

- "Warnings" and "Cautions" that require attention.
- "Notes" that might be helpful, but could be skipped.

Note how the rules (horizontal lines) also help to organize the information visually.

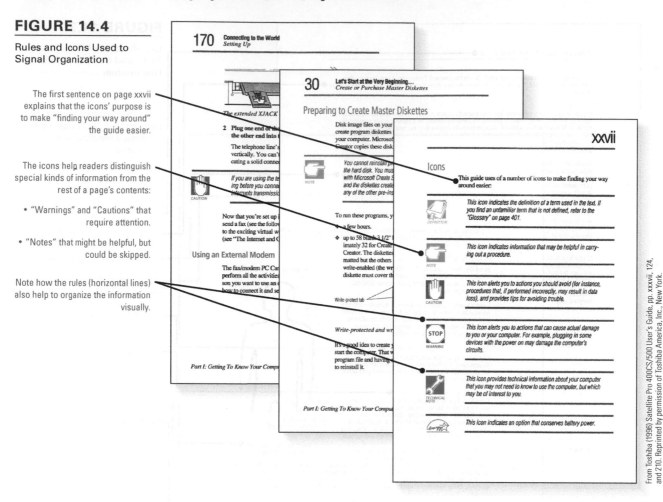

From Toshiba (1996) Satellite Pro 400CS/5001 User's Guide, pp. xxvii, 124, and 210. Reprinted by permission of Toshiba America, Inc., New York.

When you are creating a set of pages, the chief goal is to make the pages harmonize visually in a way that both supports your readers' use of them and creates an aesthetically pleasing design. To achieve this goal, use the fourth of Williams's design principles: *repetition*.

For instance, use the same grid pattern throughout. Similarly, use the same type treatment (font, size, color, etc.) for all major headings, for all second-level headings, all headers and footers, and so on. Such repetition enables your readers to "learn" the structure of your pages so that they immediately understand the organization of each new page as they turn to it. Repetition also creates visual harmony that is more pleasing to the eye than a set of inconsistent designs.

Of course, it's not possible to have exactly the same design for every page in some communications. In long reports, for instance, each chapter may begin with a page that looks different from every other page in the chapter. Instruction manuals often have troubleshooting sections that present different kinds of information than the manuals' other pages do. For such communications, follow these two strategies.

■ **Use the same design for all pages that present the same kind of information.** For instance, use one design for all pages that begin new chapters and another for all pages that don't.

■ **Even for pages that have different overall designs, use the same treatment for all elements they share.** For example, use the same style for headings wherever they appear. On all pages, stick with the same fonts and the same style for headers and footers.

Figure 14.5 shows pages from a communication that uses these strategies.

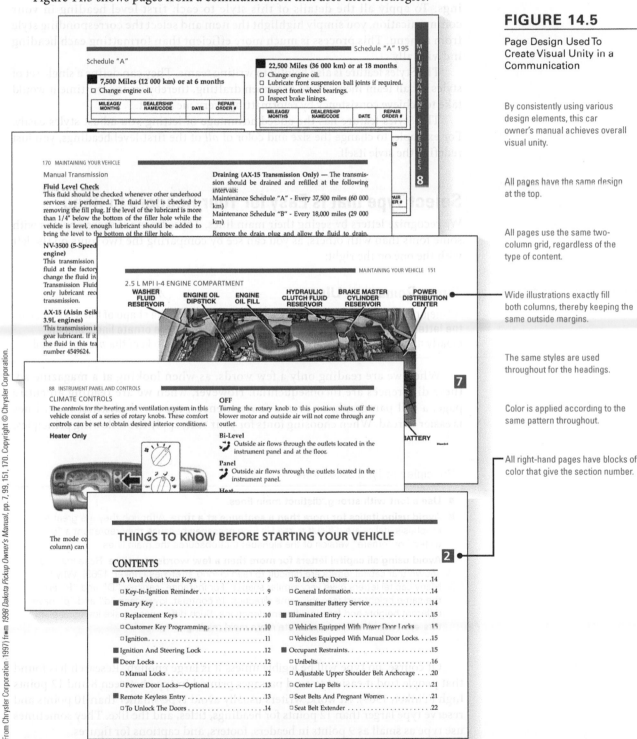

FIGURE 14.5

Page Design Used To Create Visual Unity in a Communication

By consistently using various design elements, this car owner's manual achieves overall visual unity.

All pages have the same design at the top.

All pages use the same two-column grid, regardless of the type of content.

Wide illustrations exactly fill both columns, thereby keeping the same outside margins.

The same styles are used throughout for the headings.

Color is applied according to the same pattern throughout.

All right-hand pages have blocks of color that give the section number.

Using Word-Processing Programs to Achieve Consistency

The Styles feature included with word-processing programs enables you to create consistent pages very efficiently. For example, you can name your first-level headings "Heading 1" and then use a series of menus and checkboxes to define the font, size, color, margins, indentations, and other features for these headings. To apply all the details of this style to each first-level heading in your communication, you simply highlight the item and select the corresponding style from a menu. This process is much more efficient than formatting each heading individually.

The Styles feature is also helpful to writing teams. They can define a single set of styles for all team members to use when drafting, thereby saving the time it would take to achieve consistency by reformatting everyone's draft.

The Styles feature has the added advantage of letting you adjust styles easily. For example, to change the size and color of *all* of the first-level headings, you just redefine the style itself.

Select Type that Is Easy for Your Readers to Read

We recognize letters by seeing their main lines. These lines are more obvious with some fonts than with others, as you can see by comparing the two fonts on the left with the one on the right:

Comparison of three fonts

Times Roman　　**Ultra**　　*Script*

| All major lines of all the letters are clearly defined. | The very clean, unadorned lines of all major strokes make these letters easy to recognize. | The shape of the *s* is hidden by the ornate lines, and the circle of the *p* is not closed. |

When we are reading only a few words, as when looking at a magazine ad, these differences are inconsequential. However, when we are reading an entire page, a full paragraph, or even a few sentences, type with distinct main lines is easier to read. When choosing fonts for your text, apply the following principles.

Principles of Type Selection

- **Use a font with strong, distinct main lines.**
- **Avoid using italics for more than a sentence at a time**. Although they are great for emphasis, italics are difficult to read in long passages because they represent a "distorted" or "ornate" version of the alphabet that obscures the main lines.
- **Avoid using all capital letters for more than a few words at a time.** Research has shown that capital letters are difficult to read in long stretches (Tinker, 1969). Why? Many capital letters have very similar shapes. For instance, capitals "D" and "P" are the same height and face the same direction. In contrast, lowercase "d" and "p" have different heights and face in different directions. Such differences make lowercase letters easier to identify and promote more rapid reading.

Of course, no type will be legible unless it is large enough. Research has found that people find it easiest to read passages printed in type between 8 and 12 points high (Tinker, 1969). At work, writers usually avoid type smaller than 10 points and reserve type larger than 12 points for headings, titles, and the like. They sometimes use type as small as 9 points in headers, footers, and captions for figures.

Choose the Physical Characteristics That Support Your Communication's Goals

For some communications, you will need to think about how to design not only the elements on the page but the overall product that you will provide to your readers. Here, too, consider your communication from your readers' perspective.

- **Size.** Choose a size that your readers will find convenient in the situations in which they will use your communication. For example, if you are creating instructions that readers will use at a desk already crowded by a computer and a keyboard, imitate software manufacturers who print manuals that are smaller than 8½ by 11 inches.

- **Shape.** Choose a shape that will make it convenient for your readers to carry and store. If appropriate, for example, make it a shape that will fit into a shirt pocket or a special case (such as the case for a blood glucose monitor).

- **Binding.** Some bindings stay open more easily than others. If your readers will want your communication to lie flat because they need to use both hands for something else, consider a spiral binding or three-ring binder.

- **Paper.** If you are creating a shop manual that might be spilled on or a document that might be used outdoors, consider a coated paper that resists liquids and soil.

Conclusion

Every page has some sort of page design. This chapter has explained how you can look at your pages in the way that graphic designers do, thinking about each graphic element in terms of the way it affects your readers. By following this chapter's advice, you will be able to design pages and communications that help your readers read efficiently, emphasize the important contents of your communication, and create a favorable impression.

USE WHAT YOU'VE LEARNED

EXERCISE YOUR EXPERTISE

1. Find three different page designs. (Choose no more than one from a popular magazine; look at instructions, insurance policies, leases, company brochures, technical reports, and the like.) If your instructor asks, find all the samples in documents related to your major. Photocopy one page illustrating each design. Then, for each page, describe what you think the purpose of the document is and discuss the specific features of the page design that help or hinder the document from achieving that purpose.

2. Do the Informational Page Project in MindTap.

3. Experiment with the page design of a draft or finished communication of yours that has headings and at least one list. Using your desktop publishing program's styles feature, define styles for the communication's design elements. Instead of using the program's default styles, create your own. After you have applied your styles throughout your communication, change the size, color, and fonts for the headings. Also try different designs for the other elements. (To learn how to use the program's Styles feature, use its Help feature.)

EXPLORE ONLINE

Search for "Page Design" with your web browser. Find two suggestions that you consider to be good supplements to the ones given in this chapter. Would each of them be an additional subpoint under one of the chapter's guidelines

(continued)

or would one or both of them be new guidelines? Provide a reader-centered reason for following each of them.

COLLABORATE WITH YOUR CLASSMATES

Work with another student on this exercise, which will provide you with practice at evaluating and improving a page design.

1. List the ways in which the design of the page shown in Figure 14.6 (page 277) could be improved.

2. Redesign the page using the strategies described in the Writer's Tutorial for Creating a Multicolumn Page Design available in MindTap. Specifically do the following:

   a. Create three thumbnail sketches for the page.
   b. Create a full-size mockup of the best design.

3. Display your mockup on the wall of your classroom along with the mockups prepared by the other students in your class. Decide which mockups work best. Discuss the reasons.

APPLY YOUR ETHICS

This exercise invites you to consider the role that page and screen design can play in your ability to communicate effectively on issues that are important to you for ethical reasons. Visit three or more websites whose creators are advocating a position on an ethical basis. Sites maintained by groups taking various positions on environmental policy or education policy are examples. Evaluate the ways that the screen design of each site promotes or hinders the group's ability to achieve the usefulness and persuasive goals it has for the site.

REFLECT FOR TRANSFER

1. Write a memo to your instructor describing a printed or word-processed writing project you prepared for another course or for some other purpose (in school or out) that could have been improved by following the visual design advice in this chapter. Tell what the project was about and describe three specific ways you could have made it more effective by improving the visual design.

2. Write a memo to your instructor in which you describe a printed or word-processed writing project for which you created an effective visual design. Describe three specific features of your design and explain why each improved your project. Which of these features made the most difference? Why? Did you use any visual design strategies *not* mentioned in this chapter? If so, describe one and tell how it contributed to your project's success.

FIGURE 14.6

Sample Page for Use with
Collaboration Exercise

4)

4) Flatten the clay by pounding it into the table. Remember, work the clay thoroughly!!

Slab Roller

The slab roller is a simple but very efficient mechanical device. A uniform thickness of clay is guaranteed. The ease of its operation saves countless hours of labor over hand rolling techniques.

1) The thickness of the clay is determined by the number of masonite® boards used. There are three $1/4$" boards and one $1/8$" board. They can be used in any combination to reduce or increase the thickness of the clay. If no boards are used, the clay will be 1" thick. If all the boards are used, the clay will be $1/8$" thick. So, for a $1/2$" thickness of clay, use two $1/4$" boards.

1)

2) The canvas cloths must be arranged correctly. They should form a sort of envelope around the wet clay. This prevents any clay from getting on the rollers and masonite® boards.

2)

3) Place the clay flat on the canvas near the rollers. It may be necessary to trim some of the clay since it will spread out as it passes through the rollers.

3)

4.

Courtesy of Steve Oberjohn

15

Revising Your Drafts

LEARNING OBJECTIVES

1. Identify on your own the possible ways to improve your draft.
2. Obtain truly helpful advice from people who review your draft.
3. Give truly helpful advice when you are reviewing someone else's draft.
4. Produce the maximum improvement in a limited time.

Even the most skilled writers don't always get it right the first time. In fact, the more effective the writer, the more likely he or she does the same thing your instructors have probably advised you repeatedly to do: Take time to make another draft.

On the job, revising is serious work. It usually goes far beyond the important task of proofreading. The goal is to deepen thinking, fill out ideas, refine recommendations, and clarify expression. These improvements are so important in the workplace that many employers *require* writers to revise their more critical documents, often by mandating that the writers give their drafts to managers or other co-workers for review.

The Three Activities of Revising

Revising involves three separate activities. Although they are often performed almost simultaneously, understanding that each requires a separate mental activity can greatly increase your revising expertise.

1. **Identifying possible improvements.** In this activity, you evaluate your draft from the perspective of its intended readers. Will they find it to be helpful? Persuasive? This chapter explains two methods of identifying possible improvements—checking your draft yourself and asking a co-worker to review it for you. Because you will often be asked to review other people's drafts, the chapter also explains ways to make your advice to other writers as helpful as possible. The next chapter describes a third method of identifying possible improvements: testing.

2. **Deciding which improvements to make.** At work, you will rarely have time to make all possible improvements. Deadlines will loom. Other responsibilities will demand your attention. This chapter describes ways you can choose from all the revisions you *might* make, the ones that will yield the greatest improvement.

3. **Making the selected improvements.** For good reason, this chapter offers no advice about this activity. The principles of good communication are the same whether you are creating a first draft or revising a nearly final draft. If you determine that you need to polish your writing style, go to Chapter 10, "Developing an Effective, Professional Style," for advice. Similarly, if you discover that you need to reorganize, redesign your graphics, or make any other sort of change, consult the appropriate chapters in this book.

MindTap®

Find additional resources related to this chapter in MindTap.

Identify Ways to Improve Your Draft

The first thing to know about checking your own drafts is this: It is extremely difficult to recognize problems in your own writing. Several obstacles hamper your ability to do so. The following guidelines identify and suggest ways to overcome these obstacles.

GUIDELINE 1 Check from your readers' point of view

Writers sometimes create the first obstacle by defining the focus of their checking too narrowly. They concentrate on spelling, punctuation, and grammar, forgetting to consider their communication from the readers' perspective. Always begin by reminding yourself of your communication's objectives. If you wrote them down, pull out your notes. If you recorded them mentally, review them now. What tasks do you want your communication to help your readers perform? How do you want it to alter their attitudes?

Then, with your communication's objectives fresh in mind, read your draft while imagining your readers' moment-by-moment responses. As they proceed, how usable will they find each section, paragraph, and sentence to be? How persuasive will they find each of the writing strategies by which you hope to shape their attitudes? Look especially for places where the readers may not respond in the way you desire. That's where you need to revise.

GUIDELINE 2 Check from your employer's point of view

In many working situations, you must consider your communication not only from your readers' perspective but also from your employer's.

- **How will my communication impact others in the organization?** By anticipating conflicts and objections that your communication might stir up, you may be able to reduce their severity or avoid them altogether.

- **Does my communication promise something on my employer's behalf?** For example, does it make a commitment to a client or customer? If so, ask yourself whether you have the authority to make the commitment, whether the commitment is in your employer's best interest, and whether it is one your employer wishes to make.

- **Does my communication comply with my employer's policies?** First, check your draft against your employer's policies on writing style and format. Second, determine whether it complies with the employer's regulations concerning what you can say. For example, if you are working on a patentable project or one that is regulated by a government body, such as a state Environmental Protection Agency, your employer may restrict what you reveal about it.

Things to consider when evaluating from your employer's perspective

GUIDELINE 3 Distance yourself from your draft

A third obstacle to effective checking is being too "close" to what we write. Because we know what we mean to say, when we check for errors we often see what we intend to write rather than what is actually there. A word is misspelled, but we see it spelled correctly. A paragraph is cloudy, but we see clearly the meaning we wanted to convey.

To distance yourself from your draft, try the following measures.

- **Let time pass.** As time passes, your memory of your intentions fades. You become more able to see what you actually wrote. Set your draft aside, if for only a few minutes, before checking it.

Ways to distance yourself from your draft

- **Read your draft aloud, even if there is no one to listen.** Where you stumble when saying your words, your readers are likely to trip as well. Where the "voice" you've created in your draft sounds a little off to you, it will probably sound wrong to your readers, too (see Chapter 10).

- **Let your computer read your paper aloud to you.** Mac and some PC computers come with free programs that read your writing aloud through your computer's speakers. This simple way of distancing yourself from your draft can help you detect problems.

TRY THIS Why not proofread by letting your computer read your writing aloud to you?

GUIDELINE 4 Read your draft more than once, changing your focus each time

Yet another obstacle to effective checking is summed up in the adage "You can't do two things at once." To "do" something, according to researchers who study the way humans think, requires "attention." Some activities (such as walking) require much less attention than others (such as solving calculus problems). We can attend to several more or less automatic activities at once. However, when we are engaged in any activity that requires us to concentrate, we have difficulty doing anything else well at the same time (J. R. Anderson, 2014).

This limitation has important consequences. When you concentrate on one aspect of your draft, such as the spelling or consistency of your headings, you diminish your ability to concentrate on the others, such as the clarity of your prose.

To overcome this limitation on attention, check your draft at least twice, once for *substantive* matters (such as clarity and persuasive impact) and once for *mechanical* ones (such as correct punctuation and grammar). If you have time, you might read through it more than twice, sharpening the focus of each reading.

GUIDELINE 5 Use computer aids to find (but not cure) possible problems

Desktop publishing programs offer a variety of aids that can help you check your drafts.

- **Spell checkers.** Spell checkers identify *possible* misspellings by looking for words in your draft that aren't in their dictionaries.

- **Grammar checkers.** Grammar checkers identify sentences that *may* have any of a wide variety of problems with grammar or punctuation.

- **Style checkers.** Some computer programs analyze various aspects of your writing that are related to style, such as average paragraph, sentence, and word length. Using special formulas, they also compute scores that attempt to reflect the "readability" of a draft. See Figure 15.1.

LEARN MORE To learn how to distinguish effective from ineffective uses of the passive voice, see "Use the active voice unless you have a good reason to use the passive voice" (page 198).

All three types of aids can be helpful. But use them with caution. They all have serious shortcomings. For example, spell checkers ignore words that are misspelled but look like other words, such as *forward* for *foreword* or *fake* for *rake*. Grammar checkers can't tell a good use of the passive voice from a poor one. And style checkers can't tell a clearly written long sentence from a murky short one. Some suggestions from these aids may make your writing worse. Review their suggestions carefully. And always read over your drafts yourself.

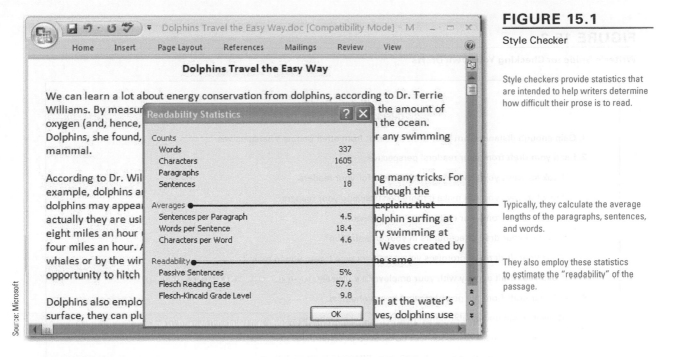

FIGURE 15.1

Style Checker

Style checkers provide statistics that are intended to help writers determine how difficult their prose is to read.

Typically, they calculate the average lengths of the paragraphs, sentences, and words.

They also employ these statistics to estimate the "readability" of the passage.

GUIDELINE 6 Take special care with social media

Because of their variety of forms and uses in the workplace, social media create a special challenge with respect to revising. Some don't require revising at all. In some organizations, short text messages that co-workers use to coordinate work, answer questions quickly, or check on the status of projects can be sent with typos and misspellings as long as the key information is accurate and clear. Content is king. On the other hand, for some organizations, polish is essential for longer messages posted on a blog or similar platform that attracts a larger audience. In these cases, it is prudent to draft messages in a word-processing program, proofread them carefully, and then cut and paste them into the social medium you are using.

As when preparing any other type of communication on the job, learn and meet the expectations of your readers and employer.

GUIDELINE 7 Ethics Guideline: Consider the stakeholders' perspective

If you followed the ethics guidelines provided earlier in this book, you will already have considered your communication from the stakeholders' point of view at least twice. Early in your work on your draft, you identified its stakeholders and learned about the impact your message might have on them. Then, while drafting, you will have kept your stakeholders constantly in mind.

However, there are so many things to think about when drafting that you may have overlooked your stakeholders' concerns in some way. Therefore, when checking a draft, review your communication one last time from their perspective.

LEARN MORE To learn more about treating your communication's stakeholders ethically, see "How to Treat Your Communication's Stakeholders Ethically" (page 64) and "Ethics Guideline: Review from the stakeholder perspective" (page 286).

Checklist for Checking

Figure 15.2 is a checklist you can use to ensure that you are checking your drafts in a reader-centered way.

FIGURE 15.2

Writer's Guide for Checking Your Own Drafts

> **Writer's Guide**
> **CHECKING YOUR OWN DRAFTS**
>
> 1. Gain enough distance from your draft to read it from other people's perspectives.
>
> 2. Read your draft from your readers' perspective.
>
>   Look for ways you can make it more useful for your readers.
>
>   Look for ways you can make it more persuasive to your readers.
>
> 3. Read your draft from your employer's perspective.
>
>   How will your draft affect others in your organization?
>
>   Does your draft make promises your employer won't want you to make?
>
>   Does your draft comply with your employer's policies?
>
> 4. Read your draft from the perspective of its stakeholders.
>
>   Does it create undesirable consequences for the stakeholders?
>
>   If so, how can these consequences be reduced or eliminated?
>
> 5. Read your draft more than once, focusing on different issues each time.
>
> 6. Check for errors missed by the computer program you used.

Obtain Truly Helpful Advice from People Who Review Your Drafts—And Give Good Advice when You Are Reviewing Someone Else's Draft

In *reviewing*, writers give their drafts to someone else to look over. You may be familiar with this process through the peer reviewing often done in classes. At work, writers are often required to have their drafts reviewed to ensure that their communications are well written, conform to the employer's policies, and say things that serve the organization's interests. For routine communications, reviews are usually conducted by the writer's manager. For sensitive communications, the reviewers may also include company executives, lawyers, public relations specialists, and many others. Commonly, the persons performing required reviews have the authority to demand changes.

Even when a review is not required, employees often ask co-workers or managers to look over their drafts. The helpfulness of the suggestions you receive will depend largely on the way you manage your interactions with your reviewers. The following six guidelines will help you develop expertise at eliciting supportive and helpful reviews of your drafts.

Several of these guidelines also include suggestions for reviewing other people's drafts effectively. If you are promoted to a supervisory or managerial position, reviewing other people's writing will be part of your job. Even if you are not a manager, you will almost surely be asked to look over someone else's draft, perhaps as early as your first months at work. Moreover, you will be evaluated on the value of the advice you offer.

GUIDELINE 1 Discuss the objectives of the communication and the review

Reviewing needs to be as reader-centered as any other writing activity. Don't just hand your draft to reviewers with a request that they "please look it over." Share the detailed information you developed while defining your communication's objectives. Describe the tasks you want to help your readers perform, the ways you want to influence your readers' attitudes, and the aspects of your readers and their situation that will affect the way they respond. Only with this information can reviewers suggest effective strategies for increasing your communication's usefulness and persuasiveness for your specific readers.

Also discuss the scope and focus of the review. As the writer, you may have a good sense of areas that most need attention—whether organization, selection of material, accuracy, tone, page design, or something else. Without discouraging your reviewers from noting other kinds of improvement, guide them to examining what you think is most critical.

Similarly, if you are the reviewer and a writer doesn't tell you his or her communications' objectives or his or her desires for the review, ask for this information. You need it to be able to give the best possible assistance.

GUIDELINE 2 Build a positive interpersonal relationship with your reviewers or writers

The relationship between writers and their reviewers is not only intellectual but also interpersonal and emotional. The quality of this human relationship can greatly affect the outcome of the review process. When writers feel criticized and judged rather than helped and supported, they become defensive and closed to suggestions. When reviewers feel their early suggestions have been rebuffed rather than welcomed, they cease giving the advice the writer needs. The following strategies will help you build positive, productive relationships with your reviewers and writers.

When You Are the Writer

Reviewers are more helpful if they believe you appreciate their comments. Throughout your discussions with them, project a positive attitude. Treat your reviewers as people who are on your side, not as obstacles to the completion of your work. Begin meetings by thanking them for their efforts. Paraphrase their comments to show that you are listening attentively, and express gratitude for their suggestions.

When your reviewers offer comments, stifle any temptation to react defensively, for instance by explaining why your original is superior to a suggested revision. Even if a particular suggestion is wrongheaded, listen to it without argument. On the other hand, don't avoid dialogue with your reviewers. When they misunderstand what you are trying to accomplish, explain your aims while also indicating that you are still open to suggestions. If your reviewers misunderstood what you were attempting to say, it's probably a sign that you aren't being clear and would benefit from their advice.

At work, drafts are sometimes reviewed in person with a supervisor or a group of co-workers.

Ways to encourage your reviewers to help as much as they can

When You Are a Reviewer

When you are the reviewer, use these strategies to build a good relationship with the writer.

■ **Begin with praise.** By doing so, you show the writer that you recognize the good things he or she has done, and you indicate your sensitivity to the

Ways to encourage openness to your suggestions

writer's feelings. In addition, some writers don't know their own strengths any better than they know their weaknesses. By praising what they have done well, you encourage them to continue doing it and you reduce the chances they will weaken strong parts of their communication by changing them.

- **Focus your suggestions on goals, not shortcomings.** For example, instead of saying, "I think you have a problem in the third paragraph," say, "I have a suggestion about how you can make your third paragraph more understandable or persuasive to your readers."

- **Use positive examples from the writer's own draft to explain suggestions.** For example, if the writer includes a topic sentence in one paragraph but omits the topic sentence in another, cite the first paragraph as an example of a way to improve the second. By doing this, you indicate that you know the writer understands the principle but has slipped this one time in applying it.

GUIDELINE 3 Rank suggested revisions—and distinguish matters of substance from matters of taste

When you review a writer's draft, one of the most helpful things you can do is rank or prioritize your suggestions. By doing so, you help the writer decide which revisions will bring the greatest improvement in the least amount of time. Also, many writers can face only a limited number of suggestions before feeling overwhelmed and defeated. Look over your suggestions to determine which will make the greatest difference. Convey those to the writer. Keep the rest to yourself, or make it clear that they are less important. Ranking also helps *you* work productively. Begin by scanning the draft to identify the issues that most need attention, then focus your effort on them.

To rank suggestions, distinguish those based on substantial principles of writing from those based on your personal taste. If you suggest changes that would merely replace the writer's preferred way of saying something with your preferred way, you will have done nothing to improve the writing—and you will almost certainly spark the writer's resentment. When you are unsure whether you are expressing a personal preference, try to formulate a reason for the change. If all you can say is, "My way sounds better," you are probably dealing with a matter of taste. If you can offer a more objective reason—for instance, one based on a guideline in this book—you are dealing with a matter of substance.

So far, the discussion of this guideline has focused on the reviewers' perspective. When you are the writer, you can be helped greatly by your reviewers' ranking of their suggestions. If your reviewers don't volunteer a ranking, request one.

GUIDELINE 4 Explore the reasons for your suggestions

The more fully writers and reviewers explore the reasons for the reviewers' suggestions, the more helpful the review will be. When you are the reviewer, explain your suggestions so the writer doesn't think you are simply expressing a personal preference and dismiss advice that would improve the communication.

Your full explanations also help the writer learn to write better. Imagine that you suggest placing an echo word at the beginning rather than the end. The writer may agree that your version is better but may not know the reason unless you explain the *principle* that you applied.

LEARN MORE To learn about the use of echo words to smooth the flow of thought through a communication, see "Smooth the flow of thought from sentence to sentence" (page 126).

Some reviewers withhold such explanations because they think the reasons are obvious. However, reasons that are obvious to the reviewer are not necessarily obvious to the writer. Otherwise, the writer would have avoided the problem in the first place.

Whenever possible, phrase your suggestions from the perspective of the intended readers. Instead of saying, "I think you should say it like this," say, "I think your intended readers will be able to understand your point more clearly if you phrase it like this." Such a strategy will help the writer take a reader-centered approach to revising the draft. It will also help you present yourself not as a judge of the writer's writing but as a person who wants to help the writer achieve his or her communication objectives.

GUIDELINE 5 Present your suggestions in the way that will be most helpful to the writer

Whenever possible, discuss reviews in person with your writer or reviewer. Conversation enables the writer to ask for clarification and the reviewer to provide full explanations. If you can't meet face to face, talk on the phone or through a free program such as Skype (www.Skype.com).

Whether you are the reviewer or writer, talk with your partner about the best way to convey written suggestions. In particular, think about the method that will make it easiest for the writer to incorporate suggested revisions into the draft being reviewed. If you are the writer, consider giving your reviewer a digital copy of your draft. Your reviewer can insert suggestions and return the draft to you. You can then use the desktop publishing program to accept or decline them one by one, thereby eliminating the need to retype suggestions you want to take (see Figure 15.3).

If more than one person will review your draft, however, it can be inefficient to have each reviewer use a desktop publishing program to enter suggestions into your draft. Each reviewer will return his or her comments to you in a separate file. You will need to read back and forth among drafts to see if different reviewers

LEARN MORE Programs like Google Docs also help teams write collaboratively. See "Use Internet and Cloud Technology for Drafts" (page 314).

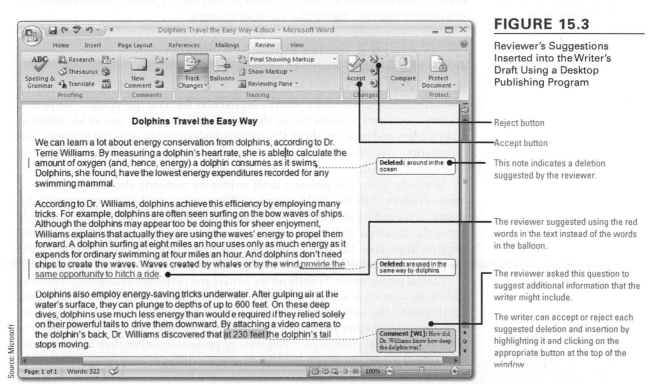

Source: Microsoft

FIGURE 15.3

Reviewer's Suggestions Inserted into the Writer's Draft Using a Desktop Publishing Program

Reject button

Accept button

This note indicates a deletion suggested by the reviewer.

The reviewer suggested using the red words in the text instead of the words in the balloon.

The reviewer asked this question to suggest additional information that the writer might include.

The writer can accept or reject each suggested deletion and insertion by highlighting it and clicking on the appropriate button at the top of the window

FIGURE 15.4

Comments that Two Reviewers Made on a Single Draft They Accessed at Google Docs Using Their Browsers

The writer clicks here to share uploaded documents with reviewers.

All reviewers add comments and suggest changes to this one copy of the writer's draft.

• The green comment is by Margie.

• This orange comment is by Orlando.

Reviewers can also suggest additions, deletions, and other changes.

The writer can see all suggestions from all reviewers at one time.

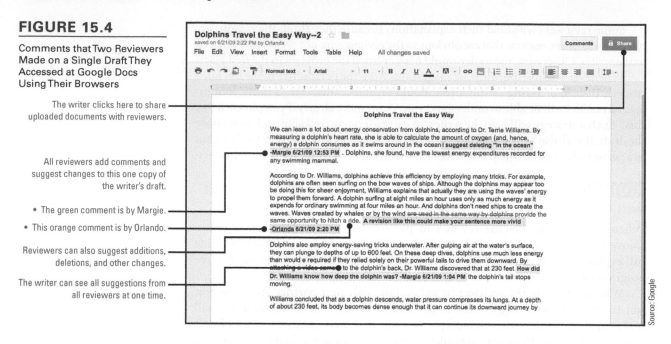

Source: Google

make different suggestions about the same issue. You will also need to go repeatedly to the different files to extract suggestions you want to accept as you create a single new draft that incorporates them all. To avoid this work, use a program like Google Docs (http://docs.google.com) that enables you to upload your draft to a central location on the Internet. All reviewers can use their browsers to access and insert comments into this single copy, greatly simplifying your work. As a bonus, because the reviewers can read each other's comments they can supplement rather than repeat one another's advice. Figure 15.4 shows an example.

GUIDELINE 6 Ethics Guideline: Review from the stakeholders' perspective

Reviewers can be as helpful in evaluating stakeholder impacts as in identifying ways to increase a communication's effectiveness for its target readers. Be sure to ask your reviewers to assist you in this way. Tell them whom you've identified as stakeholders and how you think these individuals might be affected by your communication. Ask them to help you discover other stakeholders or impacts that you might have missed.

Similarly, when you review a draft written by someone else, ask about his or her assessment of its stakeholder impacts. Then, as you read, make your own judgments. Who will be affected? How? Should the stakeholders be consulted directly about the decisions and actions discussed in this communication? Can potential negative impacts on the stakeholders be reduced or eliminated? Is it ethical to take the proposed action at all?

Ethical problems can be difficult for writers and reviewers to discuss. None of us likes to be accused of acting unethically, so a writer may quickly become defensive. Furthermore, the writer and reviewer may hold quite different ethical principles. Consequently, whether you are the writer or the reviewer, you may find it helpful to refer to the ethical code of your employer or profession rather than insist on your own ethical views. But remember that despite the difficulties of doing so, you have an obligation to raise ethical questions even if you have not been asked.

FIGURE 15.5

Writer's Guide for Reviewing Drafts

Writer's Guide
REVIEWING DRAFTS

1. When preparing for the review, do the following:
   - ☐ Discuss the objectives of the review.
   - ☐ Discuss the objectives of the communication.
   - ☐ Build a positive relationship with the writer or reviewer.

2. When discussing comments and suggestions, do the following:
   - ☐ Identify the revisions that are most important.
   - ☐ Distinguish matters of substance from matters of taste.
   - ☐ Explore the reasons for all suggestions.

3. Present your suggestions in the way that will be most helpful to the writer.

4. Review the draft from the perspective of its stakeholders.

Writer's Guide for Reviewing

Figure 15.5 is a Writer's Guide you can use to ensure that you are checking your drafts in a reader-centered way.

Produce the Maximum Improvement in Limited Time

The two preceding sets of guidelines focused on ways you can identify improvements you could make in a draft—by checking the draft yourself and by having someone else look over your draft. The following guidelines take you one step farther in the revision process: choosing which of the possible improvements you should actually make.

In some cases, of course, you won't need to make any choices. You may find that your draft (probably not your first one) is already good enough to help and influence your readers in the ways you intend. Or you will have adequate time to make all the changes that have been identified.

Often, however, writers at work face daunting decisions about which changes to make and which to forgo. The need for these decisions can arise for any of the following reasons.

- When there isn't enough time to make all the revisions that checking, reviewing, or testing suggest

- When two or more reviewers give contradictory advice

- When a reviewer, such as a boss, suggests changes that the writer believes will actually diminish the communication's effectiveness

The guidelines that follow will help you develop expertise at dealing effectively with these more complex revising situations.

GUIDELINE 1 **Adjust your effort to the situation**

This guideline suggests that you approach revising as if you were an investor. As you begin revising any communication, let your first concern be the amount of time and energy you might need to invest in it rather than in your other duties.

Some students think it is peculiar to ask how good a communication needs to be. "After all," they observe, "shouldn't we try to make everything we write as good as it can be?" On the job, the answer is "No."

Of course, all your communications should present their central points clearly and achieve an appropriate tone. At work, however, different communications need different levels of additional polish. In many organizations, printed messages are expected to be error-free, but email messages sent internally may contain spelling errors and grammatical mistakes as long as their meaning is clear and their tone inoffensive.

To determine how "good" your communication needs to be, do the following.

Ways to determine how good a communication has to be

- **Think about your communication's usefulness and persuasiveness objectives.** How carefully does your communication need to be crafted to help your readers perform their tasks? To affect your readers' attitudes in the way you want?

- **Consider general expectations about quality.** At work you usually need to polish your communications most thoroughly when addressing people at a higher level than you in the organizational hierarchy, when addressing people outside your organization rather than inside, and when trying to gain something rather than give something (for example, when writing proposals rather than reports). These general expectations can be translated into a list that ranks some typical communications according to the level of polish they usually need. Figure 15.6 illustrates a typical ranking. Of course, the ranking in your workplace may be different.

- **Look at similar communications.** Often, the level of quality needed is determined largely by what's customary. By looking at communications similar to yours, you can see what's succeeded in the past.

- **Ask someone.** Your boss or co-workers can be excellent sources of information about the level of quality needed, as can your intended readers.

There are sometimes good reasons for wanting your communication to go beyond the minimum level needed to make your communication "good enough."

FIGURE 15.6

Amount of Polish Typically Needed by Several Kinds of Communication

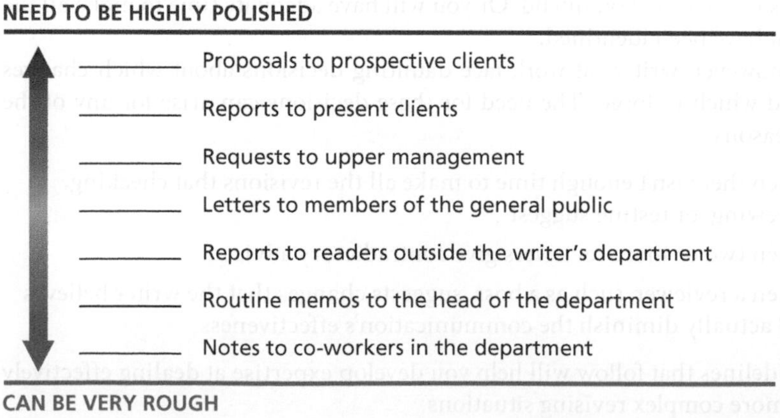

NEED TO BE HIGHLY POLISHED

_____ Proposals to prospective clients

_____ Reports to present clients

_____ Requests to upper management

_____ Letters to members of the general public

_____ Reports to readers outside the writer's department

_____ Routine memos to the head of the department

_____ Notes to co-workers in the department

CAN BE VERY ROUGH

First, your writing is a major factor on which your job performance will be evaluated. In fact, people who are not involved with your work from day to day may base their opinion of you solely on your writing. There's a real advantage to having readers say not only, "Your report provided us with lots of useful information," but also, "And you wrote it so well!"

Second, communications prepared at work often have a wider audience than the writer assumes. Notes to co-workers and your immediate manager can often be fairly rough. Sometimes, however, those people may pass your communications along to upper management or members of other departments for whom you would want to prepare more polished communications. When determining what level of quality to strive for, consider who all your possible readers may be.

Reasons for making your communications better than "good enough"

GUIDELINE 2 Make the most significant revisions first

After you've decided how much effort to invest in revising, identify the revisions that will bring the greatest improvement in the shortest time. If you spend fifteen minutes on revisions that result in small improvements rather than large ones, you have squandered your time.

Because the advice to make the most important revision first sounds so sensible, you may wonder whether anyone ever ignores it. In fact, many people do. They begin by looking at the first paragraph, making revisions there, and then proceed through the rest of the communication in the same sequential way. This procedure works well if the most serious shortcomings occur in the opening paragraphs, but not if they come in later passages the writers might not even reach before the time available for revising runs out. Also, some problems—for instance, inconsistencies and organizational problems—require simultaneous attention to passages scattered throughout a communication. A sequential approach to revising may ignore them altogether.

How to Rank Revisions

No single ranking applies to all communications. The importance of most revisions depends on the situation. For one thing, the relative importance of a certain revision is determined by the other revisions that are needed or desirable. Consider, for instance, the place of "supplying missing topic sentences" in the ranked list (Figure 15.7) that Wayne created after surveying the comments made by four reviewers of a proposal

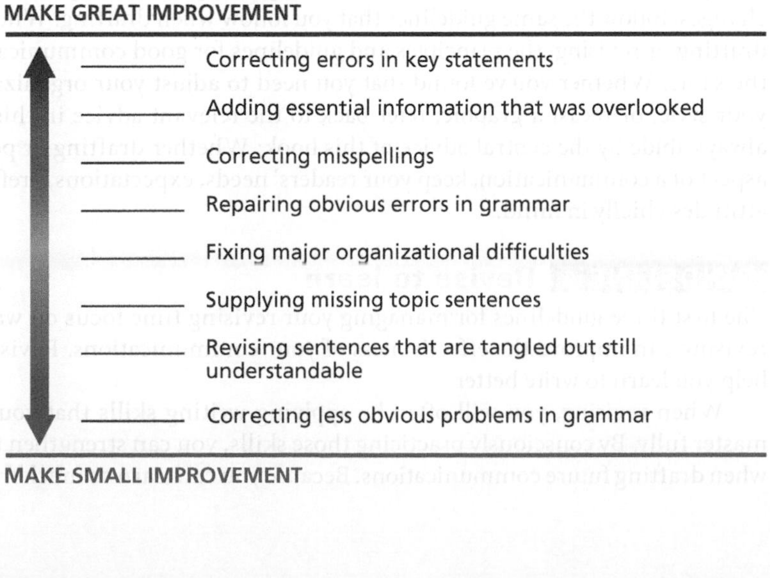

MAKE GREAT IMPROVEMENT

_____ Correcting errors in key statements

_____ Adding essential information that was overlooked

_____ Correcting misspellings

_____ Repairing obvious errors in grammar

_____ Fixing major organizational difficulties

_____ Supplying missing topic sentences

_____ Revising sentences that are tangled but still understandable

_____ Correcting less obvious problems in grammar

MAKE SMALL IMPROVEMENT

FIGURE 15.7

Amount of Improvement that Various Revisions Would Make in Wayne's Proposal he had drafted. In comparison with the other items in the list, supplying topic sentences is relatively unimportant. However, in a draft with fewer problems in more critical areas, providing topic sentences might bring more improvement than any other revision Wayne could make.

Another factor affecting the relative importance of a revision is its location. Improvements in an opening summary would increase the usefulness and persuasiveness of a long report more than would a similar revision to an appendix. Revisions in passages that convey key points have greater impact than similar revisions to passages that provide background information and explanations.

When you have limited time for revising, some of the revisions may never be made. By creating a ranked list, you ensure that the revisions you do complete are the ones that contribute most to your communication's effectiveness.

Be Sure to Correct Mechanical Problems

No matter how you rank other possible improvements, place the correction of mechanical problems—those that involve a clear right and wrong—at the top of your list, unless you are writing in a situation where you are sure that errors won't matter. Here are the major mechanical problems you should focus on:

- **Correctness.** Fix all errors in spelling, grammar, and punctuation as well as such things as the numbering of figures and the accuracy of cross-references.

- **Consistency.** Where two or more items should have the same form, be sure they do. For instance, revise parallel tables and parallel headings that should look the same but don't.

- **Conformity.** Correct all deviation from your employer's policies on such matters as the width of margins, the use of abbreviations, and the contents of title pages. Even if your employer hasn't developed written policies on these things, remember that your communication must still conform to your employer's informal policies and expectations.

- **Attractiveness.** Be sure your communication looks neat and professional.

GUIDELINE 3 To revise well, follow the guidelines for writing well

This guideline echoes an observation at the beginning of this chapter: When making changes, follow the same guidelines that you follow when drafting. Whether you are drafting or revising, the principles and guidelines for good communication remain the same. Whether you've found that you need to adjust your organization, polish your style, or recast a graphic, refer back to the relevant advice in this book. And always abide by the central advice of this book: Whether drafting or polishing any aspect of a communication, keep your readers' needs, expectations, preferences, and attitudes chiefly in mind.

GUIDELINE 4 Revise to learn

The first three guidelines for managing your revising time focus on ways in which revising can improve the effectiveness of your communications. Revising can also help you learn to write better.

When revising, you will often be applying writing skills that you have yet to master fully. By consciously practicing those skills, you can strengthen them for use when drafting future communications. Because you can learn so much from revising, it can often be worthwhile to continue polishing a draft after it is good enough to satisfy the needs and expectations of your readers and employer.

Conclusion

The complexities of working situations often require you to play several roles while revising. As an investor, you decide how much time to devote to revising a draft. As a diplomat, you negotiate with reviewers about changes they have suggested or demanded. As a student, you revise for the sake of what you can learn. The primary strategies underlying all the guidelines in this chapter are the same ones you should employ when you are planning and drafting any communication on the job: Consider the communication from the perspective of its readers, the contexts in which they will read and you are writing, and the communication's ethical impact on its stakeholders.

USE WHAT YOU'VE LEARNED

EXERCISE YOUR EXPERTISE

1. Following the advice given in this chapter, carefully check a draft you are preparing for class. Then give your draft to one or more of your classmates to review. Make a list of the problems they find that you overlooked. For three of those problems, explain why you missed them. If possible, pick problems that have different explanations.

2. The memo shown in Figure 15.8 contains eighteen misspelled words. Find as many as you can. Unless you found all the words on your first reading, explain why you missed each of the words you overlooked. Based on your performance in this exercise, state in a sentence or two how you could improve your reading for misspellings and typographical errors.

3. The memo shown in Figure 15.8 has several other problems, such as inconsistencies and missing punctuation. Find as many of those problems as you can.

EXPLORE ONLINE

1. Exchange electronic files of the drafts for your current assignment with a classmate. Using your desktop publishing program, insert comments in your partner's file, then return it to him or her. Evaluate the comments your partner provided for their clarity and ease of use, while he or she does the same for your comments. Share praise for comments that were presented effectively, and share suggestions concerning comments that could be presented more effectively.

2. Open an account at Google Docs (http://docs.google.com). Upload a short communication you've prepared for this course. Share the uploaded documents with a classmate. When you receive an email saying that a classmate has shared a document with you, go to the document and enter reviewer suggestions. Return to the document you uploaded to see the comments your classmate made in it.

COLLABORATE WITH YOUR CLASSMATES

This exercise will help you strengthen your ability to make constructive suggestions to a writer. (See this chapter's advice on reviewing other people's writing.)

1. Exchange drafts with a classmate, together with information about the purpose and audience for the drafts.

2. Carefully read your partner's draft, playing the role of a reviewer. Ask your partner to read your draft in the same way.

3. Offer your comments to your partner.

4. Evaluate your success in delivering your comments in a way that makes the writer feel comfortable while still providing substantive, understandable advice. Do this by writing down three specific points you think you handled well and three ways in which you think you could improve. At the same time, your partner should be making a similar list of observations about your delivery. Both of you should focus on such matters as how you opened the discussion, how you phrased your comments, and how you explained them.

5. Talk over with your partner the observations that each of you made.

6. Repeat steps 3 through 5, but have your partner give you his or her comments on your draft.

APPLY YOUR ETHICS

Find a print or online communication that you believe to be unethical or that you believe might possibly be unethical. Advertising is one possible source of such communications. Imagine that you are seeing it not as a finished

(continued)

communication but in draft form because its creator has asked you to review it. Plan your approach to raising and discussing the ethical issue with the person who drafted the communication. Because this person is free to accept or ignore your suggestions, you will need to convey your recommendation in a way that this person finds persuasive. Present your plan in the manner your instructor requests.

REFLECT FOR TRANSFER

1. Write a memo to your instructor describing two things that a reviewer of a draft you wrote did that you appreciated. Explain why each action assisted you. Were these actions described in this chapter? Are they actions you could take when reviewing other people's drafts on the job?

2. Write a memo to your instructor describing the three features of the communications you will write on the job that you feel you should pay the most attention to when checking your own drafts and when asking co-workers to review your drafts. Why have you put each of these features at the top of your personal priority list?

FIGURE 15.8

Memo for Use in Expertise Exercises 2 and 3

MARTIMUS CORPORATION
Interoffice Memorandum

February 19, 2017

From   T. J. Mueller, Vice President for Developmnet

To   All Staff

RE   PROOFREADING

     Its absolutely critical that all members of the staff carefully proofread all communications they write. Last month we learned that a proposal we had submited to the U.S. Department of Transportation was turned down largely because it was full of careless errors. One of the referees at the Department commented that, "We could scarcely trust an important contract to a company that cann't proofread it's proposals any better than this. Errors abound.

     We received similar comments on final report of the telephone technology project we preformed last year for Boise General.

     In response to this widespread problem in Martinus, we are taking the three important steps decsribed below.

I. TRAINING COURSE

We have hired a private consulting form to conduct a 3-hour training course in editing and proofreading for all staff members. The course will be given 15 times so that class size will be held to twelve participants. Nest week you will be asked to so indicates times you can attend. Every effort will be made to accommodate your schehule.

II. ADDITIONAL REVIEWING.

To assure that we never again send a letter, report, memorandum, or or other communication outside the company that will embarrass us with it's carelessness, we are establishing an additional step in our review proceedures. For each communication that must pass throug the regular review process, an additional step will be required. In this step, an appropriate person in each deparmtent will scour the communication for errors of expression, consistancy, and correctness.

III. WRITING PERFORMANCE TO BE EVALUATED

In addition, we are creating new personnel evaluation forms to be used at annual salary reviews. They include a place for evaluating each employees' writing.

III. Conclusion

We at Martimus can overcome this problem only with the full cooperation of every employe. Please help.

16

Testing Your Drafts for Usefulness and Persuasiveness

By following the reader-centered strategies described elsewhere in this book, you try your best to draft a communication that your readers will find highly usable and persuasive. However, you can't know for sure how your readers will respond until you actually place your draft in their hands. This chapter describes ways you can *test* a draft so that you can identify places that don't work the way you intended. With this knowledge, you can make improvements that increase the effectiveness of the final version you will deliver to your readers.

Companies such as Microsoft and Apple use testing extensively for the online and print instructions that accompany their products. By identifying problems and making improvements before products are launched, these companies reduce customer frustration and also save the significant amounts of money it would take to staff help lines or otherwise help customers solve problems that could have been avoided with clearer communication. In addition to instructions, drafts of websites, informational documents, and a wide variety of other communications are tested by both commercial organizations and government agencies. Testing is such an important activity in some organizations that they employ testing specialists and build special test facilities like the ones shown in Figures 16.1 and 16.2. Even without such facilities, you can increase the effectiveness of your communications by using the same basic strategies that testing experts employ.

The Logic of Testing

The basic logic of testing is simple: You ask one or more people, called *test readers*, to read and use your draft in the way your intended readers will read and use your finished communication. Then, by various means, you gather information from your test readers that will enable you to predict how other readers will respond. If your test is constructed effectively, the parts that work well for your test readers will probably work well for your target readers. Where the test readers encounter problems, your target audience probably will, too.

This chapter will guide you through the process of planning, conducting, and interpreting a test, whether you are testing a project for your technical writing course or a communication you are preparing on the job.

How to Define the Goals of Your Test

At a general level, all your tests will aim to answer two questions.

- How can I improve my draft? In the workplace, most tests emphasize the diagnostic goal of learning ways the draft can be improved.

LEARNING OBJECTIVES

1. Define the goals of your test.
2. Choose the people who will be your test readers.
3. Test your draft's usefulness.
4. Test your draft's persuasiveness.
5. Interpret the results of your test.
6. Test communications you write for readers in another culture.
7. Treat your test readers ethically.

MindTap®

Find additional resources related to this chapter in MindTap.

FIGURE 16.1

Layout of a Facility for Testing Computer Manuals

Test readers can use a draft manual or a website in the room on the Participant side.

On the Observer side, writers and testing specialists watch through a one-way mirror. They can also record all of the participant's actions and keystrokes.

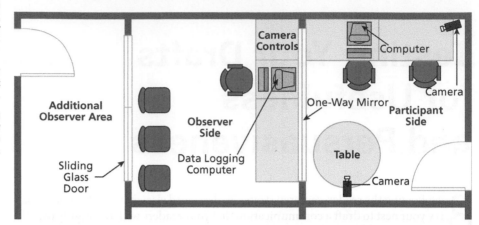

FIGURE 16.2

Professional Testing Facility

The test session is recorded on a video camera.

In the Test Room, the person with his back to the Observation Room asks readers about communications they read or view simultaneously.

The Test Room can be set up to test communications used in large spaces, such as instructions for assembling, servicing, or operating large equipment.

The observers are looking through a one-way mirror.

Spencer Grant/PhotoEdit

- Is my communication good enough? It is always helpful to learn how close your draft is to a "final" draft you can deliver to your target audience.

The first step in creating a test for answering these questions is to define its goals. Then you can tailor your test to provide the specific information you need.

Defining Your Test's Goals

1. **Review your communication's goals.** These goals, which you developed by following the guidelines in Chapter 3, tell exactly how your communication is intended to help and influence your readers. Your test should help you determine where your draft succeeds in achieving these goals and where it needs improvement.
2. **Identify the features you hope will contribute most to your communication's usefulness and persuasiveness.** For longer communications, you may not be able to test everything. Therefore, you need to focus on what you believe will have the greatest impact on your target readers' responses.
3. **Identify any features about which you feel unsure.** You can design your test to determine their effectiveness.

You can encapsulate your test's goals in a set of questions you want your test to answer. How well does my draft support my readers as they perform their first task? Their second one? How effective is it in influencing their attitudes in the way I want? How well do the headings work? Are the technical terms I used in the introduction understandable to them?

If you wish to determine whether your draft is "good enough," you may also want to establish *measurable criteria* for your test. For example, if you are drafting an instruction manual, you might decide that the manual is good enough if it enables all test readers to perform their tasks in less than two minutes or with no more than one error that is easily corrected.

However, in many cases a subjective judgment will be fully adequate: Your draft is good enough if it enables your test readers to perform their tasks without much difficulty and if they say your draft had a positive impact on their attitudes.

How to Choose Test Readers

To construct a test that accurately predicts how your target audience will respond to your finished communication, you must pick test readers who truly represent your target readers. If you are writing instructions for plumbers or computer programmers, choose plumbers or computer programmers for your test readers. If you are creating a website for adults who suffer from asthma, choose adult asthma sufferers. You could make your communication *less* effective for your intended readers if you made revisions based on responses from people who are not in your target audience.

If it is impossible to locate test readers from your target audience, find people as similar as possible to that audience. Begin by reviewing the knowledge of your readers that you developed by following the guidelines in Chapter 3. Pay special attention to your target readers' familiarity with your subject. If you select test readers who already know what your communication is intended to convey, they will probably understand passages that could baffle your target readers. Consequently, they won't help you detect needed improvements.

On the other hand, if your test readers know less than your target readers about your subject, their puzzlement could lead you to include background information and explanations your target readers neither need nor want.

How many test readers are enough? Because there is generally little variation in the ways readers respond to step-by-step instructions, two or three test readers may be sufficient for them. In contrast, people vary considerably in the ways they interpret and respond to a report or informational website. Common practice is to use about a dozen test readers for such communications. However, testing with even a single reader is infinitely better than no testing at all.

How to Test Your Draft's Usefulness

The following three guidelines explain how to test your draft's usefulness.

GUIDELINE 1 **Ask your test readers to use your draft the same way your target readers will**

Whenever you test a draft, one of your primary goals will be to evaluate its usefulness for its target readers. To learn where you need to make revisions, you must construct a test in which the test readers use your draft to perform the same tasks your target readers will want to perform with the finished communication.

The selection of test readers is a critical part of test design.

TRY THIS Using the search term "usability lab" or "usability test," find a company whose website enables people to sign up to participate in user tests. Can you find a site that conducts user tests on products or services that interest you?

Even a single test reader can provide valuable insights.

At work, most readers' tasks usually fall into three major categories.

- Perform a procedure, as in reading instructions.
- Locate information, as in looking for a certain fact in a reference manual or website.
- Understand and remember content, as when trying to learn about something through reading.

For many communications, all three types of tasks are important. For example, when people consult the owner's manual for a desktop publishing program, they usually want to *locate* directions for a particular procedure, *understand* the directions, and then use the directions as a guide while they *perform* the procedure. Although the three types of tasks are often bound together, each can be tested in a different way. Consequently, the following sections discuss separately the design of performance, location, and understandability tests.

TRY THIS Go to the usability website created by the government of any country (e.g., www.usability .gov for the United States). Find its advice about usability testing. On what points does its advice agree with the advice in this chapter? On what points does its advice disagree?

Performance Tests

To test a draft's effectiveness at helping its readers perform a procedure, make your test readers' reading situation resemble as closely as possible the situation of your target readers. The following strategies will help you plan the four major elements of *performance tests*.

Strategies for Performance Tests	
Tasks	Ask your test readers to perform the same tasks your target readers will perform.
Location	Conduct the test in the same setting as the target readers would use.
Resources	Provide the test readers with the same tools, equipment, reference materials, and other resources that the target readers would have—but not additional ones.
Information gathering	Gather information in ways that enable you to observe the details of the test readers' efforts without interfering in their use of your draft.

To see how you might apply these guidelines, consider the way Imogene tested instructions that would enable ordinary consumers to remove the radios that came with their cars and replace them with the higher-quality ones manufactured by her employer.

First, Imogene recruited two friends as her test readers. Both owned cars and neither had made a similar replacement before, so both represented her target readers. Imogene asked Rob to use her draft instructions to install the equipment in his garage at home and Janice to install hers in the parking lot of her apartment building. These are work areas that would be used by Imogene's target readers. Rob already had the necessary tools. Imogene supplied Janice with a power drill and Phillips-head screwdriver but nothing else because these might be the only tools an ordinary consumer would have.

While Rob and Janice worked, Imogene took detailed notes, using the form shown in Figure 16.3. During the test, she focused only on the left-hand column, writing down everything that seemed to indicate that Rob or Janice was having difficulty. So that she could pay full attention to the details of Rob's and Janice's efforts, she waited to fill in the other columns until after the test had been completed. To obtain a full sense of what Rob and Janice were thinking as they used her instructions, she asked them to "think aloud," verbalizing their thoughts throughout the process.

As he worked on the installation, Rob asked Imogene several questions. Instead of answering, she urged him to do his best without her help. If she had begun to provide oral instructions, she would no longer have been testing her written ones. However, she

FIGURE 16.3

Test Observation Form

After the test, you can ask the test reader to tell you what caused the problems you observed.

Observation Sheet for User Tests

Problem (What difficulty did the reader have?)	Interpretation (What might have caused the difficulty?)	Solution (What might prevent this difficulty?)
After completing a step, Rob sometimes hunted around the page for his place.	1. The steps don't stand out plainly enough.	1. Enlarge the step numbers. 2. Print the steps in bold so they are easier to distinguish from the notes and cautions.
At Step 7, Rob couldn't find the hole for the mounting bracket.	1. Rob couldn't understand the figure.	1. Simplify the figure by eliminating unnecessary details.

did assist at a point where Rob was completely stumped. Without her help, he would have had to stop work altogether, and Imogene wouldn't have been able to find out how well the rest of her instructions worked.

For some communications, you may need to create *scenarios*, or stories, that tell your test readers the tasks you want them to perform. For example, Shoon has drafted a website for a company that sells sporting goods on the Internet. In his performance test, Shoon asked his test users to imagine that they want to purchase a product for some particular activity, such as day hikes along easy trails or a weeklong ascent of a glacier-wrapped mountain peak. He then requested that they use his draft of the website to determine the appropriate products and order them.

Several other features of the test Shoon conducted illustrate variations that are possible when applying the strategies for performance tests. Because Shoon's website offered so many different kinds of products, it was not feasible for Shoon to ask his test readers to use every part of the site. Instead, he asked them to use only certain carefully selected parts of the site. Also, instead of using a worksheet to record information, Shoon placed a video camera behind his test users to record the entire test for later study. To learn what his test users were thinking as they used his website, Shoon engaged them in a conversation in which he continuously asked them what they were trying to do and how they were attempting to do it. He did not, however, tell them how to overcome problems they encountered. He chose this technique, developed by researchers M. Ted Boren and Judith Ramey (2000), because it is especially useful for testing online communications, where users often work so quickly that it is difficult to follow their actions.

Location Tests

Location tests closely resemble performance tests: You give your draft to your test readers, asking them to find specific pieces of information or the answer to a particular question as rapidly as possible. This procedure enables you to evaluate the effectiveness of the headings, topic sentences, table of contents, and other guideposts that reveal

AP Images/Ted S. Warren

Research Development & Human Factors Laboratory at the William J. Hughes Technical Center/FAA/United States Department of Transportation

Some usability labs have equipment that mirrors the test user's computer monitor (top) or detects where on the screen a test user is looking at each moment (bottom).

the organization of print communications (see Chapter 7). For websites and other online communications, location tests can help you assess the organization of your site and the navigational aids it contains.

Understandability Tests

In an *understandability test*, you ask your test readers to read your draft and then you ask them questions designed to determine whether they understood it accurately. Understandability tests often are combined with other kinds of tests. For example, because readers cannot use information unless they understand it, understandability is often an element in performance tests, such as those designed by Imogene and Shoon. Similarly, an understandability test can be combined with a location test by asking test readers to use the information they find.

If you want to test understandability in isolation, ask your test readers to read your draft and then pose questions about what they've read. Several types of questions are suggested below. Most were used by Norman, an insurance company employee, to test the following paragraph from a draft version of a new automobile insurance policy.

A passage Norman tested

> In return for your insurance payments, we will pay damages for bodily injuries or property loss for which you or any other person covered by this policy becomes legally responsible because of an automobile accident. If we think it appropriate, we may defend you in court at our expense against a claim for damages rather than pay the person making the claim. Our duty to pay or defend ends when we have given the person making a claim against you the full amount of money for which this policy insures you.

■ **Ask your test readers to recognize a correct paraphrase.**

Questions to determine whether test readers accurately recognize a correct paraphrase

> **True or false:** We will defend you in court against a claim when we believe that is the best thing to do.
>
> **Multiple choice:** When will we defend you in court? (a) When you ask us to. (b) When the claim against you exceeds $20,000. (c) When we believe that is the best thing to do.

■ **Ask your test readers to create an accurate paraphrase.**

A request that test readers create an accurate paraphrase

> In your own words, tell when we will defend you in court.

■ **Ask your test readers to apply your information.** To test your readers' ability to apply information, create a fictional situation in which the information must be used. Norman created the following question to test the understandability of one section of his draft (not the section quoted above).

A scenario used to determine whether test readers understood Norman's draft well enough to apply its information

> You own a sports car. Your best friend asks to borrow it so he can attend his sister's wedding in another city. You agree. Before leaving for the wedding, he takes his girlfriend on a ride through a park, where he loses control of the car and hits a hot dog stand. No one is injured, but the car and stand are damaged. Is the damage to the stand covered by your insurance? Explain why or why not.

GUIDELINE 2 **Interview your test readers after they have used your draft**

By interviewing your test readers after they've used your draft, you can often obtain a wealth of insights you couldn't otherwise have discovered. Here are five especially productive questions.

■ **What were you thinking when you encountered each problem?** Even when test readers try to speak their thoughts aloud as they read your draft, their account may be incomplete. For example, some test readers become so engrossed in their task that they forget to speak. Or, their thoughts fly faster than they can report them.

- **How did you try to overcome each problem?** Testing readers' strategies may suggest ways you can improve the presentation of your information.

- **How do you respond to specific elements of my communication?** Ask about any element about which you'd like feedback. Imogene might ask, "Do you find the page design appealing?" Shoon could inquire, "Do you like the website's colors?"

- **What do you suggest?** Readers often have excellent ideas about ways a communication can better meet their needs.

- **How do you feel about the communication overall?** When responding to open-ended questions about their feelings, test readers may reveal helpful information that your other questions didn't give them an opportunity to express.

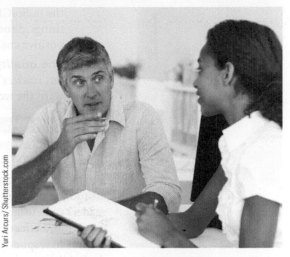

Interviewing test readers can be informative and enjoyable.

GUIDELINE 3 Minimize the impact of your presence

The value of your test depends on how closely your test readers' use of your communication matches your target readers' use of it. Anything that diminishes the closeness of this match can cause you to miss needed changes or revise in counterproductive ways.

Performance and location tests almost always involve one major difference: Unless you have someone else conduct your test, you will be right there when your test readers use your communication; however, you won't be present when your target readers use it. To gain the most value from your test, you must diminish the impact of your presence as much as possible.

Minimizing the Impact of Your Presence

- **Remain unobtrusive.** Let the test readers focus on their tasks. Be quiet. Sit to the side or behind them, out of their line of sight. If they are overly conscious of being observed, they may have difficulty concentrating.

- **Refrain from intervening unless absolutely necessary.** Even if your test readers ask for help, request that they rely only on the draft to solve problems. Intervene only if you must do so to enable them to continue with other parts of the test or if they are about to injure themselves or others.

- **Emphasize that you want to find problems and ways to improve your draft.** If they believe you will be disappointed to hear criticisms, they may refrain from pointing out some problems.

How to Test Your Draft's Persuasiveness

To test the persuasive impact of a draft, compare your test readers' attitudes before and after reading. At work, user tests usually focus on the readers' attitudes toward one or both of the following.

- **The *subject matter* of the communication.** Among other things, Imogene would look for positive changes in the readers' attitudes toward the radio/CD player and the job of installing it by themselves. Among other things, Shoon would want to determine his website's ability to prompt positive changes in his test readers' attitude toward products sold there.

■ **The *quality* of the communication.** Imogene would be concerned with test readers' attitudes toward the instructions she drafted and Shoon toward the website he created.

GUIDELINE 1 Use Likert-scale questions to evaluate persuasiveness

There are several ways to obtain information about your test readers' attitudes before and after they read your draft. Once they've finished reading, you can ask how their attitudes have changed. This approach may yield flawed results, however, because readers sometimes misremember their earlier attitudes.

A more effective method for determining attitude change is to ask your test readers to respond to the same Likert-scale question before and after they use your draft. Likert-scale questions use a numerical scale. If your test readers respond with the same numbers both times, there's been no change. If they respond with different numbers, change has occurred. Figure 16.4 shows three questions Imogene used.

Of course, there's also value in asking test readers to answer open-ended questions related to the persuasiveness of your draft, such as "How do you feel about my communication?" and "What could I do to make it more appealing or attractive?"

GUIDELINE 2 Avoid biasing your test results

Unless you are careful, your test readers may report their attitudes inaccurately. For example, if they believe you want praise more than suggestions for improvement, they will try to please you rather than help you improve your communication. You can take practical action to avoid accidentally biasing what your test readers tell you.

Avoiding Bias

■ **Emphasize your hope that the test readers will help you improve your draft.** Say this before anything else, and repeat the message from time to time.

■ **Phrase questions in a neutral way.** When drafting Likert-scale questions and interviewing test readers, avoid phrasing questions in ways that seem to steer them toward a certain answer. Your test readers may comply, thereby depriving you of accurate information needed as the basis for solid, reader-centered revisions.

■ **Welcome criticism.** If you seem disappointed or unreceptive to criticisms, test readers will stop making them. Assure the readers that all comments are helpful to you.

FIGURE 16.4

Likert-Scale Questions
Imogene Used to Test Her
Instructions' Persuasiveness

By asking her test readers to answer these questions before and after they used her instructions, Imogene could determine whether the instructions affected their attitudes.

Question	Strongly Agree	Agree	Neutral	Disagree	Strongly Disagree
I am confident that I can replace a car radio without difficulty.	1	2	3	4	5
The instructions for replacing the radio will be easy to follow.	1	2	3	4	5
Replacing a standard car radio with a high-quality one is worth the work it requires.	1	2	3	4	5

How to Interpret the Results of Your Test

As explained at the beginning of this chapter, testing involves prediction. Based on the responses of your test readers, you predict how members of your target audience will respond. However, every reader is an individual. Different persons respond in different ways to any communication.

Your challenge when interpreting test results is to determine whether test readers' responses to a particular paragraph, graphic, or design element is likely to be typical of the responses that your larger audience will have. There's no surefire way to answer that question. However, answering the following three questions can be helpful.

- **How many test readers did you have?** The more you have, the more likely that responses they share represent your target audience's likely responses.

- **In what ways do your test readers differ from your target readers?** Do they have a different level of prior knowledge about your subject? Different expectations and needs? You will need to make such comparisons in order to determine whether your target readers would respond differently than your test readers did.

- **What did you observe during the test that can help you interpret the results?** Sometimes your observations of test readers can help you interpret results. For example, if you notice that some test readers rushed through the instructions you are testing, you might conclude that their hurry, not a deficiency in your draft, may account for some of the difficulties they encountered.

How to Test Communications You Write to Readers in Another Culture

Testing communications that will be used in other cultures can present special challenges. For reasons explained throughout this book, the language, images, and other aspects of a communication that seem perfectly clear and appropriate in one's own culture can be incomprehensible, laughable, or offensive in another. When preparing communications for audiences outside your own culture, make a special effort to use test readers that represent those audiences. Readers from different cultures can approach communications from different perspectives and expectations. For example, researcher Daniel D. Ding (2003) found significant differences between instruction manuals for installing a water heater that were written in China and in the United States. The Chinese manual describes a series of ideal relationships among the parts but does not describe the specific actions the installer would take to achieve these relationships. Installers must bring their experience to the instructions in order to perform the task. Ding explains that this understanding of the way instructions should be written emerges from Chinese culture, in which context and individual objects are seen as a unity. In contrast, the U.S. instructions consist of lists of actions, described in detail, so that the installation can be accomplished independent of the installer's previous knowledge. These substantial differences illustrate how important it is for writers in one culture who are addressing readers in another culture to use test readers from that other culture.

If it is impossible for you to arrange for such test readers, it will be particularly important for you to have your communication reviewed by persons who are knowledgeable about those audiences.

How to Treat Test Readers Ethically

An ethical principle developed for research in medicine, psychology, and many other fields also applies to the testing of the drafts you write at work or in class. This principle states that the people involved should be volunteers who have agreed to participate after being fully informed about what you are asking them to do.

For certain kinds of research, this is not only an ethical requirement but also a legal one. Researchers must tell potential volunteers about the study in writing and obtain in writing the volunteers' consent to participate.

Although such legal requirements probably will not apply to the testing you do of draft communications you prepare at work, the ethical principle still does. When asking others to try out something you have drafted, let them know the following.

- The test's purpose
- The things you will ask them to do during the test

FIGURE 16.5

Sample Informed Consent Form

Using this informed consent form, a student efficiently communicated the information needed to treat his test readers ethically.

The writer tells potential volunteers what they would be doing during the test.

The writer describes the risks involved. (There are none in this case. For tests that involve risk, the statement would need to be changed.)

The writer explains the test's purpose.

He lets potential volunteers know that they would also be asked to answer questions.

The writer explains how much time he is asking potential volunteers to give.

He explains that volunteers can stop participating at any time.

The writer tells when and where the test will take place.

The writer provides a place for volunteers to confirm that they are participating voluntarily.

Informed Consent

I am seeking volunteers to test a _____ that I am creating as part of my work for_____.

If you are interested in volunteering, please read the following information carefully. If you decide to participate, please sign below.

1. In the test, you will use a draft of _____.

2. You will not encounter any risk of harm in this process.

3. The purpose of this test is to enable me to identify ways to make my draft more effective. It is not a test of your abilities.

4. I will observe you as you read and use my draft. Afterward, I will ask you questions about the draft.

5. The entire test will take between _____ and _____ minutes.

6. You may stop participating in this test at any time.

7. I will conduct the test at a time that is convenient for you.

8. The test will take place at _____.

9. I will be glad to answer any questions you have about the test.

_____

I have read and understood the description given above concerning the test. I volunteer to participate.

Name _____

Signature _____

Date _____

- The time required for the test
- Where the test will take place
- Any potential risks to them (some lab procedures, for instance, do involve risk if not done properly)
- Their right to decline to participate
- Their right, if they do volunteer, to stop participating at any time

Figure 16.5 shows a sample informed consent form.

Conclusion

Figure 16.6 shows a planning guide that will help you learn how your readers will respond to a communication you are drafting. With the insights your tests produce, you will have the surest possible guide to revising your drafts so you can create final versions that work for your readers in exactly the ways you want them to.

FIGURE 16.6

Writer's Guide for Testing Drafts

Writer's Guide
TESTING DRAFTS

GOALS OF YOUR TEST

1. What are your communication's usefulness objectives?
2. What are its persuasive objectives?
3. What parts or features do you need to test for the following reasons?
   - ❏ Because they are so crucial to your communication's success
   - ❏ Because you are unsure how well they will work

TEST READERS

1. Who are your target readers?
2. Can you find test readers from this group? If you cannot, what test readers would closely resemble your target readers? What are the important differences between your test readers and the target readers?

TESTING YOUR DRAFT'S USEFULNESS

1. What will you ask your test readers to do so you can evaluate your draft's usefulness?
2. What resources will your target readers have that you should provide to your test readers?
3. Can you conduct your test where your target readers will use your communication? If not, how can you simulate that location?
4. How will you observe what your test readers do while using your draft?
5. How will you learn what your test readers are thinking as they use your draft?
6. What questions will you ask your test readers when you interview them after they've finished using your draft?

TESTING YOUR DRAFT'S PERSUASIVENESS

1. What attitudes will you ask your test readers about?
2. How will you avoid biasing their responses?

ETHICS

1. What do you need to say in your informed consent form in order to treat your test readers ethically?

USE WHAT YOU'VE LEARNED

EXERCISE YOUR EXPERTISE

1. Explain how you would test each of the following communications:

    a. A display in a national park that is intended to explain to the public how the park's extensive limestone caves were formed.

    b. Instructions that tell homeowners how to design and construct a patio. Assume that you must test the instructions without having your test readers actually build a patio.

2. Following the guidelines in this chapter, design a test for a communication that you are now drafting.

3. Complete the project on user testing given in MindTap.

4. Design a user test for your school's website. Imagine that the target audience is high school students who are deciding what college to attend. Who would you select as test readers? How would you define the objectives of the test? What would you ask them to do, and how would you gather information from them concerning the site's usefulness and persuasiveness?

EXPLORE ONLINE

At one of its websites, the United States government provides instructions for conducting usability testing (www.usability .gov). After visiting the site, write a memo to your instructor explaining why the government feels that usability testing is so important that it publishes instructions for performing the tests.

COLLABORATE WITH YOUR CLASSMATES

Interview a classmate about a project he or she is drafting. Then design a test that will help your partner identify ways that he or she could revise the draft to make it more usable and persuasive for its target audience.

APPLY YOUR ETHICS

Following the advice given in "How to Treat Test Readers Ethically" and Figure 16.5 (pages 302–303), develop an informed consent form that you could give to people you ask to volunteer as test readers of a draft you are preparing for your class. How could you recruit volunteers so that they won't feel pressured to agree to be test readers even though they would prefer not to? Are there any aspects of your test that could conceivably harm your test readers physically, emotionally, or socially? If so, how can you alter your test to eliminate this risk?

REFLECT FOR TRANSFER

1. Write a memo to your instructor in which you address the following question: When you ask co-workers to review a draft (Chapter 15), to what extent are you asking them to play the role of a test reader? When you ask them to think about how readers might respond to your draft, are you asking them to imagine that they are the test reader so they can draw conclusions based on their own reactions? Explain your reasoning.

2. Write a memo to your instructor identifying three principles and practices of testing that you could apply in your career. Describe sample situations. Also identify three principles and practices you don't believe will be useful to you in your career. Explain why.

APPLICATIONS OF THE READER-CENTERED APPROACH

17

Creating Communications with a Team

LEARNING OBJECTIVES

1. Develop a shared understanding of team goals and procedures.
2. Make team meetings efficient and highly productive.
3. Use technology to increase the team's effectiveness.

Teamwork skills are highly valued in the workplace. As you learned in Chapter 1, when 400 employers were asked in another survey to identify the most important outcomes of a college education, they ranked *working effectively with others in teams* among their top three (Hart, 2015). The other two, you will recall, were *communicating effectively in writing* and *communicating effectively orally*, which are essential teamwork skills.

Why do employers place such a premium on teamwork skills when hiring? Both research and experience have led employers to appreciate that when people work in teams, the results are greater than the sum of what team members could produce working individually (Duhigg, 2016). However, the difference in the quality of work produced by effective teams and ineffective ones can be tremendous. And a team's effectiveness depends as much on the teamwork skills of its members as on their technical or other specialized knowledge. This chapter will help you increase your value to your employer by helping you develop the teamwork skills employers value so highly.

Varieties of Team Structures

In the workplace, you will see teams with many different structures. The structure of any particular team depends on many factors, such as the employer's preferred practices, the size and nature of the project, and the talents of the team members. Three variables are especially important for you to understand as you strive to build your expertise at teamwork.

- **Leadership.** On some teams—especially large ones—a single person is designated as the leader or manager. In other cases, team members are collectively responsible for all aspects of their work. Both arrangements have variations. For example, the person assigned as team leader may make all critical decisions personally or may facilitate discussions that produce consensus among team members.

- **Distribution of tasks.** Some teams give each member exclusive responsibility for a specific portion of the final written work. This approach can produce excellent results for written projects because each person concentrates on the sections related to his or her area of expertise. However, it risks creating redundancies and inconsistencies among sections of the final document because teams work in isolation, without knowing in detail what other team members are writing. In contrast, on some teams all members work together on every task. This arrangement can produce a highly polished outcome because all minds are applied to the entire project, but it can also be slow and inefficient.

MindTap

Find additional resources related to this chapter in MindTap.

Many teams use a hybrid model. For example, the whole team may define the communication's objectives but then assign individuals to draft different sections. The team may reassemble to edit collaboratively.

- **Technology.** Different teams use the growing array of online collaborative writing tools in different ways. Some develop online locations for storing research resources that can be accessed by all team members; many share, review, and edit one another's drafts online; some hold meetings over the Internet. Using these resources, some teams complete entire projects without ever meeting in person.

Members of a writing team can share leadership responsibilities (top) or one person can be designated as the team leader (bottom).

Keys to Team Success

No matter how they are structured, all teams depend on the same strategies to maximize their success. This chapter's goal is to supplement your instructor's guidance in helping you learn to use—and help your teams use—these strategies for maximizing team commitment and energy, developing a common understanding of the team's goal, conducting efficient and productive meetings, and using technology to increase the team's effectiveness.

Treat Other Team Members with Sensitivity and Respect

I'm guessing that as you read this advice, you were thinking that I am about to moralize rather than provide you with substantive advice. That is not the case. For five years a special research group at Google gathered extensive data on 180 teams scattered throughout its organization, trying to determine what factors made some more productive than others (Duhigg, 2016). Was it building the team exclusively with people who were the most productive as individuals? Was it having a strong leader or a more casual organization? Sticking strictly to the agenda at meetings or digressing from it? None of these and many, many other possibilities accounted for the difference.

The Google team finally had its breakthrough when it read a report by researchers at MIT, Carnegie Mellon, and Union College, whose experiments demonstrated that two factors distinguished successful teams from unsuccessful ones (Woolley, Chabris, Pentland, Hashimi, & Malone, 2010). Both involve the way team members interact with one another. First, members are sensitive and responsive to one another's feelings, often intuiting from nonverbal cues such as their expression and tone of voice. Second, every member has an equal opportunity to talk. Conversations are not dominated by one or a few members.

Analyzing the mountains of data it had collected over the years, the Google team found that the same factors distinguished the more successful from the less successful of the 180 teams it had studied. Together, these traits establish the team as a "safe place," where members can express ideas and opinions without fear of being attacked, embarrassed, or otherwise made uncomfortable.

This finding, which places interpersonal factors ahead of managerial ones, creates two questions for you as you seek to increase your teamwork skills. First, what can you do to become an even more sensitive and responsive team member than you already are? Second, because making a team a "safe place" relies on the behavior of all team members, what can you do to influence the behavior of other team members? Here are some strategies you can suggest to the other members of your teams and model for others in your interactions.

- **Get to know each other.** As the team first meets, suggest that members share information about themselves that can help others understand their goals and feelings. One constructive topic could be descriptions of good and bad experiences they've had on previous teams. What happened, and how did they feel as a result?

- **Create a team charter.** A charter is an agreement written collectively by the team that describes the ways the team will work together. Among the topics described, the ways individuals will treat one another belongs at the top of the list. Other possible topics include the team's goals, the roles of each member, and ground rules about such things as expectations for attendance and completion of tasks, as well as the team's plans for dealing with conflicts and failure to meet expectations.

- **Listen actively.** When others speak, be sure they feel that they have been heard and taken seriously. Don't interrupt. Where appropriate, ask questions to fill out your understanding of what they are communicating. Refrain from interrupting.

- **Invite quiet team members to contribute.** Ask for their suggestions or thoughts on ideas that have been generated. A simple invitation can overcome the hesitation of some people to join in.

- **Express appreciation for others' contributions.**

- **Watch nonverbal communication.** Another team member's facial expression or body language may suggest that they are feeling uncomfortable, hurt, or angry about an issue. Ask what they are thinking. Respond in a respectful and sensitive way.

- **Address conflict in a nonconfrontational way.** It is inevitable that sometimes team members will disagree. When this happens, let each person express his or her thoughts fully. Listen actively to both positions. Look for points of agreement and suggest compromises. If common agreement can't be reached, take a vote. If you are on the losing side, support the team in its decision through your actions. Research shows that if members don't compromise, a team's effectiveness plummets (Duhigg, 2016).

In sum, these suggestions are techniques for doing what common sense says: Treat team members respectfully. The more you and each of your teammates do that, the more effective your entire team will be.

Develop a Shared Understanding of Team Goals and Procedures

Teams are able to accomplish more than individuals not only because a group can accomplish more work than a single person but also because several minds working together can often produce better, smarter, and more effective results. To enable your teams to realize this strength in numbers, help them develop a shared understanding of what you collectively seek to accomplish and how you will work together to produce this result.

GUIDELINE 1 Create a shared understanding of the communication's goals

The first steps toward creating a team-authored reader-centered communication are identical to the steps you take when writing single-authored reader-centered communications.

- Learn about your readers' characteristics, situation, needs, and expectations.
- Define your communication's usability goals by identifying the tasks your readers will use the communication to perform.
- Define your communication's persuasive goals by describing how you want it to influence their attitudes and actions.

At a project's beginning, your team's members are likely to have a variety of insights and ideas about the project's objectives and readers. Some will supplement one another. Others may seem incompatible, at least at first. The following strategies can help team members achieve a common understanding of what their communication should achieve.

LEARN MORE To review the reader-centered approach to defining a communication's goals, go to Chapter 3.

LEARN MORE The "Writer's Guide: Defining Your Communication's Goals" (page 52) can be used for team projects as well as for individual ones.

Defining Goals as a Team

- Take sufficient time to explore the diverse views of all team members regarding the communication's goals. Ask questions yourself and invite questions from others.
- Keep talking until it is clear that everyone agrees.
- Record in writing the group's conclusion.
- Remain open to new insights about your readers and objectives as your team's work progresses.

GUIDELINE 2 Develop and share a detailed plan for the finished communication

When planning a communication that you'll write by yourself, you can begin with fairly general plans that become more specific as you proceed. With team projects, however, hazy plans can create multiple problems. When team members begin their individual work based on a vague plan, they often find they've invested a significant amount of time and creativity in producing drafts that later have to be modified significantly or even discarded because they don't match the more detailed plan that the team eventually develops.

Here are three actions that can help your team avoid the frayed nerves and wasted time that such misunderstandings can cause.

Planning Team Projects

- **Discuss plans as a team.** When teams discuss plans, the individual members gain a fuller understanding of what they are trying to accomplish together. In addition, the plans themselves often improve as a result of the discussion.
- **Discuss the communication's outline together.** By discussing the outline, teammates can develop a common, detailed understanding of the communication's logic and structure.
- **Create storyboards.** After your team has discussed the outline, each team member can complete a storyboard form for the sections he or she will draft. The form (Figure 17.1) asks for the section's main point, supporting points, and graphics. The team can assemble and review the storyboards, adjusting each so they fit tightly together. This storyboard review should give each writer a clear sense of what he or she will write.

LEARN MORE For information about a different variation of storyboarding used with oral presentations, see "Plan the verbal and visual parts of your presentation as a single package" (page 326).

Although teams benefit from making detailed plans early, the plans need to remain flexible. New information and ideas gained along the way can require a new organization and different presentational strategies.

FIGURE 17.1

Storyboard for Use by Writing Teams

Before anyone begins drafting, team members can use storyboards to decide together how each part will be written.

Team members sketch or make notes about the graphics that will be integrated into their sections.

Team members state the major point their sections will make

Team members identify the points that will support or explain the major point.

After all storyboards have been reviewed by the team or the team leader, each team member will have a clear idea about how to write his or her section so it will fit easily into the communication's overall structure.

Storyboard

Project _____   Writer _____

Section _____   Subsection _____

Topic _____

● Main Point   Graphics

● Major Subpoints

1.

2.

3.

GUIDELINE 3   Make a project schedule

Encourage your team to make a schedule at the very beginning of its efforts. A simple timetable, with interim deadlines, assures that the team advances steadily toward its goal and allows adequate time for all stages of its work. Each person thus knows exactly when his or her tasks must be completed. This provides motivation to complete assignments on time since it palpably demonstrates how missing a deadline will create problems for later steps in the process.

When you create a schedule, include the following elements.

Creating a Project Schedule

LEARN MORE For advice about creating schedules for client and service-learning projects, see Chapter 19, "Your Schedule" (page 340).

LEARN MORE For guidelines for creating schedule charts, see "Writer's Tutorial: Schedule Charts" (page 259).

- **Include time for defining objectives and planning.** The team will have an opportunity to act on Guidelines 1 and 2 above only if sufficient time is provided in the schedule.
- **Include milestones.** Milestones indicate the dates by which various stages of the work will be completed. These interim deadlines help teams progress at a steady pace. If a milestone is missed, the team understands it must speed up its remaining work to complete the entire project on time.
- **Create enough milestones to catch potential problems early.** For instance, in addition to a deadline for the completion of final drafts, set a date for completing rough drafts for team review. Interim deadlines enable the team to identify and address problems soon enough to complete the final draft on time.
- **Include time for editing.** No matter how much discussion and planning have occurred earlier in the process, almost every team-written communication needs editing before delivery to its readers. The sections may be written in different styles, leave gaps, or repeat some of the same information. Whether one editor is chosen or everyone edits together, time will be needed for this activity.

Make Team Meetings Efficient and Highly Productive

In addition to creating team meetings as a safe place by responding sensitively to all team members and assuring that all have a chance to talk, you can take several other steps to increase the efficiency and productivity of team meetings.

GUIDELINE 1 Set and follow an agenda

Nothing will be more precious to your communication teams than time. Each participant has other projects and responsibilities and is taking time away from them. You can help your teams spend as little time as possible in meetings by encouraging them to employ the following strategies.

Conducting Efficient Meetings

- **Prepare an agenda.** Before the meeting, have someone list the major issues to be discussed. Open the meeting by having the team review the agenda to be sure everyone agrees on what is to be accomplished. This small amount of preparation by one person can save large amounts of meeting time for many persons.
- **Ensure that agenda items are addressed.** Seeing that all agenda items are addressed by the end of a meeting is not the same as sticking strictly to the agenda. The Google research team found that some highly effective teams did, in fact, talk only about agenda items but that other highly effective teams began meetings with social chat and digressed regularly from the agenda. Whatever pattern each of your teams follows, remind them at appropriate times of the issues to be addressed.
- **Bring discussions to a close.** Communication teams sometimes keep debating a topic even after everything useful has been said. When a discussion becomes repetitive, say something like, "We seem to have explored the options pretty thoroughly. Are we ready to make a decision?" If someone objects, the team can clear up any point that still needs to be resolved.
- **Sum up.** After all the topics have been covered to everyone's satisfaction, sum up the results of the meeting verbally. Such a statement consolidates what the team has accomplished and reinforces its decisions. Follow up this summation in writing.
- **Set goals for the next meeting.** Make sure everyone knows exactly what he or she is to do before the next meeting. This strategy helps assure that when you meet again, you will have new ideas and material to discuss. Nothing is more discouraging to a team than discovering at its next meeting that it is at exactly the same point as it was at the previous meeting.

TRY THIS Think of a meeting in which you recently participated. Which of these strategies for conducting efficient meetings would have most improved that meeting?

GUIDELINE 2 Encourage discussion, debate, and diversity of ideas

As the Google research suggests, highly effective teams are so successful partly because teams work best when all members feel they can safely express their ideas—even if their ideas conflict with another team member's ideas. Consequently, one of the most important things you can do in meetings is to encourage productive debate and consideration of a rich diversity of ideas. Debate ensures that the team won't settle for the first or most obvious suggestion. It also enables your team to avoid *groupthink*, a condition in which everyone uncritically agrees at just the time when critical thinking is most needed.

Pages 307 and 308 describe ways you can treat other team members with sensitivity and respect. Treat that list as actions you can proactively take during meetings.

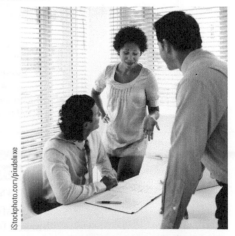

iStockphoto.com/pixdeluxe

In group meetings, active listening is as important as it is in one-to-one conversation.

GUIDELINE 3 Take special care when revising drafts

Teams can greatly increase the usability and persuasiveness of their drafts by sharing suggestions for revisions. Yet, these discussions can be the most challenging for teams to conduct in a way that creates the safe space needed for effective team functioning. Team members can be especially sensitive to comments about what they have drafted, sometimes taking what otherwise would be neutral suggestions as criticism. Being aware of their teammates' sense of vulnerability, team members may withhold suggestions that would improve the team's product. As a result, when reviewing drafts, even teams that act in highly effective ways throughout the rest of their work together may behave in the ways that characterize less-effective teams.

Chapter 15's guidelines for reviewing drafts can help teams avoid these barriers to productive reviews. Especially relevant to team writing are the following six guidelines from Chapter 15, plus two more that apply specifically to team projects.

Discussing Drafts

LEARN MORE For a fuller discussion of these strategies, see "When You Are a Reviewer" (pages 283–285).

STRATEGIES FROM CHAPTER 15

- Begin comments about a draft with praise for what's strong.
- Explain the reader-centered reasons for all suggestions. Remind the writer that your suggestions are intended to improve the communication for its intended audience.
- Avoid phrasing suggestions in ways that sound like criticisms of the writer's work.
- Rank suggested revisions beforehand and focus on the ones that will have the greatest impact.
- Distinguish matters of substance from matters of style.
- Accept all reasonable suggestions for improving your own draft and thank team members for their ideas.

SPECIAL GUIDELINES FOR TEAM PROJECTS

- **Treat drafts as team property, not individual property.** Encourage each other to give up personal "ownership" of drafts. Take pride in your contributions, but view your drafts as something you've created for the team; they have become its property.

    This doesn't mean that you should remain silent if someone suggests a change that would weaken your draft. Join in a reasoned conversation about the best way to write "our" communication rather than struggling to protect "my" writing. Your lack of defensiveness about your draft will encourage others to adopt the same attitude concerning theirs.

- **Swap responsibilities.** Responsibility for revising each section could be given to someone other than the person who drafted it. When all drafts are combined into a single communication, one person might be responsible for editing and polishing. With these arrangements, all drafts clearly become team drafts.

GUIDELINE 4 Global Guideline: Help your team work across cultural differences

Successful teamwork requires that each team member be fully engaged. One hazard for multicultural teams is that members can misinterpret one another's ways of interacting, with the result that opportunities for engagement are missed.

A first step in helping your teams work across cultural differences is to recognize the many ways these differences can impact discussions of ideas.

Cultural Differences in Discussions of Ideas

Bosley (1993) has identified four important differences in the ways that people from various cultures may interact on workplace teams.

Recognizing cultural differences

- **Expressing disagreement.** In some cultures, individuals typically express disagreement directly. In others, people say "no" only indirectly in order to save face for themselves and the other person.

- **Making suggestions.** People from some cultures offer suggestions freely. To avoid embarrassment, people from some other cultures avoid saying anything that might be interpreted as disagreement.

- **Requesting clarification.** In some cultures, individuals frequently ask others to explain themselves more clearly. People from other cultures feel that it's rude to ask for clarification; doing so would imply that the speaker doesn't know what he or she is talking about or hasn't succeeded in explaining things clearly.

- **Debating ideas.** Whereas members of work teams in some cultures debate ideas vigorously as a way of exploring ideas, people from other cultures regard such behavior as disloyal and unacceptable.

Cultural Differences in Behaviors in Conversation

The ways people interact in everyday conversation show up in the ways they converse in team meetings. In Sweden, listeners signal that they are being attentive by sitting straight and folding their arms in front of them (Rabinowitz & Carr, 2001). However, if you were raised in the United States, you might interpret the Swedish posture as signaling boredom, disagreement, or even hostility. In the United States, most people signal attentiveness by leaning forward and openness by keeping their arms spread.

Similarly, cultural differences exist concerning eye contact. People raised in the United States are likely to interpret eye contact as a signal of sincerity and interest in the other person. Eye contact is even more important in Arab cultures, where people use intense eye contact to read someone's real intentions. However, if you were raised in Korea, you might interpret constant eye contact as rudeness. In Japan, looking someone in the eye is an invasion of his or her space (Varner & Beamer, 2005).

Cultural Differences and Gender

Different cultures also have different expectations about the ways men and women will behave. For example, research shows that many men in the United States present their ideas and opinions as assertions of fact. When exploring ideas, they may argue over them in a competitive manner. In contrast, some women offer their ideas tentatively, introducing them with statements such as "I think…" or "I'm not sure about this, but …" If there is disagreement, women may support part or all of the other person's ideas and seek to reach consensus (Lay, 1989). Differences in gender expectations can lead to misinterpretations even when team members are from the same culture.

Improving Team Effectiveness across Cultural Differences

The best tools for assuring team effectiveness across cultural differences are knowledge, self-awareness, and flexibility.

■ **Knowledge.** The more members of your team know about one another's cultural expectations and behaviors, the greater their ability will be to avoid counterproductive misinterpretations. No guidebook can beat the effectiveness of conversation for developing this knowledge. These conversations can occur during team meetings or outside of them. In fact, you'll find that one of the great pleasures of working with persons from different cultural backgrounds is learning about the cultures and the people who reside in them.

■ **Self-awareness.** Also important is your own awareness of the extent to which your own cultural background influences your interpretations of your teammates' behaviors. If you grew up in a culture where people explore ideas through competitive debate, you may misinterpret the style of a team member whose culture values indirectness as signaling a lack of commitment to the team. If you are the person from a culture that values indirectness, you may misinterpret the other person's directness as rude and aggressive. Talking with other team members about your understanding of them is the best way to test your interpretations. But remember to be sensitive and cautious at first. Ask questions rather than make blunt statements.

■ **Flexibility.** Through conversation, you will likely learn that you, too, are being interpreted incorrectly by others. In addition to discussing the meaning of these behaviors in your culture and that of one or more teammates, think of ways you can modify your behavior to reduce your chances of being misunderstood.

Ultimately, each multicultural team must develop its own ways of working as productively as possible. A team can't succeed in doing that if the members merely share generalizations about their cultures. Instead, they must develop a mutual understanding that sees each member's ways of interacting as partly an expression of his or her culture. This understanding is an important element in the working relationships that team members develop.

Use Internet and Cloud Technology for Drafts

Most teams make extensive use of computer technology to support their work. You can increase your value to your teams by helping them use these resources as productively as possible.

GUIDELINE 1 **Choose the computer technology best suited to your team's project**

Many companies offer computer programs and services that support team efforts. Some are very expensive. Some are free. Each supports some team activities better than others. There's a considerable payoff from investing the time to identify the one program or group of programs that best supports your team's project.

The range of tasks that need to be supported is often more complex than might appear at first glance. One or more team members will create an initial draft or drafts, which will be distributed to the rest of the team. When reviewing these drafts, each team member will make comments, raise questions, offer suggestions, and revise by adding, deleting, or moving content. The resulting drafts will be reviewed, perhaps through many cycles. The larger the team, the more drafts to be prepared, the more difficult the work will be.

Here are three kinds of computer support, each suited to different situations.

- **Circulating drafts made with standard desktop programs.** Programs that you use every day, such as Microsoft Word and PowerPoint, include many features that support people working collaboratively. You and others on your team can insert comments, highlight changes you've made for others to review, and compare drafts readily. Drafts can be distributed as email attachments. Note, however, that if more than one of you works simultaneously on a draft, someone will need to enter both sets of changes into the same draft—and decide which version to use when they disagree. To avoid this complication, some teams pass a single draft from person to person in an agreed-upon pattern.

- **Working on a single draft in the cloud.** Rather than circulating a draft by email, you can place it at Google Drive (www.drive.google.com) or a similar site where all team members can work on it using Internet Explorer, Firefox, or another browser. Because all revisions are made to this single draft, there's no need to merge changes from separate versions. Unfortunately, some of these programs have limited word-processing options, so you can't add many of the page design features that increase a communication's usability and persuasiveness until you've downloaded it to your own computer. Also, some have trouble with tables, graphs, and similar content. However, when you are working with simple designs and straight text, they are very helpful.

- **Storing and downloading drafts on the Internet.** For documents that feature sophisticated page designs and complex content, you can also place your drafts at Google Drive, wikispaces.com, or a similar site for downloading only. If you don't open the files, the downloads will be identical to what you or your teammates uploaded.

Internet resources also enable you and your teammates to talk about drafts without being in the same room. If you are using Google Drive, for instance, to share the document, you can open the document in one window and a Skype audio or video chat in another.

LEARN MORE For more on the collaboration features of Word, see "Present your suggestions in the way that will be most helpful to the writer" (pages 285–286).

TRY THIS Set up a free account at Google Drive. Upload a draft of a project for one of your courses. Using Google Drive's Share feature, send an email to a classmate asking for suggestions. Return to your account to review his or her recommendations.

GUIDELINE 2 Use social media to your advantage

Social media can assist teams in several ways. As explained in Chapter 5, blogs, wikis, and other social media can be very helpful resources for research. Also, teams can use text messages, Twitter, and other social media for quick, convenient communication and coordination. In the workplace, many employees and organizations are finding email to be a cumbersome and time-consuming way to connect with each other for many purposes. Do you have a quick question for a teammate? Did you find a resource that other team members might find helpful? Don't wait until your next team meeting. Send a brief message using the same social media resources you would use to send a personal note.

GUIDELINE 3 **For virtual teams, foster personal relationships and conversational interchanges**

As you learned in Chapter 1, some teams, called *virtual teams*, work entirely online. Pam Brewer, an expert on international virtual teams, found that the results of Google's research are as important—or even more important—for teams that meet digitally rather than in person (Chong, 2016). Research suggests three remedies.

- **Include opportunities for social conversation.** When teams meet in person, members chat while waiting for the session to begin, during breaks, and when walking to or from the room. The personal relationships they develop motivate them to help one another by contributing to the team effort.

- **Encourage team members to be especially attuned to others' responses.** When meeting in person, you can often tell whether someone is puzzled, irritated, or reluctant to speak up by reading the person's facial expressions, body language, and other behaviors. These same signals are not easy to detect in virtual meetings. The best solution? Ask frequently for each person's thoughts.

- **Be especially alert to cultural differences.** The Internet can magnify people's impressions of others' communication styles, increasing the chances that team members from cultures with different communication customs will make negative judgments about one another. Does one team member seem to you to be overly aggressive? Perhaps the person is from a culture accustomed to exploring ideas through competitive debate. Does a team member seem to lack the conviction of points he or she seems to be making? Maybe the person is from a culture where people usually make suggestions indirectly.

Learning Team Skills through Feedback

As explained in Chapter 1, feedback from your instructor, fellow students, and others is a powerful resource as you seek to extend your communication abilities. Feedback from other team members can be especially helpful. They are able to see you in action in ways your instructor cannot. Figure 17.2 shows a feedback form that you and your teammates may complete for one another. You'll learn most if your ratings and explanations are candid rather than flattering. Also, try using the form to evaluate your own performance.

FIGURE 17.2

Collaboration Feedback Form

COLLABORATION FEEDBACK FORM

Team Member _____   Date_____

	Poor		Excellent	
Helps the team create a "safe place" for all team members to contribute				
● Sensitive and responsive to other team members' feelings	1	2	3	4
● Helps ensure that all team members have an equal opportunity to talk	1	2	3	4
Helps the team develop a shared understanding of its goals and procedures				
● Helps team members create a shared understanding of the project's objectives	1	2	3	4
● Helps the team develop a detailed plan for the finished communication	1	2	3	4
● Helps the team make a project schedule	1	2	3	4

Explanation of ratings:

	Poor		Excellent	
Helps the team have productive meetings				
● Helps the team conduct efficient meetings by helping it set agendas, address all issues, close discussions, sum up, and set goals for the next meeting	1	2	3	4
● Encourages debate and diversity of ideas	1	2	3	4
● Discusses teammates' drafts diplomatically and avoids defensiveness when others suggest revising his or her drafts	1	2	3	4
● Helps the team work across cultural differences	1	2	3	4

Explanation of ratings:

	Poor		Excellent	
Helps the team use technology to its advantage				
● Helps team choose technology and social media that best supports its project	1	2	3	4
● If the team is virtual, fosters personal relationships and conversational interchanges	1	2	3	4

Explanation of ratings:

	Poor		Excellent	
Fulfills personal responsibilities				
● Completes assignments on time	1	2	3	4
● Completes assignments thoroughly	1	2	3	4
● Meets deadlines	1	2	3	4
● Produces high-quality work	1	2	3	4
● Takes initiative	1	2	3	4
● Offers ideas and suggestions	1	2	3	4

Explanation of ratings:

USE WHAT YOU'VE LEARNED

APPLY YOUR EXPERTISE

1. Write a memo to your instructor in which you describe difficulties encountered by a team on which you have worked. Describe strategies presented in this chapter that you employed in an attempt to address these problems. Also reflect on things that you could have done differently that might have overcome the problems.

2. Write a memo to your instructor describing the style and strategies you have brought to team projects in the past. First, identify the strengths of your approach, referring to specific advice given in this chapter. Second, pinpoint ways in which you can achieve an even higher level of effectiveness.

3. If you are working on a team project in your technical communication course or another class, write a memo to your teammates suggesting three ways the team could improve its productivity. Persuasively explain the reasons for each suggestion.

EXPLORE ONLINE

1. Compare two services that enable teams to post a single draft that all team members access to add comments and revise. If you will be working on a team project in your technical communication course or another class, which one would be most helpful to your team? Why?

2. Services like Google Drive are considered to be examples of cloud computing. Using a search engine for your research, develop a brief definition of cloud computing and explain to other students how it is affecting or might affect professionals in your field of study.

3. Find and describe a college course in which students at different schools (possibly in different countries) collaborated on one or more course projects.

COLLABORATE WITH YOUR CLASSMATES

1. If you are currently involved in a team project, fill out the Collaboration Feedback Form shown in Figure 17.2 for each of the other team members.

2. If you are about to begin a group project, write a memo to your instructor detailing the plans you have made in accordance with the advice given in Guidelines 1, 2, and 3 on pages 309 and 310.

APPLY YOUR ETHICS

Various team members may bring different talents, commitments, and values to collaborative projects. What are your ethical obligations if one team member doesn't contribute, persistently completes work late, or resists compromise? What if one member is plagiarizing? Who are the stakeholders in these cases? What are your options? What is an ethical course of action? What are your obligations if members of your team ask you to contribute more, or in a manner that others consider more productive? Would any of your responses be different if the project were assigned at work rather than in school? Present your results in the way your instructor requests.

REFLECT FOR TRANSFER

1. Write a memo to your teammates describing a team project in which you felt very comfortable or uncomfortable. Referring to advice in this chapter, explain two reasons for your feelings.

2. Write a memo to your instructor describing the two ways you believe that the teams you work with in your career will differ from those in school. Describe two ways you believe they will be the same. Finally, identify the two pieces of advice in this chapter that you believe will be most helpful in your career. Explain why.

Creating and Delivering Listener-Centered Oral Presentations

18

Throughout your career, you will make many oral presentations. Sometimes you will speak to people you work with every day, sometimes to groups you haven't met previously, such as prospective clients, members of professional organizations in your field, or the general public. On occasion, your audience will be small, only one or two people. At other times, it will be large, perhaps even hundreds.

Your ability to make effective presentations under these varying circumstances can help you advance in your career. It can also help you obtain your first job after college. In response to a survey of 400 U.S. companies that employ over two million people, 95 percent said that the ability to make oral presentations is "very important" when hiring college graduates (Conference Board, 2006).

Regardless of all other variables, the audiences for the presentations you make on the job will almost always want the same thing that workplace readers will desire from your writing: information they can *use*. They will want help from you in performing a practical task, making a decision, or otherwise advancing their own work. Even the most overtly persuasive presentations, such as sales pitches to prospective customers or clients, involve practical decisions when viewed from the listener's perspective: Should they buy this product or service from your company or from a competitor?

To prepare and deliver presentations that satisfy your listeners' needs, you must polish speaking skills you already posses as well as develop some new ones. This chapter and your instructor will help you learn practical, listener-centered strategies for preparing and delivering presentations that parallel the reader-centered strategies for writing described in the rest of the book. In particular, you will learn how to define your presentation's objectives in a listener-centered way, select the oral and visual media most likely to achieve your objectives, enable your listeners to fully understand and remember your main points, maintain your listeners' good attention and goodwill, and deliver your presentations effectively.

Define Your Presentation's Objectives

Begin work on an oral presentation the same way you begin work on a written one—by defining its objectives. This means determining what will make your presentation usable and persuasive for your readers.

GUIDELINE 1 Determine who your listeners are, what task they want to perform, and what they need and expect from you

Most of what you need to know about your listeners is the same as what you need to know about the readers of your written communications. Who, exactly, are they?

MindTap®

Find additional resources related to this chapter in MindTap.

Although their audiences, topics, and goals vary considerably, oral presentations are everyday occurrences in many careers.

What task will your presentation help them perform? Make a decision? Perform a procedure? Apply the knowledge you supply to their own work?

Next, identify the facts, suggestions, and recommendations that your listeners need in order to complete their work. Find out also how much they already know about your topic. Consider, too, their professional roles. Managers, engineers, and customers may all need different information on the same subject. This information about your audience helps you see what to include in your presentation and how to present it.

To review additional information about defining your communication's goals, see Chapter 3.

In addition, find out how long your audience thinks your presentation should be. Even when you aren't given a time limit, your listeners are likely to have firm expectations about how much time you should take. Don't ruin an otherwise successful presentation by running overtime.

GUIDELINE 2 **Define your persuasive goals**

In addition to identifying your listeners' goals, define your own. Determine how you want to influence listeners' attitudes toward your topic, an activity that requires you to determine what their current attitudes are and what you want them to be after your presentation. Are your listeners prospective clients or customers who view you as an expert who can help them solve a problem they face? Do they see you as an adversary—or example, as the representative of a company that has created an environmental hazard in their community? Answers to such questions help you identify the kinds of information and ideas most likely to lead your listeners to make the decision or adopt the attitude you desire.

Select the Oral and Visual Media Most Likely to Achieve Your Objectives

When planning your presentations, you might think of yourself as having a toolbox of oral and visual media. Your success at making presentations will depend partly on your ability to pick the right tools—the most effective type of oral and visual media—for each situation.

GUIDELINE 1 **Choose the type of oral delivery by considering your audience and purpose**

At work, people generally use three forms of oral delivery: scripted talks, outlined talks, and impromptu talks. The choice among them depends largely on the kind of information you are presenting and the type of relationship you wish to establish with your listeners.

Scripted Talk

A scripted talk is a speech that you write out word for word in advance and deliver by reading the script or by reciting it from memory.

Scripted talks are ideal for presenting complex or sensitive information, where a small slip in phrasing could cause confusion, embarrassment, or harm. They can also help you keep within your time limit.

On the negative side, scripted talks take a long time to prepare, and they can be difficult to adjust if, while you are speaking, you see that your audience wants something different from what you have written. Equally important, they are difficult to compose and deliver in a natural speaking style that helps to establish a personal relationship with your audience.

Scripted talks offer security but can be rigid.

Outlined Talk

To prepare an outlined talk, do what the name implies: Prepare an outline—perhaps very detailed—of what you plan to say.

Outlined talks are easy to deliver in a "speaking voice," which helps increase listener interest and appeal. They are also very flexible. In response to your listeners' reactions, you can speed up, slow down, eliminate material, or add something that you discover is needed. Outlined talks are ideal for presentations on familiar topics to small groups, as in a meeting with co-workers in your department. Often, employees will put their outline in a series of slides, which they can refer to and that can help their listeners follow the structure of their talk.

Outlined talks offer flexibility within a general framework.

The chief weakness of the outlined talk is the same as its major strength, its flexibility. Unskilled speakers can easily run over their time limit, leave out crucial information, or have difficulty finding the phrasing that will make their meaning clear.

Impromptu Talk

An impromptu talk is one you give on the spur of the moment with little or no preparation. At most, you might jot down a few notes beforehand about the points you want to cover.

The impromptu talk is well suited to situations in which you are speaking on a subject so familiar that you can express yourself clearly and forcefully with little or no forethought.

Impromptu talks are ideal for topics you know well.

The chief disadvantage of the impromptu talk is that you prepare so little that you risk treating your subject in a disorganized, unclear, or incomplete manner. For these reasons, the impromptu talks given at work are usually short, and they are usually used in informal meetings where listeners can interrupt to ask for additional information and clarification.

GUIDELINE 2 Choose your visual medium by considering your audience, topic, and purpose

In school, you are probably most familiar with using PowerPoint or other presentation software for the visual component of presentations. At work, you will have a wider variety of media to select from, including handouts and whiteboards.

Presentation Software

Software programs such as PowerPoint and Prezi enable you to create attractive, polished slides with text, photographs, tables, movies, drawings, and many other kinds of content. Used skillfully and in the right circumstances, slides can help an audience readily grasp your meaning and appreciate the significance of your statements. A Writer's Tutorial in MindTap ("Creating a Listener-Centered Presentation") explains how to use many of PowerPoint's most useful features to quickly create an effective presentation.

TRY THIS Learn some of the criticisms of PowerPoint by searching for "Death by PowerPoint" on the Web or YouTube.

However, presentation slides are not suited for every situation and every audience. Some people and organizations complain PowerPoint and similar programs induce speakers to oversimplify complex problems into bullet lists or to provide more information than their listeners can digest. Also heavily criticized is the unskillful use of presentation software by people who, for instance, read their slides to the audience or use annoying transitions between slides. Of course, the central question when you are *planning* a presentation is whether software is the best choice for that occasion. When it is, advice given later in this chapter will enable you to use it effectively.

Handouts

When you are discussing a complex topic for which you will want your readers to examine a large table, detailed drawing, or other graphic requiring close study, handouts can be the ideal medium. Handouts also give listeners something to take away from your presentation that they can refer to later. For this reason, speakers sometimes distribute printed copies of their presentation slides. Also, your listeners can make their own notes on your handout, increasing its value to them.

However, handouts can distract your audience from what you are saying, perhaps causing them to miss the crucial points on which you want them to focus.

Whiteboards

Using a whiteboard, you can create graphics while you are speaking, enabling you to find out during your interaction with your listeners what graphics will be most helpful and persuasive for them. Whiteboards are very effective when you want to lead your audience through a process. You can add elements to your graphic as you move from one step or stage to the next. In some fields, such as computer science and software engineering, team members gather daily to discuss progress and issues. The ability to quickly sketch a concept, solution, or design while making an impromptu presentation is highly valued.

A disadvantage of whiteboards is that while you are writing on them, your listeners may lose focus or become impatient. Also, whiteboards restrict you to using words and line drawings, so you can't take advantage of the other visual representations that can be so effective when discussing certain topics.

In some presentations, you can increase your ability to help and persuade your listeners by using more than one medium. When speaking to a group of managers, a chemical engineer used PowerPoint to display photographs of the crystals grown in an experiment, a whiteboard to explain the process used, and a handout to provide detailed experimental data.

Help Your Listeners Fully Understand and Remember Your Main Points

People find it more difficult to understand what they hear than what they read.

Difficulties listeners face

- Listeners have more difficulty than readers in concentrating for extended periods. People can read for hours at a time, but many listeners have trouble concentrating for more than twenty minutes.
- A talk proceeds at a steady pace, so listeners have no chance to pause to figure out a difficult point.
- If listeners fail to understand a point or let their attention wander, they cannot flip the pages back and reread the passage. The talk goes forward regardless.

The following guidelines describe five strategies you can use to make your most important points easy for listeners to understand and remember.

GUIDELINE 1 Identify the main points you want to make

The fewer major points you make, the more likely that your listeners will remember.

Many experienced speakers limit themselves to three or, at most, four. Of course, major points may have subpoints, but their number should also be limited, and they should be linked to the main points in an easy-to-grasp, hierarchical way.

To identify your main points, answer the two central questions of the writer-centered/listener-centered approach to technical communication. Given that your presentation's purpose is to help your listeners perform some practical task, what are the most helpful things you can tell them? And, given that you want to shape your listener's attitudes and actions, what can you say that *they* will find most persuasive?

GUIDELINE 2 Create a simple structure built around your major points

The most important step you can take to make your major points stand out is to structure your presentation around them and assure that the structure is evident to your listeners. Thus, the **heart** of your presentation on solving a problem for your employer would have three parts, each with appropriate subparts. Part 1, which describes the major features of the problem, would have a subsection for each feature. To describe the major causes of the problem, Part 2 would have a subpart on each major cause. Similarly, Part 3 would have a parallel number of subparts, each describing a strategy targeting each cause.

Note that your subpoints are important elements of your presentation. Their major purpose is to help your readers understand your main points by providing the details and justification that make them useful and persuasive.

The sharply focused core of the presentation would be preceded by an **introduction** that, like the introduction to a written communication, would introduce your topic and explain its relevance to your listeners. In oral presentations, you can also almost always increase your listeners' understanding if you preview the organization of your talk. Depending on your purpose and audience, your introduction may also provide background information and state the overall point or points you want your listeners to take away.

To enhance the memorability of your presentations, end with a **conclusion** that sums up the main points. This strategy elevates the main points above the subpoints that may have taken the largest part of your presentation's time. Where appropriate, you may also indicate the next steps listeners might take to act on the information and ideas you have provided. Always, it is wise to ask for questions. Answering them gives you an additional opportunity to state your main points again.

GUIDELINE 3 Help your listeners follow the structure of your presentation

Even when you employ a very simple organization, your listeners may need help discerning it. Help them by forecasting the structure in the introduction, signaling transitions clearly, and summarizing the structure (and main points) in the conclusion. The following list describes techniques for guiding your listeners in these ways.

Signaling the Structure of Oral Presentations	
GENERAL STRATEGIES	**TECHNIQUES**
Forecast the Structure	■ In the introduction, tell what the structure will be: "In the next ten minutes, I will address the following three topics…."
	■ Show a slide that outlines the major parts of your talk.
Signal Transitions	■ Announce transitions explicitly, "Now I would like to turn to my second topic, which is…."
	■ Show a slide that announces the next topic and, perhaps, lists its subtopics.
	■ Highlight your main points. Your discussion of each main point is a major section of your talk. "I'm going to shift now to the second cause of our problem, a cause that is particularly important to understand."
	■ Pause before beginning the next topic. This pause will signal to your listeners that you have completed one part of your presentation.
	■ Slow your pace and speak more emphatically when announcing your major points, just as you would when shifting to a new topic in conversation.
	■ Move about. If you are speaking in a setting where you can move around, signal a shift from one topic to another by moving from one spot to another.
Review	■ In your conclusion, remind listeners explicitly about what you've covered: "In conclusion, I've done three main things during this talk:…."

GUIDELINE 4 Make easy-to-understand visuals

To contribute fully to a presentation's effectiveness, slides must be easy to read and understand. These two qualities depend on the way you design your slides.

Designing Simple, Effective Slides

All of Chapter 14's guidelines for designing pages apply also to the design of slides for oral presentations. The most important difference is the obvious: You can put much less information on a slide. Even so, some presenters attempt to squeeze too much information on their slides. Avoid using full sentences. Use key words instead. Keep your slides simple. Figure 18.1 shows how simplicity can be combined with the advice given in Chapter 14 to create an effective presentation.

So that the verbal and visual elements of your presentation can support one another harmoniously, distill your verbal statements into key words and phrases for the bullet lists in your slides.

Statement you will make verbally

> In this report, I will discuss three major causes of the extinction of animal and bird species in Asia: loss of habitat due to development and the harvesting of natural resources, increasing pollution from automobiles and factories, and poaching.

Content of the corresponding visual

> Causes of extinction
> ■ Loss of habitat
> ■ Pollution
> ■ Poaching

Like any other element of a communication, slides should be tested beforehand to make certain they will be as understandable and persuasive as possible. Once you've completed the set, show them to other people, preferably members of your target audience. Do this early enough to allow time to polish them, if that seems advisable, before your presentation.

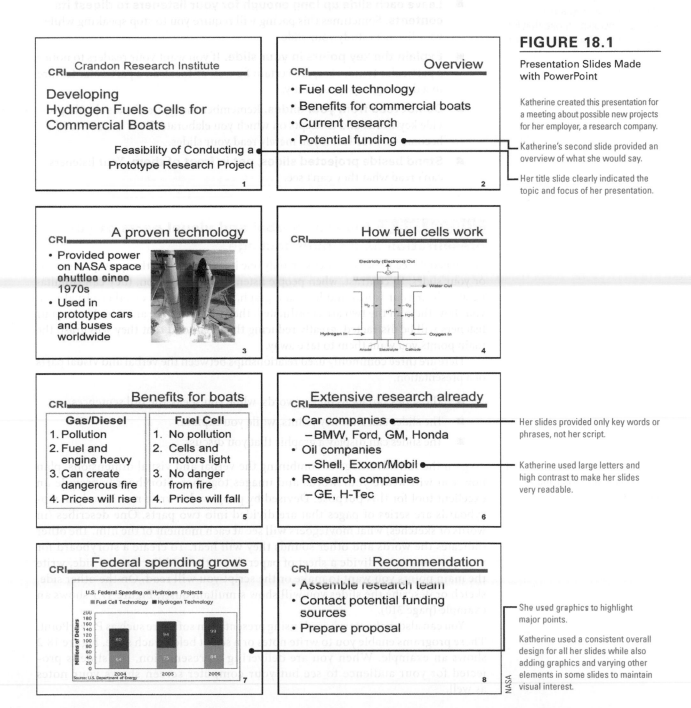

FIGURE 18.1

Presentation Slides Made with PowerPoint

Katherine created this presentation for a meeting about possible new projects for her employer, a research company.

Katherine's second slide provided an overview of what she would say.

Her title slide clearly indicated the topic and focus of her presentation.

Her slides provided only key words or phrases, not her script.

Katherine used large letters and high contrast to make her slides very readable.

She used graphics to highlight major points.

Katherine used a consistent overall design for all her slides while also adding graphics and varying other elements in some slides to maintain visual interest.

TRY THIS Presentation programs such as PowerPoint enable you to make "kiosk" presentations that run continuously on their own on a computer. Try making a presentation that, by itself, informs viewers about something important to you. Play with your program's features, including those that let you add music or record a script.

Displaying Slides Effectively

During your talk, use your slides in ways that support and reinforce your presentation rather than detract from it.

- **Display a slide only when you are talking about it.** If you talk about one thing but display a slide about something else, each person in your audience must choose whether to listen to your words or read your slide. Either way, your audience may miss part of your message.

- **Leave each slide up long enough for your listeners to digest its contents.** Sometimes this pacing will require you to stop speaking while your listeners study your slide.

- **Explain the key points in your slide.** If you want your readers to notice a particular trend, compare certain figures, or focus on a particular feature in a drawing, say so explicitly.

- **Do not read from your slides.** Remember that your visuals should provide key words and concepts on which you elaborate. Your listeners will become very restless if you merely read your slides.

- **Stand beside projected slides, not in front of them.** Your listeners can't read what they can't see.

GUIDELINE 5 Plan the verbal and visual parts of your presentation as a single package

When people read, they can focus on only one item at a time, either your paragraphs or your slides. In contrast, when people listen to a presentation, they can simultaneously hear your words and look at a graph, table, diagram, or other visual item you show them. If the two are coordinated, they will reinforce another. If not, your listeners will be distracted, greatly reducing the likelihood that they will grasp the main points you want them to take away.

Here are three commonly used relationships between the verbal and visual parts of a presentation.

- The slides can focus on key words, while you speak the full sentences.
- The slides can state key points, while you explain.
- The slides can display a graphic that you discuss.

Storyboards are an excellent tool for planning the verbal and visual together.

To gain the full advantage of combining the verbal and visual dimensions, plan how you will weave your words and images together. Storyboards provide an excellent tool for this purpose. Devised by people who write movie scripts, storyboards are series of pages that are divided into two parts. One describes (in words or sketches) what moviegoers will see at each moment of the film. The other indicates the words and other sounds they will hear. To create a storyboard for your presentations, divide a sheet of paper down the middle. On one side, write the main points you want to make or the script you will read. On the other side, sketch or describe the slides you will show simultaneously. Figure 17.1 shows an example (page 310).

You can also make storyboards using presentation software such as PowerPoint. These programs enable you to write notes or a script below each slide. Figure 18.2 shows an example. When you are delivering a presentation, the slide is projected for your audience to see but your computer screen shows your notes as well.

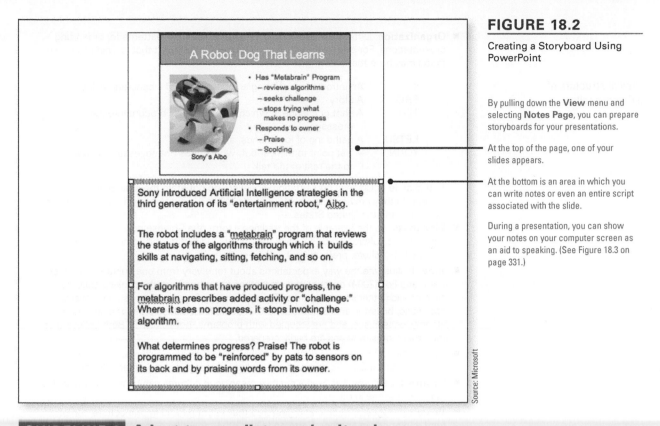

FIGURE 18.2

Creating a Storyboard Using PowerPoint

By pulling down the **View** menu and selecting **Notes Page**, you can prepare storyboards for your presentations.

At the top of the page, one of your slides appears.

At the bottom is an area in which you can write notes or even an entire script associated with the slide.

During a presentation, you can show your notes on your computer screen as an aid to speaking. (See Figure 18.3 on page 331.)

Source: Microsoft

GUIDELINE 6 Adapt to your listeners' cultural background

An audience's cultural background affects its expectations about an oral presentation in much the same way that it influences readers' expectations about written communications.

- **An audience's expectations may be shaped by several cultures.** These include the culture of their region and the national or ethnic group to which they belong, the culture of their employer's organization, and the culture of their profession, among others.

- **The members of an audience are individuals, not cultural stereotypes.** Although it can be useful to learn about your audience's culture, also find out as much as you can about the specific people you will be addressing.

Several aspects of oral presentations vary from culture to culture. Investigate your audience's expectations about each of them as you plan and prepare your presentation. You may learn that you can create a truly listener-centered presentation only if you do something different from what this chapter's other guidelines suggest.

Adapting to Your Audience's Cultural Background

- **Opening.** In some cultures, presentations typically begin with the main point. In others, they begin with formal greetings. In the United States, presentations often begin with an effort to build rapport with an audience—for instance, by telling a joke or personal anecdote. Beginning in a way that differs from what the audience expects can get a talk off to a weak start.

(Continued)

Typical structure of
a Chinese talk

- **Organization.** Different cultures have developed different patterns for organizing presentations. For example, Wolvin and Oakley (1985) report that in China a presentation may have the following structure:

KI	An introduction offering an observation of a concrete reality
SHO	A story
TEN	A shift or change in which a new topic is brought into the message
KETSU	A gathering of loose ends, a "conclusion"
YO-IN	A last point to think about, which does not necessarily relate to the rest of the talk

In this style of Chinese presentation, speakers do not state the main points explicitly nor do they preview the structure of the talk, as is customary in some European countries and the United States.

- **Directness.** As the example of the traditional form of a Chinese presentation suggests, some cultures prefer to make their points indirectly, while others, including the general U.S. culture, prefer to make them explicitly.

- **Tone.** To illustrate the way expectations about tone vary from one culture to another, Carté and Fox (2004) describe the stereotypes that two English-speaking cultures have of each other's presentations. To the British, U.S. presentations seem overly optimistic, boastful, and superficial. To U.S. audiences, British presentations seem gloomy, pessimistic, and preoccupied with problems, not solutions. Both believe that the other's presentations are badly prepared.

- **Eye contact.** Whereas eye contact is highly valued in some cultures, such as in the United States, it can seem aggressive and unwelcome in some other cultures.

- **Gestures.** Gestures that have a positive meaning in cultures can be incomprehensible, rude, or obscene in others.

- **Visuals.** Images of people and gestures that seem natural in one culture can offend audiences in another. Carté and Fox (2004) describe a presentation in which a large European company used an image of a lion tamer and lion to symbolize the way their computer systems would tame potential clients' problems. Although the presentation worked in several countries, it failed in an African country whose symbol was a lion. To this audience, the image suggested a colonial power subduing their state.

When speaking to an audience from another culture, you may also be addressing persons who are not fluent in English. In those situations, the following suggestions should help you succeed.

Speaking to Listeners Who Are Not Fluent in English

- **Use words your audience can understand.** Learning the audience's level of proficiency in English may require research. Many technical and scientific terms are the same in many languages, so that you may be able to use them even when you need to simplify your language in other ways.

- **Use slides.** Although you should be cautious about slides, people who don't understand one another's language can still understand the same drawings, graphs, flowcharts, and other visuals.

- **Provide a handout.** If you are speaking from a script, the script itself is the best handout, along with copies of your key slides. Otherwise, create a detailed outline. Consider using full sentences—even in an outline—rather than words or brief phrases. Beyond helping your listeners follow your talk, a handout provides a resource they can study later to understand points they missed.

Maintain Your Listeners' Attention and Goodwill

Oral presentations are eminently personal events. You are speaking to and with persons who are right there in front of you. Their responses to your presentation depend substantially on the quality of the relationship you establish with them. The following guidelines describe strategies you can use to maintain your listeners' attention and goodwill throughout your time with them.

GUIDELINE 1 Speak in a conversational style

For most talks, a conversational style works best. It helps you express yourself clearly and directly, and it helps you build rapport with your listeners. A more abstract, impersonal style can lead you to make convoluted statements that are difficult to understand and leave your listeners feeling that you are talking at them rather than with them. Although people often associate a conversational style with informality, the two are not synonymous. The two keys to a conversational style are (1) speaking directly to your listeners and (2) expressing yourself in the simple, natural, direct way you do in conversation. Here are suggestions for mastering this style, ones you can use whether you are giving a scripted, outlined, or impromptu talk.

Strategies for Developing a Conversational Style

- **Create your talk with your audience "present."** While preparing your talk, imagine that members of your audience are right there listening to the words you are planning to speak.
- **Use the word *you* or *your* in the first sentence.** Thank your listeners for coming to hear you, praise something you know about them (preferably something related to the subject of your talk), state why they asked you to address them, or talk about the particular goals that you want to help them achieve.
- **Continue using personal pronouns throughout.** Let the use of *you* or *your* in the first sentence establish a pattern of using personal pronouns (*I, we, our, you, your*) throughout.
- **Use shorter, simpler sentences than you might use when writing.**
- **Choose words your listeners will understand immediately.** If your audience stops to figure out the meaning of a word you have used, they will stop listening to the next point you are making.

As you strive to create a conversational style, remember that the way you use your voice can be as important as the words and phrasing you select. Listen to yourself and your friends converse. Your voices are lively and animated. To emphasize points, you change the pace, the rhythm, and the volume of your speech. You draw out words. You pause at key points. Your voice rises and falls in the cadence of natural speech.

Using your voice in the same way during your oral presentations can help you keep your listeners' interest, clarify the connections between ideas, identify the transitions and shifts that reveal the structure of your talk, and distinguish major points from minor ones. It can even enhance your listeners' estimation of your abilities. Researcher George B. Ray (1986) found that listeners are more likely to believe that speakers are competent in their subject matter if the speakers vary the volume of their voices than if the speakers talk in a monotone.

Exhibit enthusiasm for your subject. In conversation, we let people know how we feel as well as what we think. Do the same in your oral presentations, especially when you are advocating ideas, making recommendations, or promoting your employer's products and services. If you express enthusiasm about your topic, you increase the chances that your listeners will share your feelings.

Finally, use gestures. In conversation, you naturally make many movements—pointing to an object, holding out your arms to show the size of something, and so on. Similarly, when making oral presentations, avoid standing stiffly and unnaturally. Use natural gestures to help hold your listeners' attention and to make your meaning and feelings clear.

GUIDELINE 2 Establish and maintain a personal connection with your audience

In an oral presentation, you can interact in person with your audience, something you can't do through a written communication. By taking advantage of this opportunity, you can ensure that your listeners pay close attention and that you are meeting their needs.

Look at Your Listeners

LEARN MORE For more on cultural conventions concerning eye contact, see "Cultural Differences in Behaviors in Conversation" (page 313).

Benefits of maintaining eye contact with your listeners

In North America and much of Europe, the most basic way to involve listeners in your talk is make them feel that that are in a direct interchange with you—something you can accomplish simply by making eye contact with them. Through eye contact, you convey that you are interested in your listeners as individuals, both personally and professionally.

Eye contact has the additional benefit of giving your audience a more favorable impression of you. Researcher S. A. Beebe (1974) first demonstrated this effect when he asked two groups of speakers to deliver the same seven-minute talk to various audiences. One group was instructed to look often at their listeners, whereas the other group was told to look rarely. Beebe found that the speakers who looked more often at their listeners were judged to be better informed, more experienced, more honest, and friendlier than those who used less eye contact.

Eye contact also enables you to judge how things are going. You can see the eyes fastened on you with interest, the nods of approval, the smiles of appreciation, or the puzzled looks and the wandering attention. These signals enable you to adjust your talk, if necessary.

There are situations when it is appropriate for you to look briefly away from your listeners. For example, if you are using an outline or script, you may want to refer occasionally to your notes. Just don't rivet your gaze on your papers. Look back at your audience. The same is true if you look at graphics that are behind you, for instance on a chalkboard or a screen for an overhead or data projector.

If you are using a presentation program such as PowerPoint, place your computer screen so that it faces you as you look at your audience. By looking at the screen, you will see the same slide that your audience sees projected behind you. Also, if your computer supports two monitors (most laptops do), you can use the Presenter View (Windows) or Presenter Tools (Mac OS) to display helpful items on your screen that your audience won't see (Figure 18.3).

If you have difficulty looking at your listeners, try the following strategies.

Masterfile

To maintain a connection with your audience when using projected visual aids, face your listeners and look at your slides on your computer screen rather than on the large screen behind you.

Strategies for Looking at Your Listeners

- **Look around at your audience before you start to speak.** This will give you an opportunity to make initial eye contact with your listeners when you aren't also concentrating on what to say.
- **Follow a plan for looking.** For instance, at the beginning of each paragraph of your talk, look at a particular section of your audience—to the right for the first paragraph, to the left for the second, and so on.
- **Target a particular feature of your listeners' faces.** You might look at their eyes, but you could use their foreheads or noses instead. Unless they are very close, they won't notice the difference.
- **When rehearsing, practice looking at your audience.** For instance, develop a rhythm of looking down at your notes and then up at your audience—down, then up. Establishing this rhythm in rehearsal will help you avoid keeping your head down throughout your talk.
- **Avoid skimming over the faces in your audience.** To make someone feel that you are paying attention to him or her, you must focus on an individual. Try setting the goal of looking at a person for four or five seconds—long enough for you to state one sentence or idea.

Ask Your Listeners to Say or Do Something

In some oral presentations, you can ask your audience to contribute information. For example, if you are making a presentation to fellow employees about ways to improve customer satisfaction, you might ask what kinds of complaints they have heard. If your purpose is to train your listeners to perform a procedure, you may be able to ask them to perform the steps as you talk and demonstrate.

Help Your Listeners Take Notes

Note taking is another way in which listeners actively participate in a presentation. You can encourage note taking by distributing an outline at the beginning of your talk. If you are using a program such as PowerPoint, you can print out copies of the slides you will project so that members of your audience can jot notes on them.

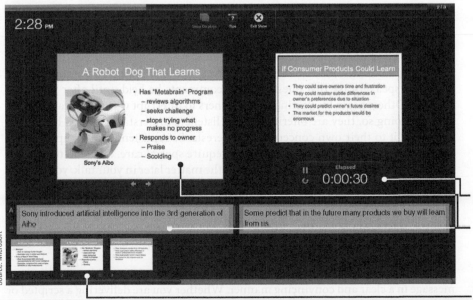

Source: Microsoft

FIGURE 18.3

Presenter Tools from PowerPoint

This figure shows the Presenter Tools in Microsoft PowerPoint 2010 for Windows. PowerPoint 2011 for Mac has a different layout but similar content.

By pulling down the View menu and choosing Presenter Tools, you can show a screen like this on your computer during your presentation.

A clock tells how many minutes have passed since you began speaking.

It shows the slide being projected and your notes about it.

At the bottom, it shows the current slide between the one before it and also the next one. You can move to any slide in your presentation by clicking on the thumbnail of it in this row.

Invite Questions

In almost every situation, it is appropriate to invite listeners to ask questions either during your presentation or after it. Guideline 3 below provides advice about ways to prepare for and respond to questions.

Give Your Listeners Something to Take Away

You may want to encourage your listeners to continue their interaction with you after your presentation. For instance, your listeners may include people who might become customers or clients of your employer's organization. If so, provide them with something to take away that includes the main points you made or supplemental information, along with your name, phone number, and email address. A handout for note taking would be perfect.

GUIDELINE 3 Respond effectively to your audience's comments and questions

Audiences at work often ask questions and make comments. In fact, most of the talks you give there will be followed by discussion periods during which members of your audience will ask you for more information, discuss the implications of your talk, and even argue with you about points you have made. This give-and-take helps explain the popularity of oral presentations at work: They permit speaker and audience to engage in a discussion of matters of common interest.

Part of preparing to deliver a talk is preparing yourself for questions and discussions. In a sense, you do this when you plan the presentation itself, at least if you follow Chapter 4's advice that you begin planning your communication by thinking about the various questions that your readers will want it to answer. Often, however, you will not have time in your talk to answer all the questions you expect listeners to ask. The questions you can't answer in your talk are ones your listeners may raise in a question period. Prepare for them by planning your responses.

When a listener asks a question—even an antagonistic one—remember that you want to maintain good relations with your entire audience. If you are speaking in a large room, be sure that everyone hears the question. Either ask the questioner to speak loudly or repeat the question yourself. If people hear only your answer, they may have no idea what you are talking about. Respond to all questions courteously. Their questions are important to the people who ask them, even if you don't see why. If you don't know how much detail the questioner wants, offer some and then ask the questioner if he or she wants more. If you don't know the answer to a question, say so.

Some speakers ask that questions and comments be held until after they have finished. Others begin by inviting their listeners to interrupt when they have a question. By doing so, they are offering to help listeners understand what is being said and relate it to their own concerns and interests. Some listeners interrupt without being invited to do so. Such interruptions require special care. Speak to the person immediately. If you are planning to address the matter later in your talk, you can ask the questioner to wait for your response, but you may want to respond right away. After you do so, remind your listeners of where you broke off: "Well, now I'll return to my discussion of the second recommendation."

GUIDELINE 4 Rehearse

All of your other good preparations can go for naught if you are unable to deliver your message in a clear and convincing manner. Whether you are delivering a scripted or outlined talk, consider the following advice.

Guidelines for Rehearsing

- **Rehearse in front of other people.** They can help you identify weak spots and make suggestions for improvement.
- **Pay special attention to your delivery of the key points.** These are the points where stumbling can cause the greatest problems.
- **Rehearse with your slides.** You need to practice coordinating your slides with your talk.
- **Time your rehearsal.** Speak at the same pace you will use in your actual presentation so you know how long your talk will require.

GUIDELINE 5 Accept your nervousness—and work with it

A final piece of advice about the nervousness you will almost certainly experience as the time approaches for you to deliver your presentation: Accept it. It's natural. Even practiced speakers with decades of experience sometimes feel nervous when they face an audience. If you fret about being nervous, you merely heighten the emotional tension. Keep in mind that your nervousness is not nearly so obvious to your listeners as it is to you. Even if they do notice that you are nervous, they are more likely to be sympathetic than displeased. Furthermore, a certain amount of nervousness can help you. The adrenaline it pumps into your system will make you more alert and more energetic as you speak.

Here are some strategies for reducing your nervousness and controlling the pacing, fidgeting, and other undesirable mannerisms it can foster.

Strategies for Controlling Nervousness

- **Arrive early.** Avoid rushing from a previous activity to your talk. Give yourself plenty of time to set up and look around before you begin.
- **Devote a few minutes before your talk to relaxing.** Take a walk or spend a quiet moment alone.
- **Speak with audience members before your presentation begins.** Doing this enables you to make a personal connection with at least some of your listeners before you begin.
- **Remind yourself that your listeners are there to learn from you, not judge you.**
- **When it's time to begin, pause before you start your talk.** Look at your audience, say "Hello," and take out your outline or notes as you accustom yourself to standing before your listeners.

Make Effective Team Presentations

At work, team presentations are as common as team writing. They occur, for example, when a project team reports on its progress or results, when various departments work together on a joint proposal, or when a company is selling a technical product (such as customized software) that requires the participation of employees from several units of the company.

Team presentations are so common because it is usually more effective to have several members make a presentation than to have one person speak for all team members. In a team presentation, each topic can be discussed by the person who

LEARN MORE The guidelines in Chapter 17, "Creating Communications with a Team," are as helpful to teams creating oral presentations as they are to teams creating written communications. Turn to page 306.

is most expert in it. Moreover, the use of a variety of speakers can help retain the audience's attention.

The following sections provide advice for creating and delivering effective team presentations.

GUIDELINE 1 Plan thoroughly as a team

When developing a team presentation, plan as carefully as you would while preparing a team-written document. Devoting a team meeting to making plans can be very helpful. Decide which topic each team member is to discuss, which points are to be made, and how each part fits with the others. Set a time limit for each part so that the total presentation doesn't run too long. Also, decide whether the team is going to ask the audience to hold questions until the end or invite the audience to interrupt the speakers with questions. In either case, provide time for the interchange between the team and its listeners.

GUIDELINE 2 Maintain overall consistency while allowing for individual differences

In a team-written project, the goal is usually to produce a document with a single voice for the entire document. In team presentations, however, each speaker can speak in his or her own style and voice, provided that the general tone of the presentation is relatively consistent.

GUIDELINE 3 Make smooth transitions between speakers

To help the audience discern the overall structure of your presentation, switch speakers where there is a major shift in topic. Also, tell your listeners how the two parts of the presentation fit together. For example, the speaker who is finishing might say, "Now, Ursula will explain how we propose to solve the downtime problem I have just identified." Or, the next speaker might then say, "In the next few minutes, I'll outline our three recommendations for dealing with the downtime problem Jefferson has described."

GUIDELINE 4 Rehearse together

Rehearsals are crucial to the success of your team presentations. You can help one another polish your individual contributions and ensure that all the parts are coordinated in a way that the audience can easily understand. Rehearsal also enables you to see that the entire presentation can be completed in the time allotted. Running overtime is a common and serious problem for groups that have not rehearsed together.

Conclusion

Making oral presentations can be among your most challenging—and rewarding—experiences at work. By taking a listener-centered approach that is analogous to the reader-centered approach described elsewhere in this book, and by striving to communicate simply and directly with your audience, you will prepare talks that your listeners will find helpful, informative, interesting, and enjoyable. The Writer's Guide shown in Figure 18.4 will help you succeed in doing so.

FIGURE 18.4

Writer's Guide for Creating and Delivering Oral Presentations

Writer's Guide

CREATING AND DELIVERING ORAL PRESENTATIONS

1. Define your presentation's objectives.

    Who are your listeners?

    What task will your presentation help them perform?

    What do they need from you in order to perform that task?

    How do you want to influence their attitudes and actions?

    What do they expect?

2. Choose the media you will use.

    What type of oral delivery is most likely to achieve your objectives?

    What type or types of visuals are most likely to achieve your objectives?

3. Plan your presentation.

    What three for four points will your listeners find most helpful and persuasive?
    (Organize the body of your presentation around them.)

    What will you say in your introduction?

    What will you say in your conclusion?

    How will you coordinate the verbal and visual elements of your presentation?

    How will you adapt your presentation to your listeners' culture (if their culture is different than yours)?

    What strategies will you use to maintain your listeners' attention and goodwill?

4. Rehearse—and time your rehearsal to stay within your time limit.

USE WHAT YOU'VE LEARNED

EXERCISE YOUR EXPERTISE

1. Outline a talk based on a written communication that you have prepared or are preparing in one of your classes. The audience for your talk will be the same as for your written communication. The time limit for the talk will be ten minutes. Be ready to explain your outline in class and be sure that it indicates the following.

    - The way you will open your talk.
    - The overall structure of your talk.
    - The main points from your written communication that you will emphasize.
    - The slides you will use.

2. Imagine that you must prepare a five- to ten-minute talk on some equipment, process, or procedure. Identify your purpose and listeners. Then write a script or an outline for your talk (whichever your instructor assigns). Be sure to plan what graphics you will use and when you will display each of them.

3. Do one of the Oral Briefing Projects in MindTap.

4. Using PowerPoint or a similar program, prepare a brief set of slides for the talk you will give in one of the preceding exercises.

EXPLORE ONLINE

1. Using a search engine, find an online source for advice about oral presentations. Compare the advice given there with the advice you have found in this chapter.

2. Locate two PowerPoint presentations online that are about a topic of interest to you. View them in html and then download them for viewing on your computer. Which of the two is most effective when viewed in html? Why? Which of the two is most effective when

(continued)

viewed on your computer? Why? Write a memo to your classmates explaining what you've learned about the Internet and about PowerPoint. To find PowerPoint presentations using Google, click on Advanced Search. On the page that opens, enter your topic. From the dropdown menu for File Type, choose Microsoft PowerPoint.

COLLABORATE WITH YOUR CLASSMATES

At work, you will sometimes be asked to contribute to discussions about ways to make improvements. For this exercise, you and a classmate are to deliver a five-minute impromptu talk describing some improvement that might be made in some organization you are familiar with. Topics you might choose include ways of improving efficiency at a company that employed you for a summer job, ways of improving the operation of a club you belong to, and ways that a campus office can provide better service to students.

Plan your talk together, but divide your content so that you both talk. In your talk, clearly explain the problem and your solution to it. Your instructor will tell you which of the following audiences you should address in your talk.

- Your classmates in their role as students. You must try to persuade them of the need for, and the reasonableness of, your suggested action.
- Your classmates, playing the role of the people who actually have the authority to take the action you are suggesting. You should take one additional minute at the beginning of your talk to describe these people to your classmates.

APPLY YOUR ETHICS

Prepare a brief (three- to five-minute) presentation concerning an ethical issue that you or someone you know has confronted on the job. Describe the situation, identify the stakeholders, tell how the stakeholders would be affected, and tell what was done by you or someone else. If nothing was done, tell why. Be sure to keep within your time limit. Depending on your instructor's assignment, give your report to the entire class or to two or three other students.

Managing Client and Service-Learning Projects

19

I n almost every field, from engineering to medicine and physics to environmental sciences, some companies earn at least some of their income by working under contract for corporate or government organizations. Within many companies, some departments routinely conduct projects at the request of other departments in the same organization who are, in essence, their clients.

To prepare you for the likelihood that you will write for clients in your career, your instructor may ask you to prepare a communication on behalf of a company, agency, or university office. Perhaps you will write a procedure manual, improve a website, or study the feasibility of a new venture that is being considered. In a variation on this type of assignment, your instructor may ask you to use your talents to assist a nonprofit such as United Way or a volunteer or community organization such as a neighborhood health clinic. Projects of the latter type are often called service-learning projects.

Whatever the details, a client project can be one of the most satisfying assignments of your college years. You get to produce a communication that has real, practical results, work with people who greatly appreciate your assistance, and develop valuable project management skills.

Project Management and Client Communication

This chapter's topic is *managing* client and service learning projects. You will quickly see, however, that managing these projects primarily involves *communicating effectively* with your client.

In this communication, you have two goals. The first

Alexander Raths/Shutterstock.com

Martin Shields/Science Source

image-egami/Shutterstock.com

Client projects give students an opportunity to research practical problems and challenges and then develop effective solutions.

LEARNING OBJECTIVES

1. Establish a detailed, mutual understanding of all important aspects of the project.
2. Maintain a productive relationship with your client throughout your project.
3. Hand off your project in a way that helps your client use your results effectively.

MindTap®

Find additional resources related to this chapter in MindTap.

is to develop and maintain a close, cooperative, productive partnership with your client. The success of client projects often depends largely on the quality of the relationship you build with your client. The second goal is to ensure that you and your client have a common understanding of what you will provide one another during the project. Without such an agreement, one or both of you could be very disappointed at the project's end.

This chapter and your instructor will help you apply to a client project the reader-centered skills you have already learned in your course, with a special focus on establishing a common understanding of the project, maintaining a productive relationship with your client, and handing off your final product to your client in a helpful way.

Establish a Detailed, Mutual Understanding of All Important Aspects of the Project

Think of your first steps in a client project in the same way you think of the first steps in creating a reader-centered communication: Learn your client's goals and preferences, plan ways you can help the client achieve what the client wants, and then test out your plan by presenting it to your client in a project proposal.

GUIDELINE 1 Determine what your client wants and why

The best way to learn your client's goals and desires is to ask. If possible, arrange a face-to-face meeting. Bring a list of questions you need to have answered in order to create a "client-centered" plan. In the meeting, however, let your client speak first. Then ask any questions from your list that remain unanswered. The Writer's Guide for Defining Your Communication's Objectives (Figure 3.1, page 52) can provide a helpful starting point for your list. Be sure to include questions on the following topics.

LEARN MORE Review Chapter 3 for advice about defining a communication's objectives.

By interviewing your client, you gain information needed to develop a detailed, effective project plan.

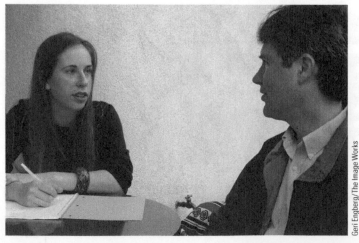

Geri Engberg/The Image Works

- **Client's organization.** Learn about the organization's products or services, as well as its goals and values. In addition, inquire about the events that led the client to request your assistance. The story of these events can help you understand what the client wants your communication to achieve. It can also reveal a wealth of information about factors you need to consider as you proceed.

- **Readers.** Determine whom the client sees as the target audience. Who will use the communication report, instructions, or website you create?

- **Usefulness and persuasive objectives.** Learn what the client wants the communication to enable the readers to do. How does the client want it to affect the readers' attitudes?

- **Stakeholders.** Learn who, besides the target readers, will be affected by the communication. This information will help you understand the ethical dimensions of the project.

- **Deadline.** Determine when the client wants you to complete the project.

- **Preferences and requirements.** Find out what characteristics your client wants the communication to have. Should it be written in a certain style? If it will be printed, is there a page limit?

- **Resources.** Learn which members of the client's staff will answer your questions and when they will be available to you. What hardware, software, or other equipment will the client provide?

- **Budget.** If you will be paid or if production expenses are involved, ask how much the client has allocated for the project.

GUIDELINE 2 Develop your own assessment of the situation

The next step in preparing your proposal is to deepen your understanding of the communication's goals by seeking your own answers to each of the questions you asked your client. When describing projects, clients sometimes forget to mention important facts. Sometimes they don't realize the importance of certain kinds of information that you know to be critical. As you size up the situation, focus on gathering information about the readers, their tasks, and their attitudes. Talk with some of them, if possible, or with other people knowledgeable about them.

Through this process, you may reach some conclusions that differ from your client's. For example, you may reach a different understanding of what the readers want, what will inspire them to change their attitudes in the desired way, and even who the readers will be. Discuss these differences with your client as soon as possible.

GUIDELINE 3 Define what you will do for your client and how you will do it

After you've developed a full understanding of the project, define what you'll do for your client and explain how you'll produce that result. In the plan, describe in detail each of the following key elements of the project: the **deliverable** (the finished communication you will deliver to your client at the end of the project), the **resources** you'll need from the client to complete the project, and the **schedule** you and your client will follow as you work together.

Your Deliverable

Describe your deliverable in as much detail as you can without prematurely making decisions that are best made as the project unfolds. Here are three major topics to include.

- **Size.** You and your client will both be unhappy with the results of your project if one of you is expecting the final product to be twice the size the other is expecting. Use numbers when describing size rather than adjectives. Instead of saying that the document will be "short," decide whether you are going to deliver 10, 20, or 50 pages. Rather than saying the website you are creating will "provide all the information visitors might desire," say whether it will have approximately 12 screens or 120.

- **Overall content and major features.** Although you should remain open to new ideas and insights as you work on the project, your proposal should describe the general nature of the communication you will deliver. Be careful not to promise more than you can prepare in the time available and (if you are being paid) for the price you are quoting. Don't agree to create features that require you to learn new skills unless you have the time necessary to do so.

- **Technical aspects.** If you are creating an online communication, describe its functionality. Will it be created only for a Windows or MacOS platform, or for both? If it is a website, which versions of which browsers will it display in? If the communication will be printed, will it be in black and white or full color?

Your Client's Contributions

In addition to describing your deliverable, identify what you will need from your client. In most cases, this will include information as well as access to people who can answer the questions you encounter along the way. Perhaps you'll also need access to certain equipment or facilities or the use of certain software. If you are going to test for usefulness and persuasiveness, you may need your client to provide you with a certain number of persons from the target audience to serve as test users. Certainly, you'll need responses to outlines, drafts, and other materials you provide for review.

Your Schedule

A detailed project schedule will serve you in many ways. It will enable you to determine how many hours, days, and weeks it will take you to complete the project and will assure your client that you have a workable plan. It will also tell your client when to expect drafts, progress reports, and other communications from you. And it will guide your activities, telling you when you need to complete each task in order to complete your deliverable on time.

First, list all major activities you will perform. Begin with the framework provided in this book: defining objectives (including learning in detail about the communication's readers and purpose), researching, drafting, and revising (including testing for usefulness and persuasiveness, if appropriate). Next, identify the specific tasks you will need to perform as part of each of these major activities. Will your research include interviewing subject matter experts as well as reading additional information? While drafting, will you create charts, drawings, and other graphics from scratch as well as acquire existing ones?

Once you've identified the activities you'll perform, create a detailed timetable. Work backwards from the final deadline, allotting a reasonable proportion of the available time to each activity. Establish milestones, or mini-deadlines, to tell when each task needs to be accomplished so that adequate time remains for the rest of the work to be completed. For instance, indicate the date by which research needs to be done so that you can begin drafting.

In most projects, several tasks can be performed simultaneously. For example, research can be conducted for one section of a website while another section is being built. In these cases, identify the sequence of tasks that determines the quickest time the project can be completed. This is the sequence along which the addition of one hour or day to any task postpones completion of the entire project. It is called the *critical path*. For large projects, you may find it helpful to use project management software such as Microsoft Project.

In your schedule, identify the dates when you and your client will communicate with one another. Include the deadlines for each draft and progress report you will submit. Also, schedule interactions at the points where you need to check your results with your client. For example, if your work includes audience analysis, set a time to go over the results to be sure that your client agrees with your analysis. Disagreement here could lead to very different expectations about what you'll produce. Other times to check with your client are after you've created a detailed outline of the communication and when you have completed an early draft that shows how you will implement your communication strategies.

In your schedule, include your client's deadlines as well as your own.

Finally, indicate when you will need things from your client. Specify not only the dates you will submit drafts or other material, but also the dates you need to hear back from the client. There is a great difference between waiting two days or two weeks for the approval or comments you need before you can proceed to the next task.

LEARN MORE For advice about schedule charts, see Ch. 13's "Writer's Tutorial: Schedule Charts" (page 259).

GUIDELINE 4 Give your client a written proposal— and ask for a written agreement

A written proposal is an excellent tool for reaching a common understanding with your client as well as a valuable guide to your ongoing work together. It can serve two important purposes.

- **To build your client's confidence in your ability to do an excellent job.** A well-written proposal shows the client that you thoroughly understand the client's goals and the process and product needed to achieve these goals.

- **To ensure that you and your client reach substantial agreement about the project before you begin work on it.** A well-written proposal provides explicit details about the final product you will deliver to your client, the things your client will provide you, and the schedule and other arrangements for your work together. Also, the proposal can provide a framework that helps both of you plan your time and other commitments. Finally, it can protect you both from the problems that might arise if you had only an oral agreement and later discovered that you recollected some significant aspect of that agreement differently.

The two purposes of a proposal

Ask for a written response to the proposal. If the client agrees to the proposal as you've written it, then a one-sentence memo will do. If not, then your proposal provides the perfect basis for detailed, specific negotiations. In either event, it provides you—and your client—with an invaluable measure of protection.

Even though the proposal is primarily a formal agreement designed to protect you and your client, remember that its overall goal is to create a cordial, cooperative, mutually satisfying partnership. Be careful to avoid taking on a demanding tone when describing what your client will give you. Write in a cordial, businesslike manner. Convey enthusiasm for working with the client on the project.

Figure 19.1 shows a sample client proposal written by students.

LEARN MORE Chapter 23 provides detailed advice for writing proposals.

FIGURE 19.1

Client Proposal Written by a
Student Team

October 9, 2017

Dr. G. F. Hargis, Curator
Wright Zoology Museum
Crandell University
Iowa City, IA 52240

Dear Dr. Hargis:

The student team opens by continuing to build a positive relationship with their client, Dr. Hargis.

> We are very excited about the opportunity to work with you to develop a website for the Wright Zoology Museum. As you know, our work on the site will serve as a project for our technical communications course with Professor Dellapiana.

The students tell what they want Dr. Hargis to do with the letter: determine whether he agrees with what they say.

> In this letter, we explain our understanding of your objectives for the site and our proposed strategy for creating it. If you approve of what we say, please send us a memo authorizing us to begin work. If you want to discuss the project further, please contact Tim at T_Banner@Crandell.edu.

Background

The students briefly summarize information they've heard from Dr. Hargis to demonstrate that they've understood him accurately.

> We understand that you wish to increase greatly the number of visits the museum receives from elementary school classes in grades three through six. You have recently developed new programs for this purpose. As you've also explained, very few teachers in the region know that the museum exists because it has been used primarily as a resource for students here at the university. Also, you have reported that some teachers who have brought their classes to the museum have been disappointed because its programs have been designed for older students.

Objectives

The students state their understanding of Dr. Hargis's objectives.

> Through talking with you, we have established the following objectives for the website.

1. To introduce the museum's programs in a way that encourages teachers to visit.

2. To enable teachers to pick the programs best suited to their classes.

3. To enable teachers to request additional information or arrange a class trip online.

FIGURE 19.1

(*Continued*)

Dr. G. F. Hargis   2   October 9, 2017

We also have these additional objectives.

1. To provide a site that you or your graduate students can easily manage and update.

2. To help you develop a plan for publicizing the site.

Final Product

We propose to achieve these objectives by creating a site that has the following features.

1. An attractive home page that provides easy navigation to the information teachers most want.

2. Detailed information about each of your programs.

3. Lesson plans teachers can use to prepare their students for a visit.

4. Information about arranging visits and the downloadable forms and email links that will enable teachers to make their arrangements online.

5. Photographs, drawings, and other images that make the museum and its programs seem attractive.

We will build the site in Dreamweaver for use on the museum's server. The site will include approximately 60 pages and about 45 images.

Process for Developing the Website

While creating the site, we will focus on five types of activities.

   Research. We will study the museum's new programs, interview teachers from your target audience, and visit websites for small scientific museums that offer similar programs for elementary schools. We will learn what types of information teachers will want from the site and what type of presentation they would find usable and persuasive. We will report our research results to you in a memo, asking you to add your own comments and insight.

   Planning. We will develop three alternative screen designs for the home page and three designs for the other pages in the site. After you select the elements you like from the alternatives, we will combine them into a final design. We will also develop a site map, which we will submit for your approval.

   Building. We will build the site on developmental server space arranged by Professor Dellapiana.

By adding two objectives that Dr. Hargis had not mentioned but that will benefit the museum, the students demonstrate their knowledge of web design and emphasize that their goal is to help the museum.

The students begin their description of what they will deliver to Dr. Hargis by describing their proposed site's major features. Each feature addresses one or more of the major goals Dr. Hargis has for the site.

The students provide specific details to prevent the client from expecting more than the students will be able to do.

The students describe their process in ways that establish confidence in their ability to successfully complete the project.

The students highlight the times they will submit their ideas and results to the client for review and approval.

FIGURE 19.1

(*Continued*)

By describing the user testing they will perform, the students assure Dr. Hargis that the site they propose will succeed in achieving the museum's objectives.

The schedule identifies key interim deadlines for the student team and the client.

The students use bold type to help Dr. Hargis spot the deadlines for steps the museum will perform.

Dr. G. F. Hargis   3   October 9, 2017

User testing. When planning the site, we will ask selected teachers to respond to paper copies of our alternative designs. Besides asking how well each design appeals to the teachers, we will ask them what they would expect to see if they clicked on the links we plan to include. Through these and similar questions, we will learn how to increase the site's appeal and usability. After we've built the site, we will ask users to test it online. We will send the user test results to you, along with a list of our proposed revisions for your approval.

Technical testing. We will test the site's performance when displayed on a variety of computers using a variety of web browsers.

Schedule

The following schedule is our plan to ensure timely and accurate completion of the project. Significant deadlines for you are in bold.

Activity	Date of Completion
Conduct research	October 16
Deliver report	October 20
Receive your response	**October 23**
Create and test site map	October 23
Create and test screen designs	October 23
Submit alternatives	October 27
Receive your decisions	**October 30**
Build site	October 30–November 26
Show the site to you	November 29
Receive your comments	**November 31**
Revise site in light of your comments	December 1–December 2
Conduct user test online	December 5
Submit user test report	December 11
Receive your response	**December 13**
Test site on several platforms and browsers	December 11
Submit a report to you	December 13
Receive your response	**December 15**
Revise the site	December 15–21
Deliver the final site	December 22

FIGURE 19.1

(*Continued*)

Dr. G. F. Hargis   4   October 9, 2017

Qualifications ●————

We possess the skills needed to create an effective website for the museum. Sheila, a computer science major, has created several websites using Adobe Dreamweaver, a professional-level design program. Knowledge Tim gained in his marketing minor will help us design a website that will entice teachers to bring their classes to the museum. Using information management skills learned during an internship at the Houston Environmental Office, Vijah will organize our research efforts.

> The students cite specific knowledge and skills they have that are required to create a website that will achieve Dr. Hargis' objectives.

Museum's Contributions ●————

To meet all of your deadlines, we will need your assistance. First, we will need you to provide program descriptions and materials, as well as photographs and other images for the site. To help us meet the deadlines above, we will need in three days the responses to the questions we ask you. We will also need to use your digital camera to take pictures needed by the site.

> The students detail the things the client must contribute.

Conclusion ●————

We are eager to begin working with you on the Wright Museum's website. Please let us know as soon as is convenient whether the arrangements we described in this letter are acceptable to you.

Sincerely,

Tim Banner   *Sheila Esterbook*   *Vijah Singe*

Tim Banner   Sheila Esterbook   Vijah Singe
T_Banner@Crandell.edu   E_Esterbook@Crandell.edu   V_Singe@Crandell.edu

> The students conclude with enthusiasm. By saying that they "are eager to begin working with" the client, they reinforce the point that they want to be partners with the client.

Maintain a Productive Relationship with Your Client throughout Your Project

Your client's satisfaction with your work will depend not only on the quality of your final product but also the quality of the relationship you maintain with them. The following guidelines explain the two most important strategies for effective client communication as your work on the project advances.

GUIDELINE 1 Communicate candidly with your client during the project

How often should you communicate with your client while working on a project? Too many contacts can annoy. Too few can frustrate the client and even lead to misunderstandings. Certainly, you should communicate at the times specified in the project schedule, such as times when you committed to providing a progress report or something else for your client to review.

In addition, contact your client promptly when you require some information, need to make a basic decision about how to proceed, or encounter a problem the client should know about. The two most common problems involve a project's schedule and its scope. Schedule problems usually arise because you are taking longer than anticipated to complete some part of the project or because your client is taking longer than you expected to provide needed information or feedback on a draft or other material. Scope problems can arise when either you or the client begins to enlarge the project beyond what you both agreed to in your proposal. Sometimes the size of a project increases in small increments, a situation called *scope creep*. In other situations, an entirely new task is contemplated. The key to completing your project successfully is to bring all problems, major decisions, and significant changes in scope to your client's attention so that the two of you can determine together how best to address them.

Whenever you communicate with your client, whether at a prearranged time or because a need has arisen, follow these practices.

Communicating with Your Client

- **Be candid.** Don't hide or minimize problems. Clients need to know about difficulties so they can work with you to solve them and make contingency plans, if necessary.

- **Be specific.** The more precisely you describe the progress you've made, the information you need, the problems you've encountered, or the strategy you are recommending, the better you are preparing your client to help you complete the project successfully. Similarly, when submitting materials for review, let your client know the specific issues you'd like comments on—overall strategy? phrasing? color choice?—while also inviting comments on any topic.

- **Communicate constructively.** Even if you are discussing a problem caused by your client, remember that your goal is to prepare a communication that your client approves. You stand the best chance of achieving this goal if you always maintain a positive relationship.

- **Respect your client's time.** Collect questions so you can ask many of them at once or incorporate them in a regularly scheduled progress report. Find out whether your client prefers email, paper mail, or another form of communication.

GUIDELINE 2 Advocate and educate, but defer to your client

As you share your strategies, outlines, drafts, and other materials with your client, you may discover that you and your client disagree about some large or small aspect of the communication you are preparing. These differences may concern content, writing style, page design—any aspect of the communication whatsoever.

If this happens, explain to your client the reader-centered reasons for the choices you think best. Using your knowledge of what makes a communication useful and persuasive, educate your client and advocate for your positions.

If you disagree with your client, explain why.

In the end, however, remember that the communication you are preparing is not your communication but the client's. As communication specialist Tony Marsico (1997) says, "The client may not always be right, but the client is always the client." The only exception is a situation where you believe it would be unethical to proceed. If your client is not persuaded by your ethical arguments, your best option is to talk with your instructor about resigning from the project.

Remember that the client is always the client.

When educating and advocating, take a strategic approach. Distinguish large matters—those that have the greatest impact on the communication's effectiveness— from small ones. Then focus first on the large issues, postponing discussion of small ones or letting them go entirely. Also, know when to quit. As long as your client is listening, help the client make good decisions. But once the client's decision has been made and you have no new perspectives to offer, stop. If you continue, you will only irritate and harm your relationship with him or her.

Hand off Your Project in a Way Your Client Will Find Helpful

When you have completed your deliverable, you are not necessarily finished with the project. In the workplace, almost all client projects are accompanied by a transmittal letter or oral presentation that reviews the assignment, describes important features of the deliverable, and closes in a cordial fashion. In some cases, you can best serve your client by preparing a separate document to provide additional information, advice, and assistance. For example, if you have created a website, you may be able to include instructions for updating the content, recommendations for drawing people to the site, and information about sources for future assistance with it. You may also have generated ideas about additional ways the site could be improved in a future project. By carefully planning how you will hand off your project, you provide added assistance and gain added praise for your work.

LEARN MORE For advice about writing transmittal letters, see "How to Write a Reader-Centered Transmittal Letter" (page 212).

James Marshall/The Image Works

At the end of client projects, students often meet with their clients to summarize their results, present their findings, and answer questions.

Conclusion

When working for a client, you must be a manager as well as a communicator. This chapter's guidelines have focused on the most important strategies for managing both the project and your relations with your client: Reach a written agreement about the project before you begin, work carefully with your client throughout the process, and hand off your deliverable in a helpful way. Figure 19.2 provides a Writer's Guide for Managing Client and Service-Learning Projects.

FIGURE 19.2

Writer's Guide for Managing Client and Service-Learning Projects

---

Writer's Guide
MANAGING CLIENT AND SERVICE-LEARNING PROJECTS

DETERMINING WHAT'S NEEDED

1. Learn about your client's organization.
2. Learn the client's perception of the readers, communication objectives, and stakeholders.
3. Learn the client's deadline, requirements for the communication, and budget.
4. Learn what information and other resources the client will provide to you.
5. Develop your own assessment of the situation.

CREATING A PROJECT PLAN

1. Determine the size, overall content, major features, and technical aspects of your deliverable.
2. Create a project schedule that includes project milestones, dates for meeting with or submitting material to your client, and dates for your client's responses.
3. Include your client's contributions.

WRITING A PROPOSAL

1. Be as specific as possible to protect your client and yourself.
2. Ask for a written agreement from your client.

CONDUCTING THE PROJECT

1. Monitor your progress continuously by comparing it against your project schedule.
2. Talk with your client about changes to your original agreement.
3. Communicate with your client in ways that are candid, specific, and respectful of your client's time.
4. Advocate for your positions, but ultimately defer to your client when necessary.
5. Hand off your project with a transmittal letter and helpful information and advice.

USE WHAT YOU'VE LEARNED

EXERCISE YOUR EXPERTISE

1. If your instructor has assigned a client or service-learning project, list the specific questions you will need to ask your client in order to learn about the organization, the communication, and the communication's objectives. Report the results in a memo to your instructor.

2. Draft a project schedule for your project.

EXPLORE ONLINE

If your instructor has assigned a client or service-learning project, use the Internet to learn about the organization's mission and goals and its structure. Based on what you learn, speculate about the style and features you believe the client would want used in the communication you will prepare. Identify other information you believe will enable you to plan successful strategies for establishing a successful partnership with this organization. If you can't find information about this specific organization, study ones that are similar to it. Report the results in a memo to your instructor.

COLLABORATE WITH YOUR CLASSMATES

If you are part of a team that is working on a client or service-learning project, decide how you will assign responsibilities within your team. Consult Chapter 17's guidelines for creating communications with a team as well as this chapter's advice for establishing productive relations with clients. Report your results in a memo to your instructor.

APPLY YOUR ETHICS

As a service-learning project for their course, Celia and Stephen have been working together for the last six weeks to create a new website for a social services agency in their community. At the beginning of the project, they used a thorough proposal to develop mutually agreeable commitments with the director of the agency. They have worked very hard, and they have had good cooperation from the agency, which has devoted many hours to working with them. However, the project has turned out to be larger than they anticipated, so they have been unable to complete the project by the end of the term. Their instructor understands the circumstances and will allow the work they have completed to serve as their project

so that they will be able to finish the course. However, the agency will not have the new website it had counted on launching. For each of the following circumstances, discuss the ethical obligations Celia and Stephen would have to complete the project during the upcoming vacation or during next semester, when they will both have a full load of courses.

- If they were unable to complete the project this term because they promised features they didn't know how to implement.
- If Celia and Stephen's inability to complete the project resulted because the agency director thought they had promised to do a larger site than they thought they had agreed to provide.
- If the delay was caused by the director's wanting very substantial revisions when Celia and Stephen submitted drafts of the site map and sample sections to the director for review.
- If the delay was caused by the director's illness, which delayed the agency's response to drafts.

20

Creating Reader-Centered Websites and Professional Portfolios

LEARNING OBJECTIVES

1. Develop reader-centered content for a website.
2. Help your readers quickly and easily find what they are looking for.
3. Design web pages that are easy to read and attractive.
4. Design a website for diverse readers.
5. Follow ethical and legal practices concerning your website's content.

MindTap®

Find additional resources related to this chapter in MindTap.

"Writer's Tutorial: Creating a Website Using Tables" (pages 352–354) gives step-by-step directions for creating a website using Microsoft Word.

In this chapter, you will learn how to design and construct effective websites, as well as how to help others evaluate and devise strategies for improving their websites.

Many employers in engineering, scientific, and other specialized fields use websites for purposes that differ significantly from the purposes of the entertainment, shopping, and marketing sites with which we are all familiar. For example, some use websites to provide their clients with detailed technical information about their products and services, share proprietary information among their employees while keeping it hidden from competitors, or advocate on technical, scientific, or environmental issues. This chapter focuses on the reader-centered approach to creating these specialized sites, although the strategies presented are equally helpful when creating any other kind of website, including a professional portfolio for your job search.

How to Gain the Most Value from This Chapter

If you've already developed websites or designed web pages at a social media site, you may be asking, "Do I really need to read this chapter?" On the other hand, if you haven't created anything for the web before, you may be skeptical that one small chapter could enable you to build a website from scratch.

Together with your instructor, this chapter has a lot to teach you in either case. If you are a novice, you will find all the information needed to construct your first website with free, easy-to-use programs already on your computer. In fact, the Writer's Tutorial on pages 352–354 explains the simple process of making one with Microsoft Word. If you have plenty of web experience, you can skip this information to focus on what matters most: the reader-centered strategies people in engineering, scientific, and other specialized fields use to make their websites effective.

This Chapter's Example: Digital Portfolio Websites

To help you learn strategies for creating effective websites, this chapter guides you through the process of creating one: a professional portfolio website that could be a powerful aid in your search for an internship or a job. A résumé and job application letter can only tell employers what you can do. In a portfolio, you can show them the impressive results of your work.

Figure 20.1 shows two pages from one portfolio. The reader-centered strategies used to create it will be just as useful for any other website your instructor may assign or that you may create in your career.

Define Your Website's Goals

A critically important point to remember about workplace websites is that although the medium is very different from paper and ink, the goals are the same: to help readers perform some task and to influence readers' attitudes and actions in ways that you desire.

Consequently, the first step in developing a reader-centered website is to determine what it needs to help your readers do and how you want it to influence their attitudes and actions. Chapter 3 provides detailed advice about identifying the readers' tasks and attitudes, as well as the contextual factors that may influence your readers' responses. Chapter 2 explains how to apply this approach to résumés and job application letters. Their advice is equally valid for professional portfolios. To refresh your memory, you may want to review those chapters.

LEARN MORE Start by following Chapter 3's guidelines for defining your communication's goals.

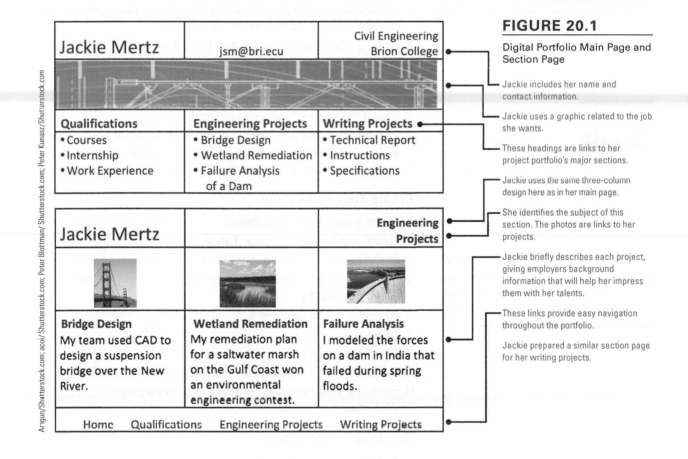

FIGURE 20.1

Digital Portfolio Main Page and Section Page

Jackie includes her name and contact information.

Jackie uses a graphic related to the job she wants.

These headings are links to her project portfolio's major sections.

Jackie uses the same three-column design here as in her main page.

She identifies the subject of this section. The photos are links to her projects.

Jackie briefly describes each project, giving employers background information that will help her impress them with her talents.

These links provide easy navigation throughout the portfolio.

Jackie prepared a similar section page for her writing projects.

CREATING A WEBSITE USING TABLES

This tutorial shows how to create a website using Word for Mac 2016, which is virtually identical to the latest versions of Word for Windows. If you use a different word processor or get stuck, click on your program's **Help** menu for assistance. To illustrate the process, the tutorial shows how to create a digital portfolio website, but you can create a website on any other topic using these instructions.

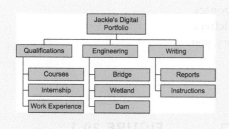

Source: Microsoft

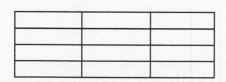

Define Your Website's Goals

1. Pick a topic or use "My Professional Portfolio."
2. Identify your target readers.
3. Define your website's usefulness and persuasive goals.

Plan Your Website

1. Determine your website's topics and subtopics, each identified by one word.
   - If you are making a professional portfolio, the major topics might identify your areas of expertise.
2. Draw a site map that organizes the topics and subtopics.

Make Your Website Folder

1. Make a new folder.
2. Title it "Website," followed by your initials.
3. In the folder, put all files that will be displayed at your website.
   - For a professional portfolio, these would be projects and other evidence of your abilities and accomplishments.

Create Your Home Page

1. Open a new Word document.
2. Under the **Insert** tab, click **Table**.
3. In the box that appears, choose 3 columns and 4 rows.
4. From the **File** pulldown menu at the top of your screen, choose **Save As**.
5. Navigate to the website folder you created.
6. Open the folder.
7. For **Save As**, type "Index."
   - "Index" is the name browsers look for at websites.
8. From the **File Format** menu, choose **Web Page**.
9. Click **Save**.

Create One Other Page

1. **Copy** the table from Index.
2. **Open** a new Word document.

3. **Paste** the table into the new document.
4. **Save** the page with the file name 'Page 2' by using Steps 4 to 9 in Create Your Home Page.

Open Your Home Page for Editing

1. From Word's **File** menu at the top of your screen, choose **Open**.
   - To edit a page you have saved as a web page, you must open it from within Word. If you open the file by clicking on it in a Finder window, the page will open in your browser, so you cannot edit it in Word.
2. Navigate to "Index.htm."
3. Click **Open**.

Enter Information into Your Home Page

1. In Row 1, enter heading text for your website.
   - For a professional portfolio, you might enter your name, contact information, major, and college.
2. In Row 3, enter the words that identify your website's topics.
3. In Row 4, enter information related to each topic.
   - For a professional portfolio, you might enter your projects' names.
4. In Row 2, highlight all the cells.
5. Under the **Table Layout** tab, click **Merge Cells**.
6. In Row 2, **Insert** a graphic related to your website's subject.
7. Change the color, type size, and alignment of the items to guide the reader's eye and make the page attractive.
8. **Save** your home page.

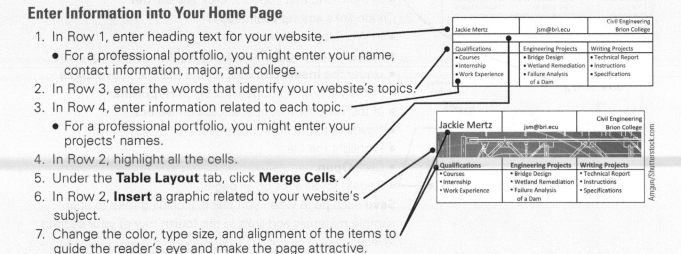

Enter Information into Your Topic Pages

1. In Word, open the second page you created.
2. In Row 1 of the table, enter heading information.
3. In Row 3, enter the titles of your website's topics.
4. In Row 3, you may also enter a description of each topic.
5. In Row 2, enter an appropriate image for each topic.
6. In Row 4, highlight all the cells.
7. Under the **Table Layout** tab, click **Merge Cells**.
8. In Row 4, type "Home" and the names of all your topics.
   - Later, you will make these words into links that promote easy navigation at your site.
9. Change the color, type size, and alignment of all items to harmonize with those used in your home page.
10. **Save** the page.

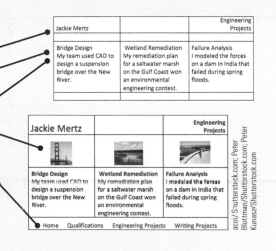

WRITER'S TUTORIAL, continued

CREATING A WEBSITE USING TABLES

11. Make one copy of this page for each of your other topics.
12. Create each topic's page by replacing the appropriate items.

Finish Your Pages

1. On all pages, eliminate the table's grid lines.
   - In Word, open each page in turn.
   - Highlight the entire table.
   - Under the **Table Design** tab, click the triangle above **Borders**.
   - In the menu that appears, click **No Border**.
2. Create links among your pages.
   - In Word, open Index.html.
   - In the home page's Row 3, highlight the text in one cell.
   - Under the **Insert** tab, click on the triangle above **Links**.
   - In the window that opens, click on **Hyperlink**.
   - In the window that opens, click on **Select**.
   - Navigate to the appropriate file.
   - Highlight the file.
   - Click **Open**.
   - Repeat for all links on all pages.
3. **Save** each page when you finish, including links to your sample materials and links in the fourth row of your second-level pages.

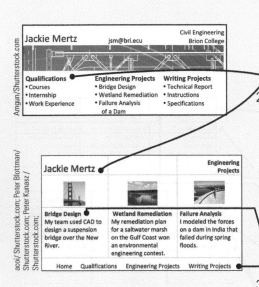

Test Your Website

1. Open a web browser.
2. From the **File** menu, select **Open File**.
3. Navigate to Index.
4. Click **Open**.
5. Test all links.
6. Ask other people to review your portfolio.
7. Revise.
8. Test all links again after revising.

Research to Identify Useful, Persuasive Content

Researching the content for a website is the same as researching for a print communication. See Chapters 4 and 5. You may wish to review those chapters. One caution: Avoid assuming that because you are researching for a website, the Internet should be your primary or only research resource.

When researching for your professional portfolio, look through your projects for the ones that best display the qualifications wanted by the employer to whom you are applying. Remember that you can include many kinds of material—text, photos, movies, and so on—some of which you may want to polish or create for your portfolio. In a portfolio, you also have a chance to explain the important points about your projects that you want to emphasize, either in text or in audio files.

A professional portfolio provides a good example of the way to develop reader-centered content. The portfolio's reader has a specific task to do—namely, to find the right person to fill a specific position. He or she wants your portfolio, along with your résumé and job application letter, to indicate how well your qualifications match the employer's needs. To be *useful* to this reader, your portfolio must help him or her find the information related to those qualifications about you as quickly as possible. To be *persuasive*—to influence the reader to invite you for an interview or give you a job offer—your portfolio must explain convincingly that your qualifications match the employer's needs better than the other applicants'. Thus, when choosing the projects and accomplishments you will include, pick ones that demonstrate that you possess the knowledge and skills the employer wants.

TRY THIS List the qualifications desired by employers for the kind of job you would like. This list can guide your selection of samples to include in a digital portfolio.

Organize to Help Your Readers Quickly Find What They Are Looking for

When people visit workplace websites, they are usually looking for something in particular. A major way to increase your site's usefulness is to help your readers find what they want quickly, without making guesses that sometimes send them traveling down dead ends. In fact, the travel metaphor is helpful. Think of each piece of information in your website as a destination. To organize an effective website, you need to help readers reach their destinations rapidly.

For a website, the most useful organization is one that matches your reader's intuitive strategy for finding what he or she wants. For example, Jackie could have organized her digital portfolio around the courses in which she produced the projects she displayed. Or, she could have organized under headings that indicated when she worked on the projects: First-Year Projects, Second-Year Projects, and so on. However, she wisely organized under headings for Engineering Projects and Writing Projects, knowing that her reader's basic task was to learn whether she had the engineering and communication abilities the employer was looking for. Note that organizing around her reader's search strategy helped Jackie make her digital portfolio persuasive. The headings for her two major sections made persuasive claims: I have the engineering and communication expertise you are seeking. The sample projects she included became her supporting evidence (see the discussion of effective persuasive arguments in "Reason Soundly," page 171).

Often, you can use your readers' questions to identify their search strategies. Figure 20.2 shows the structure of a website on bird migration. Imagining that their readers would want to know how birds prepare to fly hundreds, even thousands, of miles without stopping, the creators organized some of their information under a heading that announces the answer to that question: "How Birds Prepare." Knowing

FIGURE 20.2

Website Map with Hierarchical Organization and Intuitive Categories

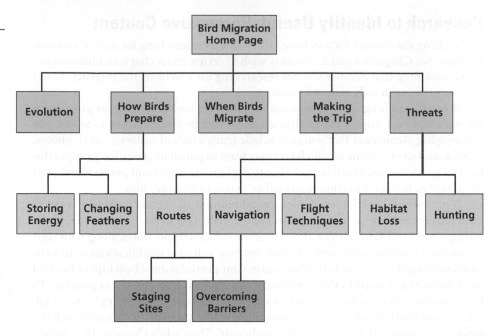

TRY THIS What would be a good site map for your digital portfolio? Try sketching one based on your list of qualifications desired by employers you would like to work for (see Try This on page 355).

that some readers would be curious about dangers birds encounter during migration, they organized a section labeled "Threats," which clearly says, "Here's your answer."

The more clicks readers must make to travel from a website's home page to the answers or information they want, the less helpful the site's organization is. A goal for many workplace websites is to enable readers to reach their destinations in three or fewer clicks from the home page. Achieving a three-click structure is rarely a problem for small websites. You may even be able to put links to every other page on the home page. However, if you are creating a larger site, count the clicks your readers will need to make. Try to simplify the structure if it will take them more than three.

Draft Your Website To Be Easy To Use, Persuasive, and Attractive

Drafting for websites is much like drafting for print communications. However, two differences often emerge. First, the visual design is typically much more important at the website, both as an aid to readers searching for specific information and as a way of maintaining readers' interest. Second, the reader is not likely to read the contents sequentially. In fact, there is no single preferred sequence. Readers follow their interest. The following guidelines will help you draft websites and a professional portfolio that are easy to use, persuasive, and attractive.

GUIDELINE 1 **Include a variety of navigational guides in your web pages and website**

After you have organized your site in a reader-centered way, you have done half the work needed to help your readers locate the information they want. The other half is to design navigational aids that enable readers to find their way quickly from the home page to their destinations. The following strategies, which are illustrated in Figure 20.3, can help readers quickly find what they want.

Helping Readers Navigate Your Site

- **Include the main menu on every page.** Enable your readers to start a new "drill down" from whatever page they are on.
- **On every page, provide a link to the home page.**
- **On each page, provide a menu for the category in which that page is located.** Once readers are in a category, they may want to explore the various types of information you provide in it.
- **Place navigational aids in the same location on every page.** Once readers have learned where your navigational aids are, let them use that knowledge throughout the time they spend at your site.

(Continued)

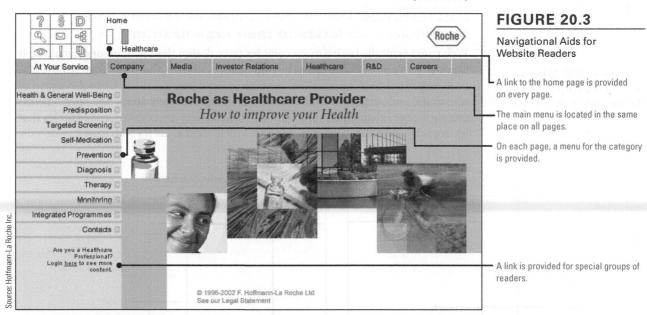

FIGURE 20.3

Navigational Aids for Website Readers

A link to the home page is provided on every page.

The main menu is located in the same place on all pages.

On each page, a menu for the category is provided.

A link is provided for special groups of readers.

Source: Hoffmann-La Roche Inc.

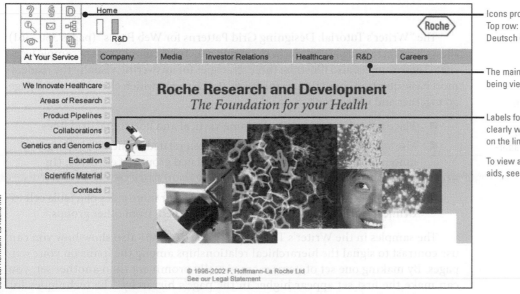

Source: Hoffmann-La Roche Inc.

Icons provide part of the main menu. Top row: Help, Legal Statement, Deutsch (German version of the site).

The main menu highlights the category being viewed.

Labels for the menu items tell readers clearly what they will find if they click on the links.

To view another set of navigational aids, see Figure 20.5 on page 363.

- **Use clear, specific labels for all menu items.** Links are helpful only if readers can accurately predict what they will find if they click on them.
- **Make clear, informative in-text links.** When creating links within your text, signal clearly what your readers will find there.
  - □ **Highlight only the major words.** If all the words in the following sentence were highlighted, readers wouldn't know whether the link leads to information on whales, the Bering Sea, or krill. Highlighting only krill makes clear what the topic of the link is.

    ▌ When whales reach the northern Bering Sea, they feed on the plentiful **krill**. ▌
  - □ **Supply explanations of the links.** These are especially helpful for links included in a list.

    ▌ **International projects**—Read about our current projects in Europe and Asia. ▌

Blue type indicates the link.

TRY THIS The sample web page in Figure 20.1 (page 351) places navigational links at the bottom of the page. Where else could the links be placed that would be handy and attractive?

LEARN MORE To learn more about grid design, see "Create a grid to serve as the visual framework for your page" (Chapter 14, page 265).

GUIDELINE 2 **Design your pages so readers can immediately understand their organization**

Web pages typically have a more complex page design than printed ones, making it especially important that you help your readers understand immediately how you have organized your information. As with printed pages, your basic tool is a grid of horizontal and vertical lines that helps you communicate visually the relationships among a page's elements. In fact, almost all web pages are constructed as tables that form grids—quite literally. Here are two widely used grids for web pages.

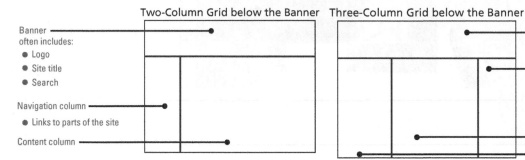

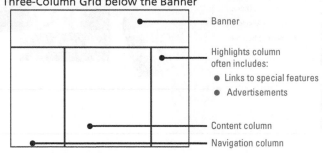

LEARN MORE To learn more about the use of alignment and visual grouping, turn to Chapter 14, "Designing Reader-Centered Pages and Documents," pages 261–277.

The "Writer's Tutorial: Designing Grid Patterns for Web Pages" (pages 360–361) shows how these two patterns form the major structural components of sample websites. These samples also illustrate three strategies for answering visually the readers' most basic question about the organization of web pages: which elements on the page go together and which don't.

- Place related items against the same vertical and horizontal grid lines.
- Use small amounts of white space between related elements and larger amounts of white space to separate unrelated groups of elements. (In websites, "white" space is often colored.)
- Add borders—black or colored lines—to one or more sides of a table cell to define groups of related items and separate them from other groups.

The samples in the Writer's Tutorial on pages 360–361 also show how you can use contrast to signal the hierarchical relationships among the items on your web pages. By making one set of items more visually prominent than another set, you can make the first set appear higher in the page's hierarchy. The techniques for giving greater visual prominence to some items are the same for web pages as for printed ones.

- Use larger type for the more major items.
- Put the larger items farther to the left by indenting less major ones.
- Use bold for the more major items.
- Use a more prominent color for the more major items.

TRY THIS Is the web page design used in the Writer's Tutorial on pages 360–361 the best possible one? Following the advice for this guideline, sketch a different design that works as well or better.

GUIDELINE 3 Make your text easy to read

A page is useful only if readers can easily read its contents. The following strategies will help you create legible web pages. Figure 20.4 (page 362) highlights the use of these strategies on effective web pages.

Making Web Pages Easy to Read

- **Use typefaces that are easy to read on screen.**
- **Make sure your text is legible.**
  - **Use dark letters against a light background (or vice versa) to produce a strong contrast between the two.** White is always a good choice for the background.
  - **Use large enough text for legibility.**
  - **Avoid intricate backgrounds that make it difficult to pick out the letters of the text.**
  - **Avoid italics.** These ornate letters can be very difficult to read on screen.
- **Present your information in small, visually distinct chunks.**
  - **Keep paragraphs short.** Break longer paragraphs into smaller ones.
  - **Use white space (blank lines) to separate paragraphs and other blocks of text,** keeping in mind that the "white space" on a web page may actually be any background color.
  - **Edit your text for conciseness.** The fewer words, the better.
  - **Use plenty of headings.** They not only help readers locate information but also break the screen into meaningful chunks.
- **Limit scrolling (it greatly increases the reading difficulty).**
  - **Avoid vertical scrolling** except on pages where users will be looking for in-depth information.
  - **Avoid horizontal scrolling altogether.**

GUIDELINE 4 Unify your website verbally and visually

Visual and verbal unity among a website's pages helps readers use the site efficiently, and it makes the site more appealing to them.

We've all had the experience of going to a website for the first time and needing to take more time than it seems reasonable to find the search box or a link we want. When designers fail to use the same overall framework throughout their sites, they require their readers to repeat that hunt each time they click to a new page. By using a consistent design, you save your readers from that time-consuming and frustrating work.

Consistent page design also creates a visual harmony that people find attractive. The positive impact of consistency is especially great in the workplace, where it communicates that the organization, team, or individual who created the site is capable of producing polished and professional work.

Figure 20.5 shows how one company uses the same page design, colors, and typefaces to make its website both functional and attractive. Note, however, that having a consistent framework doesn't mean that every page is exactly the same. In their details, the two pages shown in Figure 20.1 vary considerably, as do the pages shown in Figure 20.4.

TRY THIS In the "Writer's Tutorial for Creating a Website Using Tables", the directions for creating topic pages (page 353) shows one possible design for the topic pages that use the home page's grid pattern. Can you create another design for the topic pages that uses the home page's grid pattern?

DESIGNING GRID PATTERNS FOR WEB PAGES

Basic Designs

National Park Service

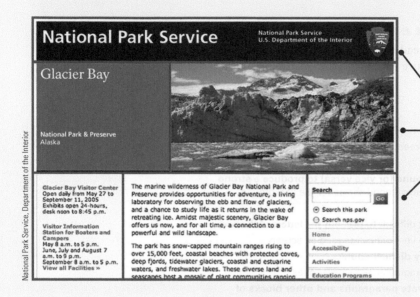

The National Park Service designed a grid with three horizontal areas.

Top: Undivided
Middle: Divided into two sides
Bottom: Divided into three parts

The design includes a very large area for the photograph to emphasize the beauty of Glacier Bay National Park and Preserve.

National Academy of Engineering

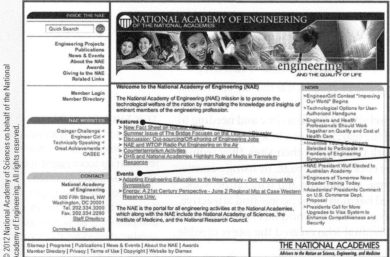

By devoting a small area for the photographic collage, the National Academy of Engineering created a grid that provides a great deal of space for links to the information readers want.

The headings that group related links also help readers find information quickly.

IBM Japan

For its Japanese website, IBM created a design whose central area (white) has five columns. Two of the grid lines from that area continue into the bottom area to divide its three cells.

Source: © IBM Corporation, 2012.

Elaborate Designs

National Aquarium in Baltimore

Where Do Angles and Curves Come From?

If all web pages are constructed from rectangular table cells (see page 450), how do designers make pages with angles and curves? Using specialized graphics programs, they create the overall image and then cut it into rectangles. The three smaller images above are among the twenty rectangles used to build the upper right-hand part of the National Aquarium in Baltimore website.

Source: National Aquarium

Source: National Aquarium

FIGURE 20.4

Readable Web Pages

In-text links are clear and informative.

- Key words highlighted
- Underlining only for links

Text is chunked.

- Short paragraphs
- Many headings
- Use of a list
- White space separating blocks of text

Text is legible.

- Dark letters against a light background
- Text large enough to be readable
- No italics

Writing is clear and concise.

No scrolling required.

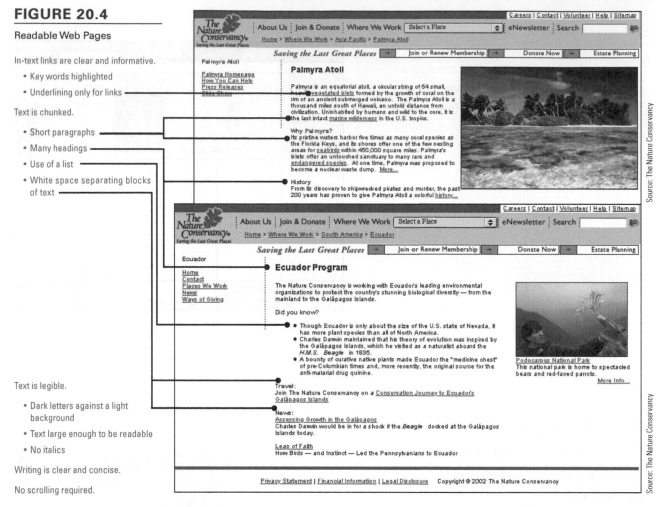

Source: The Nature Conservancy

Because you create each page in a website separately, it is easy to change the writing style unintentionally as you move from topic to topic. The change might appear in the way information is presented or in the phrasing of links and headings. You can avoid such shifts by planning the style you want to achieve before you begin drafting. You can check changes that have slipped in by making consistency of style one of the features to be examined when you and others review your site.

Design Your Website for Diverse Readers

A truly reader-centered website can be used by every reader who wants the information it provides, including people from any culture and people with disabilities. Using a small number of simple strategies, you can greatly increase your website's ability to serve all readers.

GUIDELINE 1 Ethics Guideline: Construct your website for use by readers with disabilities

Countries as diverse as Canada, Japan, Philippines, and Spain have established guidelines for making websites more accessible and easier to use for people with disabilities. In 2001, the U.S. government mandated that all federal agencies and federally affiliated institutions (including almost all U.S. colleges and universities) make their websites and other electronic and information technology accessible

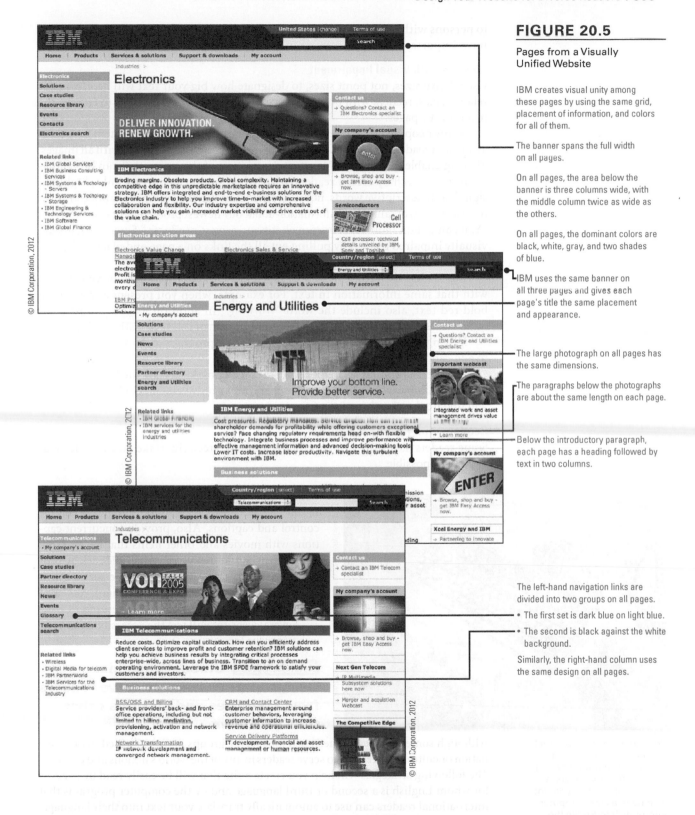

FIGURE 20.5

Pages from a Visually Unified Website

IBM creates visual unity among these pages by using the same grid, placement of information, and colors for all of them.

The banner spans the full width on all pages.

On all pages, the area below the banner is three columns wide, with the middle column twice as wide as the others.

On all pages, the dominant colors are black, white, gray, and two shades of blue.

IBM uses the same banner on all three pages and gives each page's title the same placement and appearance.

The large photograph on all pages has the same dimensions.

The paragraphs below the photographs are about the same length on each page.

Below the introductory paragraph, each page has a heading followed by text in two columns.

The left-hand navigation links are divided into two groups on all pages.

- The first set is dark blue on light blue.
- The second is black against the white background.

Similarly, the right-hand column uses the same design on all pages.

to persons with disabilities (Roach, 2002). The following strategies are among the most important ones you can use to make your site more inclusive.

Readers with Visual Impairment

Use relative sizes, not point sizes, to designate how big your text will be. If you use relative sizes, readers with poor vision can use their browser settings to enlarge the text on your pages.

Many people who can't see well use screen readers, which convert text to speech. To create a reader-centered site for these persons, include an <alt> tag with all images. This tag enables you to provide a text description of the image that can be synthesized into speech by screen readers. Also, when you create in-text links, highlight significant, descriptive words, not such phrases as "Click here." For example, instead of writing, "To obtain more information about our newest products, click here," write, "You can obtain more information about our newest products." Screen readers aid visually impaired readers by pulling out all the links on a page and placing them in a list. A list in which every link says "Click here" will be useless.

For readers with color blindness, supplement all information conveyed in color with that same information in text. For example, when you present a warning in bold red text, also include the word "*Warning.*" In addition, because most color blindness is red–green, avoid using this combination of colors to distinguish content.

When you include movies or animations, use a program such as QuickTime Pro to add captions for visually impaired readers.

If these and similar strategies don't succeed in making your site fully accessible to visually impaired readers, create a separate text-only version. Update it whenever you update the visual version of your site.

Readers with Limited Mobility

Some readers have difficulty controlling a mouse precisely. Make all clickable areas as large as possible.

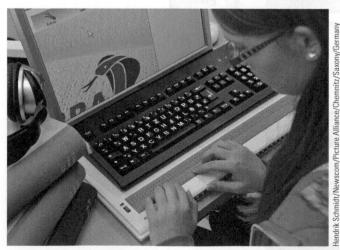

Some persons with a visual impairment supplement a regular keyboard for typing with another that enables them to rapidly navigate the web page as the computer interprets into Braille the headings, text, links, and image tags.

Hendrik Schmidt/Newscom/Picture Alliance/Chemnitz/Saxony/Germany

Readers with Hearing Impairments

If you include videos with sound, provide a volume control and captions. Also, provide concurrent captions with movies and other items that have sound. Supplement sound that gives verbal feedback by providing visual information of the same sort. For instance, if your site provides a verbal "Thank you," also include a pop-up window with that message.

Several websites will test the accessibility of your web pages for free.

GUIDELINE 2 Global Guideline: Design your website to serve readers from countries and cultures other than your own

Although some websites are designed for specific groups of users located in only one nation or culture, others aim to serve readers in any of the world's nations and cultures. The following strategies will help you create a site that will be easily read by persons for whom English is a second or third language and by the computer programs that international readers can use to automatically translate your text into their language. If your employer decides to make a second version of your site in another language, these strategies will also ease that work.

Text

Write simply. Avoid slang, colloquial expressions, and metaphors that refer to sports, occupations, or other things specific to your nation. Also, as advised in Chapters 7, 9, and 10, consider your readers' cultural background when drafting your text. Check your draft with persons knowledgeable about the cultures of your likely readers.

Images

As Chapter 12 explains, images that are meaningful to persons in your culture may be incomprehensible—even offensive—to readers in other cultures. Follow that chapter's advice when selecting your photographs, drawings, icons, and other images. Also, avoid including words in your images. Computerized translators cannot read images. And don't forget that persons creating a version of your site in another language will need to remake all images, including buttons in menus that have words.

Page Design

Many English words are translated into much longer words in other languages. Similarly, the same statement may take as many as 30 percent more words in other languages. If you expect your site to be translated into other languages, design the size of your buttons, menus, and text areas to accommodate these differences.

Follow Ethical and Legal Practices Concerning Your Website's Content

When creating a website, you will encounter two sorts of ethical issues. They concern what you borrow from other people and what you offer visitors to your site.

GUIDELINE 1 Observe copyright law and acknowledge your sources

It's very easy to download images and text from someone else's website for incorporation into your own site. Some items available on the web fall into the public domain, meaning that no one has a copyright on them. You can use them freely without permission.

However, much of what's on the web is copyrighted. This includes not only text but also technical drawings, photographs, and frames from your favorite comic strip. Many sound files, animations, and videos available on the web are also copyrighted. When you are thinking of using images from another site, check to see whether the site includes a copyright notice.

Whenever you are unsure about whether permission is needed to use something, check with your instructor or other knowledgeable person—or simply request permission.

Even when permission isn't needed, acknowledge your source when you use something that clearly represents someone else's intellectual or creative effort. Appendix A tells how to cite web sources.

GUIDELINE 2 Provide thorough, accurate information that won't harm others

When you post something on the web, keep in mind that your readers rely on your expertise and integrity. Be sure that you have checked your facts and that your information is up to date. Remember also that all the other ethical considerations discussed in this book also apply to web communications. Don't include anything that will harm other people.

In a digital portfolio, as in your résumé and job application letter, represent your work ethically by following the advice given in "Ethical Issues in the Job Search"

(Chapter 2, page 45). If you are including group projects, indicate clearly what you contributed. Ability to work effectively in groups is, by itself, an important qualification to employers.

Writer's Guide and Other Resources

Figure 20.6 presents a "Writer's Guide for Creating Websites" that you can use in your course and on the job.

FIGURE 20.6

Writer's Guide for Creating Websites

Writer's Guide
CREATING WEBSITES

DEFINE YOUR WEBSITE GOALS

1. Determine what your readers need in order to perform their tasks. (page 351)

2. Identify content that will influence their attitudes and actions in the ways you desire. (page 351)

RESEARCH TO FIND USEFUL, PERSUASIVE CONTENT

1. Review advice from Chapters 4 and 5. (page 355)

2. Use a wide range of sources, not just the Internet. (page 355)

ORGANIZE TO HELP YOUR READERS QUICKLY FIND WHAT THEY ARE LOOKING FOR

1. Organize to match your readers' search strategy. (page 355)

1. Build a three-click hierarchy. (page 356)

DRAFT YOUR WEBSITE TO BE EASY TO USE, PERSUASIVE, AND ATTRACTIVE

1. Include a variety of navigational aids on in your web pages and website. (page 356)

2. Design your pages so readers can immediately understand their organization. (page 358)

3. Make your text easy to read. (page 359)

4. Unify your website verbally and visually. (page 362)

DESIGN YOUR WEBSITE FOR DIVERSE READERS

1. Include features that make your site useful for readers with disabilities. (page 364)

2. Create your text, images, and page design for international and multicultural readers. (page 364)

FOLLOW ETHICAL AND LEGAL PRACTICES WITH YOUR WEBSITE CONTENT

1. Observe copyright law and acknowledge your sources. (page 365)

2. Provide thorough, accurate information that won't harm others. (page 365)

USE WHAT YOU'VE LEARNED

Instructors: For a website assignment, consider the Informational Website project in MindTap.

EXERCISE YOUR EXPERTISE

Visit a website, perhaps one related to your major or to a project you are preparing in this course. Begin by identifying the site's target readers and its usefulness and persuasive goals. Next, evaluate one of the site's pages in terms of this chapter's guidelines for designing effective web pages. Finally, evaluate the site's overall design in terms of this chapter's guidelines for building effective websites. Report the results of your analysis in a memo to your instructor.

EXPLORE ONLINE

1. Using a search engine, locate four websites that provide advice for designing websites. Compare their advice to the advice given in this chapter. Report your results in a memo to your instructor.

2. At the website of an international corporation (e.g., Sony), link to the organization's sites in three other languages. Discuss four ways in which these sites differ from the corporation's site for your country.

3. Evaluate the website for your university (or another site identified by your instructor) in light of this chapter's strategies for creating sites that can be used by people with disabilities.

4. Create your own website, following the guidelines given in the Informational Website project in MindTap.

COLLABORATE WITH YOUR CLASSMATES

Working with two or three other students, evaluate the usability and persuasiveness of your school's website from a web page for use by students in your major at your school that provides links to information on ethical issues that are relevant to them. Include links to your school's policies concerning academic integrity, plagiarism, and related issues as well as information from professional organizations. Make the page visually appealing, informative, and easy to use.

REFLECT FOR TRANSFER

1. Write a brief memo to your advisor in your major. Tell what you would include in a professional portfolio as you look for an internship or job. Also, guess what you might include in a portfolio you created after your first year on the job. If your instructor asks, share your memo with your advisor, asking for his or her comments.

2. Write a brief note to yourself, one you can read a few years in the future, identifying the two pieces of advice that you think will be most important for you to follow when you create a website in your career. These may be points from the chapter or advice you gained somewhere else, but identify the source of each one.

21

Writing Reader-Centered Correspondence: Letters, Memos, and Emails

In the workplace, people transact much of their business using short forms of communication that many of us use outside of work almost every day: email, instant messages, blog entries, letters, and so on. However, at work people usually write these forms in much different ways than they (and we) do in our social and personal lives.

These communications fall into two groups: correspondence and social media. Although the boundary between them is fuzzy, the two forms are different enough to be treated separately. Therefore, this chapter provides advice for workplace correspondence, and the next discusses the use of social media on the job.

Of course, the fundamental strategy for using any of these forms is to take a reader-centered approach. This chapter will remind you of the reader-centered strategies that are most important in correspondence and explain additional reader-centered strategies for correspondence. As you read the chapter and discuss it with your classmates and instructor, focus on learning how to follow its advice while working on a project your instructor has assigned or is planning to assign.

Use the Appropriate Level of Formality

When you begin a new job or internship, one of your greatest challenges will be figuring out what level of formality to use in your correspondence. Your new co-workers, managers, and clients will make many assumptions about you based on the formality or informality of your writing. That's because these relatively short forms can seem much more like personal, face-to-face communications than the longer documents they read. If they expect formality and your style is informal, they may view you as an immature person lacking seriousness about your work. On the other hand, if they expect informality and your style is very formal, they may view you as remote, someone who is not committed to working with the team.

So what can you do? The answer is to conduct a little research. Ask your supervisor and colleagues. Look at the communications sent to you. And keep in mind that while there is often a general level of formality (or informality) at many organizations, the individuals to whom you write might have strong preferences themselves.

Here are a few generalizations you might take as starting points and test through your observations and inquiries. Letters are usually more formal than memos, which are usually more formal than emails—although the degrees of difference can be very small or very large, depending on the organization. Communications written to managers and persons in other parts of the organization are often more formal than ones addressed to co-workers with whom the writer interacts daily. If you need to guess, it is usually better to lean toward formality.

Take a Reader-Centered "You Attitude"

You can also significantly increase readers' positive emotional response to your messages by adopting what specialists in business correspondence call the "you attitude." This strategy involves focusing your sentences on your readers' needs, desires, and situation rather than on you or your subject matter. By crafting sentences in this reader-centered way, you reinforce all of the other reader-centered work you perform when planning and drafting your letter or memo. The following examples show how to rephrase sentences to convey the "you attitude," not the self-centered "me attitude."

TRY THIS Find a letter or memo from an office or instructor at your school, or from a company or organization, that is written in an impersonal or writer-centered way. Create a new version using the "you attitude."

- Focus your sentences on your reader, not yourself. Sentences that use *I* as the grammatical subject appear to show that the writer is more concerned with himself or herself than with the reader.

    | I have arranged to have equipment needed for your fieldwork delivered to the Boise office. | Lacks you-attitude |

    | You can pick up the equipment needed for your fieldwork at the Boise office. | Has you-attitude |

- Phrase sentences so they create positive feelings in the reader. For instance, where the message is positive, use *you*.

    | We will reimburse you fully for the expenses resulting from this error. | Lacks you-attitude |

    | You will receive full reimbursement for your expenses resulting from this error. | Has you-attitude |

- Phrase sentences so they avoid creating negative feelings in the reader. For instance, avoid *you* when conveying criticism and other negative messages

    | Your conclusions are inaccurate because you failed to account for all variables. | Lacks you-attitude |

    | Because all variables were not accounted for, the conclusions are inaccurate. | Has you-attitude |

Apply Reader-Centered Advice from Other Chapters to Your Correspondence

Some of the advice given in other chapters is especially helpful when you are writing correspondence. Here are some reminders, along with the places in this book where you will find fuller explanations.

GUIDELINE 1 State your main point up front—unless you expect your readers to react negatively

What readers usually want most from a short communication is your main point. Do you have recommendations to make? Questions to ask? Important news to convey? Tell them right at the start.

When you lead off with your main point, you are using the direct pattern of organization. Knowing how much this pattern contributes to usefulness and persuasiveness, some writers even include the heart of their message in the subject line of their memos: "Recommendation for Replacing the Afterburner," "Phoenix Site Is the Best Location for Building Our New Factory."

At times, however, you increase your chances of writing successfully by employing a different pattern of organization—for example, when you expect the reader to have an immediate negative response to your main point, as might happen when you are refusing a request or communicating some other decision or action that will not please the reader. For more information on direct and indirect organizational patterns, see "Choose between direct and indirect organizational patterns" (Chapter 9, pages 181–182).

GUIDELINE 2 Keep it short

Few things are as effective as brevity at helping readers find what they want. Without cutting substance, shorten your message in the following ways.

- **Delete unnecessary words.** You will find strategies for finding and deleting unnecessary words in "Simplify your sentences" (see Chapter 10, pages 196–197).

- **Stick to the point.** No matter how interesting a fact is to you or how proud you are of some accomplishment, don't mention it unless your readers will find the information useful or persuasive.

- **Provide only as much detail as your readers need.** Offer to give them more information if they want it, but in your correspondence include only what readers really need.

GUIDELINE 3 Use headings and lists

LEARN MORE For advice on using headings and bullets, see "Use Headings" (Chapter 7, pages 128–132).

Headings and bulleted lists, which enable readers to quickly grasp the organization and main points of long documents, are as helpful in short messages, including email, as they are in long communications.

GUIDELINE 4 Avoid communication clichés

Some customs are better avoided because they hinder good communication. For example, at some workplaces writers are in the habit of peppering their correspondence with a variety of wordy and inelegant expressions that they do not use in conversation or other kinds of writing. Here are some examples of these correspondence clichés.

Replacing correspondence clichés with plain language

CORRESPONDENCE CLICHÉ	PLAIN LANGUAGE
As per your request	As you requested
Prior to completion	Before completing
Contingent on	Depending on
In lieu of	Instead of
It has come to our attention	We learned
Enclosed please find	We've enclosed; Here is
Please be aware that	Please note that

GUIDELINE 5 Global Guideline: Learn the customs of your readers' culture

Customs concerning correspondence vary from culture to culture. In Japan, letters customarily begin with a reference to the season. In some Spanish-speaking cultures, letters begin with statements that seem florid by the more straightforward conventions—for instance, in the United States.

GUIDELINE 6 Follow format conventions and other customs

One characteristic of reader-centered writing is that it conforms to the conventions with which your readers are familiar. You gain credibility by doing so because you demonstrate to your readers that you know "how things are done here," that you belong. By following the conventions you also benefit from the many writers and readers in the past who have found that these conventions promote effective communication from the perspective of both groups.

Here, for instance, are three conventions that letters, memos, and email share.

- Use single spacing.
- Separate paragraphs by a blank line.
- Use short paragraphs.

There are, of course, additional conventions specific to each of these forms of correspondence. Some concern format and some concern conventional ways to present your message within the formats. The rest of this chapter describes these conventions and highlights special considerations for each type of correspondence.

Writing Reader-Centered Letters

At work, you will usually use letters when writing relatively short messages to readers outside of your employer's organization. These readers may include customers, clients, suppliers, and government agencies, among others.

Conventional Format for Letters

Conventional formats for letters are described in the Writer's Tutorial on Writing Letters (see pages 372–373).

Special Considerations for Writing Letters

The following considerations supplement the general guidelines for writing reader-centered correspondence given earlier in this chapter.

- Follow the conventions for opening and closing a letter. As noted in "Global Guideline: Learn the customs of your readers' culture" (see page 370), conventions about the opening of letters vary considerably among cultures. In the United States and Canada, letters are generally more formal than memos and emails. Avoid slang and jargon. Choose a formal vocabulary—but avoid the correspondence clichés and bureaucratese described on page 370.
  - The *introductory paragraph* typically indicates the letter's topic, explains its purpose, and indicates, perhaps implicitly, its relevance to the reader. Often an introductory paragraph expresses gratitude or appreciation of the reader—for instance, by thanking the reader for a letter to which the writer is responding.
  - *Closing paragraphs* often make a social gesture—example, by expressing pleasure in working with the reader, offering to assist the reader, and the like.
- **Use a formal greeting in the salutation.** Consistent with the relative formality of letters, it is customary to use the reader's last name in the salutation: "Dear Mr. Takjeki" or "Dear Dr. Reynolds." However, if you call the

WRITER'S TUTORIAL

WRITING LETTERS

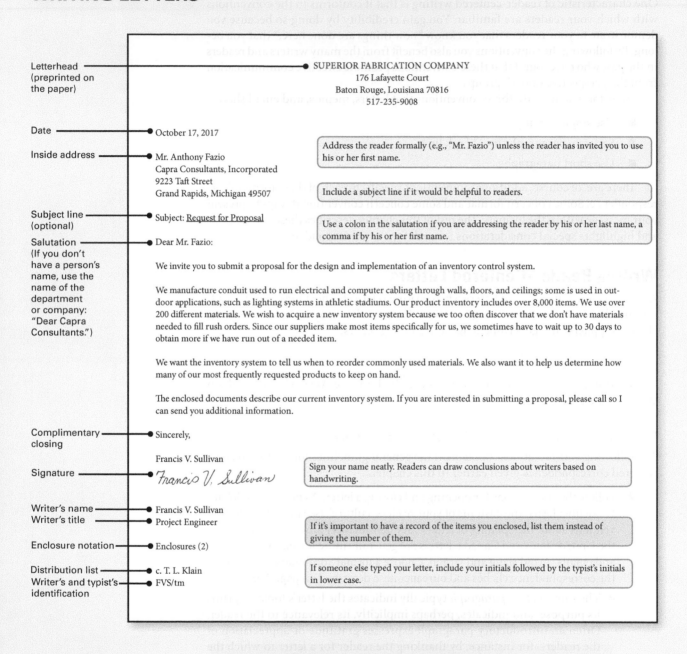

Letterhead (preprinted on the paper)

> SUPERIOR FABRICATION COMPANY
> 176 Lafayette Court
> Baton Rouge, Louisiana 70816
> 517-235-9008

Date — October 17, 2017

Inside address —
Mr. Anthony Fazio
Capra Consultants, Incorporated
9223 Taft Street
Grand Rapids, Michigan 49507

> Address the reader formally (e.g., "Mr. Fazio") unless the reader has invited you to use his or her first name.

> Include a subject line if it would be helpful to readers.

Subject line (optional) — Subject: Request for Proposal

> Use a colon in the salutation if you are addressing the reader by his or her last name, a comma if by his or her first name.

Salutation (If you don't have a person's name, use the name of the department or company: "Dear Capra Consultants.") — Dear Mr. Fazio:

We invite you to submit a proposal for the design and implementation of an inventory control system.

We manufacture conduit used to run electrical and computer cabling through walls, floors, and ceilings; some is used in outdoor applications, such as lighting systems in athletic stadiums. Our product inventory includes over 8,000 items. We use over 200 different materials. We wish to acquire a new inventory system because we too often discover that we don't have materials needed to fill rush orders. Since our suppliers make most items specifically for us, we sometimes have to wait up to 30 days to obtain more if we have run out of a needed item.

We want the inventory system to tell us when to reorder commonly used materials. We also want it to help us determine how many of our most frequently requested products to keep on hand.

The enclosed documents describe our current inventory system. If you are interested in submitting a proposal, please call so I can send you additional information.

Complimentary closing — Sincerely,

Francis V. Sullivan

Signature — *Francis V. Sullivan*

> Sign your name neatly. Readers can draw conclusions about writers based on handwriting.

Writer's name — Francis V. Sullivan
Writer's title — Project Engineer

Enclosure notation — Enclosures (2)

> If it's important to have a record of the items you enclosed, list them instead of giving the number of them.

Distribution list — c. T. L. Klain
Writer's and typist's identification — FVS/tm

> If someone else typed your letter, include your initials followed by the typist's initials in lower case.

Tips for Creating Reader-Centered Letters

- For letters that don't fill the page, place the content so its middle is slightly above the middle of the page.
- On letterhead paper, type your message with the same margins that the letterhead uses. Keep the same margins on subsequent pages.

Letter Printed on Plain Paper (Not Letterhead)

Place the return address 1 inch below the top on either the right or left of the page.

Type the date one return below the inside address.

Start the inside address at least two returns below the date.

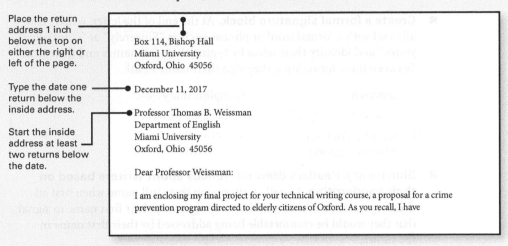

Box 114, Bishop Hall
Miami University
Oxford, Ohio 45056

December 11, 2017

Professor Thomas B. Weissman
Department of English
Miami University
Oxford, Ohio 45056

Dear Professor Weissman:

I am enclosing my final project for your technical writing course, a proposal for a crime prevention program directed to elderly citizens of Oxford. As you recall, I have

LEARN MORE For additional advice, see "Writing Reader-Centered Letters" (pages 371–374).

Heading for Second and Subsequent Pages

Print second and subsequent pages on blank paper.

If a paragraph continues from one page to the next, leave at least two lines on each page.

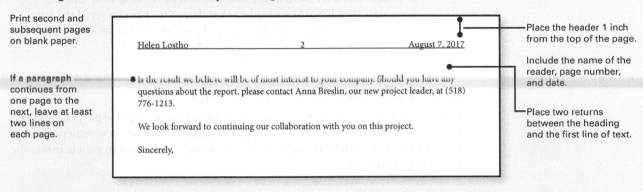

Helen Lostho   2   August 7, 2017

Is the result we believe will be of most interest to your company. Should you have any questions about the report, please contact Anna Breslin, our new project leader, at (518) 776-1213.

We look forward to continuing our collaboration with you on this project.

Sincerely,

Place the header 1 inch from the top of the page.

Include the name of the reader, page number, and date.

Place two returns between the heading and the first line of text.

Envelopes

Return address: down 1/2 inch

over 1/2 inch

Address over 5 inches

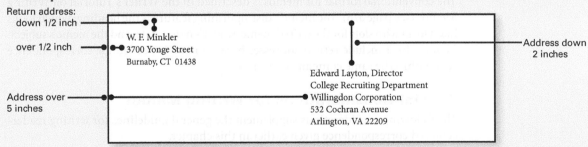

W. F. Minkler
3700 Yonge Street
Burnaby, CT 01438

Edward Layton, Director
College Recruiting Department
Willingdon Corporation
532 Cochran Avenue
Arlington, VA 22209

Address down 2 inches

LEARN MORE To see additional sample letters, see Figures 2.8 (page 43), 3.3 (page 63), 9.4 (page 178), and 19.1 (pages 342–345).

person by his or her first name in conversation or if the reader has signed a letter to you with only a first name, then it is acceptable (and expected) that you will use the person's first name in the salutation.

■ **Create a formal signature block.** At the end of the letter, writers usually end with a formal word or phrase, such as "Sincerely" or "Cordially yours," and identify themselves by typing their full names and their title. Between these formalities, they sign their name in ink.

Sincerely,	Complimentary close
Ibrahim Ettouney	Writer's signature
Ibrahim Ettouney	Writer's name
Process Engineer	Writer's title

■ **Sign neatly. Readers draw inferences about writers based on their handwriting.** Usually writers sign their full name when first addressing a reader. Afterward, they may sign only their first name to signal that they would be comfortable being addressed by their first name in the future.

■ **List the people to whom you are sending copies.** As a courtesy to your readers, let them know who else will be reading your letter. Sometimes the list assures readers that the writer has conveyed the letter's information to all who should have it. It also lets the reader know who might ask them about the letter's contents.

Writing Reader-Centered Memos

Like letters and emails, memos are used for almost any workplace purpose that can be accomplished in a relatively short space, generally between a few sentences and a few pages. Unlike letters, though, memos are usually addressed to people inside the writer's organization.

Memo Format

The conventional format for memos is described in the Writer's Tutorial on Writing Memos (see page 376). The memo's distinguishing feature is the heading, which looks like a form with slots for the writer's name, reader's name, date, and the memo's subject. Memos don't include return addresses because none is needed for correspondence sent within the same company or agency.

Special Considerations for Writing Memos

The following considerations supplement the general guidelines for writing reader-centered correspondence given earlier in this chapter.

LEARN MORE For more on the direct and indirect patterns of organization, see "Give the bottom line first" (pages 110–111) and "Organize to Create a Favorable Response" (pages 180–183).

■ **Provide a specific, informative subject line.** A heading that says "Results" doesn't tell the reader what the results are from. One that says "Gallagher Project" doesn't communicate whether the memo asks questions, provides answers, makes a proposal, or reports progress. Much more informative for current readers and future readers looking for information about the completed project is "Results of Turbulence Tests for Gallagher Project."

- **State the purpose and main point in the opening sentences.** Memos usually get off to a faster start than letters. Use the direct pattern of organization, as described in "Apply Reader-Centered Advice from Other Chapters to Your Correspondence" (see page 369)—unless you have a very strong reason for using the indirect pattern.

- **Initial the memo.** In most organizations, a writer's initials substitute for a full signature. Put your initials by your name in the "From" line.

- **List the people to whom you are sending copies.** As in a letter, you are treating your readers courteously when you let them know who else will be reading your letter. You assure your readers that you have given the letter's information to all who should have it. You have also alerted them to the possibility of receiving questions about your memo from the individuals listed.

Meanep/Shutterstock.com

Brevity is important in online correspondence because so many of these communications are read on very small screens.

Writing Reader-Centered Email

Even more than for letters and memos, customs for email vary widely. In many organizations, email is written in a formal manner—much like a formal memo that appears on-screen rather than on paper. The "Writer's Tutorial on Writing Email" (see page 377) provides advice about writing this type of email effectively. In other organizations, email content can be very informal—more like a long tweet.

In the workplace, people also reply to email in a variety of ways. Many click on "Reply" and then type their response at the top of the message they received. Others integrate their message into the original one—for instance, by inserting comments at the end of each paragraph of the original message. In a variation of that method, some writers copy parts of the original message into their own, and they follow the copied parts with their comments. To help readers distinguish their comments from the original message, they change the color of what they've copied or of their comments or they put their comments in all capital letters.

No matter how you construct your email, provide informative subject lines. Many readers use subject lines to determine whether to even open messages in their overstuffed in-boxes. Your subject line can also help your reader find your earlier email that they now want to review.

WRITING MEMOS

The heading and these words are printed on the company's memo paper.

Natalie provides a very specific subject line.

Her opening states the major finding of the tests.

Natalie provides background information she knows her reader will desire.

Natalie provides specific data from the tests, together with other information needed to interpret the results.

Natalie answers the major question she anticipates her reader will have after reading the test results: "What should we do?"

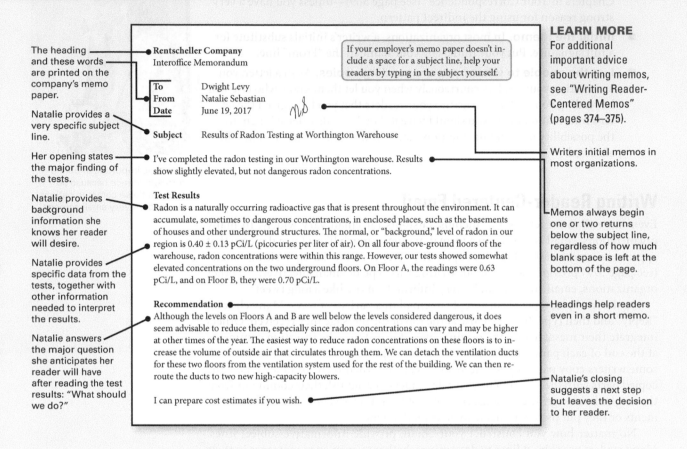

Rentscheller Company
Interoffice Memorandum

To	Dwight Levy
From	Natalie Sebastian
Date	June 19, 2017

Subject   Results of Radon Testing at Worthington Warehouse

I've completed the radon testing in our Worthington warehouse. Results show slightly elevated, but not dangerous radon concentrations.

Test Results

Radon is a naturally occurring radioactive gas that is present throughout the environment. It can accumulate, sometimes to dangerous concentrations, in enclosed places, such as the basements of houses and other underground structures. The normal, or "background," level of radon in our region is 0.40 ± 0.13 pCi/L (picocuries per liter of air). On all four above-ground floors of the warehouse, radon concentrations were within this range. However, our tests showed somewhat elevated concentrations on the two underground floors. On Floor A, the readings were 0.63 pCi/L, and on Floor B, they were 0.70 pCi/L.

Recommendation

Although the levels on Floors A and B are well below the levels considered dangerous, it does seem advisable to reduce them, especially since radon concentrations can vary and may be higher at other times of the year. The easiest way to reduce radon concentrations on these floors is to increase the volume of outside air that circulates through them. We can detach the ventilation ducts for these two floors from the ventilation system used for the rest of the building. We can then reroute the ducts to two new high-capacity blowers.

I can prepare cost estimates if you wish.

If your employer's memo paper doesn't include a space for a subject line, help your readers by typing in the subject yourself.

LEARN MORE
For additional important advice about writing memos, see "Writing Reader-Centered Memos" (pages 374–375).

Writers initial memos in most organizations.

Memos always begin one or two returns below the subject line, regardless of how much blank space is left at the bottom of the page.

Headings help readers even in a short memo.

Natalie's closing suggests a next step but leaves the decision to her reader.

Second and Subsequent Pages

Place the header 1 inch from the top of the page.

Include the name of the reader, page number, and date.

Place two returns between the heading and the first line of text.

Hasim K. Lederer   2   November 15, 2017

with many benefits for our company and our customers. These improvements can be made with minimal disruption to our employees, suppliers, and technical sales force if we schedule the transition carefully. The most important factors will be notifying our

Print second and subsequent pages on blank paper.

If a paragraph continues from one page to the next, leave at least two lines on each page.

LEARN MORE To see additional sample memos, see Figures 1.4 (page 10), 7.3 (page 130), 9.6 (page 180), 11.1 (page 213), 23.1 (395), and 26.3 (page 455).

WRITING EMAIL

Provide an informative, specific subject line. Readers use this line to decide whether to open an email and to find emails they want to review after reading them the first time.

Highlight main points and make reading easier by using bullets.

Use short paragraphs and place blank lines between them to promote reading ease.

Tell readers how to contact you by using your email system's "signature" feature.

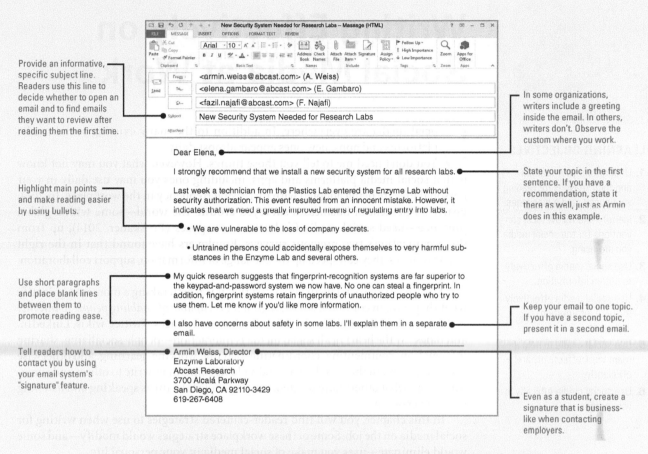

In some organizations, writers include a greeting inside the email. In others, writers don't. Observe the custom where you work.

State your topic in the first sentence. If you have a recommendation, state it there as well, just as Armin does in this example.

Keep your email to one topic. If you have a second topic, present it in a second email.

Even as a student, create a signature that is business-like when contacting employers.

Email content shown:

New Security System Needed for Research Labs — Massage (HTML)

From: <armin.weiss@abcast.com> (A. Weiss)
To: <elena.gambaro@abcast.com> (E. Gambaro)
Cc: <fazil.najafi@abcast.com> (F. Najafi)
Subject: New Security System Needed for Research Labs

Dear Elena,

I strongly recommend that we install a new security system for all research labs.

Last week a technician from the Plastics Lab entered the Enzyme Lab without security authorization. This event resulted from an innocent mistake. However, it indicates that we need a greatly improved means of regulating entry into labs.

• We are vulnerable to the loss of company secrets.

• Untrained persons could accidentally expose themselves to very harmful substances in the Enzyme Lab and several others.

My quick research suggests that fingerprint-recognition systems are far superior to the keypad-and-password system we now have. No one can steal a fingerprint. In addition, fingerprint systems retain fingerprints of unauthorized people who try to use them. Let me know if you'd like more information.

I also have concerns about safety in some labs. I'll explain them in a separate email.

Armin Weiss, Director
Enzyme Laboratory
Abcast Research
3700 Alcalá Parkway
San Diego, CA 92110-3429
619-267-6408

Tips for Creating Reader-Centered Emails

■ For headings, use all caps and underlining. Otherwise they may not stand out in your reader's email program.

■ Don't use all caps for an entire sentence or paragraph. All caps are difficult to read when they are used for more than several words together.

■ The more briefly you write, the more forcefully your points will come across to your readers.

22

Writing Effectively on Social Media at Work

LEARNING OBJECTIVES

1. Understand your readers and their reasons for reading your social media messages.
2. Identify and follow local conventions for the social media you are using.
3. Use social media effectively to obtain information.
4. Use social media effectively to share information and ideas.
5. Use social media to help a team work effectively and efficiently.
6. Use social media ethically in your career.

Social media are everywhere. In addition to the many existing social media platforms and apps, new ones appear almost daily.

You don't need me to tell you those things. However, what you may not know is that social media platforms and apps, including ones you may use daily in your personal life, are being adopted with increasing frequency in the workplace. In 2014, nearly 90 percent of businesses surveyed around the world—some with a global presence—used social media for business purposes (Proskauer, 2014), up from the organization's two previous surveys. Employers have found that in the right circumstances, they can speed up the exchange of information, support collaboration, and foster innovation.

The variety of social media is so vast that it is worth taking a moment to identify what they have in common: All are *digital tools capable of enabling people to interact instantaneously.* Examples include Facebook, text messages, Twitter, wikis, LinkedIn, and blogs. At the heart of all social media is interaction—people socializing, sharing information, commenting, coordinating, playing, working, learning, and otherwise acting together. In these exchanges, individuals sometimes write to others and sometimes read what others have written, just as people take turns speaking and listening in a conversation.

In this chapter, you will find reader-centered strategies to use when writing for social media on the job. Some of these workplace strategies would modify—and some would eliminate—uses you make of social media in your personal life.

Kinds of Social Media Used at Work

Different social media perform different functions, as you know. Each has distinctive structural features that lend it to some uses in your personal life but not others. Facebook is not as efficient as a text message for letting friends know that you will be late for a get-together, but it is very good at letting you share and comment on personal news among a group of friends.

The same is true in the workplace uses of social media, except that there is a consistent and more limited set of common workplace uses. This chapter provides advice for three (see Figure 22.1).

- Social media for quick, short exchanges
- Social media for self-presentation by individuals and groups that allow responses by readers
- Social media for collaborative development of communications and knowledge

FIGURE 22.1

Examples of Social Media Used in the Workplace

SOCIAL MEDIA FOR QUICK, SHORT EXCHANGES
- **Text Messages.** Short messages between individuals and small groups, usually using smartphones. Readers can reply instantly. Messages are preserved only on the digital devices of the sender and receivers.
- **Twitter.** Sort messages sent to a potentially large group from cell phones and other digital devices, such as computers and tablets. Readers can reply instantly. All messages not designated as private are preserved at a central site so they can be searched and read by anyone.

SOCIAL MEDIA FOR SELF-PRESENTATION BY INDIVIDUALS AND GROUPS THAT ALLOW RESPONSES BY READERS
- **Social Networking Sites and Apps such as Facebook.** Preformatted pages that allow persons and organizations to provide profiles of themselves that they can continuously update by adding new posts. Notice of new postings can be commented on by people the writer has accepted as "friends."

SOCIAL MEDIA FOR COLLABORATIVE DEVELOPMENT OF COMMUNICATIONS AND KNOWLEDGE
- **Wikis.** Allow two or more people to work on a single copy of a communication. They can draft and comment on one another's suggestions and changes. Co-authors can be as small as a group of two scientists or engineers, or as large as anyone on the web (as with Wikipedia).
- **Blogs.** Blogs (short for web logs) are sites with strings of entries on a particular topic, with the most recent entry usually appearing at the top. Some blogs have many subtopics, called "threads." Blogs can be accessed by anyone on the web or hidden from all but a group selected by the owner. The owner can also determine who can comment (perhaps a few people, perhaps anyone on the web).

In addition, other chapters describe three other work-related uses of social media: finding a job, co-op, or internship (Chapter 2, pages 21, 22, and 39); conducting research (Chapter 5, pages 94–95); and creating team-written communications (Chapter 17, pages 315–316).

Reader-Centered Guidelines for Writing Effectively on Social Media

Regardless of the specific social media platform or app you use, the following guidelines will help you create effective, reader-centered messages at work.

GUIDELINE 1 Understand your readers and their reasons for reading

People will read your workplace social media writing for the same reasons they will read any other communication you write on the job. They will look for information and ideas that can help them perform their tasks. Sometimes they will want to learn how to perform a physical task, sometimes they will want advice that will help them make a decision, and sometimes they will want to expand their knowledge. To meet your readers' needs and desires, you will need to know what they are. And you will need to know how readers will use your communication, so you can construct it to support their efforts.

Sometimes, readers are very clear about their needs, as when they ask a question on a community blog. Figure 22.2 shows a blog that lets anyone ask a question or suggest answers. Readers can vote for the best answer. Many companies, including Microsoft, Apple, and Best Buy, maintain such blogs for customers, calling them "forums" and "communities." In other situations, you may need to develop the well-rounded portrait

FIGURE 22.2

Blog for Requesting Help that
Lets Responders Rank the
Answers

Question writer provides specific
information in title of post.

Question writer provides details of
problem, including the brand and color
of the furniture she is talking about and
solutions he or she has thought of.

Best response because it has received
the most points (votes) by other users
of the blog (9)

Indented response is a comment on the
response above it.

Response receiving the second highest
number of votes (3)

Blog page continues with all of the other
answers (not shown)

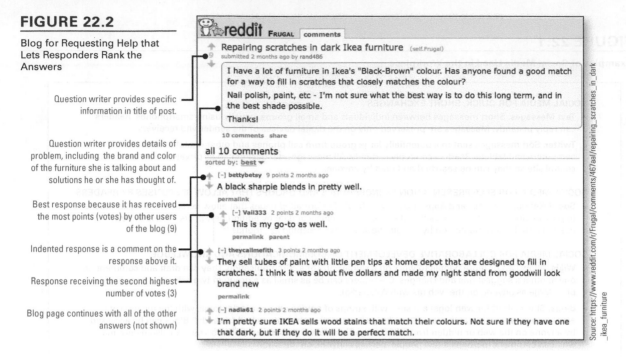

Source: https://www.reddit.com/r/Frugal/comments/467aal/repairing_scratches_in_dark_ikea_furniture

of your reader that you learned to create in Chapter 3, including their knowledge of
your subject and their attitude toward your subject, your employer, and you—among
other characteristics.

GUIDELINE 2 Follow local conventions for the social media you are using

The conventions for use of social media vary from one context to another: workplace to workplace, profession to profession, and even occasion to occasion. At some
workplaces, text messages and tweets may include the same abbreviated language,
emoticons, and emojis that you use in your personal texts and tweets. In others,
social media writing is expected to be formal. Violating your readers' expectations
can damage the clarity of your messages and reduce your credibility.

Applying the General Guidelines for Three Major On-the-Job Writing Tasks

Keeping in mind the two general guidelines you've just read, let's explore the ways
you can use social media effectively for three major writing activities.

Writing to Obtain Information

There are two situations in which you might use social media to learn something that
is important to your own work. In the first, you search through social media resources
such as blogs and wikis (e.g., Wikipedia) for the information. In these situations,
you are a reader, not a writer. The use of social media in this manner is described in
Chapter 5. In the second situation, you write a question and post it on one or more
social media outlets, such as the one illustrated in Figure 22.2. For instance, if you are
having trouble performing some action with a computer program, you might post an

inquiry on a blog maintained by the company that makes the program. If you have a question about a technical feature of a new product your employer is developing, you might post it on an internal blog, one that is open only to the organization's employees. As mentioned, some blogs allow many persons to provide answers or ask for more information from the person posting the original question. Some blogs also let readers vote on which response provides the best answer to the question.

When writing a question or writing for help, it's just as important to take a reader-centered approach as when you are writing any other workplace communication. Some items in the following list of reader-centered strategies are based on advice that Microsoft (2012) gives to those who wish to request help on a Microsoft Forum.

How to Request Information in a Reader-Centered Way

1. **Choose the right social media for your purpose.** Which ones will be read by the people best able to help you?

2. **Do your own background research.** So that you don't waste the time of people who read the social medium you are using, first look for the answer on your own in the resources that are available to you. If you are using a blog or wiki, use the search function to determine whether someone else has already asked the same question—and received a response that includes the information you need.

3. **State a precise description of your problem.** If you write, "I need help sending scans" rather than "I cannot email scans from my Emerson WorkPro 3640 printer," you may not catch the attention of the person who can most help you.

4. **Give all relevant information.** For instance, if you are having a computer problem, describe your system (e.g., operating system version, CPU or processor, video card, system memory [RAM] size, etc.). Also describe recent changes in your software or hardware. But avoid giving irrelevant information.

5. **Tell what you have already done to find the answer.** Doing so saves your readers the unnecessary work of suggesting something you've already tried. Be specific so readers don't have to ask follow-up questions about what you mean.

6. **Include a thank you.**

Sharing Information

Just as you can use social media to gain information and ideas, you can use social media to share information and ideas. For instance, you might be the person who has received a text message requesting help. Or, you may want to elaborate or correct information you find on an internal or external blog or wiki. The following guidelines suggest ways you can create reader-centered communications when your goal is to help others benefit from your knowledge and creativity.

How to Share Information in a Reader-Centered Way

1. **Understand what your readers want.** Any topic can be written about in a variety of ways. Which way will best help your readers gain what they are seeking?

2. **Adjust the level of your explanation to your readers' level of knowledge.** Writing too technically for nonspecialists or too simply for specialists can frustrate readers more than help them.

3. **Use headings, lists, and other visual cues to the organization of the information you provide.** When using social media such as blogs and wikis that permit longer messages, use the same visual cues you would use for a print communication to help your readers understand what you are saying.

4. **Provide links, if possible, to additional sources of information.** One way to avoid overly long messages, whether in a tweet or a blog, is to include links to other sources that your readers might find helpful.

5. **Use keywords that people might use when searching for the information you are providing.** The knowledge you share cannot benefit your potential readers if they cannot find it. In your messages or message titles, include the words they would use in a search.

6. **Especially if the communication is being shared with an external person, present contents in a way that is favorable to your employer.** Adopt a positive, professional tone.

Using Social Media to Help a Team Work Effectively and Efficiently

By definition, teamwork involves social interaction, most obviously at team meetings. Chapter 17 provides suggestions about conducting meetings, working together on drafts, and other activities that are often accomplished when team members gather together. But most teams alternate between meetings and individual work, with more time given to the latter. Social media help teams continue to "work together" when the individual team members are working alone.

How to Write Reader-Centered Messages that Support Teamwork

1. **Use social media to coordinate work.** For simply finding out how well each person is doing, what help he or she might need, or reminding people of meetings, text messages and tweets can be very efficient. With text messages, a team can save itself the trouble of typing in the phone number of the person addressed by first initiating a message sent to all team members, and then all team members can reply to that message, which is automatically distributed to everyone.

2. **Use social media to monitor team progress.** For tracking overall progress, teams can set up a schedule in Word or Excel that records all tasks, when they are expected to be finished, and when they are finished.

3. **Use social media to focus team energies where they are most needed at the moment.**

4. **Use social media to maintain group cohesion and enthusiasm.** When teams don't meet in person, it can be especially difficult to maintain the personal connections and a sense of responsibility that is necessary for teamwork to succeed.

Ethical Guidelines for Using Social Media

Balanced against the many ways that social media support and speed work, they can also work against an organization's interests. Here are some ways to ensure that your use is ethical.

1. **Use social media at work in only the approved ways.** Check and abide by your employer's policies. More than 70 percent of the employers

responding to the survey mentioned above (Proskauer, 2014) took disciplinary action against employees who misused social media. These policies vary widely, with some permitting even personal use of social media at work whereas others prohibit use of certain platforms and apps. In addition, the companies are refining the categories of behavior that they consider unacceptable: misuse of confidential information (80 percent), misrepresenting the views of the business (71 percent,) inappropriate non-business use (67 percent), making disparaging remarks about the business or employees (64 percent), and harassment (64 percent) (Proskauer, 2014).

2. **Don't add "likes" and positive comments on your employer's products or services without disclosing your relationship to the company.** Sometimes employers ask employees to do this. Sometime enthused employees are tempted to do it on their own. Readers depend on comments for unbiased evaluations.

3. **Be sure that you aren't sharing information your employer considers proprietary.** In your career, you will sometimes work on projects that involve information your employer wants to be held confidential within the organization, at least for the time being. Avoid harming your employer by telling others. When unsure, check with someone else.

USE WHAT YOU'VE LEARNED

EXERCISE YOUR EXPERTISE

Find a group of Twitter messages, a blog, and a website that provide information about a topic related to your major. Which makes it easiest to find the most recent information on your topic? Which seems to provide the most reliable information?

EXPLORE ONLINE

Find two social media platforms or apps that are used in the workplace. What workplace functions do they support? Which advice in this chapter would be most important for you to apply when writing for each?

COLLABORATE WITH YOUR CLASSMATES

Interview two of your classmates about the social media they have used while working on team projects in one or more of their courses. For classmates who haven't used any, find out which they think would be most helpful in future team projects—and why.

APPLY YOUR ETHICS

Identify two things it would be unethical for employees to do on social media, either on their own or at an employer's request.

REFLECT FOR TRANSFER

1. Write a memo to your classmates describing two ways you would want to adapt your social media writing when using social media on the job. For each, explain why. Also, describe two ways you would not need to adapt your current social media writing at work.

2. If you have work experience in an internship, co-op, or permanent position, draft an email or memo to your employer describing one or two specific ways it might use social media to increase organizational efficiency or effectiveness.

23

Writing Reader-Centered Proposals

LEARNING OBJECTIVES

1. Describe the two major goals of a proposal.
2. Describe the major questions asked by proposal readers.
3. Describe the superstructure for proposals, including the ways its parts correspond with readers' questions.
4. Adapt the reader-centered writing process to the special goals of proposals.
5. Write a cohesive and persuasive proposal.

MindTap®

Find additional resources related to this chapter in MindTap.

I n a proposal, you make an offer and try to persuade your readers to accept it. You say that, in exchange for money or time or some other consideration, you will give your readers something they want, create something they desire, or do something they wish to have done.

Throughout your career, you will have many occasions to make such offers. You may think up a new product you could develop—if your employer will give you the time and funds to do so. Or you may devise a plan for increasing your employer's profits—if your employer will authorize you to put the plan into effect. You may even join one of the many companies that sell their products and services by means of proposals (rather than through advertising) so that *your* proposals are the means by which your employer generates income.

The Variety of Proposal-Writing Situations

The proposals you write may vary in many ways.

- You may write to employees in your own organization or in other organizations.
- You may write to readers who have invited you to submit a proposal, or you may submit it on your own initiative.
- Your proposal may be in competition against others, or it may stand or fall on its own merits.
- You may have to obtain approval from people in your organization before you submit it to your readers, or you may be authorized to submit it directly yourself.
- You may have to follow regulations governing the content, structure, and format of your proposal, or you may be free to write your proposal as you think best.
- After you have delivered your proposal, your readers may follow any of a wide variety of methods for evaluating it.

To illustrate some of these variables in actual proposal-writing situations, the following paragraphs describe the circumstances in which two successful proposals were written. The information provided here will be useful to you later in this chapter, where advice is explained through the examples of these proposals.

Example Situation 1

Helen wanted to develop a customized computer program that employees at her company could use to reserve conference rooms. On several occasions, she arrived at a conference room she had reserved only to find that someone else had reserved it also. Because she is employed to write computer programs, she is well qualified to write this one. However, she cannot work on this special project without permission from her manager and her manager's boss. She wrote her proposal to gain their permission.

Helen's proposal-writing situation was uncomplicated. She had two internal readers. Because her employer had no guidelines for the way internal proposals should be written, she could use the content, structure, and format she thought most effective. She did not need anyone's approval to submit her proposal to her managers, and she did not have to worry about competition from other proposals. Hers would be evaluated strictly on its own merits.

Proposals for many kinds of projects are evaluated by experts in the area of research or other activity for which funds are available. Sometimes the reviewers from various parts of a country are flown to a central location to read and discuss the proposals. They determine which proposals will be funded.

Example Situation 2

The second proposal was written under very different circumstances. To begin with, it was written by three people, not just one. The writers were a producer, a scriptwriter, and a business manager at a public television station that seeks funds from nonprofit corporations and the federal government to produce television programs. The writers had learned that the U.S. Department of Education was interested in sponsoring programs about environmental concerns. To learn more about what the department wanted, the writers obtained copies of its "request for proposals" (RFP). After studying the RFP, they decided to propose developing a program that high school teachers and community leaders could use to teach about hazardous wastes.

In their proposal, the writers addressed an audience very different from Helen's. The Department of Education receives about four proposals for every one it can fund. To evaluate these proposals, the department follows a procedure widely used by government agencies and other organizations. It sends the proposals to experts around the country. These experts, called *reviewers*, rate and comment on each proposal. Every proposal is seen by several reviewers. Then the reviews for each proposal are gathered and interpreted by staff members at the department. The proposals that have received the best reviews are funded. To secure funding, the writers of the environmental proposal needed to persuade their reviewers that their proposed project came closer to meeting the objectives of the Department of Education than at least three-quarters of the other proposed projects.

Moreover, before the writers could even mail their proposal to the department, they had to obtain approval for it from several administrators at the television station. That's because the proposal, if accepted, would become a contract between the station and the Department of Education.

How Readers Use and Evaluate Proposals

Despite the many variables among proposals, almost all proposal readers use and evaluate proposals in the same way: as investors who have limited resources. Deciding whether to approve a proposed product, service, or project, they are determining whether it is better to invest some of their limited resources in it or in something else. To grant Helen two weeks to create a new scheduling program, her readers must agree that she would not spend that time on other work. To support the television team's environmental project, the Department of Education would have to turn down other proposals seeking the same dollars.

Proposal readers are investors.

To evaluate whether a proposal is worth their investment, almost all proposal readers consider the same basic questions. These are the ones your instructor and this chapter will help you to learn to answer persuasively.

■ **Does your proposal address a problem or goal that is important to us?** To write a successful proposal, you need to take the reader-centered approach of persuading your readers that you will help them gain something *they* want.

■ **Would your proposal solve our problem or achieve our goal?** If your readers are persuaded that your proposal addresses an issue important to them, they then want to be assured that the project you propose will produce the desired result. Often, proposal writers answer this question in two parts. They identify the specific objectives that any successful solution would achieve, and then they describe in detail the ways that their solution will achieve those objectives.

■ **Are you capable of producing the result you describe?** For proposal readers, a highly undesirable outcome is that a proposed project they have supported fails because the proposers lack the ability needed to bring the project to a successful conclusion. Especially for large projects, you may need to reassure your readers by persuasively describing your method, schedule, and project management plan, among other proofs of your ability to successfully complete the project.

■ **Is your cost reasonable?** Like all investors, proposal readers agree to invest only if they judge that the costs are reasonable.

Superstructure for Proposals

The superstructure or genre for proposals provides a framework you can use to answer those questions. It has ten topics. In some proposals, you may need to include information on all ten, but in others you will need to cover only some of them. Even in the briefest proposals, however, you will probably need to treat four: introduction, problem, solution, and costs.

LEARN MORE For more on superstructures, see "Adapt an appropriate superstructure or other pattern familiar to your reader" (pages 112–113).

Superstructure for Proposals

Topics	Readers' Questions	Your Persuasive Points
*Introduction	What is this communication about?	Briefly, I propose to do the following.
*Problem	Does your proposal address a problem or goal that is important to us?	The proposed project addresses a problem, need, or goal that is important to you.
Objectives	What must a solution achieve in order to succeed?	A successful solution will achieve these objectives.
*Solution	How will your proposed solution achieve these objectives?	Here's how my proposed project will achieve the objectives.
Method Resources Schedule Qualifications Management	Are you going to be able to deliver what you describe here?	Yes, because I have a good plan of action (method); the necessary facilities, equipment, and other resources; a workable schedule; appropriate qualifications; and a sound management plan.
*Costs	What will it cost?	The cost is reasonable.

*Topics marked with an asterisk are important in almost every proposal, whereas the others are needed only in certain ones.

As you can see, the ten topics of a proposal cover a lot of territory. In addition to discussing each in a thorough, reader-centered way, you need to coordinate your discussions of these diverse topics so they flow together in a single, smooth, and cohesive whole.

There is no guarantee, of course, that your readers will read your proposal from front to back or concentrate on every word. Long proposals usually include a summary or abstract at the beginning. Instead of reading the proposal straight through, many readers will read the summary, perhaps the first few pages of the body, and then skim through the other sections.

Even when readers do not read your proposal straight through—and even when your proposal is only a page or two long—the relationships among the parts will help you write a tightly focused proposal in which all the parts support one another. The next section provides advice for writing each of the ten proposal sections and making them fit together.

Crafting the Major Elements of a Proposal

All the detailed advice that Chapters 3 through 16 provide about creating workplace communications applies to proposals. The following sections explain ways you can use some of this advice most productively when writing proposals. As you read, keep in mind that the conventional superstructure is only a general plan. You must use your imagination and creativity to adapt it to particular situations.

In addition, as you plan and write proposals, remember that the ten topics identify the kinds of information you need to provide, not necessarily the titles of the sections you include. In brief proposals, some parts may take only a sentence or a paragraph, or several sections may be grouped together. For instance, writers often combine their introduction, their discussion of the problem, and their explanation of their objectives under a single heading, which might be "Introduction," "Problem," or "Need."

Introduction

At the beginning of a proposal, you want to do the same thing that you do at the beginning of anything else you write on the job: Tell your readers what you are writing about. In a proposal, this means announcing what you are proposing.

How long and detailed should the introduction be? In proposals, introductions are almost always rather brief. By custom, writers postpone the full description of what they are proposing until later, after they have discussed the problem their proposal will help solve.

You may be able to introduce your proposal in a single sentence, as Helen did.

> I request permission to spend two weeks writing, testing, and implementing a program for scheduling conference rooms in the plant.

Helen's introduction

When you are proposing something more complex, your introduction will probably be longer. You may want to provide background information to help your readers understand what you have in mind. Here, for example, is the introduction from the proposal written by the employees of the public television station to the Department of Education.

> Chemicals are used to protect, prolong, and enhance our lives in numerous ways. Recently, however, society has discovered that some chemicals also present serious hazards to human health and the environment. In the coming years, citizens will have to make many difficult decisions to solve the problems created by these hazardous substances and prevent future problems from occurring.

Introduction to the television writers' proposal

LEARN MORE For more ideas on writing introductions, see Chapter 7.

To provide citizens with the information and skills they will need to decide wisely, WPET Television proposes to develop an educational package entitled "Hazardous Substances: Handle with Care!" This package, designed for high school students, will include five 15-minute video programs and a thorough teacher's guide.

Problem

Once you've announced the project you are proposing, you must persuade your readers that it addresses some problem, need, or goal that is significant to them. As explained earlier in this chapter (page 386), you cannot hope to win approval for your project unless you persuade your readers they want the outcome your project will produce.

Often this requires considerable creativity and research. The following paragraphs offer advice that applies to each of three situations you are likely to encounter on the job: when your readers define the problem for you, when your readers provide you with a general statement of the problem, and when you must define the problem yourself.

When Your Readers Define the Problem for You

When your readers have told you in detail, perhaps in an RFP, what they want, your task is to persuade them that you understand their desire in the same way they do. In your problem or goal section, describe the problem in a way that corresponds with and builds on the readers' description but doesn't merely repeat it. You can also build confidence in your understanding of the problem by using, with precision, the technical or specialized terms the readers use.

When Your Readers Provide a General Statement of the Problem

At times, your readers may describe the problem only vaguely. This happened to the television team that submitted the environmental education proposal. In its RFP, the Department of Education provided only the general statement that it perceived a need for "educational resources that could be used to teach about the relation of the natural and artificial environments to the total human environment." Each group that wished to submit a proposal had to identify a more specific problem that the Department would find important.

Finding a problem that someone else thinks is important can be difficult. Often, it will require you to hunt for clues to your readers' values, attitudes, concerns, and opinions. You may have to interview people (including your target readers, where possible), visit the library, or search the web. The television team found its major clue in the following statement, which appeared elsewhere in the RFP: "Thus, environmental education should be multifaceted, multidisciplinary, and issue-oriented." This statement suggested to the writers that they should define the problem as the need experienced by some group of people to gain the multifaceted, multidisciplinary knowledge that would enable them to make good decisions about a practical environmental issue.

But who would that group be? And what is the environmental issue this group needs to address? First, the television writers selected an issue faced by many communities: how to handle and regulate hazardous substances that are produced or stored within their boundaries. This issue seemed ideal because citizens can make good decisions about hazardous substances only if they possess "multifaceted, multidisciplinary" knowledge from such fields as health, economics, and technology. Next, the writers also decided to target their project to high school students. These students seemed like a perfect group to address because they soon would be eligible to vote on issues involving hazardous substances in their communities. Finally, through research, the writers determined that, in fact, no such educational materials were available for classroom use. They really were needed.

Thus, by working creatively from the Department of Education's vague statement, the television writers identified a specific problem that their proposed project would help solve: the lack of educational materials that high school teachers can use to provide their students with the multifaceted, multidisciplinary knowledge the students need in order to make wise decisions, as voters, on the hazardous substance issues facing their communities.

When You Must Define the Problem Yourself

In other situations, you may not have the aid of explicit statements from your readers to help you formulate the problem. This is most likely to happen when you are preparing a proposal on your own initiative, without being asked by someone else to submit it. Describing the problem in such situations can be particularly challenging because the arguments that will be persuasive to your readers might be entirely distinct from your own reasons for writing the proposal. Helen developed the idea of writing a program for scheduling conference rooms in response to the personal frustration she experienced repeatedly when arriving at rooms she had reserved, only to find that they had been double-booked. No matter how intense her frustration, her managers were not likely to approve her project simply to make her feel better. For her proposal to win their approval, Helen had to persuade them that the program would satisfy a need or goal of their own.

The reasons you offer your readers for supporting your proposal may be different from your reasons for writing it.

Try this approach: Learn about or review your readers' overall goals, looking for ones you can link to your project. What goals or responsibilities do your readers have that your proposal will help them with? What concerns do they typically express that your proposal could help them address?

You can also gain ideas about ways to link your proposed project to your readers' needs and desires by speaking with the readers. This conversation can have two advantages. First, it will let you know whether your proposal has at least some chance of succeeding. If it doesn't, it's best to find out before you invest time writing it. Second, by talking with people who will be readers of your proposal, you can find out how they view the problem you are thinking of describing in your proposal.

When Helen spoke to her boss, she learned something she hadn't thought of before: Sometimes the rooms are used for meetings with customers. When customers see confusion arising over something as simple as meeting rooms, they may wonder whether the company has similar problems that would affect its products and services.

Objectives

When using the conventional superstructure for proposals, writers usually state the objectives of their proposed solution after describing the problem they are proposing.

Your statement of objectives plays a crucial role in the logical development of your proposal: It links your proposed action to the problem by telling how the action will solve the problem. To make that link tight, you must formulate each of your objectives so that it grows directly out of some aspect of the problem you describe. Here, for example, are three of the objectives that the writers devised for their proposed environmental education program. To help you see how these objectives grew out of the writers' statement of the problem, each objective is followed by the point from their problem statement that serves as the basis for it. (The writers did not include the bracketed sentences in the objectives section of their proposal.)

Objectives of the
environmental education
proposal

1. To teach high school students the definitions, facts, and concepts necessary to understand both the benefits and risks of society's heavy reliance upon hazardous substances. [This objective is based on the evidence presented in the problem section that high school students do not have that knowledge.]

2. To employ an interdisciplinary approach that will allow students to use the information, concepts, and skills from their science, economics, government, and other courses to understand the complex issues involved with our use of hazardous substances. [This objective is based on the writers' argument in the problem section that people need to understand the use of hazardous substances from many points of view to be able to make wise decisions about them.]

3. To teach a technique for identifying and weighing the risks and benefits of various uses of hazardous substances. [This objective is based on the writers' argument in the problem section that people must be able to weigh the risks and benefits of the use of hazardous substances if they are to make sound decisions about that use.]

The writers created similar lists of objectives for the other parts of their proposed project, each based on a specific point made in their discussion of the problem.

Like the writers at the television station, Helen built her objectives squarely on her description of the problem. She also took into account something she learned through her research: Even with a computerized system, managers wanted a single person in each department to coordinate room reservations. They didn't want each individual employee to be able to enter changes into the system.

Similarly, you will need to consider the concerns of your readers when framing the objectives for your proposals. Here are two of Helen's objectives.

Helen's objectives

1. To provide a single, up-to-date room reservation schedule that can be viewed by the entire company. [This objective corresponds to her argument that the problem arises partly because there is no convenient way for people to find out what reservations have been made.]

2. To allow a designated person in each department to add, change, or cancel reservations with a minimum of effort. [This objective relates to Helen's discovery that the managers want a single person to handle room reservations in each department.]

In proposals, writers usually describe the objectives of their proposed solution without describing the solution itself at all. The television writers, for example, wrote their objectives so that their readers could imagine achieving them through the creation of a textbook, an educational movie, a series of lectures, or a website—as well as through the creation of the video programs that the instructors proposed. Similarly, Helen described objectives that might be achieved by many kinds of computer programs. She withheld her ideas about the design of her program until the next section of her proposal.

The purpose of separating the objectives from the solution is not to keep readers in suspense. Rather, this separation enables readers to evaluate the aims of the project separately from the writers' particular strategies for achieving those aims.

Proposal writers usually present their objectives in a list or state them very briefly. For instance, Helen used only one paragraph to present her objectives. The members of the television group presented all their objectives in three pages of their 98-page proposal.

Solution

When you describe your solution, you describe your plan for achieving the objectives you have listed. For example, Helen described the various parts of the computer program she would write, explaining how they would be created and used. The television writers described each of the four components of their environmental educational package. Scientists seeking money for cancer research would describe the experiments they wish to conduct.

When describing your solution, you must persuade your readers of two things.

- **That your solution will successfully address each of the objectives.** For instance, to be sure that the description of their proposed educational program matched their objectives, the writers included detailed descriptions of the following: the definitions, facts, and concepts that their video programs would teach (Objective 1); the strategy they would use to help students take an interdisciplinary approach to the question of hazardous waste (Objective 2); and the technique for identifying and weighing risks and benefits that they would help students learn (Objective 3).

- **That your solution offers a particularly desirable way of achieving the objectives.** For example, the television writers planned to use a case study approach in their materials, and in their proposal offered this explanation of the advantages of the case study approach they planned to use.

> Because case studies represent real-world problems and solutions, they are effective tools for illustrating the way that politics, economics, and social and environmental interests all play parts in hazardous substance problems. In addition, they can show the outcome of the ways that various problems have been solved—successfully or unsuccessfully—in the past. In this way, case studies provide students and communities with an opportunity to learn from past mistakes and successes.

Television writers' argument for the special desirability of their solution

Of course, you should include such statements only where they won't be perfectly obvious to your readers. In her proposal, Helen did not include any because she planned to use standard practices whose advantages would be perfectly evident to her readers, both of whom had several years' experience doing her kind of work.

The description of your proposed solution is a place where you must be very careful to protect yourself and your organization from seeming to promise more than you can deliver. The surest protection is to be very precise. For instance, Helen included specific details about the capabilities of her program. It would support the scheduling of six conference rooms (not more), it would provide reservation capabilities up to six months in advance (not longer), and it would operate on an NT server (not necessarily other servers). If you want to identify things you might be able to do, explicitly identify them as possibilities, time and opportunity permitting, not as promises.

LEARN MORE When describing your solution, you may find it helpful to use the strategies discussed in "Describing an Object (Partitioning)," (pages 155–156) and "Describing a Process (Segmenting)," (pages 156–158).

Method

Readers of proposals sometimes need to be assured that you can, in fact, produce the results that you promise. That happens especially in situations where you are proposing to do something that takes special expertise.

To assure themselves that you can deliver what you promise, your readers will look for information about several aspects of your project: your method or plan of action for producing the result; the facilities, equipment, and other resources you plan to use; your schedule; your qualifications; and your plan for managing the project. This section is about method; the other topics are discussed in the sections that follow.

To determine how to explain your proposed method, imagine that your readers have asked you, "How will you bring about the result you have described?"

In some cases, you will not need to answer that question. For example, Helen did not talk at all about the programming techniques she planned to use because her readers were already familiar with them. On the other hand, they did not know how she planned to train people to use her program. Therefore, in her proposal she explained her plans for training.

In contrast, the television writers needed to describe their method in detail in order to persuade their readers that they would produce effective educational materials. An important part of their method, for instance, was to use three review teams: a team of scientists to assess the accuracy of the materials they drafted, a team of filmmakers to advise about plans for making the videos, and a team of specialists in high school education to advise about the effectiveness of the script. In their proposal, they described these review teams in great detail, emphasizing the way each would enhance the effectiveness of the programs the writers were proposing.

Resources

By describing the facilities, equipment, and other resources to be used for your proposed project, you assure your readers that you will use whatever special equipment is required to do the job properly. If part of your proposal is to request that equipment, tell your readers what you need to acquire and why.

If no special resources are needed, you do not need to include a section on resources. Helen did not include one. In contrast, the television writers needed many kinds of resources. In their proposal, they described the excellent library facilities that were available for their research. Similarly, to persuade their readers that they could produce high-quality programs, they described the video editing software they would use.

Schedule

Proposal readers have several reasons for wanting to know what your schedule will be. First, they want to know when they can enjoy the final result. Second, they want to know how you will structure the work so they can be sure that the schedule is reasonable and sound. In addition, they may want to plan other work around the project: When will your project have to coordinate with others? When will it take people's attention from other work? When will other work be disrupted and for how long? Finally, proposal readers want a schedule so they can determine if the project is proceeding according to plan.

The most common way to provide a schedule is to use a schedule chart, which is sometimes accompanied by a prose explanation of its important points.

In many projects, your ability to complete your work on time depends on your getting timely responses and cooperation from other people, including the client or other person or organization who approved your proposal. When discussing your schedule, be sure to include the dates when you will need something from your client, such as approvals of your drafts and information you must have in order to proceed with your work.

LEARN MORE For information on creating a schedule chart, see "Writer's Tutorial: Schedule Charts" (page 259).

Qualifications

When they are thinking about investing in a project, proposal readers want to be sure that the proposers have the experience and capabilities required to carry it out successfully. For that reason, a discussion of the qualifications of the personnel involved with a project is a standard part of many proposals. In a section titled "Qualifications," the television writers presented the qualifications of each of the eight key people who would be working on the project. In addition, they included a detailed résumé for each participant in an appendix.

In other situations, much less information might be needed. For instance, Helen's qualifications as a programmer were known to her readers because they were employing her as one. If that experience alone were enough to persuade her readers that she could carry out the project successfully, Helen would not have needed to include any section on qualifications. However, her readers might have wondered whether she was qualified to undertake the particular program she proposed because different kinds of programs require specialized knowledge and skills. Therefore, Helen wrote the following.

> As you know, I am thoroughly familiar with the kind of database design that would be used in the reservation system. I recently completed an evening course in interactive media where I learned how to program dynamic web pages that can be used to request a database search and display the results. In addition, as an undergraduate I took a course in scheduling and transportation problems, which will help me here.

Helen's statement of her qualifications

In some situations, your readers will want to know not only the qualifications of the people who will work on the proposed project, but also the qualifications of the organization for which they work.

To protect yourself and your organization, guard against implying that you and those working with you have more qualifications than you really have. Time spent learning new programs, gaining new knowledge, or otherwise gaining background needed to complete a project adds to the time (and money) the project costs your employer.

Management

When you propose a project that will involve more than about four people, you can make your proposal more persuasive by describing the management structure of your group. Proposal readers realize that on complex projects even the most highly qualified people can coordinate their work effectively only if a well-designed project management plan is in place. In projects with relatively few people, you can describe the management structure by first identifying the person or persons who will have management responsibilities and then telling what their duties will be. In larger projects, you might need to provide an organizational chart for the project and a detailed description of the management techniques and tools that will be used.

LEARN MORE For information on creating an organizational chart, see "Writer's Tutorial: Organizational Charts" (page 258).

Because her project involved only one person, Helen did not establish or describe any special management structure. However, the television writers did. Because they had a complex project involving many parts and several workers, they set up a project planning and development committee to oversee the activities of the principal workers. In their proposal they described the makeup and functions of this committee, and in the section on qualifications they described the credentials of the committee members.

Costs

As emphasized throughout this chapter, when you propose something, you are asking your readers to invest resources—usually money and time. Naturally, then, you need to tell them how much your proposed project will cost.

One way to discuss costs is to include a budget statement. Sometimes, a budget statement needs to be accompanied by a prose explanation of any unusual expenses and the method used to calculate costs.

In proposals where dollars are not involved, information about the costs of required resources may be provided elsewhere. For instance, in her discussion of

LEARN MORE To create a budget, follow the guidelines for creating a table. See "Writer's Tutorial: Tables" (page 242–243).

the schedule for her project, Helen explained the number of hours she would spend, the time that others would spend, and so on.

In some proposals, you may demonstrate that the costs are reasonable by also calculating the savings that will result from your project.

Bringing It All Together

After you have completed a full draft of your proposal, review it with special emphasis on three pieces of advice from Chapter 9's discussion of persuasion.

LEARN MORE For more see "Create a tight fit among the parts of your communication" (page 182).

- ■ **Create a tight fit among the parts.** There's a tight fit between the proposal's problem and solution when the solution described addresses every aspect of the problem. Nothing is missed. Similarly, the solution and costs have a tight fit if there is a cost for every component of the solution—and no costs that aren't attached to the solution. Create such obvious, direct connections among all the elements you include from the superstructure. As described earlier in this chapter, being careful investors, proposal readers check for this kind of correspondence among parts to assure themselves that the proposed project is carefully planned.

LEARN MORE For more on providing sufficient, reliable evidence, see "Reason Soundly," (pages 171–176).

- ■ **Provide sufficient, reliable evidence to support all of your persuasive claims.** As discussed in Chapter 9 (see "Present sufficient and reliable evidence," page 174), the amount and kinds of information readers find to be sufficient and reliable vary from one reader to another and one situation to another. The insights you gather about your specific readers while defining your proposal's objectives will help you determine what they will want. At the very least, give specific details rather than general statements. Chapter 10's discussion of using concrete, specific words will also help you persuade your readers that there is good reason to trust your proposal.

LEARN MORE For more see "Respond to—and learn from—your readers' concerns and counterarguments" (page 175).

- ■ **Address likely counterarguments.** Being careful investors, proposal readers also look for counterarguments to the various claims you make in your proposal, such as the claim that your method will succeed in producing the solution you've described. Conduct research that will help you predict what these counterarguments might be. Even if you are proposing a project for which research won't help, consider your claims and assumptions with the same critical eye your readers will cast on them. Then address the counterarguments your readers are likely to raise.

Sample Proposals

It is much easier to understand writing advice if you can see how another person has successfully applied it. Take a moment to look at two sample proposals and the marginal notes that point out some of their writers' major strategies. Figure 23.1 shows Helen's proposal, so you can see how she made all the parts fit together. You will find another sample proposal, this one written by students, in Chapter 19 (see Figure 19.1, pages 342–345).

Writer's Guides and Other Resources

Figure 23.2 (pages 400–402) presents a "Writer's Guide for Revising Proposals" that you can use in your course and on the job.

Note to the instructor For a proposal-writing assignment, consider the "Formal Report or Proposal" project in MindTap. Other projects and cases that involve proposal writing are included in MindTap. To help students reflect for transfer, ask them to complete one of the exercises in Appendix B when they finish their project.

FIGURE 23.1

Sample Proposal in the Memo Format

PARKER MANUFACTURING COMPANY

Memorandum

TO: Floyd Mohr and Marcia Valdez
FROM: Helen Constantino
DATE: July 15, 2017
RE: Proposal to Write a Program for Scheduling Conference Rooms

I request permission to spend two weeks writing, testing, and implementing a program for scheduling the buliding's conference rooms in the plant. This program will eliminate several problems with conference room schedules that have become acute in the last three months.

In the introduction, Helen tells what she is proposing and what it will cost.

Present System

At present, the chief means of coordinating room reservations is the monthly "Reservations Calendar" distributed by Peter Svenson of the Personnel Department. Throughout each month, Peter collects notes and phone messages from people who plan to use one of the conference rooms sometime in the next month. He stores these notes in a folder until the fourth week of the month, when he takes them out to create the next month's calendar. If he notices two meetings scheduled for the same room, he contacts the people who made the reservations so they can decide which of them will use one of the other seven conference rooms in the new and old buildings. He then emails the calendar to the heads of all seventeen departments. The department heads usually give the calendars to their administrative assistants.

Someone who wants to schedule a meeting during the current month usually checks with the department's administrative assistant to see whether a particular room has been reserved on the monthly calendar. If not, the person asks the assistant to note his or her reservation on the department's copy of the calendar. The assistant is supposed to call the reservation in to Peter, who will see whether anyone else has called about using that room at that time.

Because her introduction includes much background information, Helen makes a separate section for it; she labels the section "Present System" rather than "Background" to provide her readers with a more precise heading.

(continued)

FIGURE 23.1

(*continued*)

For her problem section, Helen also uses a more precise heading. ———●

Problems with the Present System

The present system worked adequately until about four months ago, when two important changes occurred:

- The new building was opened, bringing nine departments here from the old Knoll Boulevard plant.
- The Marketing Department began using a new sales strategy of bringing major customers here to the plant.

Helen pinpoints the problem. ———●

These two changes have greatly complicated the work of scheduling rooms. With the greatly increased use of the conference rooms, employees often schedule more than one meeting for the same time in the same room. If one of the meetings involves customers brought here by the Marketing Department, we end up giving a bad impression of our ability to manage our business.

Helen describes the consequences of the problem that are important to her readers. ———●

Objectives

To solve our room scheduling problems, we need a reservation system that will do the following:

- Provide a single, up-to-date room reservation schedule that can be viewed by the entire company, thereby ending the confusion caused by having a combination of departmental and central calendars

Helen ties the objectives of her proposed project directly to the points she raised when describing the problem. ———●

- Allow a designated person in each department to add, change, or cancel reservations with a minimum of effort
- Show the reservation priority of each meeting so that people scheduling meetings with higher priority (such as those with potential customers) will be able to see which scheduled meetings they can ask to move to free up a room

2

FIGURE 23.1

(*continued*)

Proposed Solution

I propose to solve our room-scheduling problems by creating a web-based system for reserving our six conference rooms up to six months in advance. A designated person in each department will be able to enter, alter, or cancel reservations quickly and easily through his or her web browser. Located on an NT server, the program will automatically revise the reservation list every time someone makes an addition or other change so people anywhere in the company can view a completely updated schedule at any time. When they enter a reservation, people will include information about the priority of their meeting, so that other individuals will be able to view this information if they are having trouble finding a place for their own meeting. The program will also be able to handle reservations for projection equipment, thereby ending problems in that area. This web-based program will have the added advantage of freeing Peter Svenson from the time-consuming task of manually maintaining the monthly reservations calendar.

> After stating the general nature of her proposed project, Helen links its features to the objectives she identified earlier.
>
> Helen protects herself from misinterpretations of what she will produce by precisely describing the capabilities of her proposed program.

Method for Developing the Program

I propose to create and implement the program in three steps: writing it, testing it, and training people in its use.

> Helen provides an overview of her method, then explains each step in a way that shows how each will contribute to a successful outcome.

Writing the Program

The program will have three features. The first will display the reservations that have been made. When users access the program, the system will prompt them to tell which day's schedule they want to see and whether they want to see the schedule organized by room or by the hour. The calendar will display a name for the meeting, the name of the person responsible for organizing it, and the projection equipment needed.

The second feature will handle entries and modifications to the schedule. When users call up this program, they will be asked for their company identification number. To prevent tampering with the calendar, only people whose identification number is on a list given to the computer will be able to proceed. To make, change, or cancel reservations, users will simply follow prompts given by the system.

3

(*continued*)

FIGURE 23.1

(*continued*)

Once a user completes his or her request, the system will instantly update the calendar that everyone can view. In this way, the calendar will always be absolutely up to date.

Third, when a higher priority meeting displaces one already scheduled, the program will automatically email all persons listed as planning to attend the lower-priority meeting. This is among the features of my program that are not available on commercially available, freestanding scheduling programs.

Testing the Program

Helen has already planned the testing.

I will test the program by having the administrative assistants in four departments use it to create an imaginary schedule for one month. The assistants will be told to schedule more meetings than they usually do to be sure that conflicts arise. They will then be asked to reschedule some meetings and cancel some others.

Training

Training in the use of the program will involve preparing a brief user's guide and conducting training sessions. I will write the user's guide, and I will work with Joseph Raab in the Personnel Department to design and conduct the first training session. After that, he will conduct the remaining training sessions on his own.

Resources Needed

Helen tells what resources she already has (people's time, in this case).

She protects herself by identifying the resources her readers must provide (again, people's time).

To write this program, I will need the cooperation of several people. I have already contacted four people to test the program, and Vicki Truman, head of the Personnel Department, has said that Joseph Raab can work on it because that department is so eager to see Peter relieved of the work he is now having to do under the current system. I will also need your authorization to use a total of about four hours of our server administrator's time.

4

FIGURE 23.1

(*continued*)

Schedule

I can write, test, and train in eight 8-hour days, beginning August 15.

Task	Hours
Designing Program	12
Coding	24
Testing	8
Writing User's Manual	12
Training First Group of Users	8
Total	64

In her schedule, Helen also tells the cost (in hours of work). She protects herself by identifying all the costs.

The eight hours estimated for training include the time needed both to prepare the session and to conduct it one time.

Qualifications

As you know, I am thoroughly familiar with the kind of database design that would be used in the reservation system. I recently completed an evening course in interactive media where I learned how to program dynamic web pages that can be used to request a database search and display the results. In addition, as an undergraduate I took a course in scheduling and transportation problems, which will help me here.

Helen focuses on her qualifications that relate directly to the project she is proposing.

Conclusion

I am enthusiastic about the possibility of creating this much-needed program for scheduling conference rooms and hope that you are able to let me work on it.

5

FIGURE 23.2

Writer's Guide for Revising Proposals

Writer's Guide
REVISING PROPOSALS

Does your draft include each of the elements needed to create a proposal that your readers will find to be useful and persuasive? Remember that you may organize the elements in various ways to increase your proposal's usefulness and persuasiveness. Also, some elements may not be needed to achieve your specific purpose with your specific readers.

INTRODUCTION

❏ Tells clearly what you propose to do.

❏ Provides background information the readers will need or want.

❏ Forecasts the rest of your proposal, if this would help your readers.

PROBLEM

❏ Explains the problem, need, or goal of your proposed action.

❏ Persuades your readers that the problem, need, or goal is important to them.

OBJECTIVES

❏ Relates your objectives directly to the problem, need, or goal you described.

❏ Presents your objectives *without* naming your solution.

SOLUTION (OFTEN THE LONGEST SECTION OF A PROPOSAL)

❏ Describes your solution in a way that assures your readers can understand it.

❏ Persuades that your solution will achieve each of the objectives you described.

❏ Persuades that your solution offers an especially desirable way of achieving the objectives.

❏ Protects you and your employer by clearly promising only what you and your employer want to deliver to your readers.

METHOD

❏ Describes clearly the steps you will follow in preparing the solution.

❏ Persuades that the method you plan to use for creating the solution will work.

RESOURCES

❏ Persuades that you have or can obtain the needed resources.

❏ Protects you and your employer by clearly identifying any resources your readers must supply.

SCHEDULE

❏ Tells when your project will be completed.

❏ Persuades that you have scheduled your work reasonably and soundly.

FIGURE 23.2

(continued)

Writer's Guide
REVISING PROPOSALS
(continued)

☐ Protects yourself and your employer by clearly stating what your readers must do in order for you to be able to meet your deadlines.

☐ Includes a schedule chart, if one would make your proposal more helpful and persuasive.

QUALIFICATIONS

☐ If necessary, persuades that you have the ability to complete the project successfully.

MANAGEMENT

☐ If your project is large, persuades that you will organize the people working on it effectively.

☐ Includes an organizational chart, if one would make your proposal more useful and persuasive.

COSTS

☐ Persuades that you have presented all the costs.

☐ Persuades that the costs are reasonable.

☐ Protects you and your employer by including all your costs in your budget.

☐ Includes a budget table, if one would make your proposal more useful and persuasive.

CONCLUSION

☐ Summarizes your key points.

☐ Concludes the proposal on a positive note that builds confidence in your ability to do a good job.

REASONING (SEE CHAPTER 9)

☐ States your claims and conclusions clearly.

☐ Provides sufficient evidence, from the readers' viewpoint.

☐ Provides evidence your readers will find to be reliable.

☐ Explains, if necessary, the line of reasoning that links your facts and your claims.

☐ Addresses any counterarguments or objections that your readers are likely to raise at any point in your report.

☐ Avoids making false assumptions and overgeneralizing.

☐ Achieves a tight fit among its parts.

PROSE (SEE CHAPTERS 8 AND 10)

☐ Presents information in a clear, useful, and persuasive manner.

☐ Uses a variety of sentence structures and lengths.

☐ Flows in a way that is interesting and easy to follow.

☐ Uses correct spelling, grammar, and punctuation.

(continued)

FIGURE 23.2

(*continued*)

Writer's Guide
REVISING PROPOSALS
(*continued*)

GRAPHICS (SEE CHAPTER 12)

☐ Included wherever readers would find them helpful or persuasive.

☐ Looks neat, attractive, and easy to read.

☐ Referred to at the appropriate points in the prose.

☐ Located where your readers can find them easily.

PAGE DESIGN (SEE CHAPTER 14)

☐ Looks neat and attractive.

☐ Helps readers find specific information quickly.

ETHICS

☐ Treats all the report's stakeholders ethically.

☐ Presents all information accurately and fairly.

Writing Reader-Centered Empirical Research Reports

I n empirical research, investigators gather information through carefully planned, systematic observations or measurements. When scientists send a satellite to investigate the atmosphere of a distant planet, when engineers test jet engine parts made of various alloys, and when pollsters ask older citizens what kinds of outdoor recreation they participate in, they all are conducting empirical research. In your career, you will almost certainly perform empirical research—and report on it in writing. This chapter guides you through the process of writing an empirical research report.

Typical Writing Situations

Empirical research has two distinct purposes. Most aims to help people make practical decisions. The engineers who test jet engine parts are trying to help designers determine which alloy to use in a new engine. Similarly, the researchers who study older persons' recreational activities are trying to help decision makers in the state park system determine what sorts of services and facilities to provide for senior citizens.

A smaller portion of empirical research aims not to support practical decisions, but rather to extend human knowledge. Here, researchers set out to learn how fish remember, what the molten core of the earth is like, or why people fall in love. Such research is usually reported in scholarly journals, such as the *Journal of Chemical Thermodynamics,* the *Journal of Cell Biology,* and the *Journal of Social Psychology,* whose readers are concerned not so much with making practical business decisions as with extending the frontiers of human understanding.

These two aims of research sometimes overlap. Some organizations sponsor basic research in the hope that what is learned can later have practical applications. Likewise, some practice-based research produces results that help explain something basic about the world in general.

How Readers Use and Evaluate Empirical Research Reports

Workplace readers typically take a skeptical approach to empirical research reports. They want to be sure that the actions they take and the new knowledge they accept is based on credible research. Consequently, they will carefully evaluate what your reports say about every step of your research process. If they believe they have found a flaw with even one step, they may be unwilling to act on or accept your results. No skill with words will make up for a research design that your readers don't consider to be credible.

Similarly, even if you have conducted your study flawlessly, a slip-up in your description of your process may also lead your readers to distrust, disregard, and

24

LEARNING OBJECTIVES

1. Describe the typical situations in which empirical research reports are written.
2. Describe the major questions asked by readers of empirical research reports.
3. Describe the superstructure for empirical research reports, including the ways its parts correspond with the readers' questions.
4. Adapt the reader-centered writing process to the special goals of empirical research reports.
5. Write an effective empirical research report.

Empirical research reports propel advances in engineering, science, and medicine.

dismiss your results. This chapter assumes that your research will always be perfect. The chapter's goal is to enable you to persuasively answer your readers' probing questions so that they don't mistakenly believe your work was less than outstanding.

From situation to situation, readers' basic questions remain the same. Here are their typical questions, the ones your instructor and this chapter will help you learn to answer persuasively.

- **Why is your research important to us?** Readers concerned with solving specific practical problems want to know what problems your research will help them address. Readers concerned with extending human knowledge want to know what your research contributes to that pursuit.

- **Did your research ask the right questions?** A well-designed empirical research project is based on carefully formulated research questions that the project will try to answer. Readers want to know what those questions are so they can determine whether they are significant.

- **Was your research method sound?** Unless your method is appropriate to your research questions and intellectually sound, your readers will not place any faith in your results or your conclusions and recommendations.

- **What results did your research produce?** Naturally, your readers want to learn what results you obtained.

- **Did you interpret the results correctly and usefully?** Your readers want you to interpret your results in ways that are intellectually sound and meaningful to them.

- **What is the significance to us of those results?** How do your results relate to the problems your research was to help solve or to the area of knowledge it was meant to expand?

- **What do you think we should do?** Readers concerned with practical problems want to know what you advise them to do. Readers concerned with extending human knowledge want to know what you think your results imply for future research.

Superstructure for Empirical Research Reports

The superstructure for empirical research reports contains seven elements. Each element responds to one of the decision makers' seven basic questions.

LEARN MORE For more on superstructures, see "Adapt an appropriate superstructure or other pattern familiar to your reader" (pages 112–113).

Superstructure for Empirical Research Reports

Report Element	Readers' Question
Introduction	Why is your research important to us?
Objectives of the research	Did you ask the right questions?
Method	Was your research method sound?
Results	What results did your research produce?
Discussion	Did you interpret your results correctly and usefully?
Conclusions	What is the significance to us of those results?
Recommendations	What do you think we should do?

Crafting the Major Elements of an Empirical Research Report

All the detailed advice that Chapters 6 through 16 provide about drafting workplace communications applies to empirical research reports. The following sections tell how you can apply some of this advice most productively when writing this type of report. Much of the advice is illustrated through the use of two sample reports. The first is presented in full in Figure 24.1 (pages 411–425). Its aim is practical. It was written by engineers who were developing a satellite communication system that will permit companies with large fleets of trucks to communicate directly with their drivers at any time. In their report, the writers tell decision makers and other engineers in their organization about the first operational test of the system, in which they sought to answer several practical engineering questions.

The aim of the second sample report is to extend human knowledge: a famous study by Robert B. Hays (1985) of the ways people develop friendships. Selected passages from this report, which appeared in the *Journal of Personality and Social Psychology*, are quoted throughout this section.

As you read the advice below, keep in mind that the conventional superstructure, or genre, is only a general plan. You must use your imagination and creativity to adapt it to particular situations.

In addition, as you plan and write empirical research reports, remember that the seven elements identify the kinds of information readers expect, not necessarily the titles of the sections you include. In brief reports, some parts may take only a sentence or a paragraph, or several sections may be grouped together. For instance, writers often combine their introduction and explanation of the objectives of the research or include the results and discussion in a single section.

Introduction

In the introduction to an empirical research report, you should seek to answer the readers' question, "Why is this research important to us?" Typically, writers answer that question in two steps: They announce the topic of their research and then explain the importance of the topic to their readers.

Announcing the Topic

You can often announce the topic of your research simply by including it as the key phrase in your opening sentence. For example, here is the first sentence of the report on the satellite communication system.

> For the past eighteen months, the Satellite Products Laboratory has been developing a system that will permit companies with large, nationwide fleets of trucks to communicate directly to their drivers through a satellite link.

Topic of report

Here is the first sentence of the report on the way people develop friendships.

> Social psychologists know very little about the way real friendships develop in their natural settings.

Topic of report

Explaining the Importance of the Research

To explain the importance of your research to your readers, you can use either or both of the following methods: Tell how your research is relevant to your organization's goals or how it expands existing knowledge on the subject.

LEARN MORE For additional advice on writing an introduction, see Chapter 7 ("Drafting Reader-Centered Paragraphs, Sections, Chapters, and Communications," pages 121–146).

- **Relevance to organizational goals.** In reports written to readers in organizations (whether your own or a client's), you can explain the relevance of your research by relating it to some organizational goal or problem. Sometimes, in fact, the importance of your research to the organization's needs will be so obvious to your readers that merely naming your topic will be sufficient. At other times, you will need to discuss at length the relevance of your research to the organization. In the first paragraph of the satellite report, for instance, the writers mention the potential market for the satellite communication system they are developing. That is, they explain the importance of their research by saying that it can lead to a profit. For detailed advice about how to explain the importance of your research to readers in organizations, see "Write a Beginning that Motivates Your Reader To Read" (pages 135–138).

- **Expansion of existing knowledge.** A second way to establish the importance of your research is to show the gap in current knowledge that it will fill. The following passage from the opening of the report on the friendship study illustrates this strategy.

The writer tells what is known about his topic.

The writer identifies the gaps in knowledge that his research will fill.

> A great deal of research in social psychology has focused on variables influencing an individual's attraction to another at an initial encounter, usually in laboratory settings (Bergscheid & Walster, 1978; Bryne, 1971; Huston & Levinger, 1978), yet very little data exist on the processes by which individuals in the real world move beyond initial attraction to develop a friendship; even less is known about the way developing friendships are maintained and how they evolve over time (Huston & Burgess, 1979; Levinger, 1980).

The writer continues this discussion of published research for three paragraphs. Each follows the same pattern: It identifies an area of research, tells what is known about that area, and identifies gaps in the knowledge—gaps that will be filled by the research the writer has conducted. These paragraphs serve an important additional function performed by many literature reviews: They introduce the established facts and theories that are relevant to the writer's work and necessary to an understanding of the report.

Objectives of the Research

Every empirical research project has carefully constructed objectives. These objectives define the focus of your project, influence the choice of research method, and shape the way you interpret your results. Readers of empirical research reports want and need to know what the objectives are.

The following example from the satellite report shows one way you can inform your readers about your objectives.

Statement of research objectives

> In particular, we wanted to test whether we could achieve accurate data transmissions and good-quality voice transmissions in the variety of terrains typically encountered in long-haul trucking. We wanted also to see what factors might affect the quality of transmissions.

When reporting research that employs statistics, you can usually state your objectives by stating the hypotheses you tested. Where appropriate, you can explain these hypotheses in terms of existing theory, again citing previous publications on the subject. In the following passage, the writer who studied friendship explains some of his hypotheses. Notice that he begins by stating the research's overall goal.

The goal of the study was to identify characteristic behavioral and attitudinal changes that occurred within interpersonal relationships as they progressed from initial acquaintance to close friendship. With regard to relationship benefits and costs, it was predicted that both benefits and costs would increase as the friendship developed, and that the ratings of both costs and benefits would be positively correlated with ratings of friendship intensity. In addition, the types of benefits listed by the subjects were expected to change as the friendships developed. In accord with Levinger and Snoek's (1972) model of dyadic relatedness, benefits listed at initial stages of friendship were hypothesized to be more activity centered and to reflect individual self-interest (e.g., companionship, information) than benefits at later stages, which were expected to be more personal and reciprocal (e.g., emotional support, self-esteem).

Overall goal

First objective (hypothesis)

Second objective (hypothesis)

Third objective (hypothesis)

Method

Readers of your empirical research reports will look for precise details concerning your method. Those details serve three purposes. First, they let your readers assess the soundness of your research design and its appropriateness for the problems you are investigating. Second, the details enable your readers to determine the limitations that your method might place on the conclusions you draw. Third, they provide information that will help your readers repeat your experiment if they wish to verify your results or conduct similar research projects of their own.

The nature of the information you should provide about your method depends on the nature of your research. The writers of the satellite report provided three paragraphs and two tables explaining their equipment (truck radios and satellite), two paragraphs and one map describing the eleven-state region covered by the trucks, and two paragraphs describing their data analysis.

The friendship study's writer began his method discussion this way.

At the beginning of their first term at the university, first-year students selected two individuals whom they had just met and completed a series of questionnaires regarding their relationships with those individuals at 3-week intervals through the school term.

In the rest of that paragraph, the writer explains that the questionnaires asked the students to describe such matters as their attitudes toward each of the other two individuals and the specific things they did with each of them. However, that paragraph is just a small part of the researcher's account of his method. He then provides a 1,200-word discussion of the students he studied and procedures he used.

How can you decide which details to include? The most obvious way is to follow the general reporting practices in your field. Find some research reports that use a method similar to yours and see what they report. Depending on the needs of your readers, you may need to include some or all of the following topics.

LEARN MORE Often a description of empirical research methods uses the pattern for organizing a description of a process (see "Describing a Process (Segmenting)," pages 156–158).

- Every procedural decision you made while planning your research.
- Every aspect of your method that your readers might ask about.
- Any aspect of your method that might limit the conclusions you can draw from your results.
- Every procedure that other researchers would need to understand in order to design a similar study.

LEARN MORE For information on using tables and other graphics for reporting numerical data, see Chapter 12, "Creating Reader-Centered Graphics" (pages 224–241).

Results

The results of empirical research are the data you obtain. Although the results are the heart of any empirical research report, they may take up a very small portion of it. Generally, results are presented in one of two ways.

- **Tables and graphs.** The satellite report, for instance, uses two tables. The report on friendship uses four tables and eleven graphs.
- **Sentences.** When placed in sentences, results are often woven into a discussion that combines data and interpretation, as the next paragraphs explain.

Discussion

Sometimes writers briefly present all their results in one section and then discuss them in a separate section. Sometimes they combine the two in a single, integrated section. Whichever method you use, your discussion must link your interpretative comments with the specific results you are interpreting.

One way of making that link is to refer to the key results shown in a table or other graphic and then comment on them, as appropriate. The following passage shows how the writers of the satellite report did that with some of the results they presented in one of their tables.

Writers emphasize a key result shown in a table. Writers draw attention to other important results.

Writers interpret those results.

> As Table 3 shows, 91% of the data transmissions were successful. These data are reported according to the region in which the trucks were driving at the time of transmission. The most important difference to note is the one between the rate of successful transmissions in the Southern Piedmont region and the rates in all the other regions. In the Southern Piedmont area, we had the truck drive slightly outside the ATS-6 footprint so that we could see if successful transmissions could be made there. When the truck left the footprint, the percentage of successful data transmissions dropped abruptly to 43%.

When you present your results in prose only (rather than in tables and graphs), you can weave them into your discussion by beginning your paragraphs with general statements that serve as interpretations of your data. Then you can cite the relevant results as evidence in support of the interpretations. Here is an example from the friendship report.

General interpretation

Specific results presented as support for the interpretation

> Intercorrelations among the subjects' friendship intensity ratings at the various assessment points showed that friendship attitudes became increasingly stable over time. For example, the correlation between friendship intensity ratings at 3 weeks and 6 weeks was .55; between 6 weeks and 9 weeks, .78; between 9 weeks and 12 weeks, .88 (all $p < .001$).

In a single report, you may use both of these methods of combining the presentation and discussion of your results.

Conclusions

Besides interpreting the results of your research, you need to explain what those results mean in terms of the original research questions and the general problem you set out to investigate. Your explanations of these matters are your conclusions.

For example, if your research project is focused on a single hypothesis, your conclusion can be brief, perhaps only a restatement of your chief results. However, if your research is more complex, your conclusion should draw all the strands together.

In either case, the presentation of your conclusions should correspond very closely to the objectives you identified toward the beginning of your report. Consider, for instance, the correspondence between objectives and conclusions in the satellite study. The first objective was to determine whether accurate data transmissions and good-quality voice transmissions could be obtained in the variety of terrains typically encountered in long-haul trucking. The first of the conclusions addresses that objective.

> The Satellite Products Laboratory's system produces good-quality data and voice transmissions throughout the eleven-state region covered by the satellite's broadcast footprint.

Conclusions from the satellite report

The second objective was to determine what factors affect the quality of transmissions, and the second and third conclusions relate to it.

> The most important factor limiting the success of transmissions is movement outside the satellite's broadcast footprint, which accurately defines the satellite's area of effective coverage.

> The system is sensitive to interference from certain kinds of objects in the line of sight between the satellite and the truck. These include trees, mountains and hills, overpasses, and buildings.

The satellite research concerns a practical question. Hence, its objectives and conclusions address practical concerns of particular individuals — in this case, the engineers and managers in the company that is developing the satellite system. In contrast, research that aims primarily to extend human knowledge often has objectives and conclusions that focus on theoretical issues.

For example, at the beginning of the friendship report, the researcher identifies several questions that his research investigated, and he states what answers he predicted his research would produce. In his conclusion, then, he systematically addresses those same questions in terms of the results his research produced. Here is a summary of some of his objectives and conclusions. (Notice how he uses the technical terminology commonly employed by his readers.)

OBJECTIVE	CONCLUSION
As they develop friendships, do people follow the kind of pattern theorized by Guttman, in which initial contacts are relatively superficial and later contacts are more intimate?	Yes. "The initial interactions of friends . . . correspond to a Guttman-like progression from superficial interaction to increasingly intimate levels of behavior."
Do *both* the costs (or unpleasant aspects) and the benefits of personal relationships increase as friendships develop?	Yes. "The findings show that personal dissatisfactions are inescapable aspects of personal relationships and so, to some degree, may become immaterial. The critical factor in friendships appears to be the amount of benefits received. If a relationship offers enough desirable benefits, individuals seem willing to put up with the accompanying costs."

Conclusions from the friendship report

Are there substantial differences between friendships women develop with one another and the kind men develop with one another?	Apparently not. "These findings suggest that—at least for this sample of friendships—the sex differences were more stylistic than substantial. The bonds of male friendship and female friendship may be equally strong, but the sexes may differ in their manner of expressing that bond. Females may be more inclined to express close friendships through physical or verbal affection; males may express their closeness through the types of companionate activities they share with their friends."

Typically, in presenting the conclusions of an empirical research project directed at extending human knowledge, writers discuss the relationship of their findings to the findings of other researchers and to various theories that have been advanced concerning their subject. The writer of the article on friendship did that. The preceding table presents only a few snippets from his overall discussion, which is several thousand words long and is full of thoughts about the relationship of his results to the results and theories of others.

In the discussion section of their empirical research reports, writers sometimes discuss any flaws in their research method or limitations on the generalizability of their conclusions. For example, the writer on friendship points out that all of his subjects were college students and that most lived in dormitories. It is possible, he cautions, that what he found while studying this group may not be true for other groups.

Recommendations

The readers of some empirical research reports, especially those directed at solving a practical problem, want to know what, based on the research, the writer thinks should be done. Consequently, such reports usually include recommendations.

For example, the satellite report contains three. The first is the general recommendation that work on the project be continued. The other two involve specific actions that the writers think should be taken: design a special antenna for the trucks, and develop a plan that tells what satellites would be needed to provide coverage throughout the forty-eight contiguous states, Alaska, and southern Canada. As is common in research addressed to readers in organizations, these recommendations concern practical business and engineering decisions.

Even in reports designed to extend human knowledge, writers sometimes include recommendations. These usually convey ideas about future studies that should be made, adjustments in methodology that seem to be called for, and the like. At the end of the friendship report, for instance, the writer suggests that researchers study different groups (not just students) in different settings (not just college) to establish a more comprehensive understanding of how friendships develop.

Sample Empirical Research Reports

It is much easier to understand writing advice if you see sample communications that follow that advice. Take a moment to look at the sample in Figure 24.1 (pages 411–425) and the marginal notes that point out some of the writer's major strategies. This figure shows the full report on the truck-and-satellite communication system. To see examples of empirical research reports presented as journal articles, consult journals in your field.

Writer's Guides and Other Resources

Figure 24.2 (pages 426–427) presents a Writer's Guide for Revising Empirical Research Reports that you can use in your course and on the job.

Note to the Instructor   For an assignment involving an empirical research report, you can ask students to write on research they are conducting in their majors, or you can adapt the "Formal Report or Proposal" project given in MindTap. You can also use the "User Test and Report" project, also in MindTap. To help students reflect for transfer, ask them to complete one of the exercises in Appendix B when they finish their project.

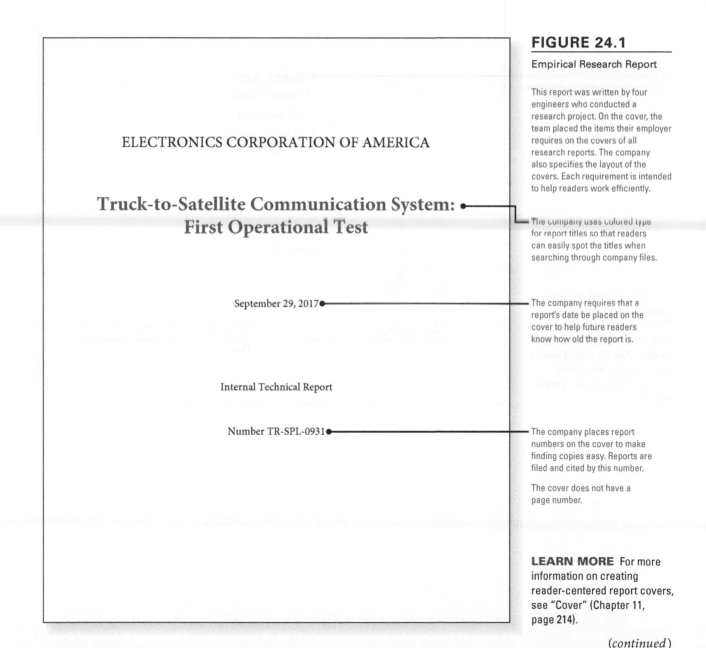

FIGURE 24.1

Empirical Research Report

This report was written by four engineers who conducted a research project. On the cover, the team placed the items their employer requires on the covers of all research reports. The company also specifies the layout of the covers. Each requirement is intended to help readers work efficiently.

The company uses colored type for report titles so that readers can easily spot the titles when searching through company files.

The company requires that a report's date be placed on the cover to help future readers know how old the report is.

The company places report numbers on the cover to make finding copies easy. Reports are filed and cited by this number.

The cover does not have a page number.

LEARN MORE   For more information on creating reader-centered report covers, see "Cover" (Chapter 11, page 214).

ELECTRONICS CORPORATION OF AMERICA

**Truck-to-Satellite Communication System:
First Operational Test**

September 29, 2017

Internal Technical Report

Number TR-SPL-0931

(continued)

FIGURE 24.1

(*continued*)

Like the cover, the title page contains exactly the information specified by the company for all research reports.

As is typical in many organizations, the title page includes more information than the cover does.

This company permits the researchers to be named. In some companies, a report's writers are not named.

The company requires that all research reports be reviewed and approved by the laboratory director. She examines drafts carefully and usually requires changes before giving approval.

LEARN MORE For more information on approvals, see "Obtain Truly Helpful Advice from People Who Review Your Drafts—And Give Good Advice When You Are Reviewing Someone Else's Draft" (pages 282–287).

ELECTRONICS CORPORATION OF AMERICA

Truck-to-Satellite Communication System: First Operational Test

September 29, 2017

Research Team

Margaret C. Barnett

Erin Sanderson

L. Victor Sorrentino

Raymond E. Wu

Internal Technical Report
Number TR-SPL-0931

Read and Approved:

Beverly Fisher

September 29, 2017
Date

FIGURE 24.1

(*continued*)

The researchers provide a brief executive summary in order to help their readers quickly understand the report's main points. The summary also enables readers to determine whether they could meet their practical needs by reading the entire report.

Every piece of information the researchers include in the summary is also in the body of the report.

In the first sentence, the researchers identify the topic of their report.

In one sentence, the researchers summarize their research method, providing specific details about the data, the satellite used, the number of trucks, and region covered.

In precise, succinct statements, the researchers detail the results that are most relevant to their readers' major question: "Does the prototype design work?"

Their conclusion is the answer to their readers' question: "Yes, the design does work."

The researchers use a bulleted list to highlight their two major recommendations for continued work on the satellite project.

LEARN MORE For more information on executive summaries, go to "Writing Reader-Centered Summaries," (pages 216–219), and see Figure 11.6.

EXECUTIVE SUMMARY

For the past eighteen months, the Satellite Products Laboratory has been developing a system that will permit companies with large, nationwide fleets of trucks to communicate directly to their drivers at any time through a satellite link. Costs would be lower than for cell phones, and a GPS system mounted in each truck would enable a company to determine the exact location of all its vehicles at one time. During the week of May 22, we tested our concepts for the first time, using the ATS-6 satellite and five trucks that were driven over an eleven-state region with our prototype mobile radios.

More than 91% of the 2500 data transmissions were successful and more than 91% of the voice transmissions were judged to be of commercial quality. The most important factor limiting the success of transmissions was movement outside the satellite's broadcast footprint. Other factors include the obstruction of the line of sight between the truck and the satellite by highway overpasses, mountains and hills, trees, and buildings.

Overall, the test demonstrated the soundness of the prototype design. Work on it should continue as rapidly as possible. We recommend the following actions:

- Develop a new antenna designed specifically for use in communications between satellites and mobile radios.

- Explore the configuration of satellites needed to provide thorough footprint coverage for the 48 contiguous states, Alaska, and southern Canada at an elevation of 25° or more.

(*continued*)

FIGURE 24.1

(*continued*)

In addition to telling readers where they can find particular kinds of information, the table of contents also provides them with an outline of the report.

The researchers include the major headings within their sections.

They align the page numbers on the right-hand side.

LEARN MORE For more information on writing the front matter of a report, go to Chapter 11, "Writing Reader-Centered Front and Back Matter."

They number the front matter with Roman numerals. They begin using Arabic numerals on the first page of the report's body.

TABLE OF CONTENTS

ii

FIGURE 24.1

(*continued*)

Section 1

INTRODUCTION

For the past eighteen months, the Satellite Products Laboratory has been developing a system that will permit companies with large, nationwide fleets of trucks to communicate directly to their drivers at any time through a satellite link. Costs would be lower than for cell phones, and a GPS system mounted in each truck would enable a company to determine the exact location of all its vehicles at one time. Several trucking lines and supermarket chains have expressed an interest in such a service. At present, they can communicate with their long-distance drivers only when the drivers pull off the road to phone in, meaning that all contacts are originated by the drivers, not the central dispatching service. The potential market for such a satellite service also includes many other companies and government agencies (such as the National Forest Service) that desire to communicate with trucks, cars, boats, or trains that regularly operate outside the very limited range of urban cellular telephone systems.

This report describes the first operational test of the system we have developed. Such tests were particularly important to conduct before continuing further with the development of this system because our system is much different from those currently being used with commercial satellites. Specifically, our system will transmit to mobile ground stations by using the short antennas and the low power provided by conventional terrestrial broadcasting systems.

In particular, we wanted to test whether we could achieve accurate data transmissions and good-quality voice transmissions in the variety of terrains typically encountered in long-haul trucking. We wanted also to see what factors might affect the quality of transmissions.

The test results indicate that our design is basically sound, although a new mobile antenna needs to be designed and the satellite configuration needs to be examined.

1

The researchers begin the body of their report by stating its topic.

They immediately tell their readers the significance of their research to the company.

The researchers describe the overall objectives of their test.

LEARN MORE For advice about presenting the objectives of empirical research, see "Objectives of the Research" (page 406).

They tell their readers their main finding ("the bottom line") at the very beginning of their report.

LEARN MORE For advice about writing the introduction to an empirical research report, see "Crafting the Major Elements of an Empirical Research Report" (pages 405–408). In addition, see "Write a Beginning that Motivates Your Reader To Read" (pages 135–138).

(*continued*)

FIGURE 24.1

(*continued*)

The researchers' major goal in the method section is to persuade their readers that they used sound procedures. They know that their readers will not trust the researchers' results unless they are persuaded that the research method was solid.

Their second goal in the method section is to provide details that will enable the readers to interpret the results.

The researchers begin with a one-sentence overview of their method.

In their next sentence, they forecast the organization of their method section.

The researchers organize this paragraph (and many others) so that it flows from general ideas to particular ones.

They provide precise details about the equipment they used.

In this sentence the researchers refer readers to the technical data in Table 1.

Anticipating a possible question from their readers, the researchers explain their justification for using the large ground station at King of Prussia.

Section 2

METHOD

In this experiment, we tested a full-scale system in which five trailer trucks communicated with an earth ground station via a satellite in geostationary orbit 23,200 miles above the earth. This section describes the equipment we used, the area covered by the test, and the data analyses we performed.

Equipment

The five trucks, operated by Smithson Moving Company, were each equipped with a prototype of our newly developed 806 megahertz (MHz) two-way mobile radio equipment. Each radio had a speaker and microphone for voice communications, along with a ten-key keyboard and digital display for data communications. Equipped with dipole antennas, the radios broadcast at 1650 MHz with 12 to 15 watts of power. They received signals at 1550 MHz and had an equivalent antenna temperature of 800°K, including feedline losses. Technical specifications for the receivers and transmitters of these radios are given in Table 1.

The satellite used for this test was the ATS-6, which has a larger antenna than most commercial communication satellites, making it more sensitive to the low-power signals sent from the mobile stations. Technical specifications for its receiver and transmitter are given in Table 2.

Through the ATS-6 satellite, the five trucks communicated with the earth ground station in King of Prussia, Pennsylvania. This facility is a relatively large station, but not larger than is planned for a fully operational commercial system.

Region Covered

The five trucks drove throughout the region covered by the "footprint" of ATS-6. The footprint is defined as the area in which the broadcast signals received are within at least 3 dB of the signal received at the center of the beam. In all, the trucks covered eleven states: Georgia, South Carolina, North Carolina, Tennessee, Virginia, West Virginia, Ohio, Indiana, Illinois, Iowa, and Nebraska.

2

LEARN MORE For advice about writing the method section of an empirical research report, see "Method" (page 407).

LEARN MORE For advice about integrating graphics and text, see "How To Integrate Your Graphics with Your Text" (pages 235–236).

LEARN MORE For guidelines on organizing sections and paragraphs in a reader-centered way, see "Present your generalizations before your details" (pages 124–125).

FIGURE 24.1

(*continued*)

Table 1

Specifications for Satellite-Aided Mobile Radio ●

Transmitter

Frequency	1655.050 MHz
Power Output	16 watts nominal
	12 watts minimum
Frequency Stability	±0.0002% (–30° to +60°C)
Modulation	$16F_3$ Adjustable from 0 to ±5 kHz swing FM with instantaneous modulation limiting
Audio Frequency Response	Within +1 dB and –3 dB of a 6 dB/octave pre-emphasis from 300 to 3000 HZ per EIA standards
Duty Cycle	EIA 20% Intermittent
Maximum Frequency Spread	±6 MHz with center tuning
RF Output Impedance	50 ohms

Receiver

Frequency	1552.000 MHz
Frequency Stability	±0.0002% (–30° to +60°C)
Noise Figure	2.6 dB referenced to transceiver antenna jack
Equivalent Receiver Noise Temperature	238° Kelvin
Selectivity	–75 dB by EIA Two-Signal Method
Audio Output	5 watts at less than 5% distortion
Frequency Response	Within +1 and –8 dB of a standard 6 dB per octave deemphasis curve from 300 to 3000 Hz
Modulation Acceptance	±7 kHz
RF Input Impedance	50 ohms

3

With this table, the writers provide detailed data in a highly useful manner.

The researchers give their table a title that briefly but precisely tells their readers what the table shows.

Because the table contains text rather than numbers, the researchers have aligned the contents of their columns on the left.

LEARN MORE The "Writer's Tutorial on Tables" (pages 242–243) provides guidelines for creating reader-centered tables.

(*continued*)

FIGURE 24.1

(*continued*)

Table 2

Performance of ATS-6 Spacecraft L-Band Frequency Translation Mode

Receive

Receiver Noise Figure (dB)	6.5
Equivalent Receiver Noise Temperature (°K)	1005
Antenna Temperature Pointed at Earth (°K)	290
Receiver System Temperature (°K)	1295
Antenna Gain, peak (dB)	38.4
Spacecraft G/T, peak (dB/°K)	7.3
Half Power Beamwidth (degrees)	1.3
Gain over Field of View (dB)	35.4
Spacecraft G/T over Field of View (dB/°K)	4.3

Transmit

Transmit Power (dBw)	15.3
Antenna Gain, peak (dB)	37.7
Effective Radiated Power, peak (dBw)	53.0
Half Power Beamwidth (degrees)	1.4
Gain over Field of View (dB)	34.8
Effective Radiated Power over Field of View (dBw)	50.1

4

FIGURE 24.1

(*continued*)

Within this region, the trucks drove through the kinds of terrain usually encountered in long-haul trucking, including both urban and rural areas in open plains, foothills, and mountains.

Data Analysis

All test transmissions were recorded at the earth station on a high-quality reel-to-reel tape recorder. The strength of the signals received from the trucks via the satellite was recorded on a chart recorder that had a frequency response of approximately 100 hertz. In addition, observers in the trucks recorded all data signals received and all data codes sent. They also recorded information about the terrain during all data and voice transmissions.

We analyzed the data collected in several ways:

- To determine the accuracy of the data transmissions, we compared the information recorded by the observers with the signals recorded on the tapes of all transmissions.

- To determine the quality of the voice transmissions, we had an evaluator listen to the tape using high-quality earphones. For each transmission, the evaluator rated the signal quality on the standard scale for the subjective evaluation of broadcast quality. On it, Q5 is excellent and Q1 is unintelligible.

- To determine what factors influenced the quality of the transmissions, we examined the descriptions of the terrain that the observers recorded for all data transmissions that were inaccurate and all the voice communications that were rated 3 or less by the evaluator. We also looked for relationships between the accuracy and quality of the transmissions and the distance of the trucks from the edge of the broadcast footprint of the ATS-6 satellite.

5

The researchers describe their method of analysis to persuade their readers of the soundness of their study. They know that their readers will want this information because the study results would be incorrect if a wrong or flawed method of analysis had been used.

The research team used three methods for analyzing data. Each method corresponds to one of their research objectives, which they describe in the report's introduction.

To help their readers see the correspondence between their analysis methods and their research objectives, the team describes their three analysis methods in the same order they used when naming their research objectives in the introduction.

(*continued*)

FIGURE 24.1

(*continued*)

The research team begins this section with general information.

The researchers tell their readers the key result first.

They then present other important results.

Immediately after presenting their results, the researchers interpret the results. This interpretation is one of the discussion parts of their results and discussion section.

Again, the researchers begin with the key result. In this sentence they also tell their readers the main point to draw from Table 4.

As soon as they have presented the results, the researchers interpret the results from the point of view of their readers' main question: "Does the system work enough for commercial application?"

The team summarizes the data and voice transmission results, emphasizing the main points of interest to their readers.

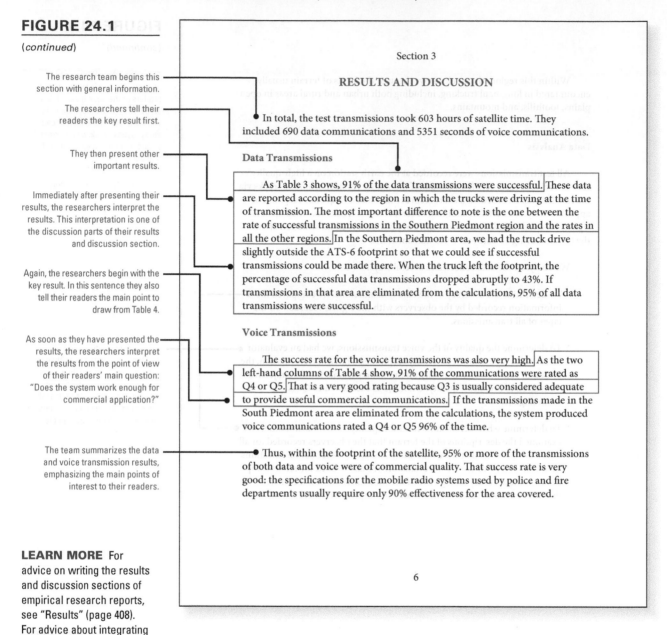

Section 3

RESULTS AND DISCUSSION

In total, the test transmissions took 603 hours of satellite time. They included 690 data communications and 5351 seconds of voice communications.

Data Transmissions

As Table 3 shows, 91% of the data transmissions were successful. These data are reported according to the region in which the trucks were driving at the time of transmission. The most important difference to note is the one between the rate of successful transmissions in the Southern Piedmont region and the rates in all the other regions. In the Southern Piedmont area, we had the truck drive slightly outside the ATS-6 footprint so that we could see if successful transmissions could be made there. When the truck left the footprint, the percentage of successful data transmissions dropped abruptly to 43%. If transmissions in that area are eliminated from the calculations, 95% of all data transmissions were successful.

Voice Transmissions

The success rate for the voice transmissions was also very high. As the two left-hand columns of Table 4 show, 91% of the communications were rated as Q4 or Q5. That is a very good rating because Q3 is usually considered adequate to provide useful commercial communications. If the transmissions made in the South Piedmont area are eliminated from the calculations, the system produced voice communications rated a Q4 or Q5 96% of the time.

Thus, within the footprint of the satellite, 95% or more of the transmissions of both data and voice were of commercial quality. That success rate is very good: the specifications for the mobile radio systems used by police and fire departments usually require only 90% effectiveness for the area covered.

6

LEARN MORE For advice on writing the results and discussion sections of empirical research reports, see "Results" (page 408). For advice about integrating graphics and text, see Chapter 12 (pages 235–236).

FIGURE 24.1

(*continued*)

Table 3

Success in Decoding DTMF Automatic Transmitter

REGION	STATES	ATS TRANSMISSIONS		ELEVATION
		Sent by Vehicle	Received and Decoded Correctly	Angle to Satellite (°)
Open Plains	Indiana Ohio Nebraska Illinois Iowa	284	283 (100%)	17–26
Western Appalachian Foothills	Ohio Tennessee	55	53 (96%)	15–19
Appalachian Mountains	West Virginia	112	93 (83%)	15–17
Piedmont	Virginia North Carolina	190	178 (97%)	11–16
Southern Piedmont	Georgia South Carolina	49	21 (43%)	17–18
TOTAL		690	628 (91%)	

7

(*continued*)

FIGURE 24.1

(*continued*)

To help their readers use this table without confusion, the research team placed vertical lines between the columns, which are very close together. Such lines are not needed in the other tables, where the space between the columns is greater.

Table 4
Quality of Voice Communication Signal[1]

Area	Transmission Time (Secs)	No Blockage Time (Secs)		Trees			Mountains and Hills			Overpasses (Momentary Dropouts)			Buildings		
		Q5	Q4	Q3	Q2	Q1	Q3	Q2	Q1	Q3	Q2	Q1	Q3	Q2	Q1
Open Plains	2481	2334	73	1	1	0	4	1	3	13	2	20	15	10	4
Western Foothills	344	322	2	0	0	0	6	12	13	0	0	0	0	0	0
Appalachian Mountains	1037	614	267	42	17	0	20	31	37	0	2	7	0	0	0
Piedmont	1219	481	623	5	15	19	25	16	10	4	1	4	8	4	1
Southern Piedmont	270	0	149	109	3	12	0	0	0	0	0	2	0	0	0
TOTAL TIME (seconds)	5351	3751	1114	157	36	31	55	60	63	17	5	33	23	14	5

[1]Total times in seconds for each quality of received signals. Q5 is excellent; Q1 is unintelligible.

8

FIGURE 24.1

(*continued*)

Factors Affecting the Quality of Transmissions

The factor having the largest effect on transmission quality is the location of the truck within the footprint of ATS-6. The quality of transmissions is even and uniformly good throughout the footprint, but almost immediately outside of it the quality drops well below acceptable levels.

> In this section, too, the researchers begin with the most important point.

Several factors were found to disrupt transmissions even when the trucks were in the satellite's footprint. Table 4 shows what these factors are for the 4% of the transmissions in the footprint that were not of commercial quality. In all cases the cause is some object passing in the line of sight between the satellite and the truck.

> The team refers to Table 4 again. This time, however, they want their readers to notice something different—and they tell the readers what that is.

The disruption caused by a single tree created only a very brief and usually insignificant dropout of one second or less. Only driving past a group of trees will cause a significant loss of signal. Yet this happened often in the terrain of the Appalachian Mountains and the South Piedmont.

> The researchers continue to present the information that is most important to their readers before the information that is less important to them. In this section, they start with the factor that causes the most disruptions: trees.

We believe we could eliminate many of the disruptions caused by trees if we developed an antenna specifically for use in communications between satellites and mobile radios.

> The researchers devote this paragraph to answering the question they imagine their readers will ask: "What can be done about the trees?"

Mountains and hills caused 36% of the disruptions. That happened mostly in areas where the satellite's elevation above the horizon was very low. Otherwise a hill or mountain would have to be very steep to block out a signal.

Highway overpasses accounted for 11% of the disruptions, but these disruptions had little effect on the overall quality of the broadcasts. The only serious disruption was a one-second dropout while a truck was directly under an overpass. This dropout was so brief that it did not cause a significant loss of intelligibility in voice communications.

Finally, buildings and similar structures accounted for about 8% of the disruptions. These were experienced mainly in large cities. Isolated buildings usually caused only brief disruptions. However, when the trucks were driving down city streets lined with tall buildings, they were unable to obtain satisfactory communications until they were driven to other streets.

9

(*continued*)

FIGURE 24.1

(*continued*)

The researchers present their conclusions in an easy-to-read bulleted list.

Each conclusion corresponds to one of the research objectives that the team named in the introduction.

LEARN MORE For advice on writing the conclusions section of an empirical research report, see "Conclusions" (page 408). In addition, see "How to Write Endings that Support Your Communication's Goals" (pages 140–143).

Section 4

CONCLUSIONS

This test supports three important conclusions:

- The Satellite Products Laboratory's system produces good-quality data and voice transmissions through the eleven-state region covered by the satellite's broadcast footprint.

- The most important factor limiting the success of transmissions is movement outside the satellite's broadcast footprint, which accurately defines the satellite's area of effective coverage.

- The system is sensitive to interference from certain kinds of objects in the line of sight between the satellite and the truck. These include trees, mountains and hills, overpasses, and buildings.

10

FIGURE 24.1

(*continued*)

Section 5

RECOMMENDATIONS

Based on this test, we believe that work should proceed as rapidly as possible to complete an operational system. In that work, the Satellite Products Laboratory should do the following things.

1. **Develop a new antenna designed specifically for use in communications between satellites and mobile radios.** Such an antenna would probably eliminate many of the disruptions caused by trees, the most common cause of poor transmissions.

2. **Define the configuration of satellites needed to provide service throughout our planned service area.** We are now ready to determine the number and placement in orbit of the satellites we will need to launch in order to provide service to our planned service area (48 contiguous states, Alaska, southern Canada). Because locations outside of the broadcast footprint of a satellite probably cannot be given satisfactory service, our satellites will have to provide thorough footprint coverage throughout all of this territory. Also, we should plan the satellites so that each will be at least 25° above the horizon throughout the area it serves; in that way we can almost entirely eliminate poor transmissions due to interference from mountains and hills.

11

The researchers present their overall recommendation first: Work on the project should continue.

They use bold type to highlight each of their other recommendations.

They then add the details that answer their readers' question, "What would be involved in the work you recommend?"

After making their last recommendation, the research team ends its report.

LEARN MORE For advice on writing the recommendations in an empirical research report, see "Recommendations" (page 410).

FIGURE 24.2

Writer's Guide for Revising Empirical Research Reports

Writer's Guide
REVISING EMPIRICAL RESEARCH REPORTS

Does your draft include each of the elements needed to create a report that your readers will find to be useful and persuasive? Remember that some elements of the superstructure may be unnecessary for your specific readers and purpose and that the elements may be organized in various ways.

INTRODUCTION

❏ Announces the topic of research presented in the report.

❏ Persuades readers that this research is important to them.

❏ Explains the relevance of the research to the organization's goals and, if appropriate, existing knowledge on the topic.

❏ States briefly your main conclusions and recommendations, if the readers would welcome or expect them at the beginning of the report.

❏ Provides background information readers will need or want.

❏ Forecasts the rest of your report, if this would help your readers.

OBJECTIVES OF RESEARCH

❏ Describes precisely what you were trying to find out through your research.

METHOD

❏ Tells the things your readers want to know about the way you obtained the facts and ideas presented in the report.

❏ Persuades the readers that this method would produce reliable results.

RESULTS

❏ Presents in clear and specific terms the things you found out.

❏ Includes material that is relevant to your readers and excludes material that isn't.

DISCUSSION (OFTEN PRESENTED ALONG WITH THE RESULTS)

❏ Interprets your results in a way that is useful to your readers.

CONCLUSIONS (OFTEN PRESENTED ALONG WITH THE DISCUSSION)

❏ Explains the significance—from your readers' viewpoint—of your results and generalizations about them.

❏ States the conclusions plainly.

RECOMMENDATIONS

❏ Tells what you think the readers should do.

❏ Makes the recommendations stand out prominently (for instance, by presenting them in a numbered list).

❏ Indicates how your recommendations are related to your conclusions and your readers' goals.

❏ Suggests some specific steps your readers might take to act on each of your recommendations, unless the steps will be obvious to the readers.

FIGURE 24.2

(Continued)

Writer's Guide

REVISING EMPIRICAL RESEARCH REPORTS

(Continued)

REASONING (SEE CHAPTER 9)

❏ States your claims and conclusions clearly.

❏ Provides sufficient evidence, from the readers' viewpoint.

❏ Explains, if necessary, the line of reasoning that links your facts and your claims.

❏ Addresses any counterarguments or objections that your readers are likely to raise at any point in your report.

❏ Avoids making false assumptions and overgeneralizing.

PROSE (SEE CHAPTERS 8 AND 10)

❏ Presents information in a clear, useful, and persuasive manner.

❏ Uses a variety of sentence structures and lengths.

❏ Flows in a way that is interesting and easy to follow.

❏ Uses correct spelling, grammar, and punctuation.

GRAPHICS (SEE CHAPTER 12)

❏ Included wherever readers would find them helpful or persuasive.

❏ Looks neat, attractive, and easy to read.

❏ Referred to at the appropriate points in the prose.

❏ Located where your readers can find them easily.

PAGE DESIGN (SEE CHAPTER 14)

❏ Looks neat and attractive.

❏ Helps readers find specific information quickly.

ETHICS

❏ Treats all the report's stakeholders ethically.

❏ Presents all information accurately and fairly.

Writing Reader-Centered Feasibility Reports

The workplace buzzes with decisions about future action. Among the more common are decisions about whether to do something new or in a new way. Although some of these decisions are easily made, others involve substantial changes. A company that manufactures wind turbines might want to know whether it should build a second factory several states away from its original location. A hospital might want to know whether it should replace its conventional telephone system with the Internet-based technology known as Voice over Internet Protocol (VoIP) that works through computers and computer lines rather than telephone switchboards and wiring.

To determine whether such large changes are practical and desirable, decision makers usually ask consultants or some of their own employees to conduct extensive research. This research is typically called a *feasibility study*. The study's results are presented in a *feasibility report.* You are likely to be asked to prepare feasibility reports many times in your career. In the simplest situation, you will be asked to compare two alternatives: continuing to do things the way they are done now (e.g., using a conventional phone system) or doing them in some specific new way (e.g., using VoIP). In more complex situations, you will be writing for decision makers who are considering three or more alternatives.

How Readers Use and Evaluate Feasibility Reports

Despite the extensive variety of situations in which you might write progress reports in your career, your decision-making readers will almost always want your reports to answer the following general questions. They will judge the quality of your reports based on the confidence they have in using the answers you provide as a basis for the decision they are making.

- **Why is it important for us to consider these alternatives?** In some cases, you need to tell decision makers why they have to make a choice in the first place. They may need a detailed explanation of the problem before they can appreciate the urgency of considering alternative courses of action. If they are already familiar with the problem, they may need to be reassured that you understand it in the same way they do.

- **What are the important features of our alternatives?** So that they can understand your detailed discussion of the alternatives, readers want you to highlight the key features of each one.

- **Are your criteria reasonable and appropriate?** To help your readers choose between alternative courses of action, you must evaluate the alternatives in terms of specific criteria. At work, people want these criteria to reflect the needs and aims of their organization.

- **Are your facts reliable?** Evaluating alternatives involves comparing facts about them against the criteria for judging them. Your readers want to know that they can rely on the accuracy of the facts you present.

- **How do the alternatives stack up against your criteria?** The heart of a feasibility study is your evaluation of the alternatives in terms of your criteria. Your readers want to know the results.

- **What overall conclusions do you draw about the alternatives?** Based on your detailed evaluation of the alternatives, you will reach general conclusions about the merits of each. Your readers need to know what your conclusions are because these overall judgments are the basis on which they will make their decision.

- **What do you think we should do?** Because of your expertise on the subject, your readers want you to tell what you recommend. In addition, they may desire suggestions about how to proceed if they decide to follow your recommendation.

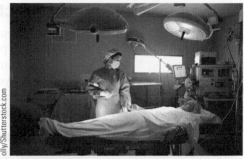

Superstructure for Feasibility Reports

The superstructure, or genre, for feasibility reports contains seven elements. Each element responds to one of the decision makers' seven basic questions.

Decisions and actions based on feasibility studies affect every aspect of our lives and environment.

Superstructure for Feasibility Reports	
Report Element	**Readers' Question**
Introduction	Why is it important for us to consider these alternatives?
Overview of alternatives	What are the important features of the alternatives?
Criteria	Are your criteria reasonable and appropriate?
Method	Are your facts reliable?
Evaluation	How do the alternatives stack up against your criteria?
Conclusions	What overall conclusions do you draw about the alternatives?
Recommendations	What do you think we should do?

Conducting Research for Feasibility Reports

Because your decision-making readers will be so deeply concerned that your feasibility reports provide them with thorough, credible information and analysis on which to base their decision, let's pause for a moment to discuss the research on which you will build your reports. Chapters 4 and 5 provide detailed advice about conducting research. Here are a few additional points that deserve special emphasis when you are preparing a feasibility report.

- **Gain a full understanding of your readers' criteria.** Begin by researching your readers. Learn how they will evaluate the alternatives. This knowledge can guide the rest of your research efforts, assuring that you gather the information your readers want. It also saves you from spending time gathering information that is irrelevant to your readers.

- **Investigate all implications of the alternatives being considered.** In particular, consider the actions that would be required to implement the alternatives. The hospital thinking of replacing its conventional phone system with Internet technology needs to consider many factors other than the features and cost of the two systems. Can its central computer server support the Internet system? What training in using that system would the doctors, nurses, and other staff members need? Who would provide it? Because the entire hospital couldn't be converted in one day, what needs to happen on the days when some departments have the new system and others have the old system?

- **Consult several kinds of sources.** Different kinds of sources can provide different perspectives and types of information. Chapter 4 suggests that you consult each of these four types (see "Identify the full range of sources and methods that may provide helpful information," pages 72–73).
    - Persons who would be affected by changing to each of the alternatives
    - Persons who would help to implement the alternatives if they were chosen
    - Other organizations or groups using the alternatives
    - Professional publications that may report on or evaluate the alternatives

- **Avoid bias in your information gathering.** Employees conducting feasibility studies sometimes start out favoring a particular alternative. They may then tend to seek out information that supports that choice, rather than gathering the full range of relevant information.

- **Consider creating tables for analyzing the information you've gathered.** Tables or matrixes can help you examine the alternatives the way your readers want to: point by point against the criteria. This approach matches the way your readers will use your report.

Organizing Feasibility Reports

As you read the advice below, keep in mind that a conventional superstructure is only a general plan. You must use your imagination and creativity to adapt it to specific situations. In particular, the five middle elements of a feasibility

---

TRY THIS Think of a change you would like to see at your school or in your community. Who could decide to bring about that change? If this person or group were to request a feasibility study related to the change, what criteria would the person or group use to evaluate the alternatives?

LEARN MORE For detailed advice about conducting research, see Chapter 4 ("Conducting Reader-Centered Research: Gathering, Analyzing, and Thinking Critically About Information") and Chapter 5 ("Using Six Reader-Centered Research Methods").

LEARN MORE For more on superstructures, go to "Adapt an appropriate superstructure or other pattern familiar to your reader" (pages 112–113).

report—the ones between the Introduction and Conclusion—may be organized in a variety of ways. To choose the best organization for a particular report, determine which will be easiest for your readers to use.

Sometimes the most effective way to organize is to create one section for each of the five middle elements of the superstructure. Figure 25.1 shows how Magnus used this organization when reporting on the feasibility of installing a heat-pump system to replace the conventional heating and air-conditioning systems at his employer's building. This organization works best when the readers have only one decision to make and the writer used only a single method to gather the information about the alternatives.

Sometimes the order of elements in the superstructure provides an effective organizational pattern.

A second organization can be more effective for readers who have two or more decisions to make. Megan and Rajiv addressed such readers in their report on the feasibility of switching to an Internet telephone system. Their readers had to consider not only the system's features, but also who might install it, how training could be handled, and so on. To help readers make these separate decisions, you can treat each decision in its own section or chapter that discusses the criteria, methods, evaluations, conclusions, and recommendations relevant to that decision. Figure 25.2 (page 433) shows the organization of Megan and Rajiv's report.

If your readers have several decisions to make, they will probably find it easiest to use a report organized around those decisions.

A third organization works well when your readers will make a single decision, but you used a different research method for gathering information relevant to each of the criteria. In such cases, readers find it helpful to have a separate chapter or section devoted to each criterion. Each chapter would include the research method, evaluations and conclusion, and recommendation related to the specific criterion the chapter addresses. Figure 25.3 (page 434) shows how Allen used this pattern when reporting on the feasibility of using a new ingredient to make paper at one of his employer's paper mills.

In other cases, organizing around the criteria helps readers.

To organize in these reader-centered ways, you need to remember that the superstructure for feasibility reports represents the elements of the report, not their outline. The elements may be combined and rearranged in many ways to meet your readers' needs.

Crafting the Major Elements of a Feasibility Report

All the detailed advice that Chapters 6 through 16 provide about drafting workplace communications applies to feasibility reports. The following sections tell how you can apply some of this advice most productively to the seven report elements, regardless of how they are organized.

Introduction

In the introduction to a feasibility report, you should answer your readers' question, "Why is it important for us to consider these alternatives?" The most persuasive way to answer this question is to identify the problem your feasibility report will help your readers solve or the goal it will help them achieve: to reduce the number of rejected parts, to increase productivity, and so on. Beyond that, your introduction should announce the alternative courses of action you studied and tell generally how you investigated them.

LEARN MORE For additional advice on writing an introduction, see Chapter 7.

Consider, for example, the way Allen, a process engineer in a paper mill, wrote the introduction of a feasibility report he prepared. (Allen's entire report appears in Figure 25.5, pages 441–446). Allen was asked to study the feasibility of substituting one ingredient for another in the furnish for one of the papers the mill produces (*furnish* is the combination of ingredients used to make the pulp for paper).

Problem

At present we rely on the titanium dioxide (TiO_2) in our furnish to provide the high brightness and opacity we desire in our paper. However, the price of TiO_2 has been rising steadily and rapidly for several years. We now pay roughly $1400 per ton for TiO_2, or about 70¢ per pound.

FIGURE 25.1

Feasibility Report Organized by Devoting One Section to Each of the Seven Elements in the Superstructure for Feasibility Reports

The organization that Magnus created for this report matches the order of the elements in the superstructure for feasibility reports.

Introduction
Magnus created a brief, one-paragraph introduction focused on the background for his feasibility study and report.

Overview of alternatives
To assure that his readers understood each of the alternatives, Magnus provided detailed explanations, including schematic diagrams of each alternative.

Criteria
Magnus used only a few sentences to describe the criteria.

Methods
He used the same methods for gathering information related to all of the criteria.

Evaluation
In the section for each criterion, Magnus presented and discussed the relevant information about all three types of systems.

Conclusions
He stated each of his major conclusions and cited the specific facts discussed in the evaluation section that supported the conclusions.

Recommendations
Magnus highlighted the main actions he believed his employer should take.

LEARN MORE For information on patterns for organizing comparisons, see "Comparing Alternatives" (pages 152–154).

Replacing the Building's Heating and Air-Conditioning System with
a Heat-Pump System: A Feasibility Study

I. Introduction

    A. The heating and air-conditioning system in our building needs to be replaced

    B. I was asked to study the feasibility of changing to a heat-pump system

II. Types of heat pumps for use in office buildings

    A. Air source (exchanges heat with air outside the building)

    B. Geothermal (exchanges heat with ground)

    C. Heat absorption (runs on gas to exchange heat with outside air)

III. Criteria

    A. Efficiency in our climate (different systems work better in different climates)

    B. Amount of damage installation would cause to our wooded property

    C. Cost (installation and operation)

IV. Research methods

    A. Consulted credible sources in print and online

        i. U.S. government resources on green technologies

        ii. Engineering journals

    B. Interviewed knowledgeable people

        i. Representatives of companies that sell heat-pump systems in our area

        ii. Owners of buildings in the area that use heat pumps

V. Evaluation

    A. Effectiveness in our area's climate

    B. Amount of damage installation would cause to our property

    C. Cost (installation and operation)

VI. Conclusions

    A. Over 6 years, total cost for all heat pumps is lower than standard heat and air

    B. Geothermal heat pumps are most effective here, though more costly and most potentially damaging to our property

VII. Recommendations

    A. Select a geothermal system

    B. Use the vertical rather than horizontal installation of the pipes to minimize damage to landscape around the building

    C. Next step: Invite proposals from companies that install heat-pump systems

Some mills are now replacing some of the TiO_2 in their furnish with silicate extenders. Because the average price for silicate extenders is only $500 per ton, well under half the cost of TiO_2, the savings are very great.

Possible solution

To determine whether we could enjoy a similar savings for our 30-pound book paper, I have studied the physical properties, material handling requirements, and cost of two silicate extenders, Tri-Sil 606 and Zenolux 26 T.

What Allen did to investigate the possible solution

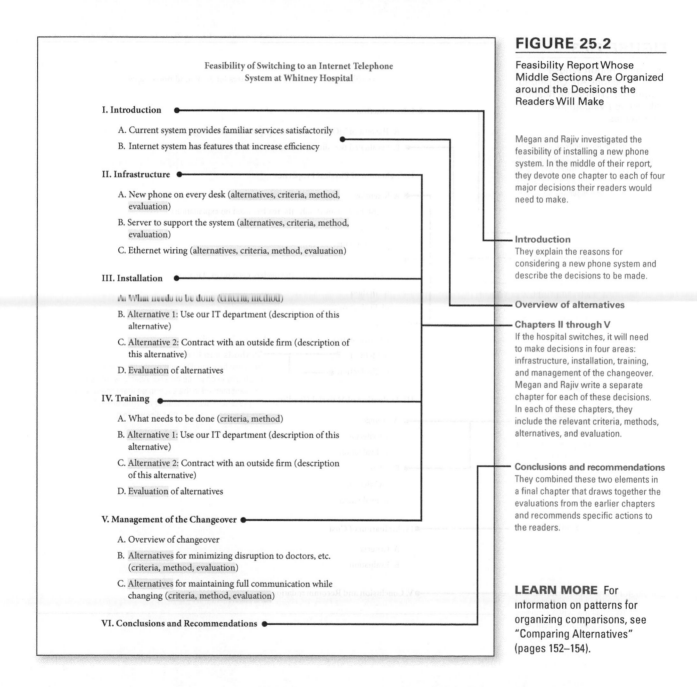

FIGURE 25.2

Feasibility Report Whose Middle Sections Are Organized around the Decisions the Readers Will Make

Feasibility of Switching to an Internet Telephone System at Whitney Hospital

I. Introduction

   A. Current system provides familiar services satisfactorily

   B. Internet system has features that increase efficiency

II. Infrastructure

   A. New phone on every desk (alternatives, criteria, method, evaluation)

   B. Server to support the system (alternatives, criteria, method, evaluation)

   C. Ethernet wiring (alternatives, criteria, method, evaluation)

III. Installation

   A. What needs to be done (criteria, method)

   B. Alternative 1: Use our IT department (description of this alternative)

   C. Alternative 2: Contract with an outside firm (description of this alternative)

   D. Evaluation of alternatives

IV. Training

   A. What needs to be done (criteria, method)

   B. Alternative 1: Use our IT department (description of this alternative)

   C. Alternative 2: Contract with an outside firm (description of this alternative)

   D. Evaluation of alternatives

V. Management of the Changeover

   A. Overview of changeover

   B. Alternatives for minimizing disruption to doctors, etc. (criteria, method, evaluation)

   C. Alternatives for maintaining full communication while changing (criteria, method, evaluation)

VI. Conclusions and Recommendations

Megan and Rajiv investigated the feasibility of installing a new phone system. In the middle of their report, they devote one chapter to each of four major decisions their readers would need to make.

Introduction
They explain the reasons for considering a new phone system and describe the decisions to be made.

Overview of alternatives

Chapters II through V
If the hospital switches, it will need to make decisions in four areas: infrastructure, installation, training, and management of the changeover. Megan and Rajiv write a separate chapter for each of these decisions. In each of these chapters, they include the relevant criteria, methods, alternatives, and evaluation.

Conclusions and recommendations
They combined these two elements in a final chapter that draws together the evaluations from the earlier chapters and recommends specific actions to the readers.

LEARN MORE For information on patterns for organizing comparisons, see "Comparing Alternatives" (pages 152–154).

Generally, the introduction to a long feasibility report (and most short ones) should also include a preview of the main conclusions and, perhaps, the major recommendations. Allen included his major conclusion.

Allen's main point

I conclude that one of the silicate extenders, Zenolux 26 T, looks promising enough to be tested in a mill run.

FIGURE 25.3

Feasibility Report Whose Middle Sections Are Organized around the Criteria the Readers Will Use to Evaluate the Alternatives

In this feasibility report, Allen organizes the central chapters around the criteria his readers will use to decide whether to change the extender in the paper mill where he works.

Introduction

Overview of alternatives

First three criteria
For each of the first three criteria, Allen describes the relevant criterion, method, and evaluation.

Fourth and fifth criteria
For the fourth and fifth criteria, the methods of gathering information are obvious, so Allen doesn't report them.

Sixth criterion

Conclusions and recommendation
Because these two elements are so brief, Allen combined them in one chapter.

Feasibility of Using Silicate Extenders for 36-Pound Book Paper

I. Introduction

A. The cost of TiO_2 is rising; some paper mills are using silicate extenders

B. I evaluated two silicate extenders: Zenolux and Tri-Sil

II. Evaluation of Physical Properties

A. Retention

i. Criterion (named only: the readers need no explanation)

ii. Method

iii. Evaluation

B. Opacity

i. Criterion (named only; the readers need no explanation)

ii. Method

iii. Evaluation

C. Brightness

i. Criterion (named only; the readers need no explanation)

ii. Method

iii. Evaluation

Methods and Evaluation
Because he used different methods to gather information related to each of the criteria, Allen described the relevant method in the section on that criterion. Likewise, he presented his evaluations in terms of each criterion in the chapter or section that discussed it.

III. Evaluation of Material Handling

A. Storage

i. Criterion

ii. Evaluation

B. Safety

i. Criterion

ii. Evaluation

IV. Evaluation of Cost

A. Criteria

B. Evaluation

V. Conclusion and Recommendation

As another example, consider the way Ellen wrote the introduction of a feasibility report she prepared for the board of directors of the bank where she works (see Figure 25.4). Ellen was asked to evaluate the feasibility of opening a new branch in a particular suburban community. She began by announcing the topic of her report.

> This report discusses the feasibility of opening a branch office of Orchard Bank in Rolling Knolls, Tennessee.

Ellen's introduction

FIGURE 25.4

How Ellen Used the Superstructure for Feasibility Reports

In her report on the feasibility of opening a new branch of her bank, Ellen mixes in several ways the elements of the superstructure for feasibility reports.

Feasibility of Opening a Branch of Orchard Bank in Rolling Knolls, Tennessee

I. Introduction

   A. Orchard Bank has built its success on carefully selecting locations for its offices

   B. A new branch office in Rolling Knolls has been suggested

   C. This report evaluates the proposed location

II. Proposed Location

   A. The branch would serve Rolling Knolls and Pickett

   B. The site selected is Lot P1-C in Rolling Knolls

III. Market Analysis

   A. This is a prosperous, growing area

   B. The population has two distinct groups

      1. Longtime residents of the area

      2. Newcomers, mainly young commuters

   C. The competition

      1. Local offices of banks and savings and loans

      2. Nashville banks used by many commuters

   D. Potential marketing strategies

      1. A two-pronged strategy for appealing to this two-part market

      2. The strategy would require Orchard to do some things it hasn't done before

IV. Financial Analysis

   A. The branch could become profitable after three years if a good marketing plan is established and followed

   B. Estimated deposits

   C. Estimated income and expenses

V. Fixed Asset Expenditures

VI. Conclusions and Recommendations

   A. Rolling Knolls offers opportunity, but also risk

   B. Recommendations

      1. Hire marketing firm to help assess the situation more precisely

      2. Take an option on the lot in Rolling Knolls

Introduction
Ellen explains the importance of considering the feasibility of opening a new branch.

Overview of alternatives
For Ellen's study there are two alternatives: Open a new branch or don't open it. Because her readers already know what it is like to have no new branch, she describes only what it would be like to open one.

Evaluation of the alternatives
Ellen divides her evaluation of the alternative of opening a new branch into three sections, one for each of her criteria:
• Marketability
• Ability to make a profit
• Costs

Criteria
Ellen identifies each of her criteria at the beginning of the appropriate section. For instance, at the beginning of her section on market analysis, she discusses her criteria related to marketability.

Method
Ellen explains each of her methods of gathering facts at the beginning of each analysis. For instance, she describes how she learned about the prosperity and growth of the Rolling Knolls area at the beginning of Section II. A.

Conclusions Recommendations
Ellen combines these two elements of the superstructure in her final section.

Then, after giving a sentence of background information about the source of the bank's interest in exploring this possibility, Ellen emphasized the importance of conducting a feasibility study.

> In the past, Orchard Bank has approached the opening of new branches with great care, which is undoubtedly a major reason that in the twelve years since its founding it has become one of Tennessee's most successful small, privately owned financial institutions.

Ellen also included her major conclusion.

> Overall, the Rolling Knolls location offers an enticing opportunity but would present Orchard Bank with some challenges it has not faced before.

She ended her introduction with a brief summary of her major recommendation.

> We should proceed carefully.

The introduction of a feasibility report is often combined with one or more of the other six elements, such as a description of the criteria, a discussion of the method of obtaining facts, or an overview of the alternatives. It may also include the various kinds of background, explanatory, and forecasting information that may be found in the beginning of any technical communication (see Chapter 7).

Criteria

Criteria are the standards used to evaluate the alternative courses of action your readers are considering. One of the most important points about writing successful feasibility studies is to be sure that you present the criteria in ways that demonstrate that you understand what your *readers'* criteria are. For instance, Ellen phrased her criteria to reflect the bank's goals, such as the possibility of attracting depositors away from competitors and the likelihood that profits at the branch would exceed the expenses of operating it. If she had found that the proposed branch failed to meet any of these criteria, she would have concluded that opening it was not feasible.

Two Ways of Presenting Criteria

There are two common ways of telling your readers what the criteria are.

- **Devote a separate section to identifying and explaining them.** Writers often do this in long reports or in reports in which the criteria themselves require extended explanation.

- **Integrate your presentation of them into other elements of the report.** Allen did this in the following sentence from the third paragraph of his introduction:

Allen names his three criteria.

> To determine whether we could enjoy a similar savings for our 30-pound book paper, I have studied the physical properties, material handling requirements, and cost of two silicate extenders, Tri-Sil 606 and Zenolux 26 T.

For each of the general criteria named in this sentence, Allen had specific criteria, which he described when he discussed his methods and evaluations. For instance, at the beginning of his discussion of the physical properties of the two extenders, he named the three properties he evaluated.

Importance of Presenting Criteria Early

Whether you present your criteria in a separate section or integrate them into other sections, you should introduce them early in your report. There are three good reasons for doing this. First, because your readers know that the validity of your conclusion depends on the criteria you use to evaluate the alternatives, they will want to evaluate the criteria themselves. They will ask, "Did you take into account all the considerations relevant to this decision?" and "Are the standards you are applying reasonable in these circumstances?"

Second, your discussion of the criteria tells readers a great deal about the scope of your report. Did you restrict yourself to technical questions, for instance, or did you also consider relevant organizational issues such as profitability and management strategies?

The third reason for presenting your criteria early is that your discussion of the alternative courses of action will make much more sense to your readers if they know in advance the criteria by which you evaluated the alternatives.

Sources of Your Criteria

You may wonder how to come up with the criteria you will use in your study and report. Often, the person who asks you to undertake a study will tell you what criteria to apply. In other situations, particularly when you are conducting a feasibility study that requires technical knowledge that you have but your readers don't, your readers may expect you to identify the relevant criteria for them.

In either case, you are likely to refine your criteria as you conduct your study. As you draft you must think in detail about the information you have obtained and decide how best to evaluate it.

You may refine your criteria while you are writing your report.

Four Common Types of Criteria

As you develop your criteria, you may find it helpful to know that, at work, criteria often address one or more of the following questions.

- **Will this course of action really do what's wanted?** This question is especially common when the problem is a technical one: Will this reorganization of the department really improve the speed with which we can process loan applications? Will the new type of programming really reduce computer time?

- **Can we implement this course of action?** Even though a particular course of action may work technically, it may not be practical. For example, it may require overly extensive changes in operations, equipment, or materials that are not readily available, or special skills that employees do not possess.

- **Can we afford it?** Cost can be treated in several ways. You may seek an alternative that costs less than some fixed amount or one that will save enough to pay for itself in a fixed period (for example, two years). Or you may simply be asked to determine whether the costs are reasonable.

- **Is it desirable?** Sometimes a solution must be more than effective, implementable, and affordable. Many otherwise feasible courses of action are rejected because they create undesirable side effects.

Ultimately, your selection of criteria for a particular feasibility study will depend on the problem at hand and on the professional responsibilities, goals, and values of the people who will use your report. In some instances, you will need to deal only with criteria related to the question, "Does it work?" At other times, you might need to deal with all the criteria mentioned above, plus others. No matter what your criteria, however, announce them to your readers before you discuss your evaluation.

Method

By explaining how you obtained your facts, you answer your readers' question, "Are your facts reliable?" That is, by showing that you used reliable methods, you assure your readers that your facts form a sound basis for decision making.

The source of your facts will depend on the nature of your study—library research, calls to manufacturers, interviews, meetings with other experts in your organization, surveys, laboratory research, and the like.

How much detail should you provide about your methods? That depends on your readers and the situation, but in every case your goal is to say enough to satisfy your readers that your information is trustworthy. For example, Ellen used some fairly technical procedures to estimate the amount of deposits that Orchard Bank could expect from a new branch in Rolling Knolls. However, because those procedures are standard in the banking industry and well known to her readers, she did not need to explain them in detail.

In contrast, Allen provided very specific details about his methods of testing the extenders. Even a small mistake could produce inaccurate results that, in turn, could lead the paper mill to make a very expensive error. He knew, therefore, that his readers would want to review for themselves each step in his test procedure.

The best place for describing your methods depends partly on how many techniques you used. If you used only one or two—say, library research and interviews—you might describe each in a separate paragraph or section near the beginning of your report, perhaps in the introduction. On the other hand, if you used several different techniques, each pertaining to a different part of your analysis, you might describe each of them at the point where you present and discuss the results you obtained.

Of course, if your methods are obvious, you may not need to describe them at all. You must always be sure, however, that your readers know enough about your methods to accept your facts as reliable.

Overview of Alternatives

To understand your detailed evaluation of the alternatives, your readers must first understand what the alternatives are. Often, you can foster this understanding with a short explanation, perhaps only a sentence long. For instance, as a consultant to a chain of convenience stores you would need only a few words to enable your readers to understand whether increasing starting pay would help your company attract and retain skilled store managers.

However, you may sometimes need to explain the alternatives in detail. Although the hospital executives were familiar with what conventional phones could do, they didn't know anything about the capabilities of an Internet telephone system, such as their ability to forward voicemail messages to the email account of the person called. Consequently, Megan and Rajiv described the Internet system in detail, highlighting features conventional telephones don't have that could help doctors, nurses, and other hospital personnel.

Evaluation of the Alternatives

The heart of a feasibility report is your evaluation of the alternatives you examined. As explained earlier in this chapter (see "How Readers Use and Evaluate Feasibility

Reports," (pages 428–429), a primary consideration is organizing your evaluation in the way that makes it easiest for your readers to use it as they make their decisions. Three additional strategies can also assist your readers.

Put Your Most Important Points First

Chapter 7's advice to begin each segment of your communication with the most important information applies to segments in which you evaluate alternatives. By presenting the most important information first, you save your readers the trouble of trying to figure out what generalizations they should draw from the details you are presenting. For instance, in her report to the bank, Ellen begins one part of her evaluation of the Rolling Knolls location like this.

LEARN MORE See Chapter 7.

> The proposed Rolling Knolls branch office would be profitable after three years if Orchard Bank successfully develops the type of two-pronged marketing strategy outlined in the preceding section.

Beginning of Ellen's evaluation

Ellen then spends two pages discussing the estimates of deposits, income, and expenses that support her overall assessment.

Similarly, Allen begins one part of his evaluation of the silicate extenders in this way.

> With respect to material handling, I found no basis for choosing between Zenolux and Tri-Sil.

Beginning of Allen's evaluation

Allen then spends two paragraphs reporting the facts he has gathered about the physical handling of the two extenders.

Use Graphics

Because tables, graphs, charts, and other graphics can display features of alternatives side by side, they help compare alternatives in a glance.

Dismiss Obviously Unsuitable Alternatives

Lengthy discussions of clearly unsuitable alternatives provide no benefit to your readers. They only increase the time required to read your report.

If you discover that an alternative fails to meet one or more of the critical criteria, explain this fact briefly, providing only the details needed to persuade your readers that you were right to drop the alternative from consideration.

Provide this explanation in the introduction (when you are talking about your report's scope) or in the overview of the alternatives. If you delay the explanation to a later section, the mistaken thought that you have missed an important alternative may distract your readers as they read the rest of your report.

Conclusions

Your conclusions constitute your overall assessment of the feasibility of the alternative courses of action you studied. You might present your conclusions in two or three places in your report. You should certainly mention them in summary form near the beginning. If your report is long (say, several pages), you might also remind your readers of your conclusions at the beginning of the evaluation segment. Finally, you should provide a detailed discussion of your conclusions in a separate section following your evaluation of the alternatives.

Recommendations

It is customary to end a feasibility report by answering the decision makers' question, "What do you think we should do?" Because you have investigated and thought about the alternatives so thoroughly, your readers will place special value on your recommendations. Depending on the situation, you might need to take only a single sentence or many pages to present your recommendations.

Sometimes, your recommendations will pertain directly to the course of action you studied: "Do this" or "Don't do this." At other times, you may perform a preliminary feasibility study to determine whether a certain course of action is promising enough to warrant a more thorough investigation. Ellen's report about opening a new bank office is of that type. She determined that there was a substantial possibility of making a profit with the new branch, but she felt the need for expert advice before making a final decision. Consequently, she recommended that the bank hire a marketing agency to evaluate the prospects.

Sometimes, you may discover that you were unable to gather all the information you needed to make a firm recommendation. Perhaps your deadline was too short or your funds too limited. Perhaps you uncovered an unexpected question that needs further investigation. In such situations, you should point out the limitations of your report and let your readers know what else they should find out so they can make a well-informed decision.

Sample Feasibility Report

It is much easier to understand writing advice if you see sample communications that follow that advice. Figure 25.5 shows the feasibility report Allen wrote to help his employer decide whether to use a silicate extender at one of its paper mills. Allen's outline for this report is shown in Figure 25.3 (page 434).

Writer's Guides and Other Resources

Figure 25.6 (pages 447–448) presents a "Writer's Guide for Revising Feasibility Reports" that you can use in your course and on the job.

Note to the Instructor For an assignment involving a feasibility report, you can ask students to write on research they are conducting in their majors, or you can adapt the "Formal Report or Proposal" project given in MindTap. To help students reflect for transfer, ask them to complete one of the exercises in Appendix B when they finish their project.

REGENCY INTERNATIONAL PAPER COMPANY

MEMORANDUM

FROM: Allen Hines, Process Engineer

TO: Jim Shulmann, Senior Engineer

DATE: December 13, 2017

SUBJECT: **FEASIBILITY OF USING SILICATE EXTENDERS FOR 30-POUND BOOK PAPER**

Summary

I have investigated the feasibility of using a silicate extender to replace some of the TiO_2 in the furnish for our 30-pound book paper. Because the cost of the extenders is less than half the cost of TiO_2, we could enjoy a considerable savings through such a substitution.

The tests show that either one of the two extenders tested can save us money. In terms of retention, opacity, and brightness, Zenolux is more effective than Tri-Sil. Consequently, it can be used in smaller amounts to achieve a given opacity or brightness. Furthermore, because of its better retention, it will place less of a burden on our water system. With respect to handling and cost, the two extenders are roughly the same.

I recommend a trial run with Zenolux.

Introduction

At present we rely on the titanium dioxide (TiO_2) in our furnish to provide the high brightness and opacity we desire in our paper. However, the price of TiO_2 has been rising steadily and rapidly for several years. We now pay roughly $1400 per ton for TiO_2, or about 70¢ per pound.

FIGURE 25.5

Feasibility Report

In this report, Allen reports on the feasibility of replacing one chemical with another in the "furnish" his employer uses to manufacture paper.

Allen includes a precise and informative subject line so his readers know exactly what his report is about.

Allen uses a 125-word summary of the entire report to enable his readers to understand the main points immediately.

In his summary, Allen emphasizes the content that will most interest his readers: his conclusion and recommendation.

In the first sentence of his introduction, Allen presents the background information his readers will need in order to understand why his report is important to them.

Allen identifies the specific problem his report addresses.

(continued)

FIGURE 25.5
(*continued*)

Allen tells his readers about a possible solution that other paper mills have used: Replace TiO$_2$ with a silicate extender.

He describes what he did to investigate the feasibility of the possible solution.

Allen states his major conclusion and recommendation.

He identifies his criteria for evaluating the physical properties of the silicate extenders.

Allen helps his readers find key information by briefly stating the overall result of his tests of physical properties before going into the details of the tests.

Allen explains the importance of testing retention.

To build his readers' confidence in his tests, Allen gives precise details about his method for testing retention.

For his evaluation, he combines his test data with his discussion of them. In this paragraph, he gives exact percentages.

Allen introduces his method for testing opacity.

Silicate Extenders   Page 2

Some mills are now replacing some of the TiO$_2$ in their furnish with silicate extenders. Because the average price for silicate extenders is only $500 per ton, well under half the cost of TiO$_2$, the savings are very great.

To determine whether we could enjoy a similar savings for our 30-pound book paper, I have studied the physical properties, material handling requirements, and cost of two silicate extenders, Tri-Sil 606 and Zenolux 26 T. I conclude that one of the silicate extenders, Zenolux 26 T, looks promising enough to be tested in a mill run.

Evaluation of Physical Properties

The three physical properties I evaluated are retention, opacity, and brightness. In all three areas, Zenolux is superior.

Retention

As with any ingredient in our furnish, we must be concerned with the proportion of a silicate extender that will be retained in the paper and the proportion that will be left in the water, where it is wasted and may cause problems in our water system. To test retention of the two silicate extenders, I made two dozen handsheets, each containing the equivalent of 3 grams of oven-dried pulp and 2 grams of oven-dried extender. By weighing the finished handsheets, I determined how much silicate extender had been lost from each.

The tests showed that the average retention for Zenolux was 75%, whereas the average retention for Tri-Sil was 51%. Higher retention should result in higher opacity and brightness because more particles remain in the furnish to prevent clumping of the TiO$_2$.

Opacity

To determine the effectiveness of each extender in preventing light from passing through the paper, I conducted a two-stage test of opacity. First, I investigated the opacity of TiO$_2$, Tri-Sil, and Zenolux when each is used alone. To do that, I made the following sets of handsheets:

Courtesy of Allen J. Hines

FIGURE 25.5

(*continued*)

Silicate Extenders   Page 3

- 8 handsheets containing each of the following loadings of TiO_2: 2%, 4%, 6%, 8%, 10%, 12%, 14%, and 16%.
- 8 handsheets containing each of the same loadings of Tri-Sil.
- 8 handsheets containing each of the same loadings of Zenolux.

I stored the handsheets at the standard conditions of 50% relative humidity and 23°C for 24 hours. Then, I found the TAPPI opacity of each handsheet. The results, given for the average opacity at each loading, are shown in Figure 1. Again, Zenolux is superior to Tri-Sil.

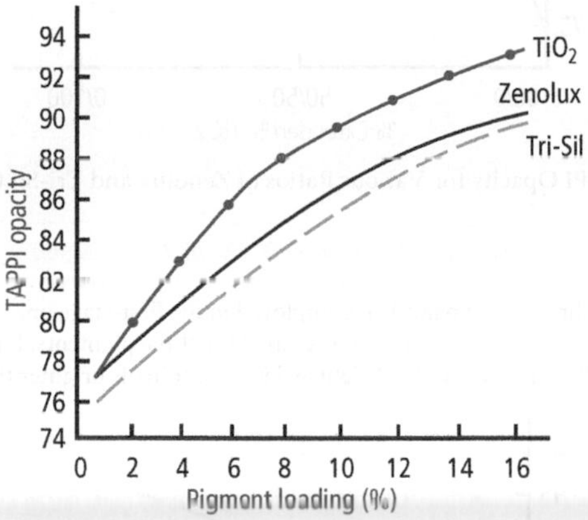

Figure 1. TAPPI Opacity for Various Loadings of TiO_2, Zenolux, and Tri-Sil.

In the second stage of the opacity test, I made two additional sets of handsheets:

- 5 handsheets with each of the following pigment loadings: 100% TiO_2 and 0% Tri-Sil; 75% TiO_2 and 25% Tri-Sil; 50% TiO_2 and 50% Tri-Sil; 25% TiO_2 and 75% Tri-Sil; and 0% TiO_2 and 100% Tri-Sil.
- 5 handsheets with each of the same proportions of TiO_2 and Zenolux.

As in his section on testing physical properties, Allen includes precise information about his testing method in order to persuade his readers that he used a sound method. To make reading easy, he presents some details in a bulleted list.

Allen refers his readers to a graph that presents his test data in a more understandable way than sentences would. However, he also states the main point he wants his readers to derive from the graph: Zenolux is superior.

He places the graph at the point in the report when his reader will want to look at it.

Again, Allen describes his method in detail, and again he uses a graph to present the test data in a reader-centered way (next page). He also states the main point he wants his readers to see when they look at the graph.

(*continued*)

FIGURE 25.5

(*continued*)

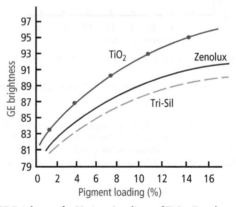

Silicate Extenders   Page 4

As Figure 2 shows, Zenolux is again superior.

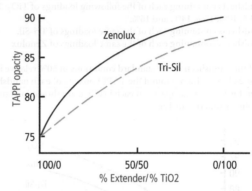

Figure 2. TAPPI Opacity for Various Ratios of Zenolux and Tri-Sil to TiO_2.

Brightness

●   Using the three sets of handsheets employed in the first-stage opacity tests, I calculated the GE brightness achieved by each of the three pigments. Figure 3 shows the results. Although not as bright as TiO_2, Zenolux is brighter than Tri-Sil.

Figure 3. GE Brightness for Various Loadings of TiO_2, Zenolux, and Tri-Sil.

When describing his tests of brightness, Allen uses the same reader-centered strategies he employed in the section on opacity:

• He briefly, but precisely, describes his method.

• He presents his test data in graphs.

• He states in a sentence the main point readers should take away from each graph.

LEARN MORE To learn the design guidelines Allen followed to create these reader-centered line graphs, see "Writer's Tutorial: Line Graphs" (pages 244–245).

Courtesy of Allen J. Hines

FIGURE 25.5

(*continued*)

Silicate Extenders   Page 5

Using the two sets of handsheets employed in the second-stage opacity tests, I examined the brightness of each of the ratios of TiO_2 to the extenders. The test data, shown in Figure 4, indicate that, as expected, TiO_2 with Zenolux is brighter than TiO_2 with Tri-Sil.

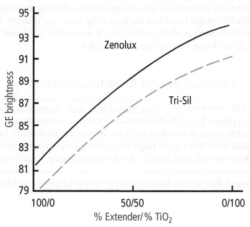

Figure 4. GE Brightness for Various Ratios of Zenolux and Tri-Sil to TiO_2.

Evaluation of Material Handling

With respect to material handling, I found no basis for choosing between Zenolux and Tri-Sil. Both are available from suppliers in Chicago. Both are available in dry form in bags and as a slurry in bulk hopper cars. Zenolux slurry is also available in 20,000-gallon tank cars, which provide a small savings, but we do not have the storage facility needed to receive it in this way.

Similarly, both silicate extenders are quite safe. Both are 100% pigment and both are chemically stable. Neither poses a hazard of fire or explosion and neither is hazardous when mixed with any other substance, whether liquid, solid, or gas. Both create the same effects in the event of overexposure: dehydration of the respiratory tract, eyes, and skin. These effects can be avoided by purchasing the extenders in slurry rather than dry form.

In the first sentence of his section on material handling, Allen states his main finding: There are no significant differences among the three extenders he tested.

Allen then provides details about each aspect of material handling that he investigated. In this way, he assures his readers that he didn't overlook any important consideration.

Courtesy Allen J. Hines

(*continued*)

FIGURE 25.5

(*continued*)

Allen realized that his readers will want to know how much a conversion to one of the silicate extenders will cost. Therefore, he provides not only the cost per ton of each item, but also considers how much of each the paper mill would need to buy in order to produce good-quality paper. In reaching this conclusion, Allen is building on evaluation of opacity and brightness.

Allen ends his report with the following items:

- His conclusion
- The evaluations that support his conclusion
- His recommendation

Silicate Extenders   Page 6

Evaluation of Costs

Either silicate extender could save us money, and they both cost about the same: $421 per ton for Zenolux and $391 per ton for Tri-Sil. The $30 difference is only 7% of the cost of Zenolux. This savings may be offset by the larger amounts of Tri-Sil needed to achieve the same brightness and opacity. Thus, based on the information available at this time, it seems that cost is not a factor that would help us choose between the two extenders.

Conclusion and Recommendation

Although the two extenders are comparable in many ways, Zenolux appears to be the better choice for us. Both extenders could save us money and both are about the same in terms of handling and cost. However, Zenolux can be used in smaller amounts to obtain a given brightness and opacity. Furthermore, its higher retention rate will place less of a burden on our aging water system.

I recommend that we test Zenolux on our 30-pound paper machine.

Courtesy of Allen J. Hines

FIGURE 25.6

Writer's Guide for Revising Feasibility Reports

Writer's Guide
REVISING FEASIBILITY REPORTS

Does your draft include each of the elements needed to create a report that your readers will find to be useful and persuasive? Remember that some elements of the superstructure may be unnecessary for your specific readers and purpose and that the elements may be organized in various ways.

INTRODUCTION

❑ Identifies the action or alternatives you investigated.

❑ Tells (or reminds) your readers why you conducted this study.

❑ Persuades readers that the study is important to them.

❑ States briefly your main conclusions and recommendations, if the readers would welcome or expect them at the beginning of the report.

❑ Provides background information the readers will need or want.

❑ Forecasts the rest of your report, if this would help your readers.

OVERVIEW OF ALTERNATIVES

❑ Presents a general description of each alternative.

CRITERIA (OFTEN INCLUDED IN THE INTRODUCTION)

❑ Identifies the standards by which the action or alternatives were evaluated.

❑ Focuses on criteria that are important to the reader.

METHODS

❑ Tells the things your readers want to know about the way you obtained the facts and ideas presented in the report.

❑ Persuades the readers that these methods would produce reliable information and data.

EVALUATION (USUALLY THE LONGEST PART OF A FEASIBILITY REPORT)

❑ Evaluates the action or alternatives in terms of the criteria.

❑ Presents the facts and evidence that support each evaluative statement.

CONCLUSIONS (OFTEN PRESENTED ALONG WITH THE DISCUSSION)

❑ Explains the significance—from your readers' viewpoint—of your facts and generalizations about them.

❑ States the conclusions plainly.

RECOMMENDATIONS

❑ Tells which course of action or alternative you recommend.

❑ Makes the recommendations stand out prominently (for instance, by presenting them in a numbered list).

❑ Indicates how your recommendations are related to your conclusions and your readers' goals.

❑ Suggests some specific steps your readers might take to act on each of your recommendations, unless the steps will be obvious to them.

(continued)

FIGURE 25.6
(*continued*)

Writer's Guide
REVISING FEASIBILITY REPORTS
(*continued*)

REASONING (SEE CHAPTER 9)

❏ States your claims and conclusions clearly.

❏ Provides sufficient evidence, from the readers' viewpoint.

❏ Explains, if necessary, the line of reasoning that links your facts and your claims.

❏ Addresses any counterarguments or objections that your readers are likely to raise at any point in your report.

❏ Avoids making false assumptions and overgeneralizing.

PROSE (SEE CHAPTERS 8 AND 10)

❏ Presents information in a clear, useful, and persuasive manner.

❏ Uses a variety of sentence structures and lengths.

❏ Flows in a way that is interesting and easy to follow.

❏ Uses correct spelling, grammar, and punctuation.

GRAPHICS (SEE CHAPTER 12)

❏ Included wherever readers would find them helpful or persuasive.

❏ Look neat and attractive, and are easy to read.

❏ Referred to at the appropriate points in the prose.

❏ Located where your readers can find them easily.

PAGE DESIGN (SEE CHAPTER 14)

❏ Looks neat and attractive.

❏ Helps readers find specific information.

ETHICS

❏ Treats all the report's stakeholders ethically.

❏ Presents all information accurately and fairly.

26

Writing Reader-Centered Progress Reports

A progress report is a report on work you have begun but not yet completed. The typical progress report is one of a series submitted at regular intervals, such as every week or month.

Typical Writing Situations

There are two basic types of progress reports. In the first, you tell your readers about your progress on one project. Lee is a geological consultant who works for a civil engineering company. He is currently assigned to evaluate the site where a city would like to build a civic center but worries that the land may not be geologically suited for this construction. Every two weeks, Lee submits a progress report to his supervisor in his own company and to the city's architect. His supervisor reads the report to make sure Lee is working in a technically sound manner. The city's architect uses the report to see that Lee is on schedule. She also looks for indications of the likely outcome of Lee's study. She could accelerate or halt other work as a result of Lee's preliminary findings.

In the second type of progress report, you describe your work on a group of projects that you or your team is working on simultaneously. A chemist, Jacqueline manages a department that develops improved formulas for the laundry detergents her company manufactures—making them clean better, smell better, more economical to produce, and safer for the environment. Every month, she prepares a report that summarizes her department's progress on approximately twelve projects. Her readers include her immediate superiors, who want to be sure that her department's work is proceeding satisfactorily; researchers in other departments, who want to see whether her staff has made discoveries that can be used in other products, such as dishwashing detergents; and corporate planners, who want to anticipate changes in formulas that will require alterations in production lines or advertising.

As the examples of Lee and Jacqueline indicate, progress reports may differ in many ways: They may cover one project or many; they may be addressed to people inside the writer's organization or outside it; and they may be used by people who have a variety of reasons for reading them.

Readers' Concern with the Future

Despite their diversity, almost all progress reports have this in common: Their readers are primarily concerned with the *future*. That is, even though most progress reports talk primarily about what has happened in the past, their readers usually want that information so that they can plan for the future.

LEARNING OBJECTIVES

1. Describe the situations in which progress reports are typically written.
2. Describe the major questions asked by readers of progress reports.
3. Describe the superstructure for progress reports, including the ways its parts correspond with the readers' questions.
4. Adapt the reader-centered writing process to the special goals of progress reports.
5. Write an effective progress report.

Some progress reports concern a single project.

Some progress reports concern all projects worked on during a single period.

Although progress reports talk about the past, they are used to make decisions about the future.

For example, from your report they may be trying to learn the things they need to know in order to manage *your* project. They will want to know, for instance, what they should do (if anything) to keep your project going smoothly or to get it back on track. The progress reports written by Lee and Jacqueline are used for this purpose by some of their readers.

Other readers may be reading your progress reports to learn what they need to know in order to manage *other* projects. Almost all projects in an organization are interdependent with other projects. For instance, suppose you are conducting a marketing survey whose results will be used by another group as it designs an advertising campaign. If you have fallen behind schedule, the other group's schedule may need to be adjusted.

Your readers may also be interested in the preliminary results of your work. Suppose, for instance, that you complete one part of a research project before you complete the others. Your readers may very well be able to use the results of that part immediately. The city engineer who reads Lee's reports about the possible building site wishes to use each of Lee's results as soon as it is available.

How Readers Use and Evaluate Progress Reports

Despite the variety of situations in which you might write progress reports in your career, your readers will almost always want your reports to answer the following questions. They will judge the quality of your progress reports based on the ease and efficiency with which they can use the reports to plan for the future of your project and other projects that depend on the outcomes of your work.

- **What work does your report cover?** To be able to understand anything else in a progress report, readers must know what project or projects and what time period the report covers.

- **What is the purpose of the work?** Readers need to know the purpose of your work to see how it relates to their own responsibilities and to the other work, present and future, of the organization. In some cases, your readers may already know the purpose of your work, so you may not need to include it in your report.

- **Is your work progressing as planned or expected?** Your readers will want to determine whether adjustments are needed in the schedule, budget, or number of people assigned to the project or projects on which you are working.

- **What results have you produced?** The results you produce in one reporting period may influence the shape of work in future periods. Also, even when you are still in the midst of a project, readers will want to know about any results they can use in other projects now, before you finish your overall work.

- **What progress do you expect during the next reporting period?** Again, your readers' interests will focus on such management concerns as schedule and budget and on the kinds of results they can expect.

- **How do things stand overall?** This question arises especially in long reports. Readers want to know what the overall status of your work is, something they may not be able to tell readily from all the details you provide.

- **What do you think we should do?** If you are experiencing or expecting problems, your readers will want your recommendations about what should be done. Your ideas about how to improve the project will also be welcome.

Superstructure for Progress Reports

The conventional superstructure for progress reports provides a very effective framework for answering your readers' questions about your projects:

Superstructure for Progress Reports

Report Element	Readers' Question
Introduction	What work does your report cover? What is the purpose of the work?
Facts and Discussion	
Past Work	Is your work progressing as planned or expected? What results have you produced?
Future work	What progress do you expect during the next reporting period?
Conclusions	In long reports—How do things stand overall?
Recommendations	What do you think we should do?

LEARN MORE Remember that a superstructure is not an outline; you may combine the elements of a superstructure in many ways to serve your readers' needs. See "Adapt an appropriate superstructure or other pattern familiar to your reader" (pages 112–113).

Crafting the Major Elements of a Progress Report

All of the advice that Chapters 6 through 16 provide about drafting workplace communications applies to progress reports. The following sections tell how you can apply some of this advice most productively to each of the elements of a progress report, regardless of how they are organized. As you read, keep in mind that the conventional superstructure, or genre, is only a general plan. You must use your imagination and creativity to adapt it to particular situations.

Introduction

In the introduction to a progress report, you can address the readers' first two questions. You can usually answer the question, "What work does your report cover?" by opening with a sentence that identifies the project or projects your report concerns and the time period it covers.

Sometimes you will not need to answer the second question—"What is the purpose of the work?"—because all your readers will already be quite familiar with its purpose. At other times, however, it will be crucial for you to tell your work's purpose because your readers will include people who don't know or may have forgotten it. You are especially likely to have such readers when your progress reports are widely circulated in your own organization or when you are reporting to another organization that has hired your employer to do the work you describe. You can usually explain your work's purpose most helpfully by describing the problem that your work will help your readers solve.

LEARN MORE For additional advice about writing an introduction, see Chapter 7.

> This report covers the work done on the Focus Project from July 1 through September 1. Sponsored by the U.S. Department of Energy, the Focus Project aims to overcome the technical difficulties encountered in constructing photovoltaic cells that can be used to generate commercial amounts of electricity.

Project and period covered

Purpose of project

Of course, your introduction should also provide any background information your readers will need in order to understand the rest of your report.

Facts and Discussion

In the discussion section of your progress report, you should answer these readers' questions: "Is your work progressing as planned or as expected?" "What results have you produced?" and "What progress do you expect during the next reporting period?"

How to Organize

For progress reports, brevity is a virtue. The goal is to take little time from your actual work on your projects and to require as little time as possible for your readers to get the information they need. To speed these processes, many organizations have developed forms for progress reports. If your employer doesn't have a form, think about the most efficient way to organize your facts and discussion.

If you are reporting on a single project, you might organize in this way.

I. Describe the major events that happened during the most recent time period
II. Describe major events you expect to occur during the next time period

Erin used this organization when reporting every two weeks on her assignment to increase efficiency at her employer's steel mills. Figure 26.1 shows the outline of her report for the two weeks in which she discovered employees at one plant were taking too long to change the blade on a hot saw, which is a machine that cuts the white-hot ingots of metal.

If you are working on two or more projects simultaneously, you can expand the organization Erin used.

I. What happened during the most recent time period
  A. Project A
  B. Project B
II. What's expected to happen during the next time period
  A. Project A
  B. Project B

A third option is to organize around your projects rather than time periods.

I. Work on Project A
  A. What happened during the last time period
  B. What I expect to happen during the next time period
II. Work on Project B
  A. What happened during the last time period
  B. What I expect to happen during the next time period

This organization works very well when you have more than a few sentences to say about one or more of your projects. It keeps all the information on each project together, making the report easy for readers to follow. Lloyd organized this way when reporting on the progress of a large project he directed, which was to introduce a new line of high-performance smartphones (Figure 26.2, page 454).

How to Indicate Whether Projects Are on Schedule

Often the work to be accomplished during each reporting period is planned in advance, so you can indicate your progress by comparing what happened with what was planned. Where there are significant discrepancies between the two, your readers will want to know why. The information you provide about the causes of problems will help your readers decide how to remedy them. It will also help you to explain any recommendations you make later in your report.

TRY THIS Imagine that you are going to write a progress report about what you've done in the past week or few weeks on a personal effort or course project. To whom would you write? How would your reader use the information in your report? What would your main points be? Alternatively, imagine that you are writing the progress report to yourself.

How much information should you include in your progress reports? Generally, readers prefer brief reports. Although you need to provide your readers with specific information about your work, don't include details unless they will help readers decide how to manage your project or unless you believe readers will be able to make immediate use of them. As you work on a project, many minor events will occur, and you will have lots of small setbacks and triumphs along the way. Avoid discussing such matters. No matter how important these details may be to you, they are not likely to be interesting to your readers. Stick to the information your readers can use.

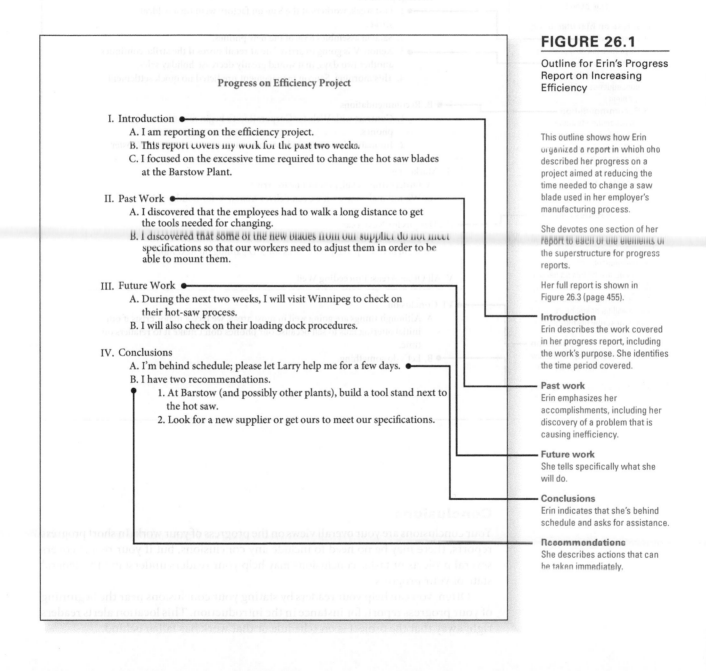

Progress on Efficiency Project

I. Introduction
   A. I am reporting on the efficiency project.
   B. This report covers my work for the past two weeks.
   C. I focused on the excessive time required to change the hot saw blades at the Barstow Plant.

II. Past Work
   A. I discovered that the employees had to walk a long distance to get the tools needed for changing.
   B. I discovered that some of the new blades from our supplier do not meet specifications so that our workers need to adjust them in order to be able to mount them.

III. Future Work
   A. During the next two weeks, I will visit Winnipeg to check on their hot-saw process.
   B. I will also check on their loading dock procedures.

IV. Conclusions
   A. I'm behind schedule; please let Larry help me for a few days.
   B. I have two recommendations.
      1. At Barstow (and possibly other plants), build a tool stand next to the hot saw.
      2. Look for a new supplier or get ours to meet our specifications.

FIGURE 26.1

Outline for Erin's Progress Report on Increasing Efficiency

This outline shows how Erin organized a report in which she described her progress on a project aimed at reducing the time needed to change a saw blade used in her employer's manufacturing process.

She devotes one section of her report to each of the elements of the superstructure for progress reports.

Her full report is shown in Figure 26.3 (page 455).

Introduction
Erin describes the work covered in her progress report, including the work's purpose. She identifies the time period covered.

Past work
Erin emphasizes her accomplishments, including her discovery of a problem that is causing inefficiency.

Future work
She tells specifically what she will do.

Conclusions
Erin indicates that she's behind schedule and asks for assistance.

Recommendations
She describes actions that can be taken immediately.

FIGURE 26.2

Outline of Lloyd's Progress Report on Introducing a New Line of Smartphones

In this report, Lloyd tells his readers about progress in several related projects, devoting a separate section to each one.

Introduction

Lloyd describes the work covered in his progress report, including the work's purpose. He identifies the time period covered.

Progress on Manufacturing

• **Past work**

Lloyd focuses on a problem.

• **Conclusion**

Lloyd explains the possible consequences of the problem.

• **Recommendation**

He recommends actions to be taken in response to the manufacturing problem he has described. If his recommendations are accepted, they will be his future work.

Progress on three other projects

Lloyd briefly describes past work on three other projects, emphasizing accomplishments. Because there are no problems, he has no need to make recommendations. Similarly, because his readers already know what he will be doing, he does not need to describe his future work.

Conclusion

Lloyd summarizes by telling how things stand overall.

Lloyd ends with a reference to his recommendations in Section II. B.

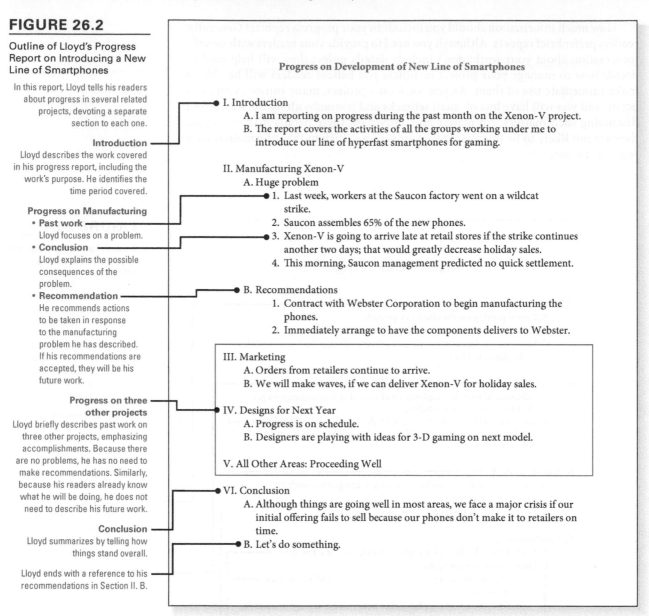

Progress on Development of New Line of Smartphones

I. Introduction
   A. I am reporting on progress during the past month on the Xenon-V project.
   B. The report covers the activities of all the groups working under me to introduce our line of hyperfast smartphones for gaming.

II. Manufacturing Xenon-V
   A. Huge problem
      1. Last week, workers at the Saucon factory went on a wildcat strike.
      2. Saucon assembles 65% of the new phones.
      3. Xenon-V is going to arrive late at retail stores if the strike continues another two days; that would greatly decrease holiday sales.
      4. This morning, Saucon management predicted no quick settlement.
   B. Recommendations
      1. Contract with Webster Corporation to begin manufacturing the phones.
      2. Immediately arrange to have the components delivers to Webster.

III. Marketing
   A. Orders from retailers continue to arrive.
   B. We will make waves, if we can deliver Xenon-V for holiday sales.

IV. Designs for Next Year
   A. Progress is on schedule.
   B. Designers are playing with ideas for 3-D gaming on next model.

V. All Other Areas: Proceeding Well

VI. Conclusion
   A. Although things are going well in most areas, we face a major crisis if our initial offering fails to sell because our phones don't make it to retailers on time.
   B. Let's do something.

Conclusions

Your conclusions are your overall views on the progress of your work. In short progress reports, there may be no need to include any conclusions, but if your report covers several projects or tasks, conclusions may help your readers understand the general state of your progress.

Often, you can help your readers by stating your conclusions near the beginning of your progress report, for instance in the introduction. This location alerts readers right away that the project is on schedule or that work has fallen behind.

Recommendations

Your readers will want your thoughts about how to improve the project or increase the value of its results. Your recommendations might be directed at overcoming a difficulty you have experienced or anticipate encountering in the future. Or they might be directed at refocusing or otherwise altering your project.

Place your recommendations where they will be most helpful to your readers. Often that location will be the end. Erin used this placement in her report on her efficiency project for the steel mill (Figure 26.3). If you are reporting on a project with parts, you may be able to help your readers by placing your recommendation in the

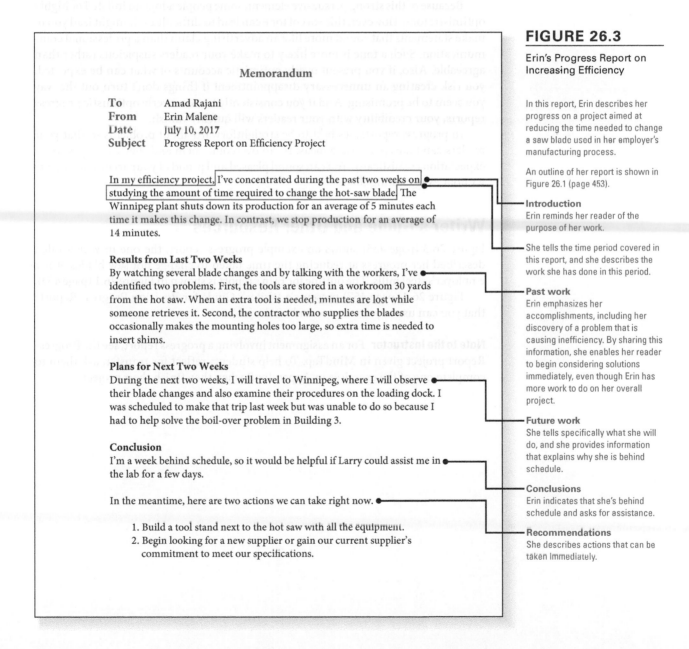

FIGURE 26.3

Erin's Progress Report on Increasing Efficiency

In this report, Erin describes her progress on a project aimed at reducing the time needed to change a saw blade used in her employer's manufacturing process.

An outline of her report is shown in Figure 26.1 (page 453).

Introduction
Erin reminds her reader of the purpose of her work.

She tells the time period covered in this report, and she describes the work she has done in this period.

Past work
Erin emphasizes her accomplishments, including her discovery of a problem that is causing inefficiency. By sharing this information, she enables her reader to begin considering solutions immediately, even though Erin has more work to do on her overall project.

Future work
She tells specifically what she will do, and she provides information that explains why she is behind schedule.

Conclusions
Erin indicates that she's behind schedule and asks for assistance.

Recommendations
She describes actions that can be taken immediately.

The text of the memorandum in the figure:

Memorandum

To   Amad Rajani
From   Erin Malene
Date   July 10, 2017
Subject   Progress Report on Efficiency Project

In my efficiency project, I've concentrated during the past two weeks on studying the amount of time required to change the hot-saw blade. The Winnipeg plant shuts down its production for an average of 5 minutes each time it makes this change. In contrast, we stop production for an average of 14 minutes.

Results from Last Two Weeks
By watching several blade changes and by talking with the workers, I've identified two problems. First, the tools are stored in a workroom 30 yards from the hot saw. When an extra tool is needed, minutes are lost while someone retrieves it. Second, the contractor who supplies the blades occasionally makes the mounting holes too large, so extra time is needed to insert shims.

Plans for Next Two Weeks
During the next two weeks, I will travel to Winnipeg, where I will observe their blade changes and also examine their procedures on the loading dock. I was scheduled to make that trip last week but was unable to do so because I had to help solve the boil-over problem in Building 3.

Conclusion
I'm a week behind schedule, so it would be helpful if Larry could assist me in the lab for a few days.

In the meantime, here are two actions we can take right now.

1. Build a tool stand next to the hot saw with all the equipment.
2. Begin looking for a new supplier or gain our current supplier's commitment to meet our specifications.

section on the part that has the problem. For example, when reporting on progress at manufacturing the new smartphone, Lloyd cited a problem with manufacturing and placed his recommendation for manufacturing immediately afterward (Figure 26.2, page 454).

Tone in Progress Reports

You may wonder what tone to use in the progress reports you prepare. Generally, you want to persuade your readers that you are doing a good job. That is especially likely when you are new on the job and when your readers might discontinue a project if they feel that it isn't progressing satisfactorily.

Because of this strong persuasive element, some people adopt an inflated or highly optimistic tone. However, this sort of tone can lead to difficulties. It might lead you to make statements that sound more like an advertising claim than a professional communication. Such a tone is more likely to make your readers suspicious rather than agreeable. Also, if you present overly optimistic accounts of what can be expected, you risk creating an unnecessary disappointment if things don't turn out the way you seem to be promising. And if you consistently turn in overly optimistic progress reports, your credibility with your readers will quickly vanish.

In progress reports, it's best to be straightforward about problems so that your readers can take appropriate measures to overcome them and they can adjust their expectations realistically. You can sound pleased and proud of your accomplishments without exaggerating them.

Writer's Guide and Other Resources

Figure 26.3 (page 455) shows an example progress report, the one in which Ellen described her progress at reducing the time taken to change hot saw blades at her employer's steel mills. The outline for Erin's report is shown in Figure 26.1 (page 453).

Figure 26.4 (page 457) presents a Writer's Guide for Revising Progress Reports that you can use in your course and on the job.

Note to the Instructor For an assignment involving a progress report, see the Progress Report project given in MindTap. To help students reflect for transfer, ask them to complete one of the exercises in Appendix B when they finish their project.

FIGURE 26.4

Writer's Guide for Revising Progress Reports

Writer's Guide
REVISING PROGRESS REPORTS

Does your draft include each of the elements needed to create a report that your readers will find to be useful and persuasive? Remember that some elements of the superstructure may be unnecessary for your specific readers and purpose and that the elements may be organized in various ways.

INTRODUCTION

❏ Identifies the work your report covers.
❏ Indicates the purpose of the work, if the readers need to be reminded.
❏ Identifies the period covered.
❏ Forecasts the rest of your report, if doing so would help your readers.

BODY OF REPORT

❏ States whether your work on each of your tasks is progressing as planned with respect to schedule, budget, or other concerns of your readers.
❏ Tells what you accomplished on each of your major tasks during the period covered by the report.
❏ Identifies planned work that is not complete.
❏ Reports results or accomplishments that your readers would like to know about immediately.
❏ Identifies any significant problems your readers would want to know about.
❏ Identifies specific tasks you will be performing during the next reporting period.
❏ Describes the progress you expect to achieve during the next period.
❏ Identifies any upcoming problems that your readers should know about.
❏ Identifies any help you feel you need.

CONCLUSION

❏ Includes a final statement that lets your readers know how things stand overall.

RECOMMENDATIONS

❏ Indicates any actions you think your readers should take.

PROSE (SEE CHAPTERS 8 AND 10)

❏ Presents information in a clear, useful, and persuasive manner.
❏ Uses a variety of sentence structures and lengths.
❏ Flows in a way that is interesting and easy to follow.
❏ Uses correct spelling, grammar, and punctuation.

GRAPHICS (SEE CHAPTER 12)

❏ Included wherever readers would find them helpful or persuasive.
❏ Look neat and attractive, and are easy to read.
❏ Referred to at the appropriate points in the prose.
❏ Located where your readers can find them easily.

PAGE DESIGN (SEE CHAPTER 14)

❏ Looks neat and attractive.
❏ Helps readers find specific information.

ETHICS

❏ Treats all the report's stakeholders ethically.
❏ Presents all information accurately and fairly.

27

Writing Reader-Centered Instructions

LEARNING OBJECTIVES

1. Describe the major questions asked by readers of instructions.
2. Describe the superstructure for instructions, including the ways its parts correspond with readers' questions.
3. Adapt the reader-centered writing process to the special goals of instructions.
4. Write effective instructions.

MindTap®

Find additional resources related to this chapter in MindTap.

Instructions come in many lengths, shapes, and levels of complexity. They range from the terse directions on a shampoo bottle ("Lather. Rinse. Repeat.") to the huge manuals that are hundreds or thousands of pages long for servicing airplane engines, managing large computer systems, and performing biomedical procedures.

Although some instructions are prepared by professional writers and editors, most other employees also create them. Whether you are developing a new procedure, training a new co-worker, or preparing to leave for vacation, you may need to provide written directions to someone else. You may write for people who will read your instructions on paper, computer monitor, smartphone screen, or other electronic device.

This chapter provides advice that you will find valuable regardless of the subject or length of the instructions you write. The advice given in the first part of this chapter applies equally to instructions written for paper and screen. A special section at the end of the chapter provides additional suggestions for instructions that will be delivered digitally as a website or video.

How Readers Use and Evaluate Instructions

Readers read instructions in many different ways. Some follow the directions meticulously, concentrating on every word. Others look at the directions only if they get stumped while relying solely on their experience and intuition. Whether they read every word or look only occasionally at instructions, the questions readers ask are almost always versions of the following six. They judge the quality of instructions based on the ease and efficiency with which they can use them to perform their tasks.

- **What will these instructions help me do?** Some readers will ask this question exactly as it reads. When others use these or similar words, they are asking, "Do I really have to read these?"
- **Is there anything special I need to know to be able to use these instructions effectively?**
- **If I'm working with equipment, where are the parts I need to use?**
- **What materials, equipment, and tools do I need?**
- **Once I'm ready to start, what—exactly—do I do?**
- **Something isn't working correctly. How do I fix it?**

Superstructure for Instructions

The superstructure, or genre, for instructions includes five elements that answer those six questions readers ask most often.

LEARN MORE For more on superstructures, see "Adapt an appropriate superstructure or other pattern familiar to your reader" (pages 112–113).

Superstructure for Instructions

Topics	Readers' Questions
Introduction	What will these instructions help me do? Is there anything special I need to know to be able to use these instructions effectively?
Description of the equipment	If I'm working with equipment, where are the parts I need to use?
List of materials and equipment needed	What materials, equipment, or tools do I need?
Directions	Once I'm ready to start, what—exactly—do I do?
Troubleshooting	Something isn't working correctly. How do I fix it?

However, not every element is in every set of instructions. The simplest instructions contain only the directions. More complex instructions contain some or all of the other four elements listed above. And some instructions also include additional elements such as covers, title pages, tables of contents, appendixes, lists of references, glossaries, lists of symbols, and indexes.

To determine which elements to include in any instructions you write, follow this familiar advice: Consider your readers' aims and needs as well as the characteristics that will shape the way they read and respond to your communication.

Tom Grill/Corbis

Guiding You through the Process of Preparing Instructions

Defining Your Instruction's Goals

Instructions have more complex goals than it might seem at first glance. Of course, they have the goal of providing exact, easy-to-understand directions. But instruction readers don't read straight through from first to last direction. They read a step, take their eyes off the page to do the step, and then turn back to the instructions to find the next step. Some readers look only occasionally at the instructions, when their intuition fails them. These readers, too, want to find quickly the information they need. Also, directions often include a mixture of graphics and text. Your instructions must help readers coordinate these two kinds of information easily.

Instructions also have persuasive goals that are less obvious than their usefulness goals but are equally important. Many people dislike reading instructions. They want to start right in on the task without taking time to read anything. Thus, one persuasive goal for instructions is simply to persuade people to read them. If your instructions are for a product made by your employer, they will have a second persuasive objective: to persuade readers to feel so good about the product that they will buy from your employer again and recommend that others do likewise.

Nigel Cattlin/Alamy Stock Photo

Courtesy of NASA

Instructions must meet the needs of readers performing vastly different tasks in a wide range of settings and circumstances. To keep both hands free, astronaut Kathryn Thornton (center photo) has her instructions strapped to her arm.

Planning

For the instructions to achieve their usefulness and persuasive goals, three features of instructions must work together harmoniously. For all three, planning overlaps with drafting.

LEARN MORE For advice about how to group the steps in your process, see "Describing a Process (Segmenting)" (pages 156–158).

- ■ **Organization of the directions.** By organizing the directions hierarchically, you can help readers find the next step as they look back at the instructions after completing the previous step. This organization can also help them find particular information when consulting the instructions as a reference document.

    To create a hierarchical organization, begin by listing all the steps in the process. Next, check the list for thoroughness. Then, group related steps under headings, such as "Preparing the Equipment," "Using the Equipment," and "Cleaning Up." If the instructions are long, shorter groups can be gathered into larger ones.

LEARN MORE For advice about creating graphics, see Chapter 12.

- ■ **Graphics.** For many purposes, well-designed graphics are even more effective than words. Words cannot show readers where the parts of a machine are located, how to grasp a tool, or what the result of a procedure should look like. Graphics are especially helpful in instructions for readers who speak languages other than your own. Sometimes, graphics alone can convey all the information your readers need (see Figure 12.1 on page 225). Look actively for places where adding a drawing, diagram, photo, or other graphic would make your directions easier for your readers to understand. Chapter 12 provides suggestions for designing graphics that your readers will welcome.

LEARN MORE For advice about designing pages and screens, see Chapter 14. Chapter 20 provides additional advice about screen design.

- ■ **Page design.** Strategically designed pages can help you and your readers in several ways. Page design can help readers find their place as they bounce back and forth between reading steps and performing them. Good page design helps readers see the connections between related blocks of information, such as a written direction and the drawing that accompanies it. An attractive design can entice readers' attention to instructions they would otherwise choose to ignore. Chapters 14 and 20 can guide you through the process of designing effective printed and on-screen pages.

Drafting and Revising

Later in this chapter, you will find suggestions for drafting each element of your instructions. The following advice applies to *all* elements: Write clearly and succinctly. Choose words that convey your meaning clearly. Construct sentences your readers will comprehend effortlessly. Use as few words as possible. More words make more work for your readers and increase the chances your readers will stop reading what you've written.

LEARN MORE For advice about writing clearly and succinctly, see Chapter 10.

When you want to find ways to revise a draft of instructions, nothing beats watching members of your target audience using the draft to perform the procedure. Where you see them succeed with some steps, you know that part of your draft is effective. Where they have problems, you have an opportunity to improve. Chapter 16 guides you through the process of planning, conducting, and interpreting the results of user tests. When testing, remember to evaluate your draft's ability to achieve its persuasive objectives as well as its usefulness objectives.

LEARN MORE To learn how to conduct user tests, see Chapter 16.

Crafting the Major Elements of Instructions

All the advice about drafting provided in Chapters 6 through 16 can help you write effective instructions. The following sections supplement that advice with suggestions for writing the five elements of the superstructure for instructions.

Introduction

An introduction should be as short as possible—or nonexistent. Many instructions don't need one. The title alone provides all the introductory information readers require. See Figure 27.5 (page 468). On the other hand, readers sometimes do need information up front. The following sections describe the eight topics most commonly included in introductions, together with suggestions for deciding whether your readers need each of them. Many are illustrated by the introduction to the Tire Uniformity Optimizer, a machine used by the manufacturers of automobile tires to ensure the quality of each tire they sell. See Figure 27.1.

Chapter 1—Introduction

This manual tells you how to operate the Tire Uniformity Optimizer (TUO) and its controller, the Tire Quality Computer (TQC). The TUO has many options. Depending upon the options on your machine, it may do any or all of the following jobs:

- Test tires
- Find irregularities in tires
- Grind to correct the irregularities, if possible
- Grade tires
- Mark tires according to grade
- Sort tires by grade

This manual explains all the tasks you are likely to perform in a normal shift. It covers all of the options your machine might have.

The rest of this chapter introduces you to the major parts of the TUO and its basic operation. Chapter 2 tells you step-by-step how to prepare the TUO when you change the type or size of tire you are testing. Chapter 3 tells you how to perform routine servicing, and Chapter 4 tells you how to troubleshoot problems with the TUO. Chapter 5 contains a convenient checklist of the tasks described in Chapter 3.

Major Parts of the Tire Uniformity Optimizer

You can find the major parts of the TUO by looking at Figure 1-1. To operate the TUO, you will use the Operator's Control Panel and the Computer Panel.

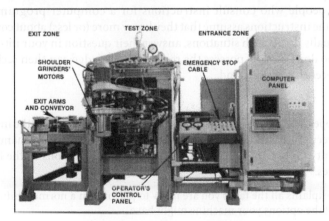

Figure 1-1. *Overview of the TUO and TQC.*

Page 1

FIGURE 27.1

Introduction to the Instruction Manual for the Tire Uniformity Optimizer

The first sentence identifies the **subject** of the manual.

The second sentence and list identify the **purposes of the procedures** that can be performed by following the instructions.

This sentence describes the **scope** of the manual: all of the procedures the reader is likely to perform during a normal shift.

This paragraph describes the **organization** of the manual.

The photograph and labels provide readers with a **description of the equipment** that enables them to locate all the major parts they will have to find while following the instructions. More detailed photos are provided later in the manual to guide the reader when using the Operator's Control Panel and other parts.

The manual presents **safety** warnings and information to **motivate** readers to follow certain parts of the procedures at the appropriate places later in the manual. This manual does not use any **conventions** that need to be explained in the introduction.

Subject

Often, the title will fully convey the subject of your instructions. In longer instructions, however, you may need to announce the subject in an introduction. Here is the first sentence from the fifty-page operator's manual for a 10-ton machine used in the manufacture of automobile and truck tires.

Opening sentence that announces the subject

This manual tells you how to operate the Tire Uniformity Optimizer (TUO).

Purpose of the Procedure

If the purpose of the procedure your instructions describe isn't obvious from the title, announce it in the introduction. You may be able to convey your instructions' aim by listing the major steps in the procedure or the capabilities of the equipment whose operation you are describing. Here is the third sentence of the manual for the Tire Uniformity Optimizer.

Depending upon the options on your machine, it may do any or all of the following jobs:

A list of the purposes for which readers can use the equipment

- Test tires
- Find irregularities in tires
- Grind to correct the irregularities, if possible
- Grade tires
- Mark tires according to grade
- Sort tires by grade

Intended Readers

When they pick up instructions, people often want to know whether the instructions are directed to them or to people who differ from them in interests, responsibilities, level of knowledge, or some other variable.

Sometimes, they can tell merely by reading the instructions' title. For instance, the operator's manual for the Tire Uniformity Optimizer is obviously addressed to people hired to operate that machine.

In contrast, people who consult instructions for a computer program may wonder whether the instructions assume that they know more (or less) about computers than they actually do. In such situations, answer their question in your introduction. Readers who don't already possess the required knowledge can then seek help or acquire the necessary background.

Scope

By stating the scope of your instructions, you help readers know whether the instructions contain directions for the specific tasks they want to perform. The manual for the Tire Uniformity Optimizer describes the scope of its instructions in the fourth and fifth sentences.

Statement of scope

This manual explains all the tasks you are likely to perform in a normal shift. It covers all of the options your machine might have.

Organization

By explaining how the instructions are organized, an introduction can help readers understand the overall structure of the tasks they will perform and locate specific pieces of information without having to read the entire set of instructions.

Often introductions explain scope and organization together. The introduction to the Tire Uniformity Optimizer devotes several sentences to explaining that manual's organization, and this information also fills out the readers' understanding of the manual's scope.

> The rest of this chapter introduces you to the major parts of the TUO and its basic operation. Chapter 2 tells you step by step how to prepare the TUO when you change the type or size of tire you are testing. Chapter 3 tells you how to perform routine servicing, and Chapter 4 tells you how to troubleshoot problems you can probably handle without needing to ask for help from someone else. Chapter 5 contains a convenient checklist of the tasks described in Chapters 3 and 4.

Paragraph describing a manual's organization

Conventions

If your instructions use abbreviations or conventions that the reader needs to know in order to interpret the directions correctly, explain them in the introduction. For instance, the introduction for a manual for operating a machine for harvesting corn reads, "Right-hand and left-hand sides are determined by facing the direction of forward travel."

Motivation

As pointed out above (and as you may know from your own experience), some people are tempted to toss instructions aside and rely on their common sense. A major purpose of many introductions is to persuade readers to read the instructions. You can accomplish this goal by using an inviting and supportive tone and creating an attractive design. You can also include statements that tell readers directly why it is important to pay attention to the instructions. The following example is from instructions for a ceiling fan that purchasers install themselves.

> We're certain that your Hampton Bay fan will provide you with many years of comfort, energy savings, and satisfaction. To ensure your personal safety and to maximize the performance of your fan, please read this manual.

Statement of scope

Motivation to read the instructions

Safety

Your readers depend on you to prevent them from taking actions that could spoil their results, damage their equipment, or cause them injury. Moreover, product liability laws require companies to pay for damages or injuries that result from inadequate warnings in their instructions.

To satisfy your ethical and legal obligations, you must provide prominent, easy-to-understand, and persuasive warnings. If a warning concerns a general issue that covers the entire set of instructions (e.g., "Don't use this electrical tool while standing on wet ground"), place it in your introduction. If it pertains to a certain step, place it before that step. The following principles apply to warnings in either location.

- **Make your warnings stand out visually.** Try printing them in large, bold type and surrounding them with a box. Sometimes, writers use the following international hazard alert symbol to draw attention to the warning.

You may also include an icon to convey the nature of the danger. Here are some icons developed by Westinghouse.

Electrical Shock　　　　　Fire　　　　　Eye Protection

- **Place your warnings so that your readers will read them before performing the action the warnings refer to.** It won't help your readers to discover the warning after they've performed the step and the damage has been done.

- **State the nature of the hazard and the consequences of ignoring the warning.** If readers don't know what could happen, they may think that it's not important to take the necessary precautions.

- **Tell your readers what steps to take to protect themselves or avoid damage.**

The box and international hazard icon draw attention to the warning.

First sentence tells what readers can do to avoid the hazard.

Second sentence describes the possible consequences of ignoring the warning.

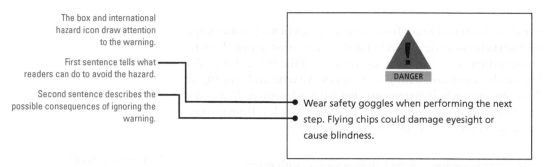

DANGER

Wear safety goggles when performing the next step. Flying chips could damage eyesight or cause blindness.

Description of the Equipment

LEARN MORE To describe equipment, see "Describing a Process (Segmenting)" (pages 156–158). To use a photograph or drawing, see "Choose the Type of Graphic Best Matched to Your Communication's Goals" (pages 226–228).

To be able to operate or repair a piece of equipment, readers need to know the location of its parts. Sometimes, they need to know their functions as well. Instructions often include a description of the equipment to be used, usually by including a labeled photograph or drawing of it. For example, the first page of the manual for the Tire Uniformity Optimizer displays a labeled photograph of the machine. In some instructions, such illustrations are accompanied by written explanations of the equipment and its parts.

List of Materials and Equipment Needed

Some procedures require materials or equipment that readers wouldn't normally have at hand. If yours do, include a list of these items. Present the list *before* giving your step-by-step instructions. This will save your readers from the unpleasant surprise of discovering that they cannot perform the next step until they have gone to the shop, supply room, or store to obtain an item that they didn't realize they would need.

Directions

At the heart of a set of instructions are the step-by-step directions that tell readers what to do. The following sections describe strategies for writing directions your readers will find easy to understand and use. Figure 27.2 illustrates much of this advice.

Write Each Direction for Rapid Comprehension and Immediate Use

Readers want to understand as quickly as possible what they should do next.

1. **In each direction, give readers only enough information to perform the next step.** If you give more, they may forget some or become confused.

2. **Present the steps in a list.** A list format helps readers see each step as a separate action.

3. **Use the active voice and the imperative mood.** Active, imperative verbs give commands: "*Stop* the engine." (This statement is much simpler than "The operator should then stop the engine.")

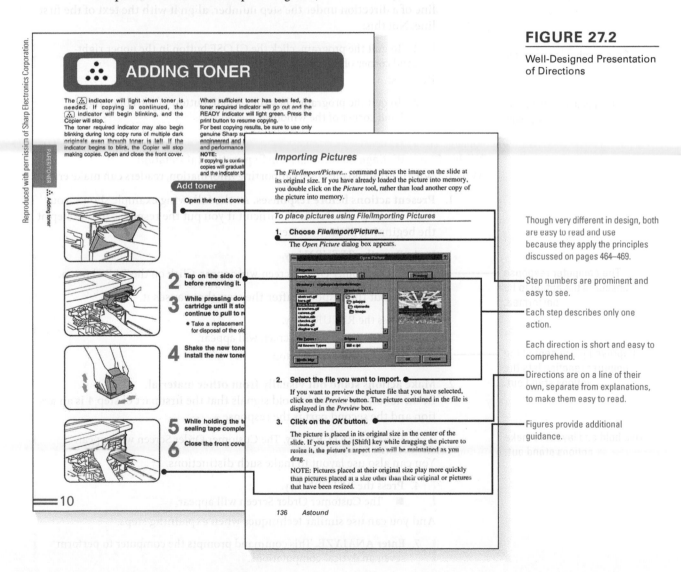

FIGURE 27.2

Well-Designed Presentation of Directions

Though very different in design, both are easy to read and use because they apply the principles discussed on pages 464–469.

Step numbers are prominent and easy to see.

Each step describes only one action.

Each direction is short and easy to comprehend.

Directions are on a line of their own, separate from explanations, to make them easy to read.

Figures provide additional guidance.

4. **Highlight keywords.** In some instructions, a direction may contain a single word that conveys the critical information. You can speed the readers' task by using bold, all-capital letters or a different typeface to make this word pop off the page. Example: Press the RETURN key.

Help Readers Locate the Next Step Quickly

There are many ways you can help your readers as they turn their eyes away from the task and back to your text:

1. **Number the steps.** With the aid of numbers, readers will not have to reread earlier directions to figure out which one they last read.

2. **Put blank lines between steps.** This white space helps readers pick out a particular step from among its neighbors.

3. **Give one action per step.** It's easy for readers to overlook a direction that is tucked in with another direction rather than having its own number.

4. **Put step numbers in their own column.** Instead of aligning the second line of a direction under the step number, align it with the text of the first line. Not this:

Step number is obscured.

> 2. To quit the program, click the CLOSE button in the upper right-hand corner of the window.

But this:

Step number is in its own column.

> 2. To quit the program, click the CLOSE button in the upper right-hand corner of the window.

Within Steps, Distinguish Actions from Supporting Information

When actions don't stand out from supporting information, readers can make errors.

1. **Present actions before responses.** As the following example shows, you make reading unnecessarily difficult if you put the response to one step at the beginning of the next step.

The computer response obscures the action to be performed.

> 4. Press the RETURN key.
> 5. The Customer Order Screen will appear. Click on the TABS button.

Instead, put the response after the step that causes it.

Improved placement of the computer reaction lets the actions stand out.

> 4. Press the RETURN key.
>   The Customer Order Screen will appear.
> 5. Click on the TABS button.

2. **Make actions stand out visually from other material.**
   In the following example, bold signals that the first part of step 4 is an action and the second part is the response.

Use bold and layout to make actions stand out.

> 4. **Press the RETURN key.** The Customer Order Screen will appear.

You can also use layout to make such distinctions.

> 4. **Press the RETURN key.**
>   ■ The Customer Order Screen will appear.

And you can use similar techniques when explaining steps.

> 7. **Enter ANALYZE.** This command prompts the computer to perform seven analytical computations.

Group Related Steps Under Action-Oriented Headings

By arranging the steps into groups, you divide your procedure into chunks that readers are likely to find manageable. You also help them *learn* the procedure so that they will be able to perform it without instructions in the future. Moreover, if you use action-oriented headings and subheadings for the groups of steps, you aid readers who need directions for only one part of the procedure. The headings enable them to locate quickly the information they require.

To create action-oriented headings, use participles, not nouns, to describe the task. For example, use *Installing* rather than *Installation* and use *Converting* rather than *Conversion*. Here are some of the action-oriented headings and subheadings from Chapter 4 of the Microsoft Windows NT manual.

> Setting Up Your Computers on Your Network
> > Connecting to Computers on Your Network
> > Sharing Your Printer
> > > Viewing Network Drives
> > > Selecting the Printer Closest to You
> > Using Peer Web Services
> > > Installing Peer Web Services
> > > Configuring and Administering Peer Web Services

The first word in each heading and subheading is a participle.

Use Many Graphics

Drawings, photographs, and similar illustrations often provide the clearest and simplest means of telling your readers such important things as the following.

1. **Where things are.** For instance, Figure 27.3 shows the readers of an instruction manual where to find four control switches.

2. **How to perform steps.** For instance, by showing someone's hands performing a step, you provide your readers with a model to follow as they attempt to follow your directions (see Figure 27.4).

3. **What should result.** By showing readers what should result from performing a step, you help them understand what they are trying to accomplish and help them determine whether they have performed the step correctly (see Figure 27.5).

LEARN MORE Chapters 12 and 13 tell how to design effective graphics for instructions.

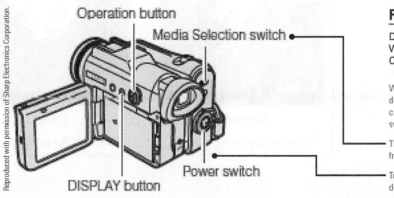

Reproduced with permission of Sharp Electronics Corporation.

Operation button

Media Selection switch

DISPLAY button

Power switch

FIGURE 27.3

Drawing that Shows Readers Where to Locate Parts of a Camcorder

Writers at Sharp Electronics used this diagram to show new owners of a camcorder the location of buttons and switches they will use.

The writers placed the labels far enough from the drawing to stand out.

To avoid ambiguity, they drew the arrows directly to the labeled part.

FIGURE 27.4

Drawings that Show How to Do Something

The writers created these instructions to tell people with diabetes how to obtain the drop of blood they need in order to check their blood glucose level.

The title provides all the information readers need to understand and use these instructions. No separate introduction is needed.

The lancet is a sharp needle used to prick the skin.

The writers use each drawing to show exactly how to hold the Gentle Touch.

In the drawing for step 4, the writers highlight the placement of the endcap against the side of a finger.

In the drawing for step 5, the writers emphasize that the drop of blood must hang from the finger so that it may be applied to a test strip (in the next part of the procedure).

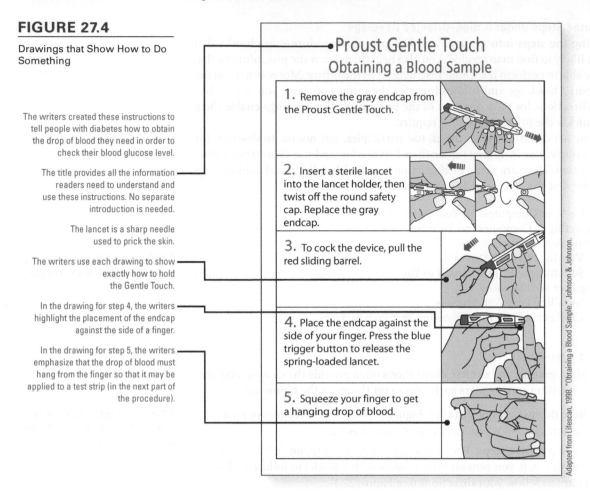

Proust Gentle Touch
Obtaining a Blood Sample

1. Remove the gray endcap from the Proust Gentle Touch.

2. Insert a sterile lancet into the lancet holder, then twist off the round safety cap. Replace the gray endcap.

3. To cock the device, pull the red sliding barrel.

4. Place the endcap against the side of your finger. Press the blue trigger button to release the spring-loaded lancet.

5. Squeeze your finger to get a hanging drop of blood.

Adapted from Lifescan, 1998. "Obtaining a Blood Sample." Johnson & Johnson.

FIGURE 27.5

Photos to Show the Successful Result of a Process

By showing both the correct and incorrect placement of the retainer clip, the National Highway Traffic Safety Administration helps readers understand the correct result of their adjustment of the retainer clip when placing children in car seats.

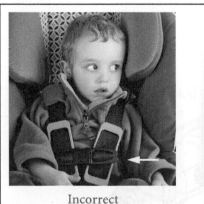

Incorrect
Retainer clip below armpit.

Correct
Retainer clip at armpit level.

Courtesy National Highway Traffic Safety Administration

Chapter 14 provides detailed advice for using page design to help readers see which figure goes with which text.

Present Branching Steps Clearly

Sometimes instructions include alternative courses of action. For example, a chemical analysis might require one procedure if the acidity of a solution is at a normal level and another if the acidity is high. In such a situation, avoid listing only one of the alternatives.

6. If the acidity is high, follow the procedure described on page 20.

Possibly confusing direction

Instead, describe the step that enables readers to determine which alternative to choose (in the example, checking the acidity is that step) and then format the alternatives clearly.

6. Check the acidity.
   - If it is high, follow the procedure described on page 20.
   - If it is normal, proceed to Step 7.

Revised direction

Follow the same logic with other places where your instructions branch into two or more directions. The following example is from instructions for a computer program.

9. Determine which method you will use to connect to the Internet:
   - If you will use PPP (Point-to-Point Protocol), see Chapter 3.
   - If you will use SLIP (Serial Line Internet Protocol), see Chapter 4.

Tell What to Do in Case of a Mistake or Unexpected Result

Try to anticipate the places where readers might make mistakes or obtain an unexpected result. Unless the remedy is obvious, tell your readers what to do.

5. Depress and release the RUN switches on the operator's panel.
   NOTE: If the machine stops immediately and the FAULT light illuminates, reposition the second reel and repeat Step 5.

Troubleshooting

In various circumstances, readers find it easier to have information about correcting mistakes or unexpected results gathered into a single section. Often, a table format works best. Figure 27.6 shows the chapter of the manual for the Tire Uniformity Optimizer that tells how to troubleshoot the TUO's Tire Quality Computer (TQC).

Physical Construction of Instructions

The physical construction of instructions is an important element of their design. Computer manuals are often printed in a small format because readers use them on crowded desktops. Cookbooks are sometimes printed on glossy paper to withstand kitchen spills. Be sure to adapt your instructions to the environment in which they will be used.

FIGURE 27.6

Troubleshooting Section
from the Manual for the Tire
Uniformity Optimizer

Chapter 4—Troubleshooting

This chapter tells you what to check when troubleshooting the TQC. It lists the problems that may occur, the probable causes, and the remedies.

The first list in this chapter consists of the error messages that appear on the CRT when a problem occurs. Next to the error messages are the causes of the problem and the possible remedies. A list of all the error messages can be found in Appendix B. The second list consists of observable phenomena that are listed in order of normal TQC operation.

One easily solved problem is caused by entering entries too quickly to the TQC through the keyboard. If the operator does not wait for the TQC to respond to one request before entering another, errors and inaccurate data will result. Make sure you allow sufficient time for the TQC to respond to your input before you press another key.

The writers use color to highlight a warning.

Warning

EXTERNAL TEST EQUIPMENT CAN DAMAGE THE TQC. If you use external equipment to troubleshoot the TQC, make sure that it does not introduce undesired ground currents or AC leakage currents.

Troubleshooting with Error Messages

Power-up Error Messages

The writers use the three-column table to enable readers to locate quickly the error message given by the TQC and read across for the relevant information and remedy.

Error Message	Probable Cause	Remedy
BACKUP BATTERY IS LOW	1. Battery on Processor Support PCB	1. Replace the battery on the Processor Support PCB.
CONTROLLER ERROR	1. PC interface PCB 2. Processor Support PCB	1. Swap the PC Interface PCB. 2. Swap the Processor Support PCB.
EPROM CHECKSUM ERROR	1. Configuration tables 2. Analog Processor PCB	1. Check the configuration tables. 2. Swap the Analog Processor PCB 88/40.
KEYBOARD MALFUNCTION: PORT	1. Keyboard or keyboard cable 2. Processor Support PCB	1. Check the keyboard and cable. 2. Swap the Processor Support PCB.
RAM FAILURE AT 0000:	1. Main Processor 86/30	1. Swap the 86/30.
RAM FAILURE AT 1000:	1. Main Processor 86/30	1. Swap the 86/30.
TIGRE PROGRAM CHECKSUM ERROR	1. TIGRE program	1. Reenter the TIGRE program or debug the program.

Table 4-1. *Power-up error messages.*

Page 59

From the Akron Standard, Operator's Manual for the Tire Uniformity Optimizer, (1986) p. 1. Used with permission of ITN Ride Quality Products.

Sample Printed Instructions

It is much easier to understand writing advice if you see sample communications that follow that advice. Take a moment to look at the sample in Figure 27.7 and the marginal notes that point out how a student followed the advice you have just read while creating instructions for a lab procedure used in paper mills. Other samples are provided throughout this chapter. For additional examples, see MindTap.

FIGURE 27.7

Instructions Written by a Student

Determining the Percentages of Hardwood and Softwood Fiber in a Paper Sample

These instructions tell you how to analyze a paper sample to determine what percentage of its fibers is from hardwood and what percentage from softwood. This information is important because the ratio of hardwood to softwood affects the paper's physical properties. The long softwood fibers provide strength but bunch up into flocks that give the paper an uneven formation. The short hardwood fibers provide an even formation but little strength. Consequently, two kinds of fibers are needed in most papers, the exact ratio depending on the type of paper being made.

Curtis explains the importance of the procedure.

To determine the percentages of hardwood and softwood fiber, you perform the following major steps: preparing the slide, preparing the sample slurry, placing the slurry on the slide, staining the fibers, placing the slide cover, counting the fibers, and calculating the percentages. The procedure described in these instructions is an alternative to the test approved by the Technical Association of the Pulp and Paper Industry (TAPPI). The TAPPI test involves counting fibers in only one area of the sample slide. Because the fibers can be distributed unevenly on the slide, that procedure can give inaccurate results. The procedure given here produces more accurate results because it involves counting all the fibers on the slide.

He provides an overview of the procedure and indicates why readers should follow it.

EQUIPMENT

He lists all of the equipment needed so readers can assemble it before starting the procedure.

Microscope	Hot plate
Microscope slide	Paper sample
Microscope slide cover	Blender
	Beaker
Microscope slide marking pen	Eyedropper
	Graff "C" stain
Acetone solvent	Pointing needle
Clean cloth	

Courtesy of Curtis J. Walor

(continued)

FIGURE 27.7

(*continued*)

Curtis creates small groups of related steps. He places them under headings that help readers understand the overall procedure and quickly locate the directions they need when referring to the instructions in the future.

Curtis explains the reason for the caution as a way of motivating readers to avoid making a mistake.

He uses bold for the action taken in each step, thereby making the action stand out from explanatory information.

He places the graphic immediately after the step it helps to explain. He keeps the graphic within the gridlines for the directions and out of the grid column for the step numbers.

2

PREPARING THE SLIDE

1. **Clean slide.** Using acetone solvent and a clean cloth, remove all dirt and fingerprints. NOTE: Do not use a paper towel because it will deposit fibers on the slide.

2. **Mark slide.** With a marking pen, draw two lines approximately 1.5 inches apart across the width of the slide.

3. **Label slide.** At one end, label the slide with an identifying number. Your slide should now look like the one shown in Figure A.

```
┌──────────┬──────────┬──────────┐
│          │          │          │
│          │          │   #1     │
│          │          │          │
└──────────┴──────────┴──────────┘
```

Figure A

4. **Turn on hot plate.** Set the temperature at warm. NOTE: Higher temperatures will "boil" off the softwood fibers that you will later place on the slide.

5. **Place slide on hot plate.** Leave the slide there until it dries completely, which will take approximately 5 minutes.

6. **Remove slide from hot plate.** Leave the hot plate on. You will use it again shortly.

PREPARING THE SAMPLE SLURRY

1. **Pour 2 cups of water in blender.** This measurement can be approximate.

2. **Obtain paper sample.** The sample should be about the size of a dime.

3. **Tear sample into fine pieces.**

4. **Place sample into blender.**

5. **Turn blender on.** Set blender on high and run it for about 1 minute.

6. **Check slurry.** After turning the blender off, see if any paper clumps remain. If so, turn the blender on for another 30 seconds. Repeat until no clumps remain.

7. **Pour slurry into beaker.**

Courtesy of Curtis J. Walor

FIGURE 27.7

(*continued*)

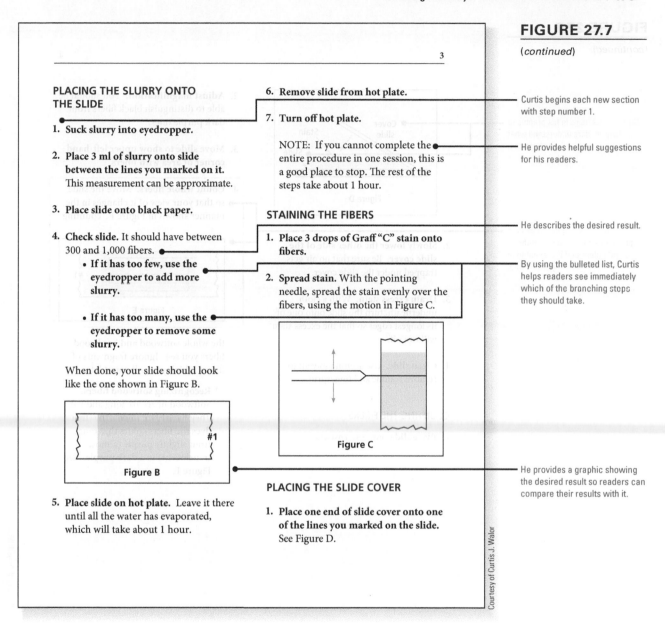

3

PLACING THE SLURRY ONTO THE SLIDE

1. Suck slurry into eyedropper.

2. Place 3 ml of slurry onto slide between the lines you marked on it. This measurement can be approximate.

3. Place slide onto black paper.

4. Check slide. It should have between 300 and 1,000 fibers.
   - If it has too few, use the eyedropper to add more slurry.
   - If it has too many, use the eyedropper to remove some slurry.

   When done, your slide should look like the one shown in Figure B.

Figure B

#1

5. Place slide on hot plate. Leave it there until all the water has evaporated, which will take about 1 hour.

6. Remove slide from hot plate.

7. Turn off hot plate.

   NOTE: If you cannot complete the entire procedure in one session, this is a good place to stop. The rest of the steps take about 1 hour.

STAINING THE FIBERS

1. Place 3 drops of Graff "C" stain onto fibers.

2. Spread stain. With the pointing needle, spread the stain evenly over the fibers, using the motion in Figure C.

Figure C

PLACING THE SLIDE COVER

1. Place one end of slide cover onto one of the lines you marked on the slide. See Figure D.

Curtis begins each new section with step number 1.

He provides helpful suggestions for his readers.

He describes the desired result.

By using the bulleted list, Curtis helps readers see immediately which of the branching steps they should take.

He provides a graphic showing the desired result so readers can compare their results with it.

Courtesy of Curtis J. Walor

(*continued*)

FIGURE 27.7

(*continued*)

4

Curtis labels items in his graphic to help readers understand what it illustrates.

Cover slide

Stain

Slide

Figure D

He uses a figure to explain a procedure that would be difficult to understand if presented in words alone.

2. **Slowly lower the other side of the slide cover.** Be sure that no air gets trapped under the slide cover.

3. **Drain excess stain.** With a cloth underneath, turn the slide onto one of its longest edges so that the excess stain will run off.

4. **Clean slide.** Use acetone solvent to remove residue and fingerprints.

COUNTING THE FIBERS

1. **Place slide onto microscope.**

2. **Adjust magnification.** You should be able to distinguish black fibers from dark purple ones.

3. **Move slide to show upper left-hand corner of area with fibers.**

4. **Count whole fibers.** Move the slide so that your view of it changes in the manner shown in Figure E, counting

#1

Figure E

the whole softwood and hardwood fibers you see. Ignore fragments of fibers, which you will count later.

* **Recognizing softwood fibers.** Softwood fibers are long and flat. They have blunt ends. The stain dyes the fibers colors that range from slightly purple (almost translucent) to a dark purple. See Figure F.

Courtesy of Curtis J. Walor

FIGURE 27.7

(*continued*)

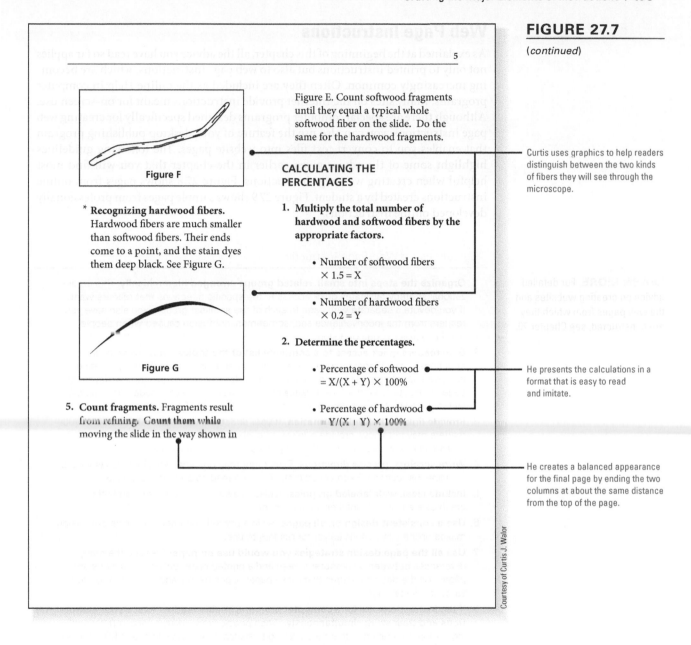

Figure F

Figure E. Count softwood fragments until they equal a typical whole softwood fiber on the slide. Do the same for the hardwood fragments.

Curtis uses graphics to help readers distinguish between the two kinds of fibers they will see through the microscope.

* **Recognizing hardwood fibers.** Hardwood fibers are much smaller than softwood fibers. Their ends come to a point, and the stain dyes them deep black. See Figure G.

Figure G

CALCULATING THE PERCENTAGES

1. **Multiply the total number of hardwood and softwood fibers by the appropriate factors.**

   - Number of softwood fibers
     $\times 1.5 = X$

   - Number of hardwood fibers
     $\times 0.2 = Y$

2. **Determine the percentages.**

   - Percentage of softwood
     $= X/(X + Y) \times 100\%$

   - Percentage of hardwood
     $Y/(X + Y) \times 100\%$

5. **Count fragments.** Fragments result from refining. Count them while moving the slide in the way shown in

He presents the calculations in a format that is easy to read and imitate.

He creates a balanced appearance for the final page by ending the two columns at about the same distance from the top of the page.

Web Page Instructions

As explained at the beginning of this chapter, all the advice you have read so far applies not only to printed instructions but also to web page instructions, which are becoming increasingly common. Often they are included as the online Help in computer programs. Many sites on the Internet provide instructions meant for on-screen use. Although there are special computer programs designed specifically for creating web page instructions, you can also use the feature of your desktop publishing program that enables you to convert text files into website pages. The following guidelines highlight some of the advice given earlier in the chapter that you will find most helpful when creating web page instructions. Figure 27.8 shows pages from online instructions created by a student. Figure 27.9 shows sample pages from professionally developed online instructions.

LEARN MORE For detailed advice on creating websites and the web pages from which they are constructed, see Chapter 20.

Guidelines for Web Page Instructions

1. **Organize the steps into small, related groups arranged hierarchically.** This organizational strategy facilitates quick access to the specific directions that readers want. If you devote a separate web page to each of the smallest groups, you also save your readers from the inconvenience and, sometimes, confusion caused when people need to scroll through directions.

2. **Give readers quick access to a complete list of the topics.** When using online instructions, many readers want to skip around among tasks. Help them by creating a complete list of items in your hierarchy so they don't have to move through layers of pages to locate what they want. Make this list available on every page, for instance, by putting a link to it in a navigation bar at the top or side of each page.

3. **Provide links to helpful information.** If your directions use a term that some of your readers may not know, provide a link to an explanation. If readers would find it helpful to learn about an alternative or related procedure, include a link to it.

4. **Write succinct, precise directions.** Even more than on paper, succinctness is valued in online instructions. Extra words make for extra reading and extra scrolling.

5. **Include clear, well-labeled graphics.** Tables, drawings, screen shots, and other graphics are just as helpful online as in print.

6. **Use a consistent design on all pages.** As in a printed document, a consistent design makes online instructions easier for readers to use.

7. **Use all the page design strategies you would use on paper.** Despite the many differences between a computer screen and a printed page, online instructions are pages. All the design features that make paper pages usable and persuasive do the same for digital pages.

8. **If the instructions are for a computer program, enable readers to see your instructions and their work simultaneously.** When people are working on a computer, nothing is more frustrating to them than having to switch between windows to follow directions. Design your web pages and their contents so that readers can keep their work open in a window beside yours.

9. **Conduct user testing.** All instructions, whether in print or online, have to be tested by members of the target audience, who try to use them under conditions identical to those of the anticipated actual use.

FIGURE 27.8

Online Instructions Created by a Student

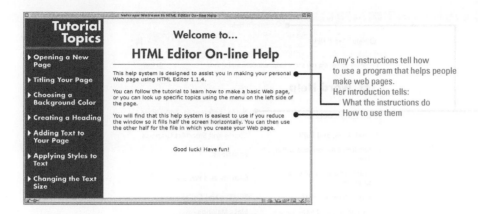

Amy's instructions tell how to use a program that helps people make web pages. Her introduction tells:
— What the instructions do
— How to use them

Amy uses the same basic design for all pages.

Most of her screens require no scrolling (though the menu on the left must be scrolled).

Note her use of figures.

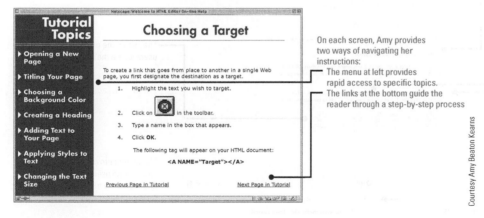

On each screen, Amy provides two ways of navigating her instructions:
— The menu at left provides rapid access to specific topics.
— The links at the bottom guide the reader through a step-by-step process

Courtesy Amy Beaton Kearns

FIGURE 27.9

Online Instructions

Writers at Microsoft create online Help pages for Word.

On the main page, they provide a table of contents plus a search option in order give users two ways to locate the information they want. Each link in the table of contents leads to a list of subtopics.

The writers use many features of effective layout that are also used in print instructions.

• Larger, bold type for main headings.

• Extra white space between steps to make reading easier.

• Bold to highlight key terms.

• Separate column for step numbers.

TRY THIS Analyze the online Help for a program you use. In what ways does it follow the guidelines provided in this book? It what ways does it deviate from them?

The writers include clear, helpful graphics to show users what they can create.

They show the program's icons to help users locate the items they needed.

The writers use color to make subheadings stand out.

The writers provide narrow Help windows so users can read the directions and also see their own work on the screen.

They use color to highlight words users can click for definitions.

They align figures in the text column, not the number column.

The writers include tables to help users see how to take alternative actions.

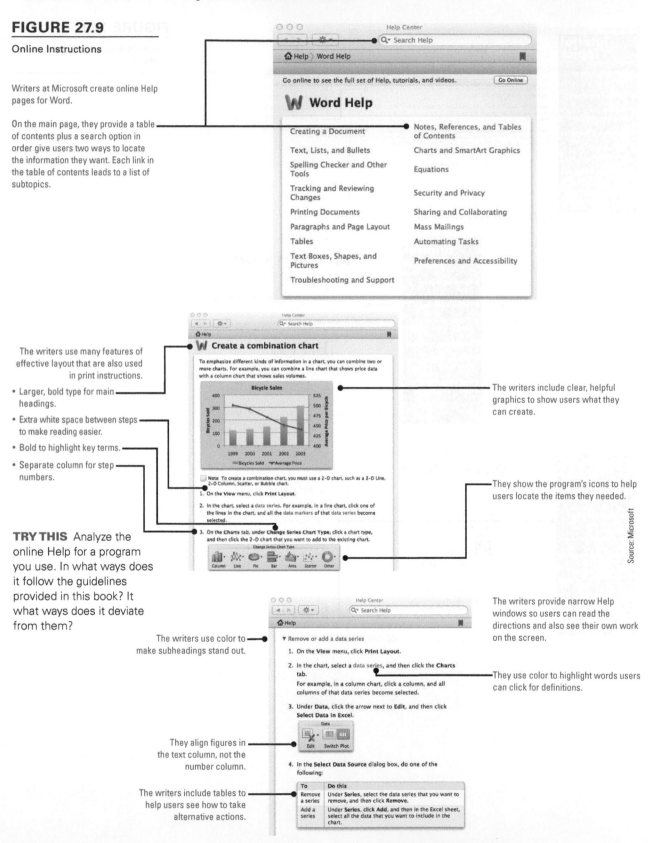

Source: Microsoft

Video Instructions

Increasingly, instructions are prepared as videos, with the text being spoken and the graphics being either a series of still images or video clips. New cars come with DVDs that provide instructions to new owners. Company websites include movies that tell people how to use the company's products. The Help feature of many computer programs, such as Microsoft Word, include videos that explain how to use some of the program's features. YouTube features an ever-growing collection of movie instructions.

Although web page instructions closely resemble printed instructions, video instructions are quite different. You couldn't create an effective script simply by reading directions prepared for print or a web page. In movie instructions, the relationship between images and script are coordinated through timing rather than placement on the page. In the movie version, you need to have images continuously. You can't leave the visual portion of the movie blank until you come to a place where you would use the next graphic in printed or web page instructions. Tone of voice and pace of reading make a great difference in the effectiveness of video instructions.

Movie-making programs are available for free on almost every computer. For Windows, you can download an easy-to-use program called Windows Movie Maker at *windows.microsoft.com*. Apple computers have iMovie. Most new computers are now equipped with cameras and microphones.

In MindTap you can read a bonus chapter about making video instructions.

Writer's Guides and Other Resources

Figure 27.10 (page 480) presents a Writer's Guide for Revising Instructions that you can use in your course and on the job.

Note to the Instructor MindTap includes a project that involves creating instructions. Instruction projects can be paired with a User Testing project, also included at MindTap. A User Testing Report would be a type of Empirical Research Report. To help students reflect for transfer, ask them to complete one of the exercises in Appendix B when they finish their project.

TRY THIS In the Help feature of a program on your computer, find a video that includes step-by-step instructions. Most desktop publishing programs, such as Microsoft Word, have them. Imagine that you are making a set of print instructions for the same procedure. Quickly draft the written instructions for the first few steps, including text and graphics. How are your print instructions like the movie instructions? How do they differ? Alternatively, write the video script, including graphics, for a simple, step-by-step procedure.

FIGURE 27.10

Writer's Guide for Revising Instructions

Writer's Guide
REVISING INSTRUCTIONS

The following headings reflect the elements of the conventional superstructure for instructions. In many situations, some of the elements are unnecessary. Include only the elements that will make your instructions useful and persuasive to your readers.

INTRODUCTION

❏ Describes the purpose, aim, or desired outcome of the procedure, if it is not clear from the title.

❏ Identifies the intended readers (their knowledge level, job descriptions, etc.).

❏ Tells the scope and organization, either separately or together.

❏ Provides motivation (reasons for following the instructions rather than using another procedure).

❏ Explains the conventions used.

❏ Includes safety information.

DESCRIPTION OF THE EQUIPMENT

❏ Shows the location of the key parts.

❏ Explains the function of the key parts.

MATERIALS AND EQUIPMENT

❏ Tells readers what materials and equipment to gather before starting the procedure.

DIRECTIONS

❏ Supports rapid comprehension and immediate use by providing only the information your readers need, listing the steps, using the active voice and imperative mood ("Do this."), and highlighting keywords.

❏ Helps readers locate the next step quickly by numbering the steps and putting blank lines between them, giving one action per step, and putting step numbers in their own column.

❏ Distinguishes action from supporting information by putting actions before responses and making actions stand out visually.

❏ Groups related steps under action-oriented headings.

❏ Uses many graphics.

❏ Presents branching steps clearly.

TROUBLESHOOTING

❏ Tells what to do if something fails or there is an unexpected result.

GRAPHICS (SEE CHAPTER 12)

❏ Includes wherever readers would find them helpful or persuasive.

❏ Looks neat, attractive, and easy to read.

❏ Refers to them at the appropriate points in the prose.

❏ Located where your readers can find them easily.

PAGE DESIGN (SEE CHAPTER 14)

❏ Looks neat and attractive.

❏ Helps readers find specific information quickly.

REVISING (SEE CHAPTERS 15 AND 16)

❏ Uses correct spelling, grammar, and punctuation.

❏ Provides directions that have been user-tested.

Documenting Your Sources

n many of the communications you will write at work, you will want to tell your readers about other sources of information concerning your subject. You may have any of the following reasons for wanting to do so:

- **To acknowledge the people and sources that have provided you with ideas and information.** For a discussion of this reason for citing your sources, see "How to Observe Intellectual Property Law and Document Your Sources" (page 82).

- **To help your readers find additional information about something you have discussed.**

- **To persuade your readers to pay serious attention to a particular idea.** By showing that an idea was expressed by a respected person or in a respected publication, you are arguing that the idea merits acceptance.

- **To explain how your research contributes to the development of new knowledge in your field.** In research proposals and in research reports published in professional journals, writers often include a literature review to demonstrate how their own research advances knowledge. For more information on using references in this way, see the discussion of expanding existing knowledge in Chapter 24 on empirical research reports (page 406).

Whenever you have determined that you should identify your sources for your readers, you must determine how to present that information. This appendix will help you by providing the following information.

MindTap®

Find additional resources related to this appendix in MindTap.

Which Documentation Style To Use

There are many formats for documentation, some very different from one another. Many organizations issue style guides that describe in detail the formats they want their employees to use. Others ask employees to follow a style guide published by a professional organization such as the American Psychological Association (APA), Institute of Electrical and Electronic Engineers (IEEE), or Modern Language Association (MLA). The rest of this appendix explains the APA, IEEE, and MLA styles. Most other documentation styles resemble one of these three.

Where to Place In-Text Citations

While documentation styles differ in many ways, most share the same two elements: a "reference" list of sources, giving full information about each one, that is placed at the end of your communication, and a "citation" placed by the statements in the text that are based on the source.

Regardless of the documentation style you use, the guidelines for placing the in-text citations are the same. For citations that pertain to a single fact, sentence, or quotation, place your citation immediately after the appropriate material (note that the APA style for formatting these citations is used in the following examples):

> According to F. C. Orley (2010, p. 37), "We cannot tell how to interpret these data without conducting further tests."

> Researchers have shown that a person's self-esteem is based upon performance (Dore, 2009), age (Latice, 1998), and weight (Swallen & Ditka, 2012).

If your citation refers to material that appears in several sentences, place the citations in a topic sentence that introduces the material. Your readers will then understand that the citation covers all the material that relates to that topic sentence. As a further aid to your readers, you may use the author's name (or a pronoun) in successive sentences:

> A much different account of the origin of oil in the earth's crust has been advanced by Thomas Gold (2008). He argues that…. To critics of his views, Gold responds….

The rest of this chapter describes the specific ways you would write the in-text citations and entries in your reference list in the APA, IEEE, and MLA styles.

How to Write APA In-Text Citations

To write an APA in-text citation, enclose the author's last name and the year of publication in parentheses inside your normal sentence punctuation. Place a comma between the author's name and the date. Use p. if you are citing a specific page and pp. if citing more than one page.

> The first crab caught in the trap attracts others to it (Tanner, 2007, pp. 33–34).

If you incorporate the author's name in the sentence itself, give only the year and pages (if any) in parentheses:

> According to Tanner (2007, pp. 33–34), the first crab caught in the trap attracts others to it.

Here are some other types of citations:

(Hoeflin & Bolsen, 2004)	Two authors.
(Wilton, Nelson, & Dutta, 1932)	First citation for three, four, or five authors.
(Wilton et al., 1932)	Second and subsequent citations for three to five authors. Omit year in subsequent citations in the same paragraph (*et al.* is an abbreviation for the Latin phrase *et alii*, which means *and others;* because it is an abbreviation, it is followed by a period).
(Norton et al., 2004)	First and subsequent citation for six or more authors.
(Angstrom, 2010, p. 34)	Reference to a particular page.
(U.S. Department of Energy, 1997)	Government or corporate author.
("Geologists Discover," 2011)	No author listed (use the first few words of the title). In the example, the words from the title are in quotation marks because the citation is to an article; when a book is cited, the words are in italics with no quotation marks.
(Justin, 1998; Skol, 1972; Weiss, 1966)	Two or more sources cited together (arrange them in alphabetical order).

In some communications, you might cite two or more sources by the same author. If they were published in *different* years, your readers will have no trouble telling which work you are referring to. If they were published in the same year, you can distinguish between them by placing lowercase letters after the publication dates in your citations and in your reference list:

(Burkehardt, 1998a)

(Burkehardt, 1998b)

How to Write an APA Reference List

The following examples illustrate and explain how to use the APA style to write reference list entries for the most common types of print, electronic, and other sources. To create entries that are not listed here, follow the logic of these examples or consult the *Publication Manual of the American Psychological Association, Sixth Edition*, available in the reference section of most libraries. Figure A.1 shows how to arrange entries in your reference list.

Print Sources

1. Book, One Author—APA

Lightman, A. P. (2005). *A sense of the mysterious: Science and the human spirit.* New York, NY: Pantheon Books.

- Give the author's last name followed by a comma and initials (not full first or middle names).
- Place the copyright date in parentheses, followed by a period.

- Italicize the title and capitalize only the first word of the title, proper nouns, and the first word after a colon or dash in the title.
- Follow the city of publication with a comma and the two-letter postal abbreviation for the state.
- Indent the second and subsequent lines by 5 spaces.

2. Book, Two or More Authors—APA

Budinski, K. G., & Budinski, M. K. (2005). *Engineering materials: Properties and selection.* Upper Saddle River, NJ: Pearson.

Adams, D. J., Dyson, P. J., & Tavener, S. J. (2004). *Chemistry in alternative reaction media.* Hoboken, NJ: Wiley.

- Give the names of all authors the same way (last name, comma, then initials).
- Use the ampersand (&) instead of the word and before the last author.

FIGURE A.1

APA Reference List

Second and subsequent lines are indented.

When the list includes two or more items by the same author, the oldest appears first.

For the second and subsequent items by the same person, the author's name is repeated.

Items by corporate and government groups are alphabetized by the groups' names (spelled out).

Items without authors are alphabetized by the title.

References

Adams, D. J., Dyson, P. J., & Tavener, S. J. (2004). *Chemistry in alternative reaction media.* Hoboken, NJ: Wiley.

Gould, S. J. (1989). *Wonderful life: The burgess shale and the nature of history.* New York, NY: Norton.

Gould, S. J. (1995). *Dinosaur in a haystack: Reflections in natural history.* New York, NY: Harmony.

IBM. (n.d.). *Introducing IBM Design Thinking: Open Possibilities.* Retrieved from http://www.ibm.com/design/.

Rethinking traditional design. (1997). *Manufacturing Engineering 118*(2), 50.

Younger manufacturing workers can teach old dogs new tricks. (2007). *Manufacturing, 86*(5), 5.

3. Anthology or Essay Collection—APA

Bodeker, G., & Burford, G. (Eds.). (2007). *Traditional, complementary and alternative medicine: Policy and public health perspectives*. London, UK: Imperial College Press.

■ If there is only one editor, use this abbreviation: Ed.

4. Second or Subsequent Edition—APA

Simmons, L. H. (2011). *Olin's construction: Principles, materials, and methods* (9th ed.). New York, NY: John Wiley.

5. Government Report—APA

Frankforter, J. D., & Emmons, P. J. (1997). *Potential effects of large floods on the transport of atrazine into the alluvial aquifer adjacent to the Lower Platte River, Nebraska* (U.S. Geological Survey Water-Resources Investigation Report 96-4272). Denver, CO: U.S. Geological Survey.

■ If the report doesn't list an author, use the agency that published it as the author. If it is a United States government agency, use the abbreviation "U.S."

■ If the report has an identifying number, place it immediately after the title in parentheses.

6. Corporate Report—APA

Daimler-Benz AG. (2015). *Sustainability report 2015*. Stuttgart, Germany: Author.

■ List the names of the individual authors rather than the corporation if the names are given on the title page.

■ If the names of the individual authors aren't given on the title page, list the corporation as the author. (In the example, "Daimler-Benz AG" is the name of a company.)

■ When the author and publisher are the same, use the word "Author" as the name of the publisher.

7. Chapter in a Book—APA

Moor, J. H. (2008). Why we need better ethics for emerging technologies. In J. Hoven & J. Weckert (Eds.), *Information technology and moral philosophy* (pp. 24–43). New York, NY: Cambridge University Press.

8. Article in an Encyclopedia, Dictionary, or Similar Reference Work—APA

Rich, E. (2003). Artificial intelligence. In *Encyclopedia Americana*. (Vol. 2, pp. 407–412). Danbury, CT: Grolier.

■ If no author is listed, begin with the article's title followed by the year.

9. Pamphlet or Brochure—APA

Ohio Department of Natural Resources. (2010). *Ring-necked pheasant management in Ohio* [Pamphlet]. Columbus, OH: Author.

■ Include the word "Pamphlet" or "Brochure" in brackets after the title.

■ If no author is listed, use the organization that published the brochure as the author.

■ When the author and publisher are the same, use the word "Author" as the name of the publisher.

10. Article in Journal that Numbers Pages Continuously through Each Volume—APA

Alonso, A., & Pulido, R. (2016). The extended human PTPome: A growing tyrosine phosphatase family. *FEBS Journal, 283,* 2197–2201.

- ■ After the journal's name (note use of capital letters), add a comma and the volume number (in italics) followed by a comma and the page numbers.

11. Article in Journal that Begins Each Issue with Page 1—APA

Bradley, J., & Soulodre, G. (2009, May). The acoustics of concert halls. *Physics World, 10*(5), 33–37.

- ■ Follow the year with a comma and the month.
- ■ After the journal title, include the volume number, followed by the issue number (in parentheses).
- ■ Put the volume number—but not the issue number—in italics.

12. Article with No Author Listed—APA

Younger manufacturing workers can teach old dogs new tricks. (2007, August). *Manufacturing, 86*(5), 5.

- ■ Begin with the article's title. This example gives the issue number in parentheses because the journal numbers its pages separately for each issue (see Example 11).

Electronic Sources

To cite online sources, begin with as much information as is available that you would include for a print source. Follow with information that enables a reader to locate the source. If a Digital Object Identifier (DOI) is available, use it. If not, use the URL. A DOI is a unique set of numbers and letters; it is preferred because, unlike urls, it does not change and can be used in browser searches.

For APA citations, if the URL leads to the item itself, precede it with "Retrieved from." If it leads to a source that tells how to obtain the item, use "Available from." Because material on the Internet often changes, APA citations do not include the date you retrieved an item. The IEEE and MLA styles have different conventions; see the sections on them for details.

13. Electronic Version of Print Book—APA

Shotton, M. A. (1989). *Computer addiction? A study of computer dependency.* New York, NY: Taylor & Francis. Available from http://www.crc.com

- ■ Give print information followed by online location.
- ■ If the book is also available in print, provide the relevant publication information (city, state, and publisher) after the title.

14. Online Journal Article that Is Not Available in Print—APA

Xu, P. (2013). Analysis of isolation effects about elastic waves by discontinuous barriers composed of rigid piles. *Electronic Journal of Geotechnical Engineering, 18,* 1–11. Retrieved from http://www.ejge.com/2013/JourTOC18A.htm

15. Online Journal Article that Is Also Available in Print—APA

Al-Taay, H. F., Mahdi, M. I., Parlevliet, D., & Jennings, P. (2017). Fabrication and characterization of solar cells based on silicon nanowire homojunctions. *Silicon*, *9*, 17–23. DOI 10.1007/s12633-015-9329-0

16. Web Page or Document that Is Not Part of a Journal—APA

IBM. (n.d.). *Introducing IBM design thinking: Open possibilities.* Retrieved from http://www.ibm.com/design/

- If the web page does not give a date, use "(n.d.)" as an abbreviation for "no date."

Additional Common Sources

17. Email—APA

The APA style includes references to emails only in parentheses in the text, not in the reference list. The parenthetical citation in the text includes the author's initials as well as his or her last name and an exact date:

(M. Grube, personal communication, December 4, 2017)

18. Letter—APA

(L. A. Cawthorne, personal communication, August 24, 2017)

- The APA style treats letters the same way it treats emails (see Example 17).

19. Interview—APA

(S. Oparka, personal communication, October 17, 2015)

- The APA style treats interviews the same way it treats emails (see Example 17).

How to Write IEEE in-Text Citations

In the IEEE style, in-text citations consist simply of a number enclosed in square brackets and placed inside the sentence punctuation. Each number refers the reader to the description of the source, which is presented in the reference list. You number the citations according to the order in which they appear in the text: the first citation is [1], the second is [2], and so on.

Wellings [1] agrees, but Astin [2] argues that the calculations are flawed.

When a source is cited for a second time, it retains its original number.

As demonstrated by Madden [7] and Thompson [8], Astin [2] was correct.

In the IEEE style, you can also treat citation numbers as nouns.

This confusion can be seen in [2]; According to [5].

How to Write an IEEE Reference List

In an IEEE reference list, arrange your sources in the order you cite them in your text—*not* in alphabetical order. The source you numbered [1] goes first, followed by [2], and so on. As shown in Figure A.2, place the numbers in square brackets and place them in a column of their own. To create entries that are not listed here, follow the logic of these examples or consult the *IEEE Editorial Style Manual (2014)*, available at www.ieee.org.

Print Sources

1. Book, One Author—IEEE

[1] A. P. Lightman, *A Sense of the Mysterious: Science and the Human Spirit.* New York, NY, USA: Pantheon Books, 2005.

- ■ Give the author's first and middle initials followed by periods, then the author's last name followed by a comma.
- ■ Italicize the title, followed by a period. Capitalize the major words.
- ■ Along with the city and state (US only), include the country in which the publisher is located.

FIGURE A.2

IEEE Reference List

Reference numbers are listed in a separate column.

The second and following lines of each entry align with the first line.

References

[1] D. J. Adams *et al., Chemistry in Alternative Reaction Media.* Hoboken, NJ, USA: Wiley, 2004.

[2] S. J. Gould, *Dinosaur in a Haystack: Reflections in Natural History.* New York, NY, USA: Harmony, 1995.

[3] S. J. Gould, *Wonderful Life: The Burgess Shale and the Nature of History.* New York, NY, USA: Norton, 1989.

[4] IBM., (n.d.). "Introducing IBM Design Thinking: Open Possibilities," n.d. [Online]. Available: http://www.ibm.com/design/. Accessed on: Aug 24, 2017.

[5] J. McLaurin and A. Chakrabartty, "Characterization of the interactions of Alzheimer β-amyloid peptides with phospholipid membranes," *European J. of Biochem.,* vol. 245, no. 5, pp. 355–363, Apr. 1997.

[6] "Rethinking traditional design," *Manufacturing Engineering,* vol. 118, no. 2, p. 50, Feb. 1997.

[7] "Younger manufacturing workers can teach old dogs new tricks," *Manufacturing,* vol. 86, no. 5, p. 5, May 2007.

- Follow the city, state (abbreviated), and country (abbreviated) of publication with a colon, the publisher, comma, and year of publication.
- Align lines after the first one with the author's first initial.

2. Book, Two or More Authors—IEEE

[2] K. G. Budinski and M. K. Budinski, *Engineering Materials: Properties and Selection.* Upper Saddle River, NJ, USA: Pearson, 2005.

[3] D. J. Adams *et al., Chemistry in Alternative Reaction Media.* Hoboken, NJ, USA: Wiley, 2004.

- If there are six or fewer authors, give the names of all in the same way (initials, then last name).
- If there are seven or more authors, use *et al.* in italics after the first author's name. Note that *et al.* includes a period that is followed by a comma (*et al.* is an abbreviation for a Latin phrase that means *and others*).

3. Anthology or Essay Collection—IEEE

[4] G. Bodeker and G. Burford, Eds., *Traditional, Complementary and Alternative Medicine: Policy and Public Health Perspectives.* London, UK: Imperial College Press, 2007.

- If there is only one editor, use the abbreviation "Ed.".

4. Second or Subsequent Edition—IEEE

[5] L. H. Simmons, *Olin's Construction: Principles, Materials, and Methods,* 9th ed. New York, NY, USA: John Wiley, 2001.

5. Government Report—IEEE

[6] J. D. Frankforter and P. J. Emmons, "Potential Effects of Large Floods on the Transport of Atrazine into the Alluvial Aquifer Adjacent to the Lower Platte River, Nebraska," U.S. Geological Survey Water-Resources Investigation Report 96-4272, 1997.

- If the report doesn't list an author, use the name of the agency that published it as the author. If it is a United States government agency, use the abbreviation "U.S."
- If the report has an identifying number, place it immediately after the title.

6. Corporate Report—IEEE

[7] Daimler-Benz AG, "Sustainability Report 2015," Daimler-Benz AG, Stuttgart, Germany, 2015.

- If the report doesn't list an author, use the name of the corporation as the author. (In the example, "Daimler-Benz AG" is a corporation.)

7. Chapter in a Book—IEEE

[8] J. H. Moor, "Why we need better ethics for emerging technologies," in *Information Technology and Moral Philosophy,* J. Hoven and J. Weckert, Eds. New York, NY, USA: Cambridge University Press, 2008, pp. 24–43.

- Capitalize only the first word of the chapter title.

8. Article in an Encyclopedia, Dictionary, or Similar Reference Work—IEEE

[9] E. Rich, "Artificial intelligence," in *Encyclopedia Americana*. Danbury, CT, USA: Grolier, 2003, vol. 2, pp. 407–412.

- If no author is listed, begin with the article's title.

9. Pamphlet or Brochure—IEEE

[10] Ohio Department of Natural Resources, "Ring-necked pheasant management in Ohio." Columbus, OH, USA: Ohio Department of Natural Resources, 2010.

- If no author is listed, use the organization that published the brochure as the author.

10. Article in Journal that Numbers Pages Continuously through Each Volume—IEEE

[11] J. McLaurin and A. Chakrabartty, "Characterization of the interactions of Alzheimer β-amyloid peptides with phospholipid membranes," *European J. of Biochem.*, vol. 245, no. 5, pp. 355–363, Apr. 1997.

- Place the article's title in quotation marks, and capitalize only the first word and proper nouns.
- Abbreviate the journal's name, if possible.
- Include the volume and issue numbers.
- Include the month of publication; abbreviate if it has more than four letters.
- In the example, the word *Alzheimer* is capitalized because it is a proper noun (a person's name).

11. Article in Journal that Begins Each Issue with Page 1—IEEE

[12] J. Bradley and G. Soulodre, "The acoustics of concert halls," *Physics World,* vol. 10, no. 5, pp. 33–37, May 1997.

- Same as for article in a journal that numbers its pages continuously through each volume. See the preceding explanation (10).

12. Article with No Author Listed—IEEE

[13] "Younger manufacturing workers can teach old dogs new tricks," *Manufacturing*, vol. 85, no. 5, p. 5, May 2007.

- Begin with the article's title if no author is listed.

Electronic Sources

13. Electronic Version of Print Book—IEEE

[14] M. A. Shotton, *Computer Addiction? A Study of Computer Dependency*, New York, NY, USA: Taylor & Francis, 1989. [Online]. Available: http://www.ebookstore.tandf.co.uk/html/index.asp. Accessed on Oct. 28, 2017.

- Include print information followed by online location.

14. Online Journal Article that Is Not Available in Print—IEEE

[15] P. Xu, "Analysis of isolation effects about elastic waves by discontinuous barriers composed of rigid piles," *Electron. J. of Geotech. Eng.,* vol. 18, no. A, pp. 1–11, 2013. [Online]. Available: http://www.ejge.com/2013/JourTOC18A.htm. Accessed on: Nov. 12, 2017.

15. Online Journal Article that Is Also Available in Print—IEEE

[16] H. F. Al-Taay, M. I. Mahdi, D. Parlevliet, and P. Jennings. "Fabrication and characterization of solar cells based on silicon nanowire homojunctions," *Silicon,* vol. 9, no. 1, pp. 17–23, Jan. 2017. [Online]. Available: DOI 10.1007/s12633-015-9329-0. Accessed on Nov. 21, 2017.

- Give print information followed by online location.
- For an explanation of DOI web locators, see the note just under the heading "Electronic Sources" in the APA section (page 486).

16. Web Page or Document That Is Not Part of a Journal—IEEE

[17] IBM. "Introducing IBM Design Thinking: Open Possibilities," n.d. [Online]. Available: http://www.ibm.com/design/. Accessed on: Aug. 24, 2017.

- If the web page does not give a date, use "n.d." as an abbreviation for "no date."

Additional Common Sources

17. Email—IEEE

[18] M. Grube, private communication, Dec. 4, 2017.

18. Letter—IEEE

[19] A. Cawthorne, private communication, Aug. 24, 2017.

19. Interview—IEEE

[20] S. Oparka, private communication, Oct. 17, 2015.

How to Write MLA In-Text Citations

A basic MLA citation contains two kinds of information: the author's name and the specific page or pages on which the cited information is to be found. Enclose these in parentheses—with no punctuation between them—and place them inside your normal sentence punctuation. If you are citing the entire work, omit the page numbers. The following citation refers the reader to pages 33 and 34 of a work by Tanner. Note that MLA style uses a hyphen to separate page ranges. In large numbers, only the last two digits are provided unless more are necessary (252-53, but 295-305).

> The first crab caught in the trap attracts others to it (Tanner 33-34).

If you incorporate the author's name in the sentence itself, give only the page numbers in parentheses:

> According to Tanner, the first crab caught in the trap attracts others to it (33-34).

Here are some other types of citations:

(Hoeflin and Bolsen 16)	Two authors.
(Wilton, Nelson, and Dutta 222)	Three authors.
(Norton et al. 776)	Four or more authors. Note that *et al.* includes a period (it is an abbreviation for a Latin phrase that means *and others*).
(U.S. Department of Energy 4-7)	Government or corporate author.
("Discover" 31)	No author listed (use the first few words of the title). In the example, the words from the title are in quotation marks because the citation is to an article; if a book is cited, the words would be in italics with no quotation marks. When there are two or more anonymous works by the same title, include in the parenthetical reference either a publication year for a book or a periodical name for an article.
(Justin 23; Skol 1089; Weiss 475)	Two or more sources cited together.

In some of your communications, you might cite two or more sources by the same author. If they were published in different years, your readers will have no trouble telling which work you are referring to. If they were published in the same year, distinguish between them by placing a comma after the author's name, followed by a few words from the title.

Whole work cited	(Burkehardt, "Gambling Addiction")
Specific pages cited	(Burkehardt, "Obsessive Behaviors," 81-87)

How to Write an MLA Works Cited List

The following sections describe how to write entries for the most common types of print, electronic, and other sources. To create entries that are not listed here, follow the logic of these examples or consult those available in the *MLA Handbook for Writers of Research Papers, Eighth Edition,* available in the reference section of most libraries. Figure A.3 shows how to arrange entries in your list of works cited.

Print Sources

1. Book, One Author—MLA

Lightman, Alan P. *A Sense of the Mysterious: Science and the Human Spirit.* Pantheon Books, 2005.

- Give the author's last name followed by a comma and then the first and middle names or initials—exactly as they appear on the title page. Follow with a period unless the author's name ends with a period after an initial.
- Italicize the title, followed by a period. Capitalize the major words.
- Give the publisher's name, comma, and publication date followed by a period.
- Indent one-half inch all lines after the first one.

FIGURE A.3

MLA List of Works Cited

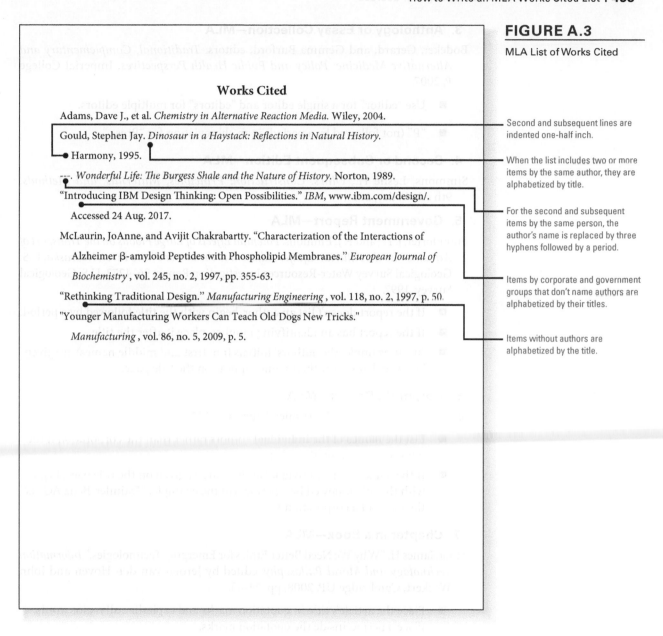

Works Cited

Adams, Dave J., et al. *Chemistry in Alternative Reaction Media*. Wiley, 2004.

Gould, Stephen Jay. *Dinosaur in a Haystack: Reflections in Natural History*.
　Harmony, 1995.

---. *Wonderful Life: The Burgess Shale and the Nature of History*. Norton, 1989.

"Introducing IBM Design Thinking: Open Possibilities." *IBM*, www.ibm.com/design/.
　Accessed 24 Aug. 2017.

McLaurin, JoAnne, and Avijit Chakrabartty. "Characterization of the Interactions of
　Alzheimer β-amyloid Peptides with Phospholipid Membranes." *European Journal of*
　Biochemistry , vol. 245, no. 2, 1997, pp. 355-63.

"Rethinking Traditional Design." *Manufacturing Engineering* , vol. 118, no. 2, 1997, p. 50.

"Younger Manufacturing Workers Can Teach Old Dogs New Tricks."
　Manufacturing , vol. 86, no. 5, 2009, p. 5.

Second and subsequent lines are indented one-half inch.

When the list includes two or more items by the same author, they are alphabetized by title.

For the second and subsequent items by the same person, the author's name is replaced by three hyphens followed by a period.

Items by corporate and government groups that don't name authors are alphabetized by their titles.

Items without authors are alphabetized by the title.

2. Book, Two or More Authors—MLA

Budinski, Kenneth G., and Michael K. Budinski. *Engineering Materials: Properties and Selection*. Pearson, 2005.

Adams, Dave J., et al. *Chemistry in Alternative Reaction Media*. Wiley, 2004.

- If a source has two authors, follow the first author's name with a comma and the word *and*. Give the second author's name in the normal order (first name first). See the first example above.

- If a source has more than two authors, follow the first author's name with a comma and "et al." Note that "et al." includes a period ("et al." is an abbreviation for a Latin phrase that means "and others"). See the second example above, which is for a book that has three authors.

3. Anthology or Essay Collection—MLA

Bodeker, Gerard, and Gemma Burford, editors. *Traditional, Complementary and Alternative Medicine: Policy and Public Health Perspectives*. Imperial College P, 2007.

- Use "editor" for a single editor and "editors" for multiple editors.

- "P" (not followed by a period) is an abbreviation for "Press."

4. Second or Subsequent Edition—MLA

Simmons, Leslie H., *Olin's. Construction: Principles, Materials, and Methods*. 9th ed., Wiley, 2011.

5. Government Report—MLA

Frankforter, J. D., and P. J. Emmons. *Potential Effects of Large Floods on the Transport of Atrazine into the Alluvial Aquifer Adjacent to the Lower Platte River, Nebraska*, U.S. Geological Survey Water-Resources Investigation Report 96-4272, U.S. Geological Survey, 1997.

- If the report doesn't list an author, begin with the title followed by a period.

- If the report has an identifying number, place it after the title.

- In the example, the authors' initials (not first and middle names) are given because that is how their names appear on the title page.

6. Corporate Report—MLA

Sustainability Report 2015. Daimler-Benz AG, 2015.

- List the names of the individual authors rather than the corporation if the names are given on the title page.

- If the names of the individual authors aren't given on the title page, begin with the title followed by a period. (In the example, "Daimler-Benz AG" is the name of a corporation.)

7. Chapter in a Book—MLA

Moor, James H. "Why We Need Better Ethics for Emerging Technologies." *Information Technology and Moral Philosophy*, edited by Jeroen van den Hoven and John Weckert, Cambridge UP, 2008, pp. 24-43.

- Place the article's title in quotation marks and capitalize all major words. Place a period inside the quotation marks.

- Use "UP" as an abbreviation for "University Press."

- Use "pp." before the chapter's page numbers. Use p. if the chapter is only one page.

8. Article in an Encyclopedia, Dictionary, or Similar Reference Work—MLA

Rich, Elaine. "Artificial Intelligence." *Encyclopedia Americana*, 2003 ed.

- If no author is listed, begin with the article's title.

- Place the article's title in quotation marks and capitalize all major words. Place a period inside the quotation marks.

- If entries in the work are arranged alphabetically, do not give the volume or page number.
- When citing familiar reference works, give the edition number (if provided) and year of publication, but not the publisher.

9. Pamphlet or Brochure—MLA

State of Ohio. Department of Natural Resources. *Ring-Necked Pheasant Management in Ohio*. State of Ohio, 1996.

- If no author is listed, begin the entry with the name of the government or other organization that published it, followed by a period and the name of the agency that issued the document.
- If the pamphlet or brochure lists no author and no publisher, begin with the document's title.

10. Article in Journal that Numbers Pages Continuously through Each Volume—MLA

McLaurin, JoAnne, and Avijit Chakrabartty. "Characterization of the Interactions of Alzheimer β-amyloid Peptides with Phospholipid Membranes." *European Journal of Biochemistry*, vol. 245, no. 2, 1997, pp. 355-63.

- Place the article's title in quotation marks and capitalize all major words.

> For page numbers larger than 99, give only the last two digits unless more are needed for clarity. Examples: "355-63" and "394-405."

11. Article in Journal that Begins Each Issue with Page 1—MLA

Bradley, John, and Gilbert Soulodre. "The Acoustics of Concert Halls." *Physics World*, vol. 10, no. 5, 2009, pp. 33-37.

- Same as for an article in a journal that numbers its pages continuously through each volume. See the preceding explanation (10).

12. Article with No Author Listed—MLA

"Younger Manufacturing Workers Can Teach Old Dogs New Tricks." *Manufacturing*, vol. 86, no. 5, 2007, p. 5.

- Begin with the article's title in quotation marks.
- This example uses the abbreviation "p." instead of "pp." because the article is only one page long.

Electronic Sources

13. Electronic Version of Print Book—MLA

Shotton, M. A. *Computer Addiction? A Study of Computer Dependency*. Taylor & Francis, 1989. *Questia*, www.questia.com/library/108876699/computer-addiction -a-study-of-computer-dependency.

- Give print information followed by online location.
- For online locations, omit "http://."
- Place a period at the end of online locations

14. Online Journal Article That Is Not Available in Print—MLA

Xu, Ping. "Analysis of Isolation Effects about Elastic Waves by Discontinuous Barriers Composed of Rigid Piles." *Electronic Journal of Geotechnical Engineering*, vol. 18, no. A, 2013, pp. 1-11. *EJGE*, www.ejge.com/2013/Abs2013.002.htm.

15. Online Journal Article That Is Also Available in Print—MLA

Al-Taay, H. F., et al. "Fabrication and characterization of solar cells based on silicon nanowire homojunctions." *Silicon*, vol. 9, no. 1, 2017, pp. 17-23. DOI 10.1007/s12633-015-9329-0.

- For an explanation of DOI web locators, see the note just under the heading "Electronic Sources" in the APA section, page 486.

16. Web Page or Document That Is Not Part of a Journal—MLA

"Introducing IBM Design Thinking: Open Possibilities." *IBM*, www.ibm.com/design/. Accessed 24 Aug. 2017.

- Give the title of the page or document in quotation marks followed by a period.

- Give the website's name in italics, followed by a the url or doi.

- This example begins with the title because no author is listed; if one were, the entry would begin with the author's name.

- Provide the access date for items that might be altered in the future so readers will know which version you used.

Additional Common Sources

17. Email—MLA

Grube, Melvin. "Effects of Sun Spots." Email received by Justin Timor, 4 Dec. 2017.

- Place the title (taken from the subject line) in quotation marks.
- Abbreviate months with five or more letters.

18. Letter—MLA

Cawthorne, Linda A. Letter to the author, 24 Aug. 2017.

19. Telephone Interview—MLA

Oparka, Sabrina. Telephone interview, 17 Oct. 2015.

Reflecting for Transfer

As you know, a major goal of your instructor and this book is to build your ability to apply what you are learning in this course to the writing situations you will encounter in your career—and also in your other courses. The key to building this ability is to develop the habit of engaging in two kinds of reflection:

- At the beginning of a project, reflecting on the knowledge and skills you already possess that you can adapt to your work on the communication you are about to write

- At the end of a project, reflecting on what you have learned and done to make the project successful that could be helpful on future projects

The assignments below ask you to engage in the second kind of reflection. Each involves writing a communication to yourself, your instructor, or another audience when you turn in one of your major course projects.

Your instructor will pick (and perhaps modify) the assignment that is best suited to each project you are completing.

For each, your instructor will specify the length of your reflection, the audience, and the form it should take (e.g., memo, email, PowerPoint, blog, etc.).

In creating these questions and the "Reflect for Transfer" exercises elsewhere in the book, I am indebted to Bransford, Pellegrino, and Donovan (2000) and Yancey, Robertson, and Taczak (2015).

Reflection 1: Transfer from the Past and into the Future

As you worked on this project, what are the three most important pieces of writing knowledge and skills that you knew before the project began that you were able to use to make your project successful? What are the three most important pieces of new knowledge and new skills that you applied? For each, cite a specific place in your project where you used the knowledge or skill you identify. In what future situations do you imagine the new knowledge and skills will be most helpful? Why did you pick those situations?

Reflection 2: Context

Describe the features of the context for your reader and communication that were the most novel or difficult for you to take into account as you worked on this project. What strategies did you use to adjust your communication to these features? Give specific examples from your communication that illustrate these adjustments.

Reflection 3: Reader-Centered Process

Describe one of the reader-centered strategies described in this book or by your instructor that you used in each of the following activities of your writing process: defining your communication's objectives, researching, organizing, drafting text and graphics, and reviewing and revising. Which of these strategies do you believe will be most helpful to you when writing on the job? Which will be least helpful? Why?

Reflection 4: The Nature of Workplace Writing

Define your understanding of the nature of workplace writing by identifying three strategies you used in this project, which simulates workplace writing, that differ most from the strategies you have used in another one of your courses. Give specific examples of places in your writing process or your finished communication where you used these strategies.

Reflection 5: Superstructure or Genre

Evaluate the ways that the superstructure (or genre) of this project helped you write a successful reader-centered communication. What specific writing choices did you make by considering the conventions of the superstructure? What features of the superstructure were most difficult for you to work with? Why? How did you overcome the difficulty?

Reflection 6: What Makes You Most Proud?

Imagine that a researcher who studies learning has asked you, "What aspect of your project and your work on it makes you most proud?" (If this reflection is used at the end of a term, the interviewer's question could be, "What aspect of your work in this course makes you the proudest?") Give specific examples, and explain why you are proud of these aspects.

Reflection 7: What Makes Writing Effective in the Workplace?

This reflection is designed to have you summarize at the end of your course what you've learned about workplace writing. Thinking back over your work on this and earlier projects in this course, describe your understanding of what constitutes effective writing in the workplace. What is your way of describing the goal of this writing? What is your way of describing the things you must do to write successfully in the workplace? Based on your experience in this course, how does workplace writing differ, and how does it resemble writing you do in other contexts, including in and out of school?

REFERENCES

Anderson, J. R. (2014). *Cognitive psychology and its implications* (8th ed.). New York, NY: Worth

Anonymous. (1994). Personal interview with corporate executive who requested that the company remain anonymous.

Axtell, R. E. (2007). *Essential do's and taboos: The complete guide to international business and leisure travel.* Hoboken, NJ: Wiley.

Beebe, S. A. (1974). Eye contact: A nonverbal determinant of speaker credibility. *Speech Teacher, 23,* 21–25. Cited in M. F. Vargas (1986). *Louder than words.* Ames, IA: Iowa State University Press.

Beer, D. F., & McMurrey, D. (2009). *A guide to writing as an engineer* (3rd ed.). New York, NY: Wiley.

Boren, M. T., & Ramey, J. (2000, September). Thinking aloud: Reconciling theory and practice. *IEEE Transactions on Professional Communication, 43(3),* 261–278.

Bosley, D. S. (1993). Cross-cultural collaboration: Whose culture is it, anyway? *Technical Communication Quarterly, 2,* 51–62.

Bransford, J. D., & Johnson, M. K. (1972). Contextual prerequisites for understanding: Some investigations of comprehension and recall. *Journal of Verbal Learning and Verbal Behavior, 11,* 717–726.

Bransford, J. D., Pellegrino J. W., & Donovan, M. S. (Eds.) (2000). *How people learn: Brain, mind, experience, and school: Expanded edition.* Washington, DC: National Academy Press.

Carté, P., & Fox, C. (2004). *Bridging the culture gap: A practical guide to international business communication.* Sterling, VA: Kogan Page.

Chong, A. (2016, March 4). "'What Google learned from its quest to build the perfect team': New York Times." IEEE Professional Communications Society Blog. Retrieved from http://sites.ieee .org/pcs/what-google-learned-from-its-quest-to-the-perfect -team-new-york-times/

Coleman, E. B. (1964). The comprehensibility of several grammatical transformations. *Journal of Applied Psychology, 48,* 186–190.

Conference Board. (2006). *Are they really ready to work? Employers' Perspectives on the Basic and Applied Skills of New Entrants to the 21st Century U.S. Workforce.* Retrieved from http://www .p21.org/storage/documents/FINAL_REPORT_PDF09-29-06.pdf

Connor, U., & Nagelhout, E. (2008). *Contrastive rhetoric: Reaching to intercultural rhetoric.* Philadelphia, PA: John Benjamins Publishing.

Cotton, G. (2013). *Say anything to anyone, anywhere: 5 keys to successful cross-cultural communication.* Hoboken, NJ: Wiley.

Covey, S. R. (2013). *The seven habits of highly effective people: Anniversary edition.* New York, NY: Simon & Schuster.

Davidson, J. (2014, October 16). The 7 social media mistakes most likely to cost you a job. *Money.* Retrieved from http://time.com /money/3510967/jobvite-social-media-profiles-job-applicants/

De Mente, B. L., & Botting, G. (2015). *Etiquette guide to Japan: Know the rules that count* (3rd ed.). North Clarendon, VT: Tuttle.

Ding, D. D. (2003). The emergence of technical communication in China. *Journal of Business and Technical Communication, 17,* 319–345.

Duhigg, C. (February 25, 2016). What Google learned from its quest to build the perfect team. *New York Times.* Retrieved from http://www .nytimes.com/2016/02/28/magazine/what-google-learned-from -its-quest-to-build-the-perfect-team.html

Hart Research Associates. (2015). *Falling short? College learning and career success: Selected findings from online surveys of employers and college students conducted on behalf of the Association of American Colleges & Universities.* Washington, DC: Author.

Harvard Business School Press. (2005). *Power, influence, and persuasion: Sell your ideas, and make things happen.* Cambridge, MA: Harvard Business School Press.

Hays, R. B. (1985). A longitudinal study of friendship development. *Journal of Personality and Social Psychology, 48,* 909–924.

Herrington, A., & Moran, C. (2005). *Genre across the curriculum.* Logan: Utah State University Press.

Herzberg, F. (2003). One more time: How do you motivate employees? [Reprinted from 1968] *Harvard Business Review, 81,* 87–96.

Hitchcock, D., & Verheij, B. (Eds.). (2010). *Arguing on the Toulmin model: New essays in argument analysis and evaluation.* New York, NY: Springer.

Hofstede, G., Hofstede, G. J., & Minkov, M. (2010). *Cultures and organizations: Software of the mind* (3rd ed.). Hightstown, NJ: McGraw-Hill.

Human Japanese. (2005). *Correspondence Japanese style.* Retrieved from http://www.humanjapanese.com/appendix/letters.htm

Kachru, Y. (2006). *Hindi.* Philadelphia, PA: John Benjamins Publishing.

Kahneman, D. (2013). *Thinking, fast and slow.* New York, NY: Farrar, Straus and Giroux.

Krug, S. (2016). *Don't make me think, revisited: A common-sense approach to Web usability* (3rd ed.). San Francisco, CA: New Riders.

Kulhavy, R. W., & Schwartz, N. H. (1981, Winter). Tone of communications and climate of perceptions. *Journal of Business Communication, 18,* 17–24.

Lay, M. M. (1989). Interpersonal conflict in collaborative writing: What can we learn from gender studies? *Journal of Business and Technical Communication, 3(2),* 5–28.

Lustig, M. W., & Koester, J. (2012). *Intercultural competence: Interpersonal communication across cultures* (7th ed.). New York, NY: Pearson.

Maslow, A. H. (2000). *Maslow business reader.* D. D. Stephens (Ed.). New York, NY: Wiley.

Meyer, E. (2014). *The culture map: Breaking through the invisible boundaries of global business.* New York, NY: Public Affairs.

Microsoft. (2012). *Suggestions for asking a question on help forums.* Retrieved from https://support.microsoft.com/en-us/kb/555375

Miller, C. R. (1984). Genre as social action. *Quarterly Journal of Speech, 70,* 151–167.

Moran, R. T., Abramson, N. R., & Moran, S. V. (2014). *Managing multicultural differences* (9th ed.). Florence, KY: Routledge.

Murray, R. L. (2003). *Understanding radioactive waste* (5th ed.). Columbus, OH: Battelle.

National Commission on Writing. (2004). *Writing: A ticket to work ... or a ticket out: A survey of business leaders. The National Commission on Writing for America's Families, Schools, and Colleges.* New York, NY: College Board.

Nishiyama, K. (1999). *Doing business in Japan: Successful strategies for intercultural communication.* Honolulu, HI: Latitude 20 Books.

Petty, R. E., & Cacioppo, J. T. (1986). *Communication and persuasion: Central and peripheral routes to attitude change.* New York, NY: Springer-Verlag.

Proskauer, R., LLP. (2014). *Social media in the workplace around the world 3.0: 2013/14 survey.* Retrieved from http://www.proskauer .com/files/uploads/social-media-in-the-workplace-2014.pdf

Rabinowitz, C. J., & Carr, L. W. (2001). *Modern day Vikings: A practical guide to interacting with the Swedes.* Yarmouth, ME: Intercultural Press.

Ray, G. B. (1986). Vocally cued personality prototypes: An implicit personality theory approach. *Communication Monographs, 53,* 266–276.

Ricks, D. A. (2006). *Blunders in international business* (4th ed.). Malden, MA: Blackwell.

Roach, R. (2002, January 31). Making web sites accessible to the disabled. *Black Issues in Higher Education, 16,* 32.

Romney, C. (2009). Conventions and writing styles of rirekisho (Japanese résumés). In A. M. Stoke (Ed.), *JALT2008 Conference Proceedings.* Tokyo: JALT. Retrieved from http://www.academia .edu/13062582/Conventions_and_writing_styles_of_rirekisho _Japanese_resumes_2009

Sageev, P., & Romanowski, C. J. (2001). A message from recent engineering graduates: Results of a survey on technical communication skills. *Journal of Engineering Education, 90,* 685–692.

Sauer, B. A. (2003). *The rhetoric of risk: Technical documentation in hazardous environments.* Mahwah, NJ: Erlbaum.

Schawbel, D. (2012). *Millennial Branding Student Employment Gap Study.* Retrieved from http://millennialbranding.com/2012 /millennial-branding-student-employment-gap-study/

Schriver, K. (1997). *Dynamics of document design: Creating text for readers.* New York, NY: Wiley.

Smith, F. (2004). *Understanding reading* (6th ed.). Hillsdale, NJ: Erlbaum.

Stanford University. (n.d.). *Copyright & fair use.* Retrieved from http:// fairuse.stanford.edu/overview

Sternthal, B., Dholakia, R., & Leavitt, C. (1978). The persuasive effect of source credibility: Tests of cognitive response. *Journal of Consumer Research, 4,* 252–260.

Suchan, J., & Colucci, R. (1989). An analysis of communication efficiency between high-impact and bureaucratic written communication. *Management Communication Quarterly, 2,* 464–473.

Thompson, M. A. (2000). *Global résumé and CV guide.* Hoboken, NJ: Wiley.

Thorell, L. G., & Smith, W. J. (1990). *Using computer color effectively: An illustrated reference.* Englewood Cliffs, NJ: Prentice Hall.

Tinker, M. A. (1969). *Legibility of print.* Ames, IA: University of Iowa Press.

van Dijk, T. (1980). *Macrostructures: An interdisciplinary study of global structure in discourse, interaction, and cognition.* Hillsdale, NJ: Erlbaum.

Varner, I., & Beamer, L. (2005). *Intercultural communication in the global workplace.* Boston, MA: McGraw-Hill.

Wells, B., Spinks, N., & Hargarve, J. (1981, June). A survey of the chief personnel officers in the 500 largest corporations in the United States to determine their preferences in job application letters and personal résumés. *ABCA Bulletin, 14*(2), 3–7.

Westinghouse Corporation. (1981). *Danger, warning, caution: Product safety label handbook.* Author.

White, J. V. (1990). *Color for the electronic age.* New York, NY: Watson-Guptil.

White, M. (2009, April 15). Handwritten evidence. *The Guardian (U.S. Edition).* Retrieved from http://www.theguardian.com/careers /handwriting-analysis

Williams, J. M., & Bizup, J. (2013). *Style: Ten lessons in clarity and grace* (11th ed.). New York, NY: Pearson.

Williams, R. (2014). *The nondesigner's design book* (4th ed.). Berkeley, CA: Peachpit.

Wolgemuth, L. (2010, May). America's Best Careers. *U.S. News & World Report, 147*(5), 20–25.

Wolvin, A. D., & Oakley, C. (1985). *Listening.* Dubuque, IA: William C. Brown.

Woolley, A. W., Chabris, C. F., Pentland, A., Hashmi, N., & Malone, T. W. (2010). Evidence for a collective intelligence factor in the performance of human groups. *Science, 330,* 686–688.

Yancey K., Robertson, L., & Taczak, K. (2015). *Writing across contexts: Transfer, composition, and sites of writing.* Logan: Utah State University Press.

Young, R. O. (2010). *How audiences decide: A cognitive approach to business communication.* New York, NY: Routledge.

INDEX

A

abstracts, 110, 387. *See also* summaries
    database, 97, 98
    example, 217
    writing guidelines for, 216–19
accuracy, word selection for, 204
acronyms, 207
actions
    goals transformed into, 65–68
    in instructions, 466
    verbal expression of, 197–99
    writing as, 8
active voice, 15, 198–99, 201, 280
activities, in résumés, 29
adaptation
    of application letters, for different employers, 45
    to readers' cultural background, 134–35
    of résumés, for different employers, 32
    of superstructure, 112–13
advisers, questions asked by, 54
advocacy, on client projects, 347
agenda
    interview, 99
    team meeting, 311
alarm bell strategy, 64
alignment, in page design, 267–268
alternating comparison pattern, 152–54
alternatives presented in feasibility reports, 430, 434
    comparing, 152–54
    dismissing unsuitable, 439

evaluation of, 438–39
    overview of, 438
ambiguity, avoiding, in survey questions, 104
American Psychological Association (APA) style guidelines
    format, 482
    in-text citations, 482–83
    references list using, 483–87
analogy, 204
analysis, evidence-based, 77–79
anchoring, in research, 81
annual report, basic design, 266
appendixes, 219–21
application of writing knowledge to new situations, 17–19, 497–98
arguments, emotional, 183–84
Aristotle, 169, 171, 177, 183
articles, bibliographic notes on, 75
attention
    oral presentations and listener, 329–33
    using color to focus reader, 232
attitudes
    graphics selection and, 226–28
    influencing readers', 168–69
    of readers, persuasive communication and, 56
    toward subject, 193
audiences
    comments and questions, in oral presentations, 332
    complex, 61, 63
    connection with, in oral presentations, 329–33

database, 97
    for oral presentation, 319–20
    participation of, in oral presentations, 331

B

background information, to aid readers, 57, 110, 111, 127, 216–218, 323, 367, 395, 436, 441
back matter
    appendixes, 219–21
    glossary and list of symbols, 221, 222
    index, 221, 223
    planning, 211–12
    references list, endnotes, bibliography, 221
    writing reader-centered, 219–23
bar graphs
    ethical use of, 237
    tutorial for creating, 246–47
    types of, 247
beginning communications
    adjusting length of, 138–39
    cultural background, adaptation to, 140
    guidelines for, 135–444
    motivation of readers in, 135–40
bias
    avoiding in feasibility reports, 430
    avoiding in test results, 300
    avoiding personal or organizational, 81
biased questions, avoiding, 102, 105
bibliographic notes, 75
bibliography, 221
binding, 137, 275